RACE, CHANGE, and URBAN SOCIETY

Volume 5, URBAN AFFAIRS ANNUAL REVIEWS

INTERNATIONAL EDITORIAL ADVISORY BOARD

RACE, CHANGE, and URBAN SOCIETY

Edited by
PETER ORLEANS
and
WILLIAM RUSSELL ELLIS, Jr.

Volume 5 URBAN AFFAIRS ANNUAL REVIEWS

SAGE PUBLICATIONS / BEVERLY HILLS, CALIFORNIA

Printed in the United States of America

For information address:

SAGE PUBLICATIONS, INC.
275 South Beverly Drive
Beverly Hills, California 90212

International Standard Book Number 0-8039-0009-0

Library of Congress Catalog Card No. 70-127992

FIRST PRINTING

Contents

Assessments of Research Trends

1

PROLOGUE

Race Research: 'Up Against the Wall' in More Ways Than One

WILLIAM RUSSELL ELLIS, Jr.
PETER ORLEANS

□ IT IS BY NOW a relatively mundane observation to say that "race" in the literature of American race research has come to refer to "nonwhites." Indeed, the term nonwhites has, for the most part, stood for black people. In similar fashion, the expressions "urban crisis" and "urban affairs" have become polite and euphemistic ways of referring to the most important facts of urban life of the past decade; namely, the takeover of many central city areas by black people; the attendant poverty of both people and central cities; and the resultant confrontations with previously unquestioned authority.

The terms race and urban society, as used in the title and explored in the contents of this Annual Review, to some degree perpetuate this dubious custom. The contributions for the most part are concerned with black people and the issues currently surrounding them. However, an important effort has been made to consider these issues as they apply among nonblacks, between class levels, and across cultural and historical boundaries.

While the concept of race has become increasingly specific, the notion of "social change" seems to have become even more general.

It is used here as the indispensable background, against which problems involving the directions of race research, the prospects of rejuvenated social theory, and the exploration of cross-national cases, are thrown into relief. Many of the chapters in this volume pay little heed to boundaries of time, discipline, tradition, or nation in their selection of problems or mode of addressing those problems.

In the decade just past, many practitioners and scholars have had their beliefs and moral stances directly tested by major political upheavals on their campuses, at their professional conventions, and in their neighborhoods. It is hardly a surprise, then, to find the consequences reflected in their examination of social life. Although major areas of social scientific investigation were unaffected by these upheavals, the 1960s witnessed a direct and profound impact on those formally expected to keep track of and interpret race relations.

It is somewhat startling to consider the rapid change implied in the contrast between the chaotic indeterminacy of the present and the liberal optimism of the recent past. It is equally startling and instructive to consider the corresponding changes which have taken place in the attitude and practice of race research. In 1963, well before the events of Watts, Newark, and Detroit, Everett Hughes addressed to his sociological colleagues his assessment of race relations research:

> Whenever before was there a race-caste of 20,000,000 people, literate, with the aspirations and basic skills of a modern industrial society, with money to spend and the tastes which make them want to spend it on the same things as do other people of highly industrialized societies, yet limited by others in their full realization of all these things; living in a society that has preached that all men are created free and equal, and has practiced it not fully, but enough so that with every increase of education, standard of living and of middle class achievement of the race-caste, the discrepancy between partial and full practice of equality becomes a deeper, more soul-searing wound. Why should we have thought, apart from the comfort of it, that the relations of the future could be predicted in terms of moderate trends, rather than by the model of the slow burn reaching the heat of massive explosion [Hughes, 1963: 889].

The political turmoil, the ameliorative legislation, the acute race consciousness, and the social strife which ensued in succeeding years have merely served to emphasize the poignancy and the irony of Hughes' observation and query. Hughes' observation is noteworthy for other reasons. It is a recognition and anticipation of the issues which were to confront the social sciences in the last part of the decade; it exemplifies the liberal intellectual melancholy which was to blossom during those years. Those who had developed their ideas and their political positions in the situation of the slow burn were to be radically tested, both as persons and professionals, in "the heat of the massive explosion" which they did not foresee.

In this volume we are not so much concerned with assessing the failure of the social sciences to anticipate the events of the sixties, as substantial and significant a failure as it was. Rather we are concerned with the consequences of that failure as they are acknowledged and manifested in current attempts by scholars and practitioners to analyze and interpret various aspects of race, change, and urban society. In our estimation it is precisely the nature of the collective response of academicians, policy makers, and laymen, that defines the contemporary concerns and the future directions of the social sciences as they attempt to interpret recent events and to develop reliable and relevant information in an effort to understand, shape, and influence future events.

The 1960s witnessed two important events in American history. During the decade just past, as a people, and perhaps for the first time, we became aware of the seemingly irreversible process of urbanization. This meant that a majority of Americans was experiencing urban life as first-generation city dwellers and suburbanites. This was occurring in the context of a culture dominated by celebration of, if not absolute veneration for, the myth of its rural origins. Our acceptance of our urban fate was characterized more by trepidation and resignation than by enthusiastic endorsement. The urban quality of American life was only finally officially and symbolically recognized with the establishment of the Department of Housing and Urban Development in 1965.

Simultaneously, the recalcitrance of our institutions, and their failure to accommodate to the course of racial integration mandated by the Supreme Court (and supported by traditional patriotic rhetoric), created the impetus for the Civil Rights Movement. Local resistance to federally sanctioned objectives merely served to emphasize the discrepancy between our professed beliefs and our predominant behavior. It was, after all, a little more than ten years that separated the 1954 landmark decision of the Supreme Court from the Kerner Commission's characterization of American society as racist.

During this same period of time the social sciences were under extraordinary pressure to demonstrate their utility. In the formative years of their development, during the early part of the twentieth century, the social sciences exhibited an overriding concern with social problems, many of which had been occasioned by the first wave of urban migration. These problems were first made evident by the increased visibility of social pathology which accompanied the unique concentration of ethnically diverse populations in the cities. The content and concerns of academic social science were yet to be precisely defined and its methodology was understandably rudimentary.

In the quest for further specification of the domains of the various disciplines, and in the concern for developing valid and reliable information and interpretation, social scientists began to differentiate their research concerns and systematically to question the epistomological foundations of their procedures. Simultaneously, they attended to the development of rigorous research techniques, aided and abetted by technological advances appropriate to the collection and manipulation of their data. By mid-century social scientists from a variety of disciplines seemingly had demonstrated the possibilities of the scientific study of human behavior, not to mention the limited, but strategically important, utility of insights gained from their research.

By 1950, seminal studies by social scientists of various disciplinary persuasions—the Western Electric research, the American Soldier studies, Kinsey's reports, work on the authoritarian personality, and Myrdal's analysis of American race relations—had made their indelible imprint (see Madge, 1962). For the most part these researches were limited in scope; they addressed manageable, though important and practical problems. However, they served to establish the possibility of rational policy development based on reliable information about human behavior.

Thus, when the Supreme Court used social science research to support the contentions contained in its school integration decision, in effect it acknowledged the legitimacy and the utility of social science research. It also created, heightened, and reinforced expectations regarding the potential role of the social sciences in assisting policy makers responsible for charting the future course of the society. As a consequence, social scientists found themselves prematurely in the difficult position of having to contend with both the further development and refinement of their disciplinary knowledge and the application of that knowledge to the practical concerns of an ongoing society.

The increasing complexity of urban life was soon to outdistance the explanatory potential of the social sciences and to throw its practitioners into an uncomfortable, untenable limbo. The quest for relevance, so characteristic of the social science enterprise during the 1960s, was anticipated in the earlier self-conscious debates concerning the role of pure and applied research in the social sciences. It developed as it became increasingly apparent that in some significant sense social scientists were out of touch with the social reality they purported to interpret. How else could they have failed to anticipate the events of the 1960s?

In an attempt to secure a firmer grasp on their enterprise, social scientists turned inward, seeking to define the boundaries of their concerns. The resulting specialization of knowledge within disciplines and fragmentation of knowledge among disciplines occurred at a time when the society itself was experiencing problems which were not amenable to the partial analyses fostered by discipline-circumscribed investigation. The search for relevance, at least in its initial stages, was not an effort to politicize the social sciences as much as it was an effort to redirect research to appropriate concerns through the development of an integrated and wholistic perspective (accomplished with limited success in research carried out by interdisciplinary teams). But, the disciplinary fragmentation of social scientific knowledge has fostered, intentionally or not, a pseudo-objectivity and an avoidance of political issues. This has become painfully obvious in recent years.

The failure of the social sciences to anticipate the events of the 1960s, no less than to understand those events once they had occurred, forced social scientists to become more realistic and modest about the limitations of their knowledge and its application. Spokesmen for the various social science associations felt constrained to explain to both their politically sophisticated younger colleagues and to the general public that the problem-solving expectations of their disciplines were too high. This persuasion effort proved exceptionally difficult since the traditional distinctions between pure and applied research commanded so little respect among the relevant audiences. At the same time there developed a tendency within the society to acknowledge or at least to defer to the expertise of the lay critic and the ordinary citizen. This tendency was perhaps sustained by the essential pragmatism of the American mind (every man a sociologist).

Moreover, it was supported by the political nostrum of participatory democracy—the idea that the beneficiaries of social policy, rather than being passive bystanders, as active knowers of their lives, should be encouraged to take an active role in the formulation of solutions to their particular problems. After all, if the experts lacked sufficient insight and knowledge to assess current events, perhaps the participants themselves had valid knowledge of their own circumstances. Of course, social scientists had not actively sought the roles of policy experts at this point. Thus, their willingness to share the expert's domain may have been as much a sharing of responsibility for potential future failures of interpretation as it was a genuine deference to the genius of popular social science.

It became evident, at any rate, that the value-neutrality of social scientists was more apparent than real, that researchers operated with implicit assumptions about social reality, and that their assumptions were for the most part no more soundly grounded than the assumptions of the people they studied. There is a sense in which Everett Hughes anticipated this trend toward the diffusion of expertise in his presidential address to the American Sociological Association when he asserted that:

> Our problem is not that we are too deeply involved in human goings-on but that our involvement is so episodic and so bound to the wheel of particular projects with limited goals; in short, that we are too professional. While professionalizing an activity may raise the competence of some who pursue it by standardizing methods and giving license only to those who meet the standard, it also may limit creative activity, by denying license to some who let their imagination and their observations run far afield, and by putting candidates for the license (PhD) so long in a straitjacket that they

> never move freely again. Our problem, as sociologists, in the next few years will be to resist the drive for professionalizing, and to maintain broad tolerance for all who would study societies, no matter what their methods [Hughes, 1963: 890].

The professionalization of social science, to use Hughes' term, has continued apace under the most difficult of circumstances. It has persisted at a time when, confronted by a crisis of confidence, social scientists have been called upon to make interpretative sense of events they had failed to anticipate, and to chart out the consequences of alternative future courses of action. Because this professionalization runs directly counter to contemporary requirements within the disciplines and demands outside, it has produced a tension which underlies the schizophrenic character of much of contemporary social science. There are many social scientists who, in a variety of ways, in different settings, and for different reasons are themselves trying to reconsider, extend, or modify, their roles as social scientists, as they attempt to examine one or another aspect of race, change, and urban society. We believe that the tension to which we have referred is clearly evident in many contributions to this Annual Review.

This tension is characterized by the conflicting needs to address simultandously the political, personal, and professional aspects of the social science enterprise. The maintenance and extension of the professional hegemony over expertise has been evidenced in the rededication of some academicians to protect themselves from the onslaught of lay insights. This is being accomplished by attempts to tighten standards of scholarship, and by the pursuit of traditionally accepted modes of inquiry and publication. For some social scientists it was evident that the expectations of the social sciences are premature and unjustified. Perhaps they reasoned that a new assessment of the work done to date might be expected to uncover errors of interpretation and omission—the missing links in past analyses. A reevaluation of prior research might then be expected to yield more refined and appropriate analyses which, in turn, could account for past failures while maximizing the likelihood of future successes. Needless to say, the pressures on such academically oriented social scientists have been extreme during the latter part of the last decade.

However, social conditions could not await the results of methodical scholarship. Ameliorative solutions, even if only temporary, often seemed to be required by the times. Such solutions had to be based on policy decisions, and so it was that social scientists were called upon to acknowledge the political implications of their work. They were, in a sense, forced to modify their value-neutral stance, to provide not only the basic information necessary to the required decisions, but to make policy recommendations, and, more importantly, to amass data to rationalize the decisions once they had been made. In short, social scientists were called upon to state a preference for one course of action as opposed to others (whether in the form of cost-benefit analyses or in the form of more obvious value judgments regarding intangible aspects of human behavior).

Whether the social scientist chooses to further the condition of man by advancing knowledge in his own discipline, or by engaging directly in policy-relevant research, it is increasingly difficult for him to remain a detached observer of the social scene. As a result, more and more social scientists have been making forays into the community from the sanctuary of the university in an attempt to bolster abstract intuition with concrete experience and evidence. Often as an almost self-conscious part of the reexamination process, social researchers have invited and have been influenced by, the intrusion of urban community spokesmen into university classrooms, research centers, and consulting firms.

For increasing numbers of academic social scientists the personalization of the research process is becoming a necessity. It is as though the attempt is being made to understand a confusing, shifting, and unanticipated social reality by confronting it directly, personally, and self-consciously. Alvin Gouldner has discussed this phenomenon with considerable insight in his most recent book. He contends that sociologists must come to terms with two kinds of reality. The first of these, the "role-reality" as he terms it, is defined by "facts" developed through research—the things sociologists learn as sociologists. This reality is counterposed with what Gouldner terms "personal reality" which is shaped by what the sociologist has himself seen, heard, been told, or read. "While these beliefs differ from 'facts' systematically gathered and scientifically evaluated, the sociologist experiences them as no less real" (Gouldner, 1970: 42).

We would suggest that it was precisely the discrepancy between the professional experience of social scientists and their personal realities, to use Gouldner's term, that has given renewed impetus to recent and increasingly widespread attempts to personalize the research process. It is as though there now exists a pressing need to bridge the disjunction between these two realities; to feel, and to cull from first-hand experience, a sense of the real world; to extend experience of the personal reality to the point where it begins to shed some light on an apparently untenable conception of the world as previously given by the role reality

Many of the new members of the social science professions (black, people, Chicanos, Puerto Ricans) have, in a sense, come prepersonalized and politicized. As they have begun to exercise their interpretive prerogatives they have affected research outcomes and have learned, simultaneously, just how problematic is the investigative enterprise of the social sciences (see Robinson et al., 1969). Their awareness serves to heighten the white researcher's felt need to reexamine his own perspectives.

Awareness of the discrepancies between their personal and role realities has led some social scientists to engage in research calculated to extend their personal experience. Others have sought to reduce the dissonance through a reassessment of prior work. Efforts in the latter direction are apparent in critical retrospectives on race history, the relatively recent and widespread development of ethnic and racial studies curricula on campuses across the country, and in the public debates over policy-relevant research documents (of which the Moynihan Report is a prime and much discussed example), which rely heavily on the extension of interpretations found in earlier studies (see Rainwater and Yancy, 1967).

Many social scientists, and especially those from minority backgrounds, have begun to question the assumptions embedded in the research and writing of their (usually white) predecessors. Current efforts of such scholars to reclaim the literature, as often as not through the process of reinterpretation or redirection, invariably stem from the recognition that in the past the predominant perspective in the social sciences was cast by those who, though often sympathetic with the underclass and the racial and ethnic castes, were nontheless not a part of them.

Although many social scientists recognize and acknowledge the existence of a multitude of social worlds defined by diverse and individualized orbits in the urban milieu, few have seriously considered the importance of common experience in the formulation of research agendas; the idea that "it takes one to know one" (see Webber and Webber, 1967). Although intrigued by the journalistic reports of writers such as Tom Wolfe and John Howard Griffin, and the literary accounts of authors such as Claude Brown and Peri Thomas (which publishing houses are only too anxious to label "social science"), few social scientists have sought to find out whether role reversal (or juxtaposition) makes a significant difference in the way in which one apprehends and understands social phenomena.

Undoubtedly, white social scientists are as capable of engaging in race research as their nonwhite colleagues even though their everyday experiences differ. However, because they come to the task with different backgrounds they are likely to see different problems and pose different questions. Their intellectual and practical concerns may overlap, yet their analyses and recommendations will almost necessarily differ insofar as these are tempered by differences in their individual sense of urgency and conception of the possible. The results of these differing perspectives for what appears as interpretations of the urban world are profound. The kinds of data and insight that have been missing and that all are attempting now to gather on ethnic and racial features of urban life are those most "vulnerable" to the peculiar effects of personalization.

Practical necessity (e.g., the attitudes of ghetto dwellers) may dictate that for the foreseeable future the domain of race research will be restricted to nonwhite professionals. If so, this may mean that social scientists will have to contend with a new set of methodological problems. Whereas in the past the white analyst engaged in race research possessed liabilities analogous to those of the anthropologist encountering an alien culture for the first time, the liabilities of the nonwhite researcher today are akin to those of any indigeneous participant as analyst. If one takes seriously the underlying thesis of this chapter, it would seem that the liabilities that stem from a professional identity are likely to supersede those which derive from a racial or ethnic status. This result, however, will depend, in some degree, on the ultimate resolution of conflict between the currently contending models of the social scientist as personally and politically "engaged" or "objective."

If a new awareness of the fallibility of social science (occasioned by a recognition of the discrepancy between role and personal realities) has impressed upon researchers the need for direct experience in the world, then reassessments of the literature reinforcing that view have begun to stimulate an increase in the incidence of personalized and particularly ethnographic research. We have already witnessed a spate of literary and ethnographic accounts, albeit accounts which are invariably partial and unsystematic, of the life situations of minority people. This trend may be expected to continue through the 1970s. It seems likely, however, that nonwhite researchers will take the lead in this effort in the years ahead.

It may be expected further that the past tendency for urban ethnography to underplay the systemic antecedents of the plight of the underclass, in favor of explanations in terms of personality and cultural deprivations and defects, already recognized as fallacious and discredited, will be altered (see Valentine, 1968). A reorientation of the interpretative mode of ethnographic study should develop from the sheer fact that the researchers, because of both their prior experience and the current concerns of the people they are studying, will be attending to previously unaddressed problems and questions.

The contributions to this Annual Review are varied, more varied indeed than our introductory remarks suggest. Here, and in the general literature of urban studies, is to be found research which clearly rests upon reasonably stable theoretical and methodological approaches of the past. But, some of these contributions, we believe, anticipate the direction of research in the 1970s. They represent a burgeoning attempt to reexamine, reconceptualize, and redirect our orientation to what remains the single most important and problematic aspect of international urban life.

REFERENCES

BROWN, C. (1965) Manchild in the Promised Land. New York: Macmillan.

GOULDNER, A. (1970) The Coming Crisis of Western Sociology. New York: Basic Books.

GRIFFIN, J. H. (1960) Black Like Me. New York: Houghton-Mifflin.

HUGHES, E. C. (1963) "Race relations and the sociological imagination," Amer. Soc. Rev. 28, 6 (December): 879-890.

MADGE, J. (1962) The Origins of Scientific Sociology. New York: Free Press.

RAINWATER, L. and W. L. YANCY [eds.] (1967) The Moynihan Report and the Politics of Controversy. Cambridge: MIT Press.

ROBINSON, A. L., C. C. FOSTER, and D. H. OGILVIE [eds.] (1969) Black Studies in the University. New Haven: Yale Univ. Press.

THOMAS, P. (1967) Down These Mean Streets. New York: Knopf.

VALENTINE, C. A. (1968) Culture and Poverty: Critique and Counter-Proposals. Chicago: Univ. of Chicago Press.

WEBBER, M. M. and C. C. WEBBER (1967) "Culture, territoriality and the elastic mile," in H. W. Eldredge (ed.) Taming Megalopolis: What Is and What Could Be. New York: Frederick A. Praeger.

WOLFE, T. (1970) Radical Chic and Mau-Mauing the Flak Catchers. New York: Farrar, Straus & Giroux.

Part I

VALUE PREMISES AND RESEARCH STRATEGIES

Race research in the United States progresses apace. As human beings alert to the fact that so much social science research proved irrelevant to the rapid and dramatic changes of the last decade, many researchers have now been moved to ask fundamental questions about the nature and basis of that research.

From a sociology of knowledge standpoint, Edward Paynter, in the opening chapter of this section, provides some historical background against which one can view the contemporary strain of value premises and dilemmas confronted elsewhere in this volume. He offers a detailed analysis of the nature-nurture controversy which so preoccupied social scientiest during the formative years of race relations research in this country. Questions of value—or perhaps more correctly, of value neutrality—still plague social scientists concerned with race research. But within the last decade the problematic character of race relations in America has created a sense of urgency within the scholarly community which has forced a rethinking of the issue, and has exposed and reinforced the political dimension of social science research. It is significant that Paynter concludes his chapter quoting, with approval, Myrdal's statement that "the scale of radicalism-conservatism has always had, and still has, strong influences upon both the selection of research problems and the conclusions drawn from research." This observation addresses, in some important degree, the political, theoretical, and methodological criticisms to which formal social studies have been increasingly subject, and not only in the area of race research. Ira Katznelson addresses himself to the peculiar strain of normative problems confronting contemporary white social scientists, involved in race research. Because, as he puts it "attempts at value free race relations scholarship are not simply absurd, but dangerous." He urges a social science of "disciplined subjectivity" and stresses the need for comparative studies of race relations focused on the fact of power, its distribution and consequences. The latter recommendation is made in an attempt to break the parochial tendency of American scholars to cast their research in conceptual terms which psychologize group relations and therefore avoid assessments of the social, political, and economic conditions which support relationships of racial inequality. The author reserves a special, if brief, critical glance at urban ethnographic studies for precisely this reason. The call for disciplined subjectivity does not simply imply thoroughgoing immersion in the subject of one's study, it involves a stance "disciplined by the canons of scholarship, yet involved with the subject of inquiry." As stated and elaborated by Katznelson, this notion represents an important, if complicated, test for the social sciences.

Katznelson's prescription for the reconceptualization of race relations research is in essence a call for the type of research exemplified in Martin Orans' comparative analysis of race and caste conflict in India, Japan, and the United States presented in the concluding chapter in this section. Although Orans' research clearly respects "the uniqueness of national historical circumstance," and in that sense conforms to one of the objectives set forth by Katznelson, he also succeeds, through a modification of his rank concession theory, to synthesize and generalize his findings.

In this latter respect, Orans' reference of race-caste conflict (in international and historic terms) to a theory of mobility and social change, shifts the conceptions of both social problems and social research to higher and comparative levels.

2

Value Premises in Race Research: The Evolution of Environmentalism

EDWARD L. PAYNTER

□ GUNNAR MYRDAL'S *An American Dilemma,* published in 1944, was probably more responsible than any other book for the ultimate acceptance of the environmentalist explanation of Negro-white differences. It was Myrdal's thesis that the "Negro problem" in the United States was really a white problem. If the Negro was a "failure," as he obviously was by every criterion that white American society recognized as legitimate, then, Myrdal insisted, he was a failure because white America had made him so.

The only other explanation of the black man's lack of success in America was that some distinctly racial peculiarity or limitation, some defect in his biological equipment, had rendered him incapable of maintaining himself on a level with his fellow Americans. But it was evident that little, if anything, could be "scientifically explained in terms of the peculiarities of the Negroes themselves." All recent attempts to reach scientific explanations of "why the Negroes are what they are and why they live as they do have regularly led to determinants on the white side of the race line" (Myrdal, 1944: lxxv).

There was, then, a "dilemma" only in the loosest sense–a complex and bitterly vexing problem for representative government. In moral terms, there was no dilemma at all. American society was presented with a clear choice between two policies toward the Negro, one of which was good according to America's liberal and egalitarian creed, and the other evil. This was hardly a novel thesis in 1944, but *An American Dilemma* not only presented a thesis, it supported it with impressive scientific documentation.

The message of the book was taken seriously not only because the "American Creed" said all men are created equal, but also because respectable scholarship now proclaimed that, so far as could be ascertained, the Negro was the equal of the white in innate mental capacity and that his shortcomings were due to the environment in which he lived. Other factors contributed to the favorable reception of *An American Dilemma:* the prestige of the Carnegie Corporation under whose auspices the study was carried out, the reputation of the many scholars who contributed to the book's preparation, and the fact that the author was a distinguished Swedish social scientist and statesman. The skill with which he presented his case and anticipated and checked arguments against it was important in assuring the volume a good reception. But it was the evidence recently gathered in the fields of anthropology, sociology, and psychology which contributed most substantially to the authority of the book. If Myrdal had begun his study in 1928 instead of 1938, and had completed it in 1933 instead of 1943, *An American Dilemma* would have been a different book.

The innate inferiority of the Negro continued to be a "fact"–or at least a hypothesis supported by a great deal of plausible evidence–well into the twentieth century. By the early 1940s, however, this racial hypothesis had been all but completely discredited, and any biologist or psychologist who continued to insist on the importance of racial differences would risk injury to his reputation as an impartial and unprejudiced scientist. A revolution occurred in the social sciences on this question of innate racial differences in the first forty years of this century. Like all intellectual revolutions, it was a long and complex process involving the refinement of many concepts and techniques, but the evidence indicates that the period of sharpest conflict was between 1925 and 1935. The change to an environmental explanation of race differences came later to psychology than to anthropology and sociology, but it was in psychology that the decisive victory was won. By the mid-thirties, almost all psychologists were admitting that if significant racial differences in mental ability did exist, it was as yet beyond the competence of the discipline to establish them. Otto Klineberg probably spoke for most psychologists when he said, "in the present state of our knowledge . . . we have no right to assume that [innate differences] exist" (Klineberg, 1935a: 345). The assumption would now be that the races were equal, although that could not be proved, either.

Reviewing this situation, Myrdal wrote (1944) that "these conclusions of psychological research have probably been more revolutionary and practically important with respect to the Negro problem than the conclusions from any other sphere of science." For even if future research should establish the existence of innate psychic differences between whites and Negroes,

> it is highly improbable that such differences would be so large that—particularly when the overlapping is considered—they could justify a differential treatment in matters of public policy, such as in education, suffrage and entrance to the various sections of the labor market [Myrdal, 1944: 149].

In this chapter, some of the more important episodes and developments in the environmental revolution will be examined in order to reveal something of its dynamics and to determine why, when it did come, it accomplished its objectives so swiftly.

THE HEREDITARIAN DOCTRINE

The status of the "nature-nurture" controversy in 1917 was accurately described by E. B. Reuter when he observed that while the weight of authority in anthropology had until recently been on the side of unequal capacity, "the consensus of scholarly opinion at the present time seems to regard the backward races, not only as not having been proven to be inferior in mental ability, but as being, in so far at least as their inherited mental capacity is concerned, substantially equal to the culture [*sic*] races." On the other hand, the "popular assumption of a difference in mental capacity in the races" seemed to have been borne out in psychology, although later studies might show "that all modifying cultural elements have not been eliminated" (Reuter, 1917: 92).

By 1917, American anthropology, largely through the influence of Franz Boas as a writer, teacher, and academic strategist, had adopted a position on the race question that psychology would not embrace for another twelve or fifteen years (Stocking, 1968: 217-219). Psychology's position on the issue of innate differences was, during these years, largely determined by the results of the intelligence test, and the test therefore came to be, as one historian has said, "the most powerful weapon of the racism of the period" (Gossett, 1963: 373). The conclusions that psychologists drew from their tests were subjected to vigorous attacks by black intellectuals and white liberals before 1925, and by an increasing number of white social scientists after that date. The Negro educator Horace Mann Bond spoke for many angry liberals, black and white, when he wrote (1924) that the dogma of race had been at the point of expiration as early as the turn of the century when the intelligence testers, by mishandling their data, had given it new life. Such recent popularizers of Nordic supremacy as Madison Grant and Lothrop Stoddard were anachronistic holdovers from the nineteenth century, Bond (1924: 197) said, and their vicious doctrines would hardly have been taken seriously but for the false evidence of the tests.

Yet the intelligence test came to be a weapon of racism in the twenties, and the psychologists most closely identified with it came to be, or to seem, racists partly by inadvertence. Not wholly so, of course, for some psychologists were convinced white supremacists who employed test results to confirm doctrine. Most, however, were neither bigots nor cynics, but simply conventional social scientists who shared the prejudices of their time. They "knew" that the races differed in mental capacity; they wanted to determine, in a scientific way, just how much. If science seemed to proclaim the existence of innate differences in the early 1920s, then science would a little later, with the assistance of more sophisticated techniques, cast serious if not conclusive doubts on the racial hypothesis.

Racism within the discipline of psychology, despite its temporarily robust appearance, was, in fact, a fragile thing, for it was not, generally speaking, an integrated and tenaciously held doctrine, but an unstable compound of potentially conflicting beliefs and loyalities. The elements of the testers' racism were (1) the narrow focus of their sector of the discipline on mental differences considered apart from environment, (2) an unconsciously opportunistic belief in the ability of their tests to isolate innate capacity, (3) a smattering of Nordic doctrine and an enthusiasm for eugenics (i.e., the improvement of racial "stocks" through selective breeding), and (4) an ignorance of the cultures of the Mexican, Oriental, American Indian, and Negro children they so doggedly tested.

Of these four elements, the second was the most important, for whether or not a psychologist subscribed to Nordic doctrine, he would, if he believed in the ability of the tests to isolate innate capacities, be drawn to hereditarian conclusions. There were strong if unconscious reasons for such a belief. If the tests were indeed capable of isolating innate capacity, the reputation of psychology as a science of man would be greatly enhanced. No longer could detractors of the discipline caricature it as a bundle of dubious theories. The empirical proof of mental differences among the races would establish psychology on a plane of equal importance with biology, anatomy, and genetics among the physical sciences. One senses the anticipation of future triumphs in Lewis Terman's statement (1916: 41) that Alfred Binet's revised test, with its concept of "mental age," was perhaps the most important contribution to knowledge "in all the history of psychology." As Thomas R. Garth (1926: 241), one of the leading testers, said,

> It must be remembered that until psychology became an experimental science it could not really claim any fellowship with the sciences, and so in the study of racial differences, the problem here in hand must be an empirical one.

Viewed one way, this was nothing more than a straightforward appeal to psychologists to avoid conclusions regarding mental processes that had not been derived from empirical observation. But the ground already gained was in Garth's case the evidence of his many tests that the races differed in their mental capacities. Few testers would easily surrender such an important discovery. In

Garth's case, it is true, ambition for the discipline was combined with a positive belief in race differences: he was a leading member of the Eugenics Research Association.

In the early 1920s, Garth's Nordic doctrines and his commitment to the tests reinforced one another. In a typical report, Garth (1923) described the manner in which he had set out to test several "principles of genetics," all of which asserted different mental capacities for different "blood" or population groups. The children tested were of mixed (white and Indian) parents, children of Mexican parents undifferentiated as to racial descent, and children of three Indian tribes of the Southwest. Garth was sufficiently aware of the criticisms of environmentalists to take care in determining the socioeconomic status and educational level of each group tested. He noted that the test scores of the different groups in his experiment were in exactly the same sequence as their estimated socioeconomic status, and, with the exception of one of the five groups, with known educational level as well. Yet he concluded that "the results indicate differences between their racial stocks as here measured."

Did Garth affirm that his tests isolated racial capacities because such a conclusion confirmed his racial theories, or did he hold to his theories because they provided the only basis, in the face of environmental evidence to the contrary, for believing that the tests isolated racial differences? The two considerations at this stage were inseparable. A decade later, when the results of Garth's experiments had been subjected to sharp attack by environmentalists, he repudiated his earlier conclusions (Garth, 1934: 323). Garth's commitment to psychology as an experimental science was stronger than his commitment to Nordic doctrine, and by 1934 a commitment to science required an admission of error.

Like Garth, Carl C. Brigham, another well-known racial tester, would ultimately repudiate his conclusions. In the early 1920s, Brigham was to all appearances an outright racist. Probably no document in the field of "race psychology" was subjected to such strenuous and deserved attack as his *A Study of American Intelligence* (1923). The book was an analysis and interpretation of the results of the U.S. Army alpha (verbal) and beta (numerical and pictorial) intelligence tests given to recruits in 1917 as a quick and reliable way of sorting out potential officer material. Brigham had helped to prepare and administer the tests as an assistant to Robert Yerkes, the chief of the division of psychology of the Office of the Surgeon General. Brigham paid tribute in his introduction to the insights of Madison Grant.

Not surprisingly, Brigham concluded that American whites were superior in intelligence to American Negroes, and that among the whites, the predominantly Anglo-Saxon old stock was superior to the eastern and southern European "new immigration" stock. He also found that within the last group, those who had come to the United States several years before were more intelligent than the most recent arrivals. Brigham's reasoning in regard to the recent arrivals is worth studying closely. He noted that they scored lower on the average than the older immigrants in both the alpha and beta tests. Did this result, he asked, "represent the growth of intelligence with increasing length of residence, does it represent an error in the method of measuring intelligence, or, looked at from another angle, does it show the gradually decreasing intelligence of the more recent immigrants examined in the army?"

Only the third interpretation, Brigham noted, was compatible with the objectives of the test: "we must," he reminded his readers, "assume that we are measuring *native or inborn intelligence*, and any increase in our test score due to any other factor may be regarded as an error." If there had been an error, Brigham thought, it would show up in a discrepancy between the results of alpha, which was indirectly a measure of education, and of beta, which did not involve the use of English or the ability to read, and therefore could not be considered an educational measure in any sense. Since Brigham found that the distribution of scores in each test was about the same, it followed that the tests accurately reflected innate mental differences[1] (Brigham, 1923: 100).

The test results confirmed Brigham's belief that America was in danger of being swamped by an increasingly stupid and unassimilable horde of Slavs and Mediterraneans. On the other hand, Brigham's confidence that the tests he had helped devise could measure "inborn intelligence" would have led him, even in the absence of race doctrine, to the conclusion that recent immigrants were innately inferior. Like Garth, Brigham abandoned his doctrines when environmentalist criticism of the Army tests forced him to do so. Intelligence testers had been guilty of a "naming fallacy," he confessed in 1930, which had enabled them to "slide mysteriously from the score in the test to the hypothetical faculty suggested by the name given to the test." Thus psychologists habitually spoke of "sensory discrimination, perception, memory, intelligence and the like while the reference is to a certain objective test situation." Brigham admitted that the early testers had made conclusions which they had no right to make on the evidence, and that he had been among the worst transgressors (Brigham, 1930: 160).

The most renowned of all testers was Professor Lewis M. Terman of Stanford, the inventor of the Stanford revision of the Binet-Simon scale, or the "Stanford-Binet," as it quickly became known. In his instruction manual, Terman referred interchangeably to the "general intelligence" of the testee as measured by the Stanford-Binet and to his "original endowment" (Terman, 1916: 45-50). Although Terman confessed that psychology could not as yet define intelligence precisely, he insisted that those in the discipline knew perfectly well what it was, for they could measure it. In this respect, they were in the same situation as the eighteenth-century scientists who could not define electricity but could nevertheless test its properties (Terman, 1916: 44).

Terman's presumptions were subjected to the scorn of Walter Lippmann in the pages of the *New Republic* in 1922. In Lippmann's most incisive statement on the matter, the subject of race was not even mentioned:

> The claim that we have learned how to measure hereditary intelligence has no scientific foundation. We cannot measure intelligence when we have never defined it, and we cannot speak of its hereditary basis after it has been indistinguishably fused with a thousand educational and environmental influences from the time of conception to the school age. The claim that Mr. Terman or anyone else is measuring hereditary intelligence has no more scientific foundation than a hundred other fads . . . and it will pass with them into that limbo where phrenology and palmistry and charactereology and the other Babu sciences are to be found [Lippmann, 1922: 248].

Lippmann was not calling the testers racists, he was calling them quacks, and the charge naturally stung Terman deeply. In a rambling and sarcastic reply he grouped Lippmann with the reactionaries who had doubted Darwin. Almost hidden in the bluster, however, was an admission: "as a matter of fact," Terman (1922: 119) said, "all the 'intelligence testers' will readily agree with Mr. Lippmann that their tests do not measure simon pure intelligence, but always native ability plus other things, with no final verdict yet as to exactly how much the other things affect the score." But not once in his four-page reply did Terman return to this matter of "other things," nor did he directly concede the importance of *environment* as a contributor to the test score, which was, after all, the whole point of Lippmann's criticism. `Terman was hardly unaware of the importance of environment: he had earlier attempted with the anthropologist A. L. Kroeber to produce a "culture-free" test (Stocking, 1968: 298). Terman's ambitions for the intelligence test were such that he could not admit its shortcomings in public print.

If the testers were only made angry or uncomfortable by flat charges of bigotry or quackery, they were sufficiently sensitive to criticisms on good methodological grounds to reexamine their beliefs and to qualify or reject them. The sin of the "race psychologists," if we are to regard the episode that way, was perhaps not racism but, as Lippmann said (1922), vanity.

ENVIRONMENTALISM COMES OF AGE

The ascendancy of environmentalism in the social sciences during the late twenties and early thirties is not easily described, for environmentalism was at once a method, a hypothesis, a moral inclination, a disciplined movement, and a drift of scholarly opinion. Yet this much may be said at the outset: insofar as it *was* a movement, environmentalism advanced in two separate columns, one composed of anthropologists and psychologists, the other of sociologists. The first group was inspired by the example and the teachings of Boas, and was composed of his students of the 1920s: Margaret Mead, Ruth Benedict, Otto Klineberg, and Melville Herskovits. The second group was benignly presided over, though hardly led, by Robert E. Park, the University of Chicago urban sociologist. This group included E. Franklin Frazier, Charles S. Johnson, John Dollard, and several others. Having said this much by way of generalization, one must rapidly qualify.

To speak of columns of advance is to employ an analogy that implies clear-cut lines of struggle dividing individuals and disciplines. The analogy is something less than perfect, for in truth the struggle was also an internal one. In the early period of this revolution—say, up to 1930—no discipline, no university department, and very few social scientists would be exempt from assumptions regarding the broad question of race and the narrower question of the American Negro which by 1940 would seem untenable. Scientists, whom we might associate with an essentially racial interpretation of human behavior, and others whom we might identify with an environmental interpretation, often held much the same views on questions of importance. "Racists" like Garth and Henry Pratt Fairchild and "environmentalists" like Boas and A. L. Kroeber all expected to find that obvious inherited differences in physical traits among the races—skin and eye color, hair texture, and so on—would be accompanied by possibly significant psychological differences as well (Garth, 1931: 207; Gossett, 1963: 386; Boas, 1911: 114-115; Kroeber, 1934: 352). Whether such differences would be important or merely marginal, whether or not they would imply an order of races in terms of intelligence, no one could say for certain. Lacking proof one way or the other, it would be as unscientific to dismiss the possibility of such differences as it would be to assert them as fact.

If those of the racial persuasion and those of the environmental persuasion held roughly the same views on many questions of importance, how does one differentiate between them? The best answer would seem to be, in terms of methodological inclination. Klineberg, in stating his own views, succinctly described the classic environmentalist stance. "The writer," he said of himself, "confesses to a preference for the environmental hypothesis on logical, or better methodological grounds. If the evidence were such that the facts could just as easily be explained by one hypothesis as the other, he would incline to regard that as a presumption in favor of environmental factors."

> For to explain observed differences on the basis of race and heredity does not seem to him to clarify the problem. Environmental

> factors—customs, culture, tradition, education, contact and relationship with other peoples—are much more tangible and definite than the factor of race, psychologically speaking . . . . Only when [the environmental explanation] fails should the racial factor be invoked—and in the present instance there seems to be no such necessity [Klineberg, 1928: 97].

Most of the racialists were so poorly grounded in the realities of which Klineberg spoke that a few dedicated environmentalists, rigorous in their methods, deft in probing for weaknesses, were able, almost from the outset, to throw the racial hypothesis into doubt.

In the cases of most scientists, ideology, in the sense of a more or less rigid set of convictions tending to induce a presumption on one side or the other of a given issue, does not provide a satisfactory explanation of the dynamics of this revolution. Rather, it happened that the accumulating evidence, usually derived from research begun from an environmental hypothesis but sometimes not, increasingly permitted environmental explanations of other problems, over an ever-widening area, and this movement progressively reinforced the inclinations of a growing number of scientists toward broad environmental conclusions.

An excellent picture of the dynamics of this revolution was provided by the psychologist Dale Yoder (1928). Yoder noted that the literature on race differences had expanded enormously in the preceding few years. He detected three distinct viewpoints. One group of social scientists accepted the *fact* of race differences. These men, now on the defensive, had lately thrown all of their efforts into adducing additional evidence to support the thesis, quite forgetting that "it is the quality rather than the quantity of evidence which is under fire." The second group, the great majority, considered inferiority possible but not adequately demonstrated, and they displayed an air of calm objectivity appropriate to their views. The third group—with whose views Yoder obviously identified his own—was small, but growing. They were "highly critical of the means used to demonstrate race inferiority and of the results so obtained and generally [insistent] on racial equality." They possessed "the fire of enthusiastic discovery" and "idealistic zeal" (Yoder, 1928).

Among those possessing the fire of enthusiastic discovery, Boas' students were easily the most important. Margaret Mead was among the earliest flatly to challenge the testers in the pages of a scholarly journal. She introduced a welcome note of open partisanship when she wrote (1926: 658) that "while the writer on general social problems has learned that the methodology of intelligence testing is still in swaddling clothes, he is not so conscious that a methodology adequate to deal with racial and nationality testing has not even been born." Melville Herskovits was one of the earliest to challenge the whole basis of the racialist consensus regarding the Negro, not just in its particulars and not merely to qualify it, but to destroy it and replace it with an egalitarian consensus. Herskovits (1926) directed his attack specifically at G. O. Ferguson's 1916 study of white and Negro school children in Virginia, and a follow-up study of white and Negro troops at Camp Lee, Virginia, in 1917. Ferguson had concluded (1916) on the basis of his own observations that the level of intelligence was in direct proportion to the amount of "white blood" contained. He calculated the racial mixtures of his Negro subjects on a sight basis, depending for the most part on skin color. He assumed that differences in intellectual capacity could be related directly to "blood" because among Negroes in general there were, he thought, no considerable distinctions based on color. Herskovits (1926) attacked the idea that ancestry could be determined on sight, and effectively challenged the assumption that social distinctions based on skin color were lacking among Negroes. He attributed test score differences to environment.[2]

Klineberg, displaying less of the disputant's zeal than Herskovits, nevertheless contributed more than any other psychologist to the decline of the racial hypothesis. His most important achievement was to cast into doubt the "selective migration" theory, one of the fundamental props of the racial persuasion. According to this view, the fact that Northern, urban Negro children scored higher on the average in intelligence tests than Southern Negro children could be explained in terms of a migratory impulse which tended to be strongest in the most fit. Klineberg demonstrated (1935b) that the impulse was indiscriminate, and drew off near-indigent Negroes as often as successful ones. The higher scores of Northern children could better be explained in terms of the greater educational opportunities the Northern environment provided.

The defeat of race psychology and the vindication of the American Negro were by no means the only concerns of Klineberg and Herskovits, and still less were they preoccupations of the Boas school as a whole. Their horizons were much broader, their objectives more fundamental. Taking the world as their field of inquiry they sought, through the study of primitive societies, to establish environmentalism as a general principle, to replace the hierarchical idea of race as a category of social investigation with the noninvidious concept of culture. Each culture, they insisted, was unique, and must be viewed in terms of its own peculiar setting and history. Above all, the student of human societies must be aware of the fact that he, fully as much as the peoples he might visit in his field trips, was "culture-bound"; he, too, was entangled in a web of customs and unarticulated beliefs. Ruth Benedict's *Patterns of Culture* (1934) which had the explicit purpose of getting Americans to think critically about their own culture and of promoting "a more intelligent and directive attitude" to social change (Benedict, 1934: 10) best exemplifies the reformist spirit that underlay much of the Columbia school's work. Race psychology was only one of many manifestations of American insularity that they sought to discredit; the American Negro was only one among many clients.

If the Boas school sought to establish environmentalism as a principle, there was another group of scholars who assumed it as a truth, and whose efforts were also directed toward intelligent social action. These were the sociologists who in one decade transformed the study of the Negro from an obscure and little regarded specialty into one of the dominant fields in American social science. Myrdal's debt to these men was profound and candidly expressed.

LITERATURE OF BLACK PROTEST

It is no more than a convenience to speak of all the social scientists who devoted themselves to the study of the Negro and his place in the American caste system as forming a "school." Such men as Sterling D. Spero, Abram L. Harris, John Dollard, Carter G. Woodson, Ralph J. Bunche, E. Franklin Frazier, Allison Davis, Charles S. Johnson, Arthur Raper, Edward R. Embree, and Horace Cayton came from a variety of backgrounds, were of different races and ages, studied at different graduate schools, and affiliated with a variety of organizations. Their motives were equally disparate, ranging from a desire to promote interracial understanding in the case of Raper to a tough and caustic militancy in the case of Bunche.

Yet, if variety defeats one sort of generalization it suggests another: with this generation of sociologists, scholarship was for the first time a racially integrated enterprise. The fact that blacks and whites could work in concert toward similarly defined goals, and occasionally collaborate on books like Spero and Harris' *The Black Worker* (1931) and Davis and Dollard's *Children of Bondage* (1940), was in itself of great importance.

This remarkable efflorescence of black studies after 1930 might best be understood as a continuation, in another form, of the literature of Negro protest which began to take shape around the turn of the century. Two considerations prompt this thesis. The first is that the two most important representatives of that protest literature, W. E. B. DuBois and James Weldon Johnson—each of whom was by turns an essayist, historian, folklorist, sociologist, and novelist—are probably the most oft-quoted of all writers whose books appear in Gunnar Myrdal's bibliography.[3] The second is that, of the post-1930 sociologists, those for whom Myrdal had the highest respect, were E. Franklin Frazier and Charles S. Johnson, both black, both scrupulous scholars, but each, after his own fashion, a militant.

One must, in fact, go back as far as 1899, to DuBois' *The Philadelphia Negro* (1967), for an earlier example of the sort of work that Johnson and Frazier were preparing in the late 1920s. In the interval there were only a few studies of consequence—the Atlanta University monographs prepared under DuBois' direction in the 1900s, Mary Ovington Hall's *Half a Man: The Status of the Negro in New York* (1911), and Ray Stannard Baker's *Following the Color Line* (1908), upon which Myrdal relied heavily for insights into Negro life during the age of progressivism. That Baker was a journalist and not an academician is both an indication of the high standards of muckraking journalism and a comment on the low level of sociological awareness in those years.

Black writers during the progressive period turned their efforts to works of propaganda, protest, and interpretation. The general intention was to get across to white society *What the Negro Thinks,* to employ the title of an excellent example of the genre, by Robert Russa Moton (1929). It must have seemed to DuBois, James Weldon

Johnson, Kelly Miller, and others, that if their voices were to be heard by whites at all, they must exhort, admonish, and accuse. As DuBois said (1940: 67) of his own state of mind when he abandoned the Atlanta studies, "one could not be a calm, cool, and detached scientist while Negroes were lynched, murdered, and starved." Even if one were of a mind to be simply a scholar, there was the bleak fact to be faced that "there was no . . . definite demand for scientific work of the sort that I was doing."

The effect of this literature of black protest on the white social scientists of the period was slight. The Columbia psychologist Robert S. Woodworth, a colleague and friend of Franz Boas, might have been expected to greet a serious work of black scholarship with some enthusiasm. Yet in a review of the current literature on race differences, Woodworth (1916: 395) found DuBois' *The Negro* (1915) merely "interesting" because it represented the "Negro point of view." However, the very qualities of the genre which guaranteed a cool reception by white social scientists—the indignant prose, the denunciations of prejudiced white historians, the appeals to Negro pride, the daring claims made on behalf of black cultural accomplishment in Africa and the United States—were precisely the qualities which Negro students and intellectuals found fresh and exciting.

In E. Franklin Frazier's career, this protest impulse was directed into activist scholarship. After his graduation from Howard in 1916, Frazier taught briefly at the Tuskegee Institute, where, using the classroom as a platform of agitation, he urged his students to take a militant attitude toward discrimination wherever it appeared. Frazier deeply resented being drafted into the Army to fight what he considered a white man's war, and regarded all talk of making the world safe for democracy as the rankest hypocrisy when Negroes were being abused, discriminated against, and murdered at home. But Frazier later recalled of the years after the war, "as I look back now, it appears to me that during this period I was developing an objective outlook on racial and other social problems which was divorced in a sense from my reaction to these problems as a person and as a member of society." In part, Frazier thought, this was due to the discipline he had developed during his graduate work at Clark University, which enabled him to study social problems at "a high intellectual level" (Odum, 1951: 234-235).

In the career of Charles S. Johnson, the offices of race leader and scholar were combined without a hint of conflicting motives or inconsistent purposes. Johnson came to the University of Chicago as a young graduate student during World War I, and stayed long enough to witness the race riots of 1919 and to collaborate on a study of them. From 1923 to 1928, Johnson edited *Opportunity*, the National Urban League monthly, in which office he "encouraged new talent . . . offered cash prizes for outstanding literary achievement," and generally "made a substantial contribution to the birth of the New Negro Movement" (Bone, 1965: 55). Johnson then took a professorship in sociology at Fisk University, and began an extremely productive career as a scholar.

The main link connecting Negro and white scholarship in the late twenties and early thirties was the sociology department at Chicago, presided over by Robert E. Park. Park was a great teacher, wrote extensively on race problems, and maintained close relationships with Negroes for many years. But he was not the influence over his black students that Boas was over the Columbia anthropologists. Doubtless such students as Johnson, Frazier, and Horace Cayton matured as sociologists under his tutelage, but on the subject of race they were ahead of Park from the beginning. In Park's writings on race relations there was an almost studied blandness—a result, perhaps, of the soothing counsels of Booker T. Washington, with whom Park had been associated at Tuskegee for several years before coming to Chicago. Violence, economic oppression, and the climate of race hatred in America hardly appeared in Park's placid essays; instead there were ruminations on "folkways" and interracial "etiquette," combined with a seemly optimism for the future.

Gunnar Myrdal was severely critical of what he termed the "do-nothing" school of sociology, of which Park and his student E.B. Reuter were leading examples (Myrdal, 1944: 1048-1057). Probably the outstanding failing of the do-nothing school, from Myrdal's point of view, was not its failure to make value premises explicit, but its apparent indifference to the moral failings of society and its acceptance of indefensible social practices as perfectly natural folkways. Examples of both the permissive attitude of the do-nothing school toward a specific social failing, race prejudice, and the anger of the reformist school in regard to the same phenomenon, may be found in essays by Park and Frazier in the late twenties.

Park (1950) thought that President John Grier Hibben of Princeton University had stated the matter in a "fundamental way" when he described prejudice as "a natural factor in any thinking, and not to be regarded in any sense as an abnormal and disturbing element." Prejudice, in Park's opinion, was "a common human weakness in which we all, more or less, share." He defined prejudice, in the broadest possible way, as any simple preference or dislike that could not be "justified on general and rational grounds," and concluded that prejudice, thus defined, could be seen as something having its "source and origin in the very nature of men and their relation to one another." "Race prejudice," he concluded, "is like class and caste prejudices–merely one variety of a species" (Park, 1950: 230-232).

This broad definition of prejudice, which could embrace anything from racial or religious bigotry to a preference for strawberry ice cream rather than vanilla, effectively guaranteed that racial oppression in the United States would not be dealt with seriously in Park's work, and it was not. Frazier approached the question from the opposite direction, emphasizing not the similarities between mild and violent forms of prejudice, but the similarities between violent prejudice and mental illness. "If," Frazier (1927) said, "we consider some of the newer conceptions of mental processes as they apply to abnormal behavior, we shall find in each case that the behavior motivated by race prejudice shows precisely the same characteristics as that ascribed to insanity." In the insane person, he continued, there was a "dissociation of consciousness," a relatively permanent severance of the mental state from objective reality. So it was with prejudiced Southerners for whom "the whole system of ideas respecting the Negro is dissociated from the normal personality and . . . this system of ideas seems exempt from the control of personality." Thus the Southerner could write and speak of the glories of democracy and the sacredness of human rights and fully believe in these things, and yet approve and even participate in the lynching and mutilation of Negroes. Frazier called this the "Negro complex" and concluded his article by likening the South to an insane asylum controlled by the inmates.

In Park's view, prejudice was not rational, but it was certainly normal. Because it was "human," nothing much could be done about it. To Frazier, certain forms and intensities of prejudice were not only irrational but sick. The comparison to mental illness carried two implicit corollaries: like mental illness, race prejudice was a terrible thing, and like mental illness, the more pathological forms of race prejudice might be susceptible to cure, making the latter explicit. Frazier said, "Race prejudice is an acquired psychological reaction, and there is no scientific evidence that it represents the functioning of inherited behavior patterns" (Frazier, 1927).

The general attitude toward the race problem implicit in Park's essay on prejudice died out a few years later because it was simply not productive of the sort of searching and critical scholarship that the problem required. Frazier's perspective, on the other hand, with its refusal to regard given social arrangements and modes of behavior as normal, and its impatience to find out why such things existed if they were not immutable, had the force and intellectual toughness to carry the study of the Negro into the productive period of the thirties.

CONVERGENCE OF THE DISCIPLINES

In every sector of this broad field of inquiry, one will find that in the late twenties and early thirties the most useful and productive work was motivated by a decided impatience with the status quo. The increasing number of intelligence studies negating earlier results was fast breaking down the racial hypothesis. But, if these studies brought into discredit traditional explanations of the Negro's backwardness, they left moot the question of its actual causes. What was needed was a picture of the Negro in his economic and social environment, and now that picture was being filled in with rapid strokes. Lorenzo J. Greene and Carter G. Woodson's *The Negro Wage Earner* (1930) surveyed the history of the Negro in various occupations since 1890, when the United States government had begun to compile occupational statistics according to race. Woodson and Greene were not just conducting a statistical survey. They were, as they said, interpreting data. Their unsurprising conclusion was that the history of the Negro wage earner had been one of economic discrimination—of differential wage rates, occupational exclusion, denial of entry into trade unions, and so forth (Greene and Woodson, 1930).

In 1931, Sterling Spero and Abram L. Harris published *The Black Worker,* a history of the Negro's experience with organized labor from the post-Civil War period to 1930. The soundness of the book's scholarship was in no way diminished by the authors' intention of showing that the AFL craft system of autonomous unions made exclusion and discrimination easy, and that the Negro's future in organized labor, if he had one, lay with industrial unionism (Spero and Harris, 1931).

All of the reformist attitudes so far discussed are found in Charles S. Johnson's *The Negro in American Civilization* (1930). The book was the outcome of the National Interracial Conference, in 1926, at which representatives of a number of organizations met to consider sponsoring a study of Negro life. The organizations included the National Urban League, the NAACP, the Federal Council of Churches, the National Catholic Welfare Conference, the American Friends Service Committee, and the Commission on Interracial Cooperation, a group of white Southern liberals formed after World War I to promote "understanding" between the races in the South. A research committee was formed which included Park and the University of North Carolina sociologist Howard W. Odum. Johnson was chosen to serve as "research secretary" and to write the book.

Mary Van Kleeck, the chairman of the 1926 meeting, indicated the reformist temper of the whole project in her introduction to the book. "The synthesis of available data," she said, "revealed a sound basis for planning programs of improvement. That the Negro is racially susceptible to disease, racially incapable of education or racially apt to violate laws, is a case not proved." The conclusion was inescapable that "there is here a task of planning and engineering. A fatalistic attitude which is skeptical of the possibility of action is not supported either by experience or by sound facts" (Johnson, 1930: vi-ix).

The study broke fresh ground in several areas, but it was particularly welcome in its focus upon Southern Negro farmers. Despite recent migrations to the North, they still formed the great majority of American blacks. Working with the Commission on Interracial Cooperation and the University of North Carolina, Johnson's staff was engaged, during 1928, in an intensive study of the social and economic conditions of Negroes in two typical Georgia black-belt counties. "So far as is known," Johnson (1930: 368) commented, "this is the first effort made to get an adequate picture of rural Negro conditions." This was indeed a pathfinding effort. In

1928 little was known in a scientific way about the rural South; by 1940, thanks to the work of Negro sociologists such as Johnson, Frazier, and Allison Davis, and their Northern white colleagues, and to the work of Odum's North Carolina group, as well as to research projects carried out by the federal government, the rural South, white and black, was as well mapped out sociologically as any region in the United States.

Johnson's book concentrated on Negro education as well. Negro educators and psychologists had long protested that the intelligence test results were indicative not of innate differences between blacks and whites but of unequal educational opportunities (Bond, 1924; Long, 1925). Such arguments, if persuasive on grounds of common sense, lacked the comprehensive statistical basis they required if they were to be listened to, much less acknowledged as true, by a largely indifferent white community. With Johnson's study, the statistical basis was established. For the first time, low educational opportunity was not simply put forward at the conclusion of an article as a possible explanation of lower Negro scores, or as an accentuating factor, but was demonstrated to be in almost exact correlation with the rate of "retardation" (Johnson, 1930: 280-287).

By the early thirties, enough had been accomplished so that work in sociology influenced results in psychology, and vice versa. Klineberg (1935a), discussing the effects of low social and economic status on intelligence testing results, observed in *Race Differences* that in the case of the Negro, "the difficulties in his way are so great that any inference from industrial status to intelligence is completely unwarranted. The kind of competition the Negro has to face in his search for a job has been carefully described by Johnson . . . and Spero and Harris" (Klineberg, 1935a: 163). Similarly, in discussing the Negro's educational opportunities, Klineberg (1935a: 176) was able to cite Johnson on "the frequent discrepancy in per capita expenditures, in equipment and facilities, in the training of the teachers, and in the length of the school year."

The best single indication of how radically the balance of forces in the nature-nurture controversy had altered by the mid-thirties is contained in the 1934 special issue of *The Journal of Negro Education* devoted to "The Physical and Mental Abilities of the American Negro."[4] In the dozen or more articles in this issue are to be found representatives of virtually every persuasion on the testing issue, from Klineberg and Herskovits, to Garth and Rudolph Pintner, one of the most dogmatic of the tests' defenders and a man possessed of an almost invincible skepticism toward evidence from the environmental side (Pintner, 1931: 41, 436).

Garth had, as early at 1931, backed off from his sweeping conclusions of the early twenties, writing in that year that there was no "indisputable evidence for belief in mental differences which are essentially racial" (Garth, 1931: 24). Nevertheless, as a conscientious eugenicist, Garth (1931: 84) had stuck by "selection" as a possible factor in the test scores, in effect holding to a kind of intermediate position between nature and nurture. Now, in 1934, he quite gracefully recanted. The tests, he wrote, were all white men's measuring devices "and naively, though rather generally, side-step the very criterion for racial use, that of universality" The fact was that these devices "make the white experimenter appear distressingly provincial" (Garth, 1934: 323). Only Pintner now held out for a *presumption* of Negro inferiority, and his conclusions were qualified and tentative (Pintner, 1934: 517).

In summary, from 1916 and the publication of Ferguson's pioneering study, to 1925 or 1926, the racial hypothesis had held the high ground in the war of the psychological journals, although toward the end of this period it was being subjected to increasingly effective attack from other quarters. From about 1925 on, however, the racial hypothesis was being attacked from within psychology itself. It is impossible to say which of these two operations was more important, though it is clear that the combination of the two reduced hereditarian doctrine to an enthusiasm of a beleaguered few.

Viewing the movements described above in terms of the change in orientation they brought about in the social sciences as a whole, it will be seen that the breakdown of the idea of "intelligence" as something autonomous and somehow independent of environment involved an accommodation by psychology to new insights provided especially by anthropology and secondarily by sociology. The concept of intelligence was only one of several related psychological concepts affected by this accommodation. More basic, perhaps, was the idea of personality, which underwent a redefinition away from that of a configuration of impulses more or less fixed at infancy to the idea of a mechanism of adaptation to a specific cultural environment.

Such a redefinition was made necessary, in part, by the discovery of the anthropologists that certain forms of behavior sanctioned and considered normal in one culture might be regarded as aberrant or pathological in another. The lesson for psychology was clear. As Ruth Benedict (1934: 236) stated it:

> If we are interested in human behavior, we need first of all to understand the institutions that are provided in any society. For human behavior will take the forms those institutions suggest, even to extremes of which the observer, deep-dyed in the culture of which he is a part, can have no intimation.

I have called this reorientation an accommodation of psychology to anthropological knowledge. It might equally be regarded as a *convergence* of anthropology and psychology upon a heretofore unoccupied area between them. Benedict dismissed as a verbal quarrel the argument which had been going on between psychologists and anthropologists over which factor, culture, or personality, was the determinant and which the effect. The relationship between the two, it was now understood, was a dynamic one.

From this convergence of the social sciences on the culture-personality nexus there would ultimately evolve the idea of "basic personality" or "group character" (Potter, 1954: ch. 2). Abram Kardiner and Lionel Ovesey, in their brilliant *The Mark of Oppression* (1951), would apply the basic personality concept to the Negro, defining it as "the mental and emotional processes shared by a group of people who live under the same institutional and environmental conditions." Thus the "subjective impact of social discrimination" was revealed—a nagging, ever-present knowledge of the contempt and hostility of whites, low self-esteem, irritability, and fear of loss of self-control, all masked by a hard-learned affability, when affability was possible (Kardiner and Ovesey, 1951: 302-304).

The Mark of Oppression was published seven years after *An American Dilemma,* but it shows the direction in which studies of the Negro were tending when Myrdal's book was written.[5] If one looks back a decade to 1932, one will find that sociology had hardly begun to evolve a notion of Negro personality cast in terms of environment. One essential condition had, however, already been met for such a development: the patent disutility of stereotype thinking had been clearly demonstrated.

This was due more to black northward migration than to an increased sophistication in the social sciences. Robert E. Park had written in 1918 that the high incidence of black folk songs in a major key tended to justify the general impression "that the Negro is naturally sunny, cheerful, optimistic." The Negro, Park thought, was by "natural disposition" neither an intellectual nor an idealist, neither a "brooding introspective" nor a pioneer. He was "primarily an artist, loving life for its own sake" (Park, 1950: 278-280). This characterization, not much different from the Southern white stereotype of the improvident but lovable "happy nigger," was just barely adequate to the realities of plantation life, but it simply would not do after the great migration. In 1923, in an essay on recent black poetry, Park recognized the existence of some surprising qualities: pride, hope, cynicism toward the white man, and anger. Now that the Negro was learning to live in cities, Park said (1950: 284-300), the traditions and habits acquired in slavery were rapidly vanishing. Some Negroes were even becoming Bolsheviks!

In short, undeniable diversity had defeated stereotype, and social scientists would soon be compelled to seek new explanations. The groundwork for such analysis was carried out during the thirties in works by Frazier (1932), Johnson (1934), Harris (1936), and Dollard (1937). All of these works were sociological in method and purpose, and were concerned with effects of the environment rather than the personality affected by it. But in studying the Negro as peasant, as capitalist, and as family man, in describing the Negro community and the relationships between that community and the white world, these scientists rendered the Negro's known or ascribed characteristics at least comprehensible without recourse to stereotype or biological theory. It was apparent from these studies that the social, economic, and psychological pressures bearing upon the Negro formed a dense, smothering net, whether he lived in a Southern town, in a black-belt cotton county, in a Northern ghetto, or even in a middle-class suburb, and that the effect of such pressures could only be the formation of "neurotic" and defeatist behavioral patterns setting the Negro off from white Americans.

Later in the decade, four studies were carried out under the auspices of the American Youth Commission (AYC) to assess the psychological effects on Negro youths of membership in a minority racial group. The publication of these titles (Johnson, 1941; Frazier, 1940; Davis and Dollard, 1940; Warner, Junker, and Adams, 1941) marked the last in a series of breakthroughs in the twenty years or so preceding the publication of *An American Dilemma.* If one conceives of the sociological study of the Negro as a building operation, as an effort to create a coherent structure of analysis, then these studies of Negro youths form a huge block directly beneath the apex, which is the Myrdal study itself.

The AYC studies were grounded in interviews with thousands of Negro youths of all social classes, locales, and conditions. Many pages were given over to the answers of youths, with the interviewers' questions deleted, so that what resulted were extended personal statements, often of real eloquence, of the kind later employed by Oscar Lewis in his studies of Mexican and Puerto Rican families. The effect was to dispel many myths, among them the belief common to whites and to genteel Negro novelists of an earlier time that the light-skinned and upper-class Negro experienced greater torment than the lower-class black because the mulatto lived "between two worlds" and was neither of one nor the other. These studies demonstrated that being fully "integrated" into a subordinate and despised caste was no guarantor of psychological health; on the contrary, a worse condition could hardly be imagined.

In the AYC studies (and in Richard Wright's *Native Son,* published the same year, 1940) lower-class Negroes who formed by far the largest part of the black population as a whole, ceased to be silent figures in the background of a pastoral tableau or dimly seen faces in the urban crowd. They came alive for the reader, spoke for themselves, and were revealed to be complex human beings. The "veil" of caste, of which DuBois had spoken for so many years, had been partially lifted.

CONCLUSIONS

With the publication of *An American Dilemma,* the environmentalist view established its paramountcy in the social sciences. Its victory may be termed decisive, inasmuch as almost all subsequent work would proceed from environmentalist premises. Whether or not one wishes to term its victory final, however, depends on the point of view. In the broadest sense, the environmental revolution was more moral than scientific. A change in white sensibility, a gradual altering of perceptions, a growing ability to see black men not as mysterious or grotesque others but as human beings—these developments were in the long run more important politically and culturally than the establishment of a scientific hypothesis.

The environmental hypothesis considered as a hypothesis never was, and perhaps never could be, established beyond doubt within psychology. In 1960, two psychologists who confessed themselves "frankly environmentalist in our bias" noted that the results of many researches might as yet "be interpreted from an hereditarian viewpoint without doing violence to them" (Dreger and Miller, 1960). In 1963, Klineberg could still say only that psychologists had "no right to assume that Negroes are, on the average, innately inferior" (Klineberg, 1963). Another environmentalist wrote the following year that it was now the "overwhelming view" that mean differences between white and black test scores "are *largely* the results of environmental, rather than genetic, factors" (Pettigrew, 1964; italics added). Strictly speaking, then, there was within psychology not victory but stalemate, and the most important fact about test results was not that they had dramatically changed, but that they were no longer news. Not until 1969 when, with the publication of Professor Arthur Jensen's deliberately contentious essay upholding the hereditarian hypothesis, was the old argument once again the subject of conversations among laymen as well as scholars (Jensen, 1969).

Jensen asserts that when all environmentalist arguments have been considered, there remain discrepancies in test scores between blacks and whites that strongly suggest significant genetic differences. He therefore proposes some sort of eugenics program to curtail lower-class black birthrates, and an educational policy under which the majority of white children would receive instruction in "cognitive learning" (problem-solving), while the majority of black children would receive drill in "associative learning" (rote), which would in fact, if not precisely in intent, mean exactly what Myrdal (1944) thought so unlikely, "a differential treatment in matters of public policy."

What is both depressing and instructive about the present "nature-nurture" controversy is that although it is now conducted on a higher technical plane than fifty years ago, and with more sophisticated weaponry, it remains, in terms of the analogies drawn, the values invoked, and the postures assumed by each side, pretty much the same. And although it may only be a declaration of the present writer's bias, this seems particularly true of the "nature" side. Jensen, like Pintner before him, tends to regard "environment" as an irritating thesis pushed by endlessly resourceful opponents instead of as the sum of the conditions under which human beings live. Like Brigham, he is alarmed at the "dysgenic" effects which must inevitably result from the breeding habits of the poor. Like Terman, he is exasperated by the objection of opponents that we do not as yet, and perhaps never will, really know what "intelligence" is. Quite independently of Terman, he even draws the same dubious analogy between intelligence and electricity, writing that intelligence, like electricity, "is easier to measure than to define" (Jensen, 1969: 5).

There are not only similarities, there are continuities as well. The authority most often cited by Jensen is the English psychologist Cyril Burt, who a half-century ago was a student of William McDougall. McDougall, who in *Is America Safe for Democracy?* (1921) wrote of the fearful consequences of continued immigration, was in turn a disciple of Sir Francis Galton, the original proponent of the "nature" explanation of class differences. One finds McDougall citing Burt's early experiments in an exchange with Walter Lippmann in 1923, just as Jensen cites Burt's late experiments in an exchange with Richard C. Lewontin in 1970 (McDougall, 1923; Jensen, 1970).

After fifty years of inconclusive struggle, one is drawn to the conviction—call it the hypothesis—that the "nature-nurture" controversy is not primarily scientific but political. As Myrdal (1944) wrote, "The place of the individual scientist in the scale of radicalism-conservatism has always had, and still has, strong influences upon both the selection of research problems and the conclusions drawn from research." In a sense, he continued, *"it is the master scale of biases in the social sciences"* (Myrdal, 1944: 1038). If in the future our philosophical premises are egalitarian, our scientific conclusions are likely to be egalitarian also.

NOTES

1. In fact, a good performance on beta depended in part on the testee's acquaintance with the artifacts and the way of life of the American middle class. Among the picture completion tests were a tennis court without a net, a dormer-roof house with the top of the chimney missing, and a victrola of the day without a megaphone. The testers were oblivious to the problem of how an Indian from a Western reservation, a tenement dweller from Little Italy, or a black sharecropper could possibly make anything of such pictures, seldom or never having seen tennis courts, houses with chimneys, or victrolas. Behind the racism of the testers, there was much naiveté.

2. Interestingly enough, the most effective challenge to the sweeping conclusions that Herskovits drew from his experiment (involving a small number of Howard undergraduates) came from Otto Klineberg (1928: 55-56). Klineberg questioned Herskovits' conclusions because of the small size of the sample group and because of Herskovits' reliance on inferential logic in lieu of precise historical and social knowledge. Klineberg said that Herskovits had not really eliminated race as a factor in intelligence. It is an indication of the confidence of the environmentalists that they could thus afford to play devil's advocate.

3. Myrdal rarely expressed complete confidence in the judgment of American scholars. But of Frazier's *The Negro Family in the United States* (1939), he wrote, "The book . . . is such an excellent description and analysis of the American Negro family that it is practically necessary only to relate its conclusions to our context and to refer the reader to it for details" (Myrdal, 1944: 931). The admiration was mutual. Frazier later recalled that he had been skeptical at the choice of a foreigner to head the study, but that his skepticism was dissipated when he came to know Myrdal. For Negroes "were simply people to him." Myrdal could "size up a Negro for what he was worth because he did not view him either

with the pathetic attitude of most friendly white Americans or through the myths that have grown up about Negroes in American life" (Frazier, 1945: 577).

4. The appearance of the Journal itself in 1931, published at Howard University, is another instance of how the new Negro protest, understood in the broad sense, had by the 1930s ascended to a higher level of sophistication, objectivity, and academic respectability. It is significant that in the early 1920s, white intelligence testers like T. H. Garth held sway in the professional journals while Negro educators who protested their results had available to them, for the most part, only magazines such as *The Crisis* and *Opportunity*. In the Journal special issue, we find Garth and Howard H. Long sharing the same forum.

5. There are, in fact, preshadowings of Kardiner and Ovesey's insights scattered throughout *An American Dilemma,* and especially in the brief chapter on "The Protest Motive and Negro Personality."

REFERENCES

BAKER, R. S. (1908) Following the Color Line: American Negro Citizenship in the Progressive Era. New York: Harper & Row.

BENEDICT, R. (1934) Patterns of Culture. New York: Houghton-Mifflin.

BOAS, F. (1911) The Mind of Primitive Man. New York: Macmillan.

BOND, H. M. (1924) "What the Army 'intelligence' tests measured." Opportunity 2.

BONE, R. A. (1965) The Negro Novel in America (revised). New Haven: Yale Univ. Press.

BRIGHAM, C. C. (1923) A Study of American Intelligence. Princeton: Princeton Univ. Press.

DAVIS, A. and J. DOLLARD (1940) Children of Bondage. Washington D.C.: Amer. Council on Education.

DOLLARD, J. (1937) Caste and Class in a Southern Town. New Haven: Yale Univ. Press.

DREGER, R. M. and K. S. MILLER (1960) "Comparative psychological studies of Negroes and whites in the United States" Psych. Bulletin 57, 5 (September).

DuBOIS, W. E. B. (1967) The Philadelphia Negro: A Social Study [1899]. New York: Schocken.

--- (1944) "The American dilemma." Phylon (Second Quarter).

--- (1915) The Negro. New York: H. Holt.

FERGUSON, G. O., Jr. (1916) "The psychology of the Negro: an experimental study." Archives of Psychology 36 (April).

FRAZIER, E. F. (1945) "Review of an American dilemma." Amer. J. of Sociology 50.

--- (1940) Negro Youth at the Crossways. Washington, D.C.: Amer. Council on Education.

--- 1939) The Negro Family in the United States. Chicago: Univ. of Chicago Press.

--- (1932) The Negro Family in Chicago. Chicago: Univ of Chicago Press.

--- (1927) "The pathology of race prejudice." Forum 77 (June).

GARTH, T. R. (1934) "The problem of race psychology: a general statement." J. of Negro Education 3 (July).

--- (1931) Race Psychology. New York: Whittlesey.

--- (1926) "Race psychology." Scientific Monthly 23 (September).

--- (1923) "A comparison of the intelligence of Mexican and mixed and full blooded Indian children." Psych. Rev. 30: 388-401.

GOSSETT, T. F. (1963) Race: The History of an Idea. Dallas: Southern Methodist Univ. Press.

GREENE, L. J. and C. G. WOODSON (1930) The Negro Wage Earner. (Reprinted 1969) New York: Russell & Russell.

HALL, M. O. (1911) Half a Man: The Status of the Negro in New York. New York: Longmans, Green.

HARRIS, A. L. (1936) The Negro as a Capitalist. Philadelphia: Amer. Academy of Political and Social Science.

HERSKOVITS, M. (1926) "On the relation between Negro-white mixture and standing in intelligence tests." Pedagogical Seminary 33: 30-42.

JENSEN, A. R. (1970) "Race and genetics of intelligence: a reply to Lewontin." Bull. of the Atomic Scientists 25, 5 (May).

--- (1969) "How much can we boost IQ and scholastic achievement?" Harvard Educational Rev. 39, 1 (Winter).

JOHNSON, C. S. (1941) Growing Up in the Black Belt. Washington, D.C.: Amer. Council on Education.

--- (1934) Shadow of the Plantation. Chicago: Univ. of Chicago Press.

--- (1930) The Negro in American Civilization. New York: Holt.

KARDINER, A. and L. OVESEY (1951) The Mark of Oppression. New York: Norton.

KLINEBERG, O. (1963) "Negro-white differences in intelligence test performance: a new look at an old problem." Amer. Psych. 18, 4 (April).

--- (1935a) Race Differences. New York: Harper & Bros.

--- (1935b) Negro Intelligence and Selective Migration. New York: Columbia Univ. Press.

--- (1928) "An experimental study of speed and other factors in 'racial' differences." Archives of Psychology 15.

KROEBER, A. L. (1934) "Cultural anthropology," in M. Bentley and E. V. Cowdry (eds.) The Problem of Mental Disorder. New York: McGraw-Hill.

LIPPMANN, W. (1922) "The mystery of the 'A' men." New Republic 32 (November 1).

LONG, H. H. (1925) "On mental tests and racial psychology—a critique." Opportunity 3 (May).

McDOUGALL, W. (1923) Letter. New Republic 34 (May 23).

--- (1921) Is America Safe for Democracy? New York. C. Scribner's.

MEAD, M. (1926) "The methodology of racial testing: its significance for sociology." Amer. J. of Sociology 32, 5 (March).

MOTON, R. R. (1929) What the Negro Thinks. Garden City, N.Y.: Doubleday, Doran.

MYRDAL, G. (1944) An American Dilemma. New York: Harper & Row.

ODUM, H. W. (1951) American Sociology. New York: Longmans, Green.

PARK, R. E. (1950) Race and Culture. Glencoe: Free Press.

PETTIGREW, T. F. (1964) "Negro American intelligence: a new look at an old controversy." J. of Negro Education 33, 1 (Winter).

PINTNER, R. (1934) "The intelligence of the Negro." J. of Negro Education 3 (July).

--- (1931) Intelligence Testing (second edition). New York: Holt.

POTTER, D. (1954) People of Plenty. Chicago: Univ. of Chicago Press.

REUTER, E. B. (1917) "The superiority of the mulatto." Amer. J. of Sociology 23, 1 (July).

SPERO, S. D. and A. L. HARRIS (1931) The Black Worker: The Negro and the Labor Movement. Port Washington, N.Y.: Kennikat.

STOCKING, G. W., Jr. (1968) Race, Culture and Evolution. Glencoe: Free Press.

TERMAN, L. M. (1922) "The great conspiracy or the impulse imperious of intelligence testers, psychoanalyzed and exposed by Mr. Lippmann." New Republic 33 (December 27).

--- (1916) The Measurement of Intelligence. New York: Houghton-Mifflin.

WARNER, W. L., B. H. JUNKER, and W. A. ADAMS (1941) Color and Human Nature. Washington, D.C.: Amer. Council on Education.

WOODWORTH, R. S. (1916) "Comparative psychology of races." Psych. Bull. 13,10 (October 15).

WRIGHT, R. (1940) Native Son. New York: Harper & Brothers.

YODER, D. (1928) "Present status of the question of racial differences." J. of Educational Psychology 29, 7 (October).

3

Power in the Reformulation of Race Research

IRA KATZNELSON

□ SICKENED BY THE VISION of "great, ugly whirlwinds of hatred, and blood and cruelty" of impending racial conflicts, W. E. B. DuBois, both deeply pessimistic and realistic, wondered in 1920, "If I cry amid this roar of elemental forces, must my cry be in vain, because it is but a cry—a small and human cry amid Promethean gloom" (DuBois, 1920: 52). A half-century later, despite massive racial upheavals and the demise of most formal, overt colonial arrangements (including the end of legally sanctioned segregation practices in the United States), the realist must remain a pessimist. Few today could convincingly take issue with Aron's prediction that "during the final third of the 20th century, ethnic conflicts over social, racial, or political dominance—in turn or simultaneously—appear more likely than the continuation of the class struggle in the Marxist sense" (Aron, 1969: 46). Paradoxically, today's "realistic pessimism" follows a long period of roughly forty years that was characterized by what Harry Hoetink has called "sociologistic optimism"; most race relations scholars, especially American scholars, were convinced that by revealing the nature of racial myths and prejudice they could administer therapy to their racist society (Hoetink, 1967).

This optimistic, hopeful vision has been mocked by recent events. In the United States the illusion of racial progress promoted by the passage of the Civil Rights Act in 1963 was shattered by widespread racial violence in the middle years of the decade. As a consequence, race relations scholars, and white scholars in particular, have been compelled to reexamine their role and the nature and direction of their research. This chapter addresses itself to these questions. Its purpose is to delineate some of the problems of the white social scientists concerned with race and to explore and propose an approach to the study of race that can, on a comparative basis, facilitate the first-order task of the identification and analysis of the structured relationships that define the limits of choice and racial justice.

THE ROLE OF THE WHITE SCHOLAR?

The relationships that link social scientists and their work to their social and political locale raise difficult and nettlesome issues. Given present patterns of racial dominance, these issues need to be raised with great urgency by white scholars whose work deals with race. The easy shibboleths about scholarly independence, appropriateness, objectivity, and value neutrality can no longer be uttered except by the most facile and insensitive. The insistent struggles by Third World peoples for decolonization and liberation have overtaken these glib formulations.

Thus, the role of the white scholar, who recognizes that to be a man is to be responsible, is an anomolous role. One is reminded of Albert Memmi's vivid characterization (1965: ch. 2, pp. 19 ff.) of the colonizer who refuses, who resists the colonial arrangements from which he benefits, who can usually only observe the nationalistically oriented struggles from afar, and who remains, despite his good intentions, politically impotent. The white race relations scholar resembles, as a result, the lonely adolescent, who, in Octavio Paz's words (1961: 9), "is astonished at the fact of his being, and this astonishment leads to reflection: as he leans over the river of his consciousness, he asks himself if the face that appears there, disfigured by the water, is his own. The singularity of his being, which is pure sensation in children, becomes a problem and a question."

In the past decade, many white social scientists have been increasingly and painfully puzzled by the issues of race. Well-intentioned liberal scholars, who have subscribed to the view that color should be irrelevant, that the political and academic worlds should be color blind, are troubled by many blacks' rejection of racial integration. In sincere but insensitive response they accord equal weight to the positions of white racists and black separatists, not realizing—as a black Englishman put it—that "white racism is at one level a matter of choice, at another a matter of privilege, but at all levels an exercise in oppression. White racism incurs somewhere down the line the denial of human dignity; black 'racism' envisages the destruction of that denial" (Sivanandan, 1968: 13).

Many Marxist and quasi-Marxist scholars, too, have had difficulty in dealing with race. They are disturbed by what they see as the deflection of class conflict and consciousness into racial conflict and consciousness, typically arguing that "anyone who assumes race exists . . . is playing the racialist game" (Harris, 1968: 24). Like the liberal's sentiments, the dogmatic Marxist's are noble, but not rooted in reality. Color has been a mark of oppression related to, yet quite independent of class. Among other indications, the blue-collar Wallace vote in 1968, and the fierce resistance by members of American construction unions to black membership, indicate that, at least for the present, color and class do not share a common denominator. The Marxist theologian might reply with appropriate arguments about false consciousness (arguments with which I would concur), yet he can no longer argue intelligently that corrosive racial distinctions, whatever their origin, have not created categories of events and behavior that must be dealt with on their own terms.

In his 1956 letter of resignation from the French Communist Party, the Martinique poet, Aimé Césaire (1956: 6, 11) affirmed the duality of class and race. He wrote Thorez:

> One fact, crucial in so far as I am concerned, is this: that we coloured men, in this specific moment of historical evolution, have consciously grasped, and grasped in its full breadth the notion of our peculiar uniqueness, the notion of just who we are and what, and

> that we are ready, on every plan and in every department, to assume the responsibilities which proceed from this coming into consciousness.
>
> The peculiarity of "our place in the world" isn't to be confused with anybody else's. The peculiarity of our problems aren't to be reduced to subordinate forms of any other problems. . . .
>
> In these conditions, it will be understood that we cannot delegate anyone to think for us; to do our searching, to make our discoveries; that we cannot henceforth accept that anyone at all, be he our best friend, answer for us.

If we take Césaire's last point seriously (as we should), then we must indicate with some precision what the role of white social scientists should be in the area of race relations.

It is possible, of course, to resolve this question by counseling white withdrawal from the field, temporarily at least. But this categorical position, in my view, confuses the tasks of black leadership and indigenous decision-making with contextual issues of black-white interaction, with those structural economic, social, and political linkages that define the limits of choice and social justice, and which, therefore, must be of as much concern to whites as to blacks. To state explicitly what should be obvious: however benevolent the intention, white scholars have no role to play in solving internal black community disputes, no role to play in leading black organizations, no role to play in defining black community goals or the direction of black political activity. Unsolicited advice concerning such issues as the relative merits of political and cultural nationalism must be shunned.

What then should be the role of the white social scientist dealing with race? The answer has a great deal to do with the appropriate task of the social scientist, whatever his subject matter, and to that I will return. But first, a discussion of two well-meaning, but misdirected approaches should help clarify the issue.

One approach to race studies now in vogue is that of urban ethnography, the attempt to apply the field techniques of the anthropologist to racial minority groups in complex industrialized societies. To date, many, though not all, urban ethnographers have focused on what Oscar Lewis has called the culture of poverty, or the cultural and personality problems of the ghetto poor (Lewis, 1966:

li-lii; also see Whitten and Szwed, 1970). Many ethnographers argue, that these patterns, which are dysfunctional for mobility in American society, must be modified so that blacks can take advantage of the *existing* opportunity structure. The direct policy implications are clear, and were spelled out, to cite a leading example, in Daniel Moynihan's report (1965) on the black family:

> What then is the problem? We feel the answer is clear enough. Three centuries of injustice have brought about deep-seated structural distortions in the life of the Negro American. At this point the present tangle of pathology is capable of perpetuating itself without assistance from the white world. *The cycle can be broken only if these distortions are set right. In a word, a national effort towards the problems of Negro Americans must be directed toward the problems of family structure.*

This approach to urban ethnography and the policy uses to which it has been put have been criticized vigorously, perhaps most convincingly by Charles Valentine in his volume *Culture and Poverty* (1969). He argues that many urban ethnographers misleadingly portray racial problems as black problems of personality and behavior, while ignoring the interdependence of systemic arrangements, cultural patterns, and debilitating effects on personalities. The life cycle of the ghetto poor can more fruitfully be seen as a cluster of related adaptations to structural-situational factors that limit the chance to choose. Thus, if society is conceptualized in terms of a hierarchy of choice behavior possibilities, the ghetto poor are at the bottom with the fewest choices. It is crucial, therefore, to analyze those structured economic, social, and political relationships that limit choice and destroy psyches.

On the face of it, this is what social scientists have been doing who have participated in the research and writing of many of the recent riot commission reports published by government agencies at the local, state, and national levels in the United States. But in this respect, Robert Lynd's classic question (1945) obtains, "Knowledge for What?" Most of the reports—the McCone Commission report on Watts (see Blauner, 1966: 3 ff.; Coser, 1968: 101 ff.) being the most glaring example—have sought to discover the means to secure the greatest level of black docility at the lowest cost to the wider society.

Put another way, many social scientists have become interested in race relations only since American blacks, using the most effective tactic available to the powerless, that of disruption, have been able to make inputs into the political system. (It is noteworthy that of the 2,611 articles published by the *American Political Science Review* between 1906 and 1963, only thirteen contained the words "Negro," "Race," or "Civil Rights" in their titles; see Matthews, 1969: 113.) Simmel (1968: 8-9) noted an analogous phenomenon when he commented on welfare assistance programs to the poor in Germany in the early years of this century:

> It becomes clear that the fact of taking away from the rich to give to the poor does not aim at equalizing their individual positions and is not, even in its orientation, directed at suppressing the social differences between the rich and the poor. On the contrary, assistance is based on the structure of society, whatever it may be. . . .The goal of assistance is precisely to mitigate certain extreme manifestations of social differentiation, so that the social structure may continue to be based on this differentiation.

Consider the best of the riot reports, the Kerner Commission's *Report of the National Advisory Commission on Civil Disorders* (1968), which dramatically indicated "white racism," concluding, "Our nation is moving toward two societies, one black, one white–separate and unequal." Even this report, as two critical observers have noted, "contained almost no criticism of established institutions or programs" (Lipsky and Olson, 1969: 17). The organizational ties of the commission membership–which included legislators, a union leader, the chairman of a corporation board, a mayor, and a police chief–made it difficult, perhaps impossible, to explore the institutional context and expression of the white racism the commission considered so central to its findings. Looked at this way, the report is no more than a placatory symbolic document that provided the Johnson Administration with the opportunity to delay implementation of its miminal programs. The report, Parenti has observed (1970),

> treats the obviously abominable living conditions as "causes" of disturbance but never really inquires into the causes of the "causes,"

> viz., the ruthless enclosure of Southern sharecroppers by big corporate farming interests, the subsequent mistreatment of the black migrant by Northern rent-gorging landlords, price-gorging merchants, urban "redevelopers," discriminating employers, insufficient schools, hospitals and welfare, brutal police, hostile political machines and state legislators, and finally, the whole system of values, material interests and public power distributions from the state to the federal capitols (sic) which gives priority to "haves" than to "have-nots," servicing and subsidizing the interests of private corporations while neglecting the often desperate needs of the municipalities.

Thus, like some urban ethnographers, the riot commission policy proposers focus on the poor, "ignoring the system of power, privilege, and profit which makes them poor. It is a little like blaming the corpse for the murder" (Parenti, 1970).

By ignoring the structure of power, riot commission studies have usually missed the political side of urban violence. Those who participated in the ghetto rebellions of the 1960s focused their attention on symbols of public authority and political legitimacy; thus the riots have been realistically compared to "native" uprisings against oppressive, colonial-like relationships (Blauner, 1966: 3ff.). Those social scientists who have serviced the needs of the government-appointed commissions at best have unwittingly played a symbolic game, leaving themselves open to the charge of having helped to perpetuate the network of structured relationships that produced the conditions they claim to abhor.

SOCIAL CONSEQUENCES OF RESEARCH

Writing about race relations research, Melvin Tumin recently observed "that never has social research been so substantial, in the first instance, and relevant in the second instance, to matters of pressing moment, and *so quickly followed up with social action and policy that run directly against the major grain of the research*" (Tumin, 1969: 254). Whether intended or unintended, research in this field clearly has social consequences. To avoid the question of consequences is to obscure a central, and inescapable dilemma, one that is characteristic of the social sciences as a whole.

The problem of intention and consequences, of course, is not new for social scientists. Marx implored us to change the world; Weber to maintain distance, to be objective.[1] Recently, Dahrendorf has made an interesting attempt to reformulate the issue by compartmentalizing the sociologist and his sociology. Dahrendorf, on the whole, accepts the canons of scientific objectivity but cautions, "Our responsibility as sociologists does not end when we complete the process of scientific inquiry." Thus "it is the sociologist's business to consider what a modern, open, civilized society might look like, and what roads might lead to it. This is the domain of theory. It is also the sociologist's business, once he is equipped with his theories, to take part in the process of changing reality, in making what is reasonable real. This is the domain of practice" (Dahrendorf, 1968: 278). Though this formulation has appeal, the neat dualism of theory and practice is unconvincing, for however detached or conventional the intent, "no two arrangements of social knowledge have quite the same uses or political implications" (Stretton, 1969: 64).

Moreover, as Dankwart Rustow has argued, the social scientist *qua* social scientist is heavily involved in the area of values whether he likes it or not as a result of his relationship, unique for the sciences, with his subject matter and his audience. "As social scientists we are cast in the awkward role of men in society trying to explain to other men in society how men live in society" (Rustow, n.d.: 9). Seen in this light, objectivity cannot possibly be achieved in the physical science sense. The professed objectivity of the social scientist can more aptly be seen as a particular solution, fulfilling particular needs for social scientists seeking a particular kind of linkage with society. Thus, Gouldner perceptively sees objectivity as "the way one comes to terms and makes peace with a world one does not like but will not oppose" (Gouldner, 1970: 103).

Counterpoised to the pose of objectivity is the more appealing approach of "disciplined subjectivity" (Erikson, 1968: 683), disciplined by the canons of scholarship, yet involved with the subject of inquiry. This approach recognizes the essential links between the scholar, subject, and audience; the social scientist is both in and of his world. This is a persuasive approach, for why study society unless you care about society? Moreover, since social science scholarship has political and moral consequences, social scientists must face up to their political and moral responsibilities (Gouldner, 1970: 410).

These considerations are magnified in the field of race relations. Given the social scientist's web of relationships, attempts at value-free race relations scholarship are not simply absurd, but dangerous. We do not work and write in a moral and political vacuum. Therefore, race relations scholarship is today, whatever our degree of regret, more than an exercise in understanding. Given the subject matter, to write is to be implicitly political. For that reason, it is crucial that we be as clear as possible as to the meaning and potential consequences of our work. And, given the burden of our racial history, white scholars must be doubly aware.

The parameters of the white social scientist's role in the area of race can be assessed by a consideration of the broader question of the social scientist's relationship to those with power. C. Wright Mills examined three alternative public roles social scientists may seek to assume. The scholar might seek political power, coveting the role of philosopher-king. Alternatively, he might counsel those with power in a subsidiary, advisory role, as in the case of those who serviced the riot commissions. Mills urged a third course, with which I concur: "It is to remain independent, to do one's own work, to select one's own problems, but to direct this work *at* kings as well as *to* publics" (Mills, 1959: 181). In this role, the social scientist imputes responsibility to those with power for the consequences of their decisions, and educates those without power by examining the consequences of public decisions for their private lives. In this role, to both those with and without political power, the social scientist clarifies the range and consequences of choice.

At its best, a social science of disciplined subjectivity adopts approaches similar to those of Gandhi's *satyagraha.* In her analysis of the Gandhian philosophy of conflict, Joan Bondurant asks, "What does it mean to be a satyagrahi?"

> Gandhi repeatedly addressed himself to this question. He not only thought, spoke, and wrote about the continuing exploration of truth, but he fashioned a method of conflict in the exercise of which a man could come to know what he is and what it means to evolve. In satyagraha, dogma gives way to an open exploration of context. *The objective is not to assert propositions, but to create possibilities.*

> In opening up new choices and in confronting an opponent with the demand that he make a choice, the satyagrahi involves himself in acts of "ethical existence." The process forces a continuing examination of one's own motives, an examination undertaken within the context of relationships as they are changed towards a new restructured, and reintegrated pattern [Bondurant, 1965: vi-vii].

Thus, a social science of disciplined subjectivity seeks to fuse moral commitment, rigorous research, and theoretical illumination to define and extend the range of choice. It is, therefore, imperative to focus on those structured relationships that limit behavioral possibilities. Stretton has put the point clearly:

> We need a science of choice and possibilities. Such a science needs moral as well as technical originality. It must invent as well as discover. Inescapably, it must include political as well as technical debate. It cannot be a science only of durable laws and extrapolated trends. . . . For us, the purpose of knowledge is often to erode constants, escape from regularities, reduce the certainty of prediction; or if the regularities seem valuable, then to replace their causes, so that men may now choose what they once suffered from necessity [Stretton, 1969: 261-262].

Given these sets of commitments, social scientists concerned with race have, as a first-order task, the identification and analysis of the relationships, linkages, and structures—social, economic, political, and normative—that define the limits of choice behavior and racial justice. The following discussion proposes an approach to the study of race that can facilitate this urgent task.

PROPOSED APPROACH TO RACIAL STUDY

In the last decade and a half, under the impact of considerable methodological development, some of political science's most exciting work has been done by students of comparative politics who have moved away from the constraints of formalist, descriptive, institutional approaches. It is an unfortunate irony that despite these gains in methodological sophistication, and despite the growing recognition that the issues of race are of international significance, to date, to their mutual impoverishment, race relations scholars have usually operated within the relatively closed world of traditional race relations literature while students of comparative politics and societies remain largely uninformed by that literature.[2]

Pierre van den Berghe's recent critique (1967) of the dominant trends in the study of race relations highlighted the major shortcomings of the literature. "Much of the literature on race relations during the last three decades has dealt with the United States and has been written by scholars who lacked comparative experience. . . . For every hundred published racial attitude studies done in the United States, we could perhaps find five to ten in all the rest of the world. This lack of comparative approach has led, among other things, to an implicit or explicit overgeneralization of American findings." He noted the relative lack of sophistication in methodological techniques in the field of comparative race relations; its low degree of integration with the social sciences; its stress on *immediate* policy payoffs; and its aversion to historical analysis. He stressed that at least in the case of most American work, the field is dominated "by a functionalist view of society and a definition of the race problem as one of integration and assimilation of minorities into the mainstream of a consensus based society," an approach complemented by a reformist meliorism that is essentially optimistic and complacent (van den Berghe, 1967: 4-8). Schermerhorn, too, has noted that much of what ails race relations research stems from the "general ignorance of related cross-cultural studies made by social scientists in other disciplines" (Schermerhorn, 1970: 5).

Traditional race relations work has largely been concerned with prejudicial attitudes and discriminatory behavior. The focus on prejudice, which dates academically from the 1930s (Schermerhorn, 1970: 6), achieved its most brilliant expression in Gunnar Myrdal's *An American Dilemma* (1944). In the keynote paragraph of his massive study (which remains invaluable to students of American race relations), Myrdal summed up his organizing thesis:

> The American Negro problem is a problem in the heart of the American. It is there that the interracial tension has its focus. It is

> there that the decisive struggle goes on. This is the central viewpoint of this treatise. Though our study includes economic, social, and political race relations, at bottom our problem is the moral dilemma of the American [Myrdal, 1944: xxi].

Besides being empirically suspect (events subsequent to the publication of *An American Dilemma* indicate that most white Americans are *not* torn by contradictions of creed and behavior), Myrdal's ethical-moral approach to prejudice begs the question: What are the social, economic, and political origins of relationships of racial inequality, and what social, economic, and political factors sustain such relationships? It is a distortion to posit the perceptual, subjective notion of prejudice as the central explanatory independent variable in analyzing social and political reality. As Schermerhorn has put it, "If we begin with the matter of prejudice, any approach to the field from this viewpoint has a subtle tendency to psychologize group relations by seeing them as personality process writ large. . . . If research has confirmed anything in this area, it is that prejudice is a product of *situations,* historical situations, economic situations, political situations; it is not a little demon that emerges in people because they are depraved" (Schermerhorn, 1970: 6). The focus on attitudes as independent variables has comparative limitations as well, not only because prejudice is difficult to measure, but because prejudicial attitudes have no meaning apart from concrete contexts. The focus on discrimination also misleads to the extent that it diverts attention from the structured differential distribution of wealth, status, and political power, and from the examination of the linkages of racial groups in society to each other and to the polity (Schermerhorn, 1970: 7).

Given the present state of the systematic study of comparative race relations, described aptly by van den Berghe as "so incoherent and embryonic that almost any cross-cultural study, however poorly designed and executed, is destined to be regarded as 'pioneering' " (van den Berghe, 1967: 5), how best to proceed? Needed now, I will argue below, are theoretical guidelines that can organize possibilities rather than assert certainties; a conceptual approach that will yield the interesting and significant questions; interrogative theory, if you like, for, as Runciman has put it, "even the most ingenious answers will prove inadequate when it is the wrong question which is being asked" (Runciman, 1963: 165). A conceptual approach productive of heuristic questions has analysis, not neoscholastic classification, as its objective.

The raison d'etre of the approach would not be a search for iron laws of race relations, but the development of propositions at the level of (what Michael Scriven has called) normic statements–not universal hypotheses or statements open to exact quantitative validation, but rather at the level of theory "which leaves open the question whether people are doing what people invariably do in these uniquely complicated circumstances or are doing one of the comparatively few things which people . . . choose to do in such circumstances" (Stretton, 1969: 327). The development of heuristic guidelines for the study of comparative race relations will be valuable to the extent that the questions asked "extricate the essential factors in the constitution of reality under study" (Goldmann, 1969: 115), and to the extent that they permit internal comprehension and external confirmation; that, in other words, they have comparative utility, and that they respect the uniqueness of national historical circumstance.

RECOMMENDATIONS

To stimulate productive research and to overcome some of the shortcomings of the literature, the following research guide proposes a number of considerations and elements that should inform the development of interrogative race relations theory. Paradoxically, despite the recognition that racial cleavages throw up categories of events and behavior that must be dealt with on their own terms, the recommendations below are not limited in their utility to the study of racial groups. Based on a conviction that the analysis of structured racial linkages is a priority task, both for normative and methodological reasons, in the development of systematic comparative race studies, I will propose the following: that race be regarded as a special case of stratification; that adherence to the Weberian tripartite approach to stratification opens up new and potential fruitful areas of inquiry; that, in the study of race, a conflict model orientation is an advance on the emphasis of consensus and value integration; that study focus on critical structural periods and decisions that define the parameters of racial group relationships; and that such relationships be evaluated along the dimensions of representativeness, responsiveness, efficacy, and on an assessment of the direction or flow of decision-making capabilities made possible by those relationships.

By themselves, the physical facts of race are of little or no analytical significance to the social scientist. Racial physical characteristics only assume social meaning when they become criteria of stratification (see van den Berghe, 1967). Thus, a racial group is defined both by its relationships to other groups and by its *potential* as well as actual consciousness. In Etzioni's terms[3] the racial group, as a collectivity, is a meaningful stratification unit whose members share positions and, therefore, objective interests, similar vantage points, and a shared consciousness, actual or potential (see Goldmann, 1969: also Etzioni, 1968a: 98, 439).[3] It follows that the analysis of objective interests should precede the analysis of subjectively perceived interests; that the analysis of structure should precede the analysis of behavior. Hence the discussion below will be limited to the first order concern with structure.

Since race relations represents a special case of stratification, I would urge that we begin by adhering strictly to Weber's multidimensional approach to the forms and elements of stratification. In his uncompleted, posthumously published essay, "Class, Status, Party," Weber implicitly took issue with Marx' view that property stratification and economic class reference is the substructure and all else is superstructure. He retained stratification as a fundamental concept, but urged that we see society as composed of discrete, separate, yet interacting spheres, each marked by the uneven distribution of available, and scarce, resources. The three major discrete spheres are the *economic,* divided into classes "as a resultant from either (1) the given distribution of property, or (2) the structure of the concrete economic order"; the *social,* divided into "status groups" marked by specific "styles of life" based on the subjective distribution of honors and prestige; and the *political,* stratified by the unequal distribution of political power based on the possession or nonpossession of political offices which have a monopoly of the legitimate means of coercion (Weber, 1958b: 180 ff.).

In a recent essay, W. G. Runciman argues convincingly that in spite of the operational difficulties in retaining these Weberian categories, stratification theories, for the sake of clarity and accuracy, should be expressed in their terms since they can be shown to be empirically as well as conceptually distinct (Runciman, 1968). The multidimensional approach, he has argued, is particularly valuable in dealing with concrete situations of inequality (race relations, of course, being a salient example):

> Any classification, of course, is legitimate if it answers to the purpose at hand. But if there are three, and only three, basic dimensions in which societies are stratified—class, status, and power—then it necessarily follows that all social inequalities (in the sense opposed to individual inequalities such as height and weight) are inequalities of one or another of these three, and only three, kinds. It is perhaps self-evident that inequalities of class, status, and power need not always coincide. But surprising as it seems, there is as far as I know no major writer on social inequality who has explicitly formulated and consistently retained the tripartite distinction. Indeed, I would go so far as to say that every one of the best known discussions of inequality, from Aristotle to Rousseau, to Tocqueville to Tawney, has been confused by the neglect of it [Runciman, 1966: 37].

Adherence to the three-dimensional approach to stratification is conceptually liberating because it permits the asking of new questions. It is precisely because "inequalities of class, status, and power need not always coincide" that we can isolate one and seek to establish its relationship, often complex, to the others. Each of society's stratified spheres is autonomous, yet in constant interaction with the others. Thus, no single developmental tendency emerges out of Weber's comprehensive picture of structured inequality. Within his systematic, schematic framework, a multitude of types of displacement, interaction, and conflict, is conceivable. It is, therefore, possible to deal, for example, with rates of mobility for each stratification dimension discretely. In doing so, Lipset and Zetterberg's complaint that political "power as a vehicle for other kinds of social mobility has so far been a neglected area of research" (Lipset and Zetterberg, 1966: 442) may be remedied.

Complementing the Weberian schema is Simmel's assumption that "typically speaking, nobody is satisfied with the position he occupies in regard to his fellow creatures; everybody wishes to attain one which is, in some sense, more favorable" (Wolff, 1959: 275). Apter makes a similar point when he asserts "that the basic motive of politics . . . is a striving motive to expand mobility opportunities, either for some special group or for large segments of the society" (Apter, 1963: 82-85). Implicit in this view is a concern with conflict and power; politics is seen as a process in which individuals and groups interact in competition for the scarce resource of political power.

Adherence to the Weberian framework might provide a corrective to the disinclination, which Joseph Roucek noted in 1956, "to view majority-minority relationships as another aspect of human power relations"; an orientation which reflects integrative theory's concern with stability and value consensus (see Dahrendorf, 1959: 161 ff.). This kind of approach, M. G. Smith has lucidly argued, is particularly inappropriate in analyzing societies that are divided by racial cleavages. "It is obvious that when societies are conceived as structural systems in equilibrium, their homogeneity is assumed, and heterogeneity is difficult to define, classify, or analyze. In consequence, one general model, namely, that of homogeneous structural systems, is applied to quite different types of society, thereby obscuring their differences, misleading their analyses, and blocking the development of social theory. . . . From such presumptions, absurdities often follow" (Smith, 1965: 77, xiii). Thus, as eminent an analyst as Robert Dahl could write in 1956 that "the full assimilation of Negroes into the normal system has already occurred in many northern states and now seems to be slowly taking place even in the South" (Dahl, 1956: 138-139). The difficulty with this construction of reality, of course, was that it bore little relation to fact.

Opposed to the severely distorting position of the integrative theorists in the area of race is the Weberian conceptualization of society as a structure of organized inequality in the distribution of wealth, status, and political power; this differential distribution being the central dynamic for conflict and change (see Dahrendorf, 1959: 161 ff.). In this view, conflict is an omnipresent characteristic of social relationships, the product of the competition for scarce resources. As Simmel noted: "If liberation from subordination is examined more closely . . . it almost always reveals itself as, at the same time, a gain in domination" (Wolff, 1950: 273). Though it is clear that conflict and consensus models of society represent two sides of the same coin, and that each is an abstraction of reality, it also is clear that in the area of race relations, the conflict model is the more compelling and useful analytical approach (see Henderson, 1964).

Power has already been referred to without elaboration; some clarification is in order. Power involves the capacity to make and carry out decisions. John Champlin (1970) has defined having power as "being in a position to get others to do what one wants them to without having to make unacceptable sacrifices." Power, then, is relational. It is positional also, in that it depends on an individual's or group's structural position in relation to others. Thus, for Gerth and Mills, the political order "consists of those institutions with which men acquire, wield, or influence distributions of power" (Gerth and Mills, 1954: 192-193).

In considering the power of a group, I will follow Etzioni who urges that to overcome the key methodological objection to the concept of power–that it can only be assessed after the event, after A gets B to do what A wants without unacceptable sacrifices to A–we should use power "not for a single exercise of it on a single issue over a single subject at one point in time, but rather to refer to a generalized capacity of an actor, in his relations with others, to reduce resistance in the course of action he prefers in a given field . . . about a set of matters over a period of time. This capacity can be anticipated with a certain degree of possibility" (Etzioni, 1968a: 314-315). To assess a group's power capacity, it is necessary to assess its assets which may be converted into power (like consciousness, power may be actual or potential). These assets or resources include money and status, but the most important political assets are positional. Thus, inevitably, there will be conflict between individuals and groups attempting to obtain the institutional positions which confer the right to legitimately make and carry out decisions, the number of such positions being limited, and, as such, being a scarce resource. But, the possession of these positions does not, by itself, fix immutably the individual's or group's power (see Bendix and Lipset, 1966: section 3). Thus, for example, one

President of the United States can be said to have been more powerful than another, even though the legal-institutional powers of the office were identical for both. Nevertheless, it is clear that to obtain the generalized capacity called political power, a person or group must have access to the resources of positions of political control. The prerequisite to that, of course, is participation in the political system (which must be fluid and open).

This view of power contrasts sharply with the use of the term by consensus-oriented systems theorists. Consider, as a leading example, Talcott Parson's essay "On the Concept of Political Power" (Parsons, 1966: 243-244). He conceives of power "as a circulating medium, analogous to money"; thus Parsons views power not as the generalized capacity of an individual or group, but as the generalized capacity of the *system* to secure outputs useful to its continuity and stability. As he states it, "Power is the generalized capacity to secure performance of binding obligations by units in a system of collective organization when the obligations are legitimized with reference to their bearing on collective goals and where in the case of recalcitrance there is a presumption of enforcement by negative situational sanctions" (Parsons, 1966: 243-244).

Parsons does not focus on who has power or on the consequences of the distribution of power (see Gouldner, 1970: 291). Not only is the implicit moral valuation repugnant, that what is legitimate is collectively good, but empirically, this formulation begs the most important questions about power–who can attain the capacity to make and carry out decisions, on behalf of whom, and to what ends. As Gouldner has noted, the Parsonian approach looks at power from the standpoint of those on top; viewed from the perspective of the subordinate, power cannot be seen as a morally neutral generalized systemic currency. It is imperative to ask "*whose* legitimate claims are being met and whose are being ignored?" As the generalized capacity of individuals and groups, power often exists quietly:

> It exists as a factor in the lives of subordinates shaping their behavior and beliefs, at every moment of their relations with those above them. Attitudes towards their superiors are continually influenced by the *awareness*–sometimes focal and sometimes only subsidiary–that superiors can give or withhold at will things that men greatly want, quite apart from their own agreement or dissent, and that

> crucial gratifications depend upon allocations and decisions by their superiors [Gouldner, 1970: 294].

To focus on the generalized power of a racial group is to focus not only on decision-making but also on what Bachrach and Baratz have called nondecision-making, or the power to keep certain issues off the political agenda, which they define as "a decision that results in suppression or thwarting of a latent or manifest challenge to the values or interests of the decision-maker . . . nondecision-making is a means by which demands for change in the existing community can be suffocated before they are even voiced; or kept covert; or killed before they gain access to the relevant decision-making arena; or, failing all these things, maimed or destroyed in the decision-implementing stage of the policy process" (Bachrach and Baratz, 1970: 44). Nondecision-making is as much of an action as positive decision-making; it is not a nonevent (Bachrach and Baratz, 1970: 46). In short, as C. Wright Mills put it, "the problem of who is involved in making [decisions] (or not making them) is the basic problem of power" [Mills, 1959: 40].

In the study of structural-racial questions, the related notions of critical structural periods and critical structural decisions are analytically useful. In 1948, Charles Beard compellingly argued that "many of our neglects, overstresses, and simplifications are due to the divorce of political science from history." This unnatural separation has had negative consequences for the uses of political knowledge. He cited John Burgess' warning at the turn of the century that "if political science, economics, law and sociology were cut entirely loose from history, they would become theoretical, superficial, and speculative, or what might be worse, merely 'practical,' that is, subservient to vested interests and politicians temporarily in power" (Beard, 1948: 220, 222). To remedy the separation of history from the social sciences, recognizing that "every social fact is a historical fact and vice versa" (Goldmann, 1969: 23; Apter, 1963: 733). Nisbet has urged that time be envisaged in the manner now largely reserved for space. Similarly, Levi-Strauss sees time not merely in mechanical or cumulative or statistical terms, but in social terms as well ("social time") as deriving its properties from concrete social phenomena (Nisbet, 1970: 99; Levi-Strauss, 1967: 281-283).

In social terms, critical structural periods are historical periods when critical structural decisions are made. Much as Barrington

Moore has examined the periods of social origin of democracy and distatorship (combining comparison with detailed investigation of individual cases), so it is important to explore the genesis of structured racial linkages that limit behavioral choice. Hypothesizing that the linkages fashioned during the early, and fluid, periods of interracial contact have a determinative impact on the future course of race relations, the examination of the formation of structured relationships by racial groups is a priority task.

Since "organization is the mobilization of bias" (Schattschneider, 1961: 71), structure not only limits but also shapes the direction of choice. Thus, critical structural decisions are those that define the structured relationships that limit and shape choice. With Apter "the structure may be defined as the relationships in a social situation which limits the choice process to a particular range of alternatives" while "the behavioral may be defined as the selection process in choice, i.e., deciding between alternatives" (Apter, 1963: 732). Structure, as we noted above, is thus the logical starting point of analysis; to begin with behavior, stressing perceptions, socialization, voting behavior, protest activity, and the like, would assume a knowledge of structure and the boundaries of choice possibilities. Without an analysis of context, an analysis of choice behavior can have little meaning. Hence complementing the Weberian approach to stratification and the concern with conflict and power, is a focus on critical structural periods and decisions. We must ask who has the power to make the critical decisions affecting structure and "who gets what and how, *and who gets left out and how"* (Bachrach and Baratz, 1970: 105) as a consequence of those structural arrangements (see Lipsitz, 1970).

Complementing the view of race as a special case of stratification, and an analytical perspective concerned with conflict and power, is the analysis of structured linkages that shape and limit choice based on the group's power or generalized capacity to make and carry out decisions. In this respect, the direction or flow of decision-making capabilities is the critical variable. We may ask, in a two-sided relationship, if the distribution of power produces unidirectional decision-making (linking actors and objects), bidirectional decision-making (linking participant-equals), or mediated or nondirectional decision-making (either linking participant-equals through buffers or the two groups exist in isolation). Manfred Halpern, in his work on political modernization, has usefully classified group relationship, or polarities, as he calls them, in this manner.

He enumerates six basic types of relationship—emanation, subjection, direct bargaining, boundary management, buffering, and isolation, as well as two other special types of linkage which he calls incoherence and transformation, and which need not detain us here (Halpern, 1968a, 1968b). Classifying his six basic types of relationship according to the direction or flow of decision-making capabilities of the linked actors yields three classes of relationships:

Class I: Unidirectional Power

In this class of relationships, the capacity to make and carry out decisions is the exclusive, or near exclusive, property of one of the linked groups. Thus, for Halpern, emanation "is an encounter in which (1) one treats the other solely as an extension of one's own personality, one's own will and power—as an embodiment of one's self—and (2) the other accepts his denial of a separate identity as legitimate because of the mysterious source or nature of the overwhelming power of the other" (Halpern, 1968b: 61). In subjection, the second polarity in this unidirectional class of relationships, the source of the power by which the actions of the other are controlled is not, as in emanation, mysterious or charismatic. Rather, the other complies simply because of the unilateral control of power. In both emanation and subjection, the relationship links superior and subordinate, not participant-equals. Clearly, decisions "reached among actors who differ significantly in their power inevitably reflects more of the needs of the mighty than those of the weak" (Etzioni, 1968b: 204).

Class II: Bidirectional Power

In this class of relationships, the capacity to make and carry out decisions is shared by the linked participants. In the linkage-form Halpern calls boundary management, each of the participant groups claims zones of autonomous jurisdiction, which can, for example, take the form of pressure group organization or direct control of geographical political units. Direct bargaining, a related form of relationship, refers to the linkage in which "individuals and groups make demands on each other and negotiate directly with each other" (Halpern, 1968b: 63-64).

Class III: Nondirectional or Mediated Power

In this class of relationships, the groups do not interact directly. Thus buffering, the fifth of Halpern's polarities, denotes an indirect encounter through a mediator, while isolation denotes the absence of communication and of a functioning relationship.

Of these three classes of linkage, only bidirectional relationships are relationships of productive justice, which refers to the way in which it is possible to participate in the making of decisions, and which is defined by Halpern in terms of the ability "to participate fully and freely and in mutuality" in relationships which attempt to "deal with the problems of society and history" (Halpern, 1968a: 2). Moreover, productive justice is usually the prerequisite for distributive justice which obtains when a community's tangible goods and services and intangible statuses as well as political rewards are shared equitably; that is, are distributed according to germane rather than capricious criteria, an example of the former being need, and of the latter being discrimination by color (Halpern, 1968a: 45).

This classification framework is only the beginning, not the end, of analysis. It should be stressed that relationships between racial (and other) groups in the "real world" are most often quite complex, do not necessarily or easily fit neatly into the social scientists' pigeonholes. More often than not they will approximate amalgams of Halpern's ideal-types. Individual linkages must be analyzed in their complexity. Evaluations of individual linkages, whether they be essentially unidirectional, bidirectional, or mediated or nondirectional in terms of power and decision-making, can be made along the three dimensions of representativeness, responsiveness, and efficacy. These three variables are especially useful in the analysis of group-polity and internal group relationships.

The most important of these dimensions is *representativeness,* a measure of personnel. We may inquire if a group has access to decision-making positions in the polity and if it has, proportional to its numbers, an equitable share of such positions. We also may ask if a group's leaders are representative of the group as a whole (what are the selection criteria? In class and status, how do they compare with the group's membership?)

The second dimension is *responsiveness,* a behavioral measure of the extent to which decision-making elites, whether representative or not, meet the group's needs and demands. Rustow concluded his review of the literature on elites by stressing the distinction between representativeness and responsiveness: "The study of the rulers' social background," he wrote, "is only a supplement, not a substitute, for the study of their performance in office—and (we may add) of the effect of that performance on the recruitment of future rulers" (Rustow, 1966: 716-717). Utilizing this distinction, conflicts about power may be conflicts about the possession of positions that confer the ability to make and carry out decisions or about the use others make of their power capacity (Stretton, 1969: 329; Runciman, 1965: 75).

The third dimension is *efficacy* an objective measure of the ability of government and political elites to produce desired effects or results. From the perspective of the ruled, elite representativeness and responsiveness without elite efficacy is clearly inadequate. The three dimensions—representativeness, responsiveness, and efficacy—yield eight possible "ideal-types" of group-polity, group-leadership relationships (see Table 1).

(1) The polity or the group's leadership is representative, responsive, and efficacious. We may call this case *clear democracy.*

(2) The polity or the group's leadership is representative and efficacious, but is unresponsive to group needs. Here the forms of democracy are intact and are seen to be operating, but the group's participating leadership elite has been coopted or bought off so that while it participates in the larger political system, it no longer speaks for those it claims to represent. We may call this case *detached oligarchy.*

(3) The polity or the group's leadership is representative, responsive, but ineffectual. We may call this case *ineffectual democracy.*

(4) In this fairly unlikely case, the polity or the group's leadership though representative is unresponsive and ineffectual. The polity or group leadership's unresponsiveness is masked by its lack of efficacy. We may, therefore, call this case *apparent democracy.*

(5) The polity or the group's leadership is responsive and efficacious but not representative. We may call this case *paternalism.*

(6) The polity or the group's leadership is unresponsive and unrepresentative, but efficacious in promoting its goals. We may call this case *efficacious autocracy.*

(7) The polity or the group's leadership, though ineffectual and not representative, does try to be responsive. We may call this case *ineffectual paternalism.*

(8) The polity or the group's leadership is not representative, as well as unresponsive and ineffectual. We may call this case *clear tyranny.*

TABLE 1

A TYPOLOGY OF GROUP-POLITY RELATIONSHIPS AND INTERNAL GROUP STRUCTURE

	Representative		Not Representative	
	Responsive	**Not Responsive**	**Responsive**	**Not Responsive**
Efficient	(1) clear democracy	(2) detached oligarchy	(5) paternalism	(6) efficacious autocracy
Not Efficient	(3) ineffectual democracy	(4) apparent democracy	(7) ineffectual paternalism	(8) clear tyranny

An analysis of critical structural periods, decisions, and linkages in the study of race relations that utilizes Weber's tripartite approach to stratification, focuses on power and conflict, that evaluate relationships along the dimensions of representativeness, responsiveness, efficacy, and assesses the direction decision-making capabilities can contribute effectively to the development of race studies that are comparative and systematic. Most important, such an analysis can enlarge the range of behavioral choice and racial justice. In short, the approach to race research sketched above is meant as a corrective to the mutual ignorance of scholars of race relations and students of comparative politics, and to behavioral approaches that ignore the first-order questions of structure. Only after racial linkages are identified and analyzed can we move intelligently to an analysis of their behavioral consequences and implications for change. The guidelines proposed above should be seen as tools to be used in the formulation of research questions that respect the uniqueness of historical circumstance while permitting comparative analysis.

RESEARCH GUIDE – AN EXAMPLE

By their nature research guides tend to be abstract; their applicability to concrete research problems can be demonstrated only by specific example. Accordingly, the following discussion indicates the kinds of questions and approaches that might inform the analysis of one topic: the politics of race in New York City, 1900 to 1930. It will not be possible here to do more than sketch the highlights of an approach to this topic and indicate the insights it might provide which have been overlooked by traditional studies (Katznelson, 1971).

In the history of American blacks, the period 1900-1930 stands out as a key structural period. In 1900, American racial contacts were limited largely to the South. There were 7,922,969 blacks, by federal government count, living in the South; only 902,025 Afro-Americans resided in all the other regions of the country combined (U.S. Department of Commerce, 1935: 9). Though the percentage of America's blacks who lived in the South had declined slightly from 90.6 in 1870 to 89.7 in 1900, most Americans, at the turn of the century, were convinced that blacks would remain a fixed element, socially as well as demographically, in Southern society (U.S Department of Commerce, 1935; 7; Gossett, 1963: 285). By the outbreak of World War I, however, this assumption had to be reassessed. Approximately 150,000 blacks left the South for the metropolitan areas of the North in the first fifteen years of the century. The movement, which came to be known as the Great Migration, rapidly accelerated during the war; by conservative estimate, almost 500,000 blacks migrated to the North in wartime (Franklin 1967: 465; U.S. Department of Labor, 1921: 10; Baker, 1917: 315; Roche, 1963: 80). In the 1920s, well over 550,000 left the South. The black population of the Middle Atlantic states alone in this decade increased (as a result of natural increase and migration) by 868,000. By 1930, the percentage of America's blacks in the South had declined dramatically to 78.7.

It is difficult to overstate the impact of this migration on the North and on the course of black history in America. Ray Stannard Baker aptly described the migration, in 1917, as "probably next to Emancipation the most noteworthy event which has happened to the Negro in America." In similar terms, August Meier has argued that the population movement was "after Emancipation . . . the great watershed in American Negro history" (Baker, 1917: 317; Meier, 1963: 170). With the exception of Chicago, no city felt the impact of the black migration from the South more than New York. In 1890, 36,617 New Yorkers (including those in the yet unconsolidated boroughs) were black; in 1930, 327, 706. In 1890, one person in seventy in Manhattan, where most of New York's blacks lived, were black; in 1930, one in nine. In 1890, Harlem was a semirural, all-white, upper-class community; in 1930, James Weldon Johnson correctly described Harlem as "*the* Negro metropolis" in America (Scheiner, 1965: 221; Haynes, 1968: 47).

In the critical and fluid period of migration, key structural-political decisions were made concerning the integration of the black population into the city's political structure. In the most widely respected history of this period, *Harlem: The Making of a Ghetto,* Gilbert Osofsky (1963) entitled the chapter on black ward politics, "A Taste of Honey," arguing that

> while the urbanization of the Negro obviously caused great difficulties, it also provided the base for significant political power–political power unprecedented in the history of the Negro in the North. As the Negro population increased in numbers, the cynical and apathetic attitudes that typified the reactions of politicians in the later nineteenth century came to an end. . . . Although the Negro's political progress in New York City was not an unqualified success, there obviously was greater advance in this sphere of community activity than in all the others. While the social

> and economic position of the city's Negroes tended to remain stable, or, with the Great Depression, even retrogressed, there was significant political mobility. In the 1890's Negroes were an almost powerless minority group; the least influential minority group in the metropolis. Their role, if any, was on the periphery of municipal affairs. Within the next generation, the generation that settled in Harlem, Negroes became an integral part of city government and politics—and politics proved a wedge for economic advancement [Osofsky, 1963: 159, 177].

The statement is internally inconsistent, arguing on one hand that blacks were not socially and economically upwardly mobile, and on the other that success at ward politics translated into economic gains for New York's blacks. More significant, however, is Osofsky's dual confusion of group with individual mobility and of ethnic recognition with ethnic decision-making power. In short, research that follows the guidelines in the previous section, as I hope to show below, asks questions that Osofsky ignores. Because he did not ask those questions, Osofsky was misled by evidence of individual success. Black political powerlessness was the norm for the generation that settled Harlem; political progress was not a wedge for economic advance because in this period neither occurred.

I agree with Osofsky's implicit Weberian model, with his implicit assumption that political mobility *may* translate into economic and social mobility as well. Thus, it can be argued that as a *result* of their political powerlessness, institutionalized in structured political linkages, the social and economic problems of the city's blacks grew to unmanageable proportions. An extended discussion of black politics in this period is not possible here. A brief exploration of black-Tammany structural relationships compared to the usual group-Tammany pattern along the lines suggested above, however, is, I think, quite revealing.

The key structural decision made by Tammany in this period was the decision to link the city's blacks to the machine through a citywide organization called the United Colored Democracy of Tammany Hall. In January 1898, after a small group of disaffected black Republicans had actively worked for the Tammany candidate for mayor in the November election, Richard Croker, Tammany's leader, appointed Edward "Chief" Lee, the black head bellman of the Plaza Hotel, to lead a new party organization, the United Colored Democracy (UCD). On paper the organization controlled all political and patronage matters concerning the city's blacks and the Democratic Party. In fact, it was a powerless segregated institution whose primary tasks were winning votes for Tammany and isolating blacks from positions of real political influence.[6]

The UCD was a citywide organization; thus blacks were unable to compete for political power on the district or ward level, and under the machine it was district political control that mattered. As a result, as John Morsell has noted, the UCD's "leaders may, from time to time, have been listened to when central decisions were made on matters affecting them—or their community's—interests, but this never approximated the system of consultation and solicitation which is dictated by custom and practicality in the case of leaders of clubs which are on a footing of recognition by the county leadership" (Morsell, 1951: 28-31). While white ethnic immigrant political leaders emerged from their neighborhood political clubs in largely homogeneous neighborhoods to participate in boundary management and direct bargaining (bedirectional relationships) with other leaders and with Tammany, black politicans were selected by Tammany's leadership, and had no community organizational base (there were no black Democratic district leaders in Harlem before 1935 though the area was over ninety percent black by 1930).

In terms of Halpern's classification typology, black-Tammany linkages were an amalgam of subjection, buffering, and isolation. In terms of power, the relationship was unidirectional. Black leaders and the United Colored Democracy acted as buffers in linking blacks to the Tammany machine; the mass of the black population was, as a result, largely isolated from the urban political system. An analysis of the composition of the UCD leadership, its aims and achievements, which cannot be detailed here for lack of space, reveals that it was neither representative, responsive, nor efficacious. Utilizing Weber's distinction between class and political power, the contrast between the white immigrant and migrant black experience in terms of mobility, is striking. For European immigrants, economic mobility was the consequence of political advance, whereas for black migrants, as Furniss (1969) recently demonstrated, political mobility for individuals usually came *after* considerable economic and social mobility.

In short, attention to the research questions proposed above leads to a concern with a key structural decision—the creation of the UCD by Tammany—a decision that shaped and limited the political choices open to the city's blacks during the period of migration, to a concern with the character and performance of black political leaders, to a comparative analysis of black and white ethnic situations, and to a focus on the relationship of political, social, and economic mobility. All of these concerns have been overlooked in the analysis of this topic in the past (of which Osofsky's work is the leading example). This neglect has produced conclusions about political success based on the study of individual leaders that are illusory and misleading.

This all-too-brief example is meant to suggest rather than exhaust the possibilities of the research schema presented above. The same questions may be applied cross-nationally to other plural situations. An inquiry along these lines concerning West Indians in Britain, for example, would reveal the emergence of racial buffers and a buffering leadership that acts much like the UCD and its leadership acted in New York (see Katznelson, 1970). Interestingly, this is the case in spite of significant political, cultural, national, temporal, and social differences.

CONCLUSION

These research proposals are meant, also, to suggest a role for the white social scientist in America. In his description of the colonizer who refuses, to whom the white race scholar was compared early in this essay, Memmi concluded (1965: 43-44):

> If he cannot stand this silence and make his life a perpetual compromise, he can end up by leaving the colony and its privileges. And if his political ethics will not permit him to "run out," he will make a fuss. He will criticize the authorities until he is "delivered to the disposal of the *metropole,*" as the chaste administrative jargon goes. By ceasing to be a colonizer, he will put an end to his contradiction and uneasiness.

But the white social scientist's reluctance to run out is more than a question of political ethics or sensibilities. Just as the colonized in America (to extend the image) cannot physically separate from the mother country—the colony being in the mother country—so the colonizer who refuses cannot resolve his contradictions by flight to the mother country. Only domestic decolonization can resolve his contradictions; he and the colonized are inextricably bound. His research that identifies and analyzes those racial structures that shape and limit choice and racial justice may create new possibilities of just restructured relationships. In this direction lie his hopes, and those of the colonized, for their mutual liberation.

NOTES

1. "The philosophers have only *interpreted* the world, in various ways; the point, however, is to *change* it" (Marx, 1969). All of his work, of course, was directed toward this end. Weber's (1958a) position is most clearly stated in "Science as a vocation." For a revisionist interpretation of Weber's position, see Alvin Gouldner (1962).

2. For a discussion, see Rose (n.d.: 40). It is noteworthy, to cite a significant recent example, that Jean Blondel's (1969) intelligent *Introduction to Comparative Government* deals at some length with class but totally ignores race.

3. Etzioni (1968a) defines a collectivity as "a macroscopic unit that has a potential capacity to act by drawing on a set of macroscopic normative bonds which tie members of a stratification category . . . the cohesion of a collectivity is not based on face to face contact. Unlike class position in the stratification structure, the position of a collectivity is not necessarily horizontal nor is its prime base necessarily economic."

4. Though this classification system is my own, I am indebted to Professor Juan Linz of Yale who stimulated my thinking along these lines by his trenchant critique of Robert Michels' classic study, *Political Parties.*

5. The term "Great Migration" has most often been used to denote the movement of blacks to the North during World War I. It is both more correct and useful to accept John Roche's and Gilbert Osofsky's revisionist use of the term; namely, to use it to describe the black migration to the North which began at the turn of the century and which continues today. Roughly speaking, there have been two periods of "Great Migration"—1900-1930 and 1940 to the present.

6. "In time, Lee was referred to as the 'Black Croker'—a name he hardly deserved. He was without voice to the councils of Tammany Hall, and only a few crumbs fell to Negroes" (Ottley, 1943: 210-211).

REFERENCES

APTER, D. (1963) "A comparative method for the study of politics," in D. Apter and H. Eckstein (eds.) Comparative Politics. New York.

ARON, R. (1969) Progress and Disillusion: The Dialectics of Modern Society. New York: New American Library.

BACHRACH, P. and M. S. BARATZ (1970) Power and Poverty: Theory and Practice. Fair Lawn, N.J.: Oxford Univ. Press.

BAKER, R. S. (1917) "The Negro goes North." The World's Work (July).

BEARD, C. (1948) "Neglected aspects of political science." Amer. Political Science Rev. (April).

BENDIX, R. and S. M. LIPSET [eds.] (1966) Class, Status, and Power. New York: Free Press.

BLAUNER, R. (1966) "Whitewash Over Watts." Trans-action (March/April).

BLONDEL, J. (1969) Introduction to Comparative Government. New York: Praeger.

BONDURANT, J. (1965) Conquest of Violence: The Gandhian Philosophy of Conflict. Berkeley.: Univ. of California Press.

CESAIRE, A. (1956) Letter to Maurice Thorez. Pamphlet. Paris: Trésence Africaine.

CHAMPLIN, J. (1970) "On the study of power." Politics and Society (November).

COSER, L. A. (1968) Continuities in the Study of Social Conflict. New York: Free Press.

DAHL, R. (1956) A Preface to Democratic Theory. Chicago: Univ. of Chicago Press.

DAHRENDORF, R. (1968) Essays in the Theory of Society. Stanford: Stanford Univ. Press.

--- (1959) Class and Class Conflict in Industrial Society. Stanford: Stanford Univ. Press.

DuBOIS, W. E. B. (1920) Darkwater: Voices from Within the Veil. New York: Schocken.

ERIKSON, E. H. (1968) "On the nature of psycho-historical evidence: in search of Gandhi." Daedalus (Summer).

ETZIONI, A. (1968a) The Active Society. New York: Free Press.

--- (1968b) "Societal guidance: a key to macro-sociology." Acta Sociologica 4.

FRANKLIN, J. H. (1967) From Slavery to Freedom. New York: Knopf.

FURNISS, G. (1969) "The political assimilation of Negroes in New York City." Ph.D. dissertation. Columbia University. (unpublished)

GERTH, H. H. and C. W. MILLS (1954) Character and Social Structure. New York: Harcourt, Brace Javanovich.

GOLDMANN, L. (1969) The Human Science and Philosophy. New York: Grossman.

GOSSETT, T. F. (1963) Race: The History of an Idea in America. Dallas: Southern Methodist Univ. Press.

GOULDNER, A. W. (1970) The Coming Crisis in Western Sociology. New York: Basic Books.

--- (1962) "Anti-minotaur: the myth of a value-free sociology." Social Problems (Winter): 199-213.

HALPERN, M. (1968a) "Conflict, violence, and the dialectics of modernization." (unpublished)

--- (1968b) "A redefinition of the revolutionary situation." J. of International Affairs (Winter): 65-75.

HARRIS, N. (1968) "Race and nation." International Socialism (Autumn).

HAYNES, G. E. (1968) The Negro at Work in New York City [1912]. New York: AMS Press.

HENDERSON, D. (1964) "Minority response and the conflict model." Phylon (Spring): 18-26.

HOETINK, H. (1967) Two Variants in Caribbean Race Relations: A Contribution to the Sociology of Segmented Societies. London: Oxford Univ. Press.

KATZNELSON, I. (forthcoming) Race, Politics and Migration. London: Oxford Univ. Press.

--- (1970) "The politics of racial buffering in Nottingham, 1954-1968." Race (April): 431-446.

LEVI-STRAUSS, C. (1967) Structural Anthropology. New York: Anchor Doubleday.

LEWIS, O. (1966) La Vida. New York: Random House.

LIPSET, S. M. and H. L. ZETTERBERG (1966) "A theory of social mobility," in L.A. Coser and B. Rosenberg (eds.) Sociological Theory. New York: Macmillan.

LIPSITZ, L. (1970) "On political belief: the grievances of the poor," pp. 142-172 in P. Green and S. Levinson (eds.) Power and Community: Dissenting Essays in Political Science. New York: Pantheon.

LIPSKY, M. and D. OLSON (1969) "Riot commission politics." Trans-action (July/August).

LYND, R. (1945) Knowledge for What?: The Place of Social Science in American Culture. Princeton: Princeton Univ. Press.

MARX, K. and F. ENGELS (1969) "Theses on Feuerbach [1845]," pp. 28-30 in Selected Works of Marx and Engels. New York: International.

MATTHEWS, D. (1969) "Political science research on race relations," in I. Katz and P. Gurin (eds.) Race and the Social Sciences. New York: Basic Books.

MEIER, A. (1963) Negro Thought in America, 1880-1915. Ann Arbor: Univ. of Michigan Press.

MEMMI, A. (1965) The Colonizer and the Colonized. Boston: Beacon Press.

MILLS, C. W. (1959) The Sociological Imagination. Fair Lawn, N.J.: Oxford Univ. Press.

MORSELL, J. (1951) "The political behavior of Negroes in New York City." Ph.D. dissertation. Columbia University. (unpublished)

MOYNIHAN, D. (1965) The Negro Family. U.S. Department of Labor, Office of Policy Planning and Research. Washington, D.C.

MYRDAL, G. (1944) An American Dilemma. [Revised ed. 1962] New York: Harper and Row.

National Advisory Commission on Civil Disorders (1968). Report. New York: 1.

NISBET, R. A. (1970) Tradition and Revolt. New York: Random House.

OSOFSKY, G. (1963) Harlem: The Making of a Ghetto. New York.

OTTLEY, R. (1943) New World A-Coming: Inside Black America. New York: Grosset and Dunlap.

PARENTI, M. (1970) "The possibilities for political change." Politics and Society (November).

PARSONS, T. (1966) "On the concept of political power," in R. Bendix and S. M. Lipset (eds.) Class, Status, and Power. New York: Free Press.

PAZ, O. (1961) The Labyrinth of Solitude: Life and Thought in Mexico. New York: Evergreen.

ROCHE, J. P. (1963) The Quest for the Dream: The Development of Civil Rights and Human Relations in Modern America. New York: Quadrangle.

ROSE, P. (n.d.) (unpublished)

RUNCIMAN, W. G. (1968) "Class, status, and power?" pp. 25-61 in J. A. Jackson (ed.) Social Stratification. Cambridge: Cambridge Univ. Press.

--- (1966) Relative Deprivation and Social Justice: A Study of Attitudes to Social Inequality in Twentieth Century England. Berkeley: Univ. of California Press.

--- (1963) Social Science and Political Theory. Cambridge: Cambridge Univ. Press.

RUSTOW, D. (1966) "The study of elites." World Politics (July).

--- (n.d.) Change as the Theme of Political Science. (unpublished)

SCHATTSCHNEIDER, E. E. (1961) The Semisovereign People. New York: Holt, Rinehart and Winston.

SCHEINER, S. M. (1965) Negro Mecca. New York: New York Univ. Press.

SCHERMERHORN, R. A. (1970) Comparative Ethnic Relations. New York: Random House.

SIMMEL, G. (1968) "The poor," in C. Waxman (ed.) Poverty, Power and Politics. New York: Grosset and Dunlap.

SIVANANDAN, A. (1968) "The revolt of the natives." Twentieth Century (Third Quarter).

SMITH, M. G. (1965) The Plural Society in the British West Indies. Berkeley.

STRETTON, H. (1969) The Political Sciences. New York: Basic Books.

TUMIN, M. (1969) "Some social consequences of research on racial relations," in H. P. Dreitzel (ed.) Recent Sociology No. 1: The Social Basis of Politics. New York: Macmillan.

U.S. Department of Commerce (1935) Negroes in the United States, 1920-1932. Bureau of the Census. Washington, D.C.

U.S. Department of Labor (1921) The Negro at Work During the World War and During Reconstruction. Washington, D.C.

VALENTINE, C. A. (1969) Culture and Poverty. Chicago: Univ. of Chicago Press.

van den BERGHE, P. L. (1967) Race and Racism. New York: John Wiley.

WEBER, M. (1958a) "Science as a vocation," pp. 129-158 in H. Gerth and C. W. Mills (eds.) From Max Weber. New York.

--- (1958b) "Class, status, party," in H. H. Gerth and C. W. Mills (eds.) From Max Weber. New York: Grosset and Dunlap.

WHITTEN, N. E., Jr. and J. F. SZWED (1970) Afro-American Anthropology. New York: Free Press.

WOLFF, K. [ed.] (1959) Georg Simmel. 1858-1918: A Collection of Essays. Columbus: Ohio State Univ. Press.

4

Caste and Race Conflict in Cross-Cultural Perspective

MARTIN ORANS

Be Blessed, Mother!
I will not send the East Wind over these sacred images as over the sands of the road.
You do not hear me when I hear you, like an anxious mother who forgets to push Button A.
But I will not efface the footprints of my father or of my father's fathers in this head open to all the winds and plunders of the North.
Mother, in this study lined with Latin and Greek, breathe the fumes of the evening victims of my heart.
May the protecting spirits save my blood from slackening like that of the assimilated and the civilized!

Though I come late, I stand upright before the Ancestors and offer one chicken without stain, so that before the milk and millet beer
May spurt upon me and my fleshy lips the warm salt blood of a bull in the prime of his life, in the splendour of his fatness!

Léopold Sédar Senghor[1]
President of Senegal

LET COME THE DAY TO SAY "ONCE IT WAS SO"[2]

I heard whispering like wind from mouth to mouth
That under each armpit I am marked, the size of an open hand.
Was it inherited from an ancient time?
My parents, I've heard, also are thus bruised by nature's brand.

Yet no memory of them affords me sight or feel of such a spot.

But in childhood I learned,
Through cruel heavy winks, instinctively to hide.
What was it I so naively wrapped with rags,
And dragged, hidden, through dark months and years?

In these concealing rags, I hid my heart.
When found, it was so bruised,
Sore, red and shrunken from the stigma I tried to lose.

Without some exposure, my songs would all be lies;
Bruises tightly bound daily increase the inner pain.

Who marked my sides? For what cause?
Why this brand upon my soul?

Even today, my ebbing thoughts,
So pale and cold, transparent as glass,
Hold me awake.

Poem by a Burakumin,
member of a Japanese pariah caste

□ SEVERAL MONTHS AGO a student with virtually no familiarity with anthropology, but with the knowledge that we often study far-out people, came before our anthropology faculty and reminded us that there were plenty of worthwhile people and customs to study right here at home. Of course we made an effort to reply by pointing out the advantages of the comparative method. We claimed the advantage of recognition by contrast of what would otherwise appear as background phenomena; we argued the necessity of cross-cultural validation to separate genuine regularities from historically connected phenomena; we even defended the investigation of the esoteric as distinct from the topically "relevant" on the ground that genuine knowledge of man in general, or in particular circumstances, might derive from the most far-out kind of study. This essay represents an effort to make good on some of these claims.

INTRODUCTION

About ten years ago I tried with considerable brashness to transform the historical experiences of a "tribe" in India, the Santal, into a processual statement of very great generality. Professor David Schneider, who was present during the defense of my thesis, was perspicuous enough to notice that if I really meant what I said, it ought to apply to the Black American experience. Unfortunately at that time I did not know enough of the related experience of any other peoples to profitably take up the comparative question raised by Professor Schneider. This essay is an effort to begin consideration on a comparative basis of the very loose theoretical construction which I had developed and which I call with some apology for yet another social science neologism, "the rank concession syndrome" (RCS; Orans, 1965).

What I have done is to compare the evidence from three ethnic groups with that of the Santal from the point of view of the RCS. These three groups are the Jatavs, an "untouchable" Indian leather-working caste, the Burakumin, an untouchable Japanese pariah caste also connected with leather working, and Black Americans. I have chosen these three groups, all of which seem to have had experiences relevant to the RCS, because the material available contains considerable treatment of the kind of evidence which I had concentrated on in dealing with the Santal. Having an Indian pariah caste to add to the Indian Santal evidence seemed also to provide an especially valuable control; in addition, the author of the Jatav material (Lynch, 1969) had himself commented on the connection between the RCS and his findings. The Japanese Burakumin are again enough like the Indian Jatavs in their general pariah situation, having even the same traditional occupation, to warrant comparison. I have found that consideration of the Black American experience has contributed significantly to my understanding of more esoteric peoples. Of course these four cases are not randomly chosen and do not provide a test in any formal sense of the claims of the RCS; but in the context of discovery they do provide the stimulus for a reformulated statement of the RCS which perhaps is a bit closer to the truth than the original.

The four groups under consideration constitute four generally poor ethnic groups, and might profitably be regarded from the point of view of "the culture of poverty" (Lewis, 1966) or from the standpoint of the critics of that approach; e.g., Valentine (1968). They might also be analyzed according to arguments based upon cross-cultural theories of caste, e.g., Berreman (1966a, 1966b) and others, whose arguments already incorporate some of the RCS material in a very pervasive context. And as I implied some time ago, much of the RCS might profitably be restated in the language of reference group theory as suggested to me by Fred Eggan and Milton Singer when I first presented my thesis. Indeed, Lynch has chosen to present his Jatav material largely in the terms of reference group theory with, I think, very good results. All of these already available cross-cultural approaches are pertinent and deserve some attempt at integration. Though I have given some considerable thought to the matter, I have chosen, for the most part, to stick to the RCS.

A brief statement of the Santal material and the original presentation of the RCS begins the essay. With a few minor changes, this theoretical material is taken directly from my book, *The Santal.* The Jatavs, the Burakumin, and the Blacks are then treated in that order, each suggesting various modifications of the RCS. In a concluding section, the various modifications of the RCS are summarized and a few general findings suggested by comparison are discussed.

Since this essay deals with topical material, I do not suppose that it will escape interpretation as a charter for this or that social action. Nevertheless, it has not been my intention to help or hinder any particular social policy, but rather to follow wherever the truth seems to lead. I do not pretend to neutrality nor do I deny that my preferences have influenced my interpretations in spite of my good intentions. Though I have tried not to allow these preferences to sway me, let me state that my preference is for democratic socialism and concern for what a man is rather than with his genetic connections. As a rootless cosmopolitan atheistic Jew, professing a belief in democratic socialism, it is with some envy and some trepidation that I view all social and cultural movements based upon ethnicity. I am on the side of the poor and exploited, and however I may feel about any particular response to this condition, it is the rich and powerful whom I hold responsible.

THE SANTAL: A THEORY OF THE RANK CONCESSION SYNDROME*

Less than two decades ago the anthropologists Redfield and Singer (1954) offered their colleagues in the social sciences a set of constructs to aid in understanding the part cities play in the development of culture. From that offering the most salient constructs, judging by usage, have been the concepts of the "great tradition" and the "little tradition." (Great traditions were regarded as the stuff of civilization and as having the qualities of explicitness, systematization, and reflectiveness. Little traditions were conceived of as the product of folk societies and as relatively lacking in these qualities.) But such constructs as great and little traditions are mainly aids to thought, and one hardly expects to encounter them as objects of experience, particularly in the very form in which they have been postulated. Imagine then my surprise at encountering a self-conscious effort to create a great tradition, precisely as Redfield and Singer define it, among a tribal people well known for having long managed to preserve a considerable degree of distinctiveness from the dominant surrounding civilized society.

The Santal study was in large measure an effort to understand the development and characteristics of this emergent great tradition. In seeking such understanding, attention was focused on the shifting historical pressures for and against acculturation and assimilation, and on the impact of a spreading market economy, industrialization, and political democracy. Though it was not at all a monograph purporting to tell all about a particular people, the events described were largely the experience of a single tribe, the Santal, who live primarily on the Chotanagpur Plateau of Bihar, West Bengal, and northern Orissa in India. Nevertheless, the experiences of the Santal suggested certain widely prevalent processes which I call the "rank concession syndrome." These processes were speculatively held to pertain to all part-societies which have conceded rank to a dominant surrounding society. I have tested that speculation in the present essay. In particular the pressures for and against assimilation are regarded as crucial determinants and these in turn are seen to depend significantly upon shifting avenues of mobility. Movements of

*Much of the material in this section is reprinted from *The Santal,* 1965, by Martin Orans by permission of the Wayne State University Press.

self-conscious cultural creativity are viewed in relation to the conflicting stimuli generated by pressures for and against assimilation.

There are adequate if not compelling theoretical and practical reasons for making the Santal the object of such a study. The characteristically primitive world-view and social organization of traditional Santal culture provides a sufficiently contrasting background against which to view the impact of such modern conditions as the market, industrial employment, and political democracy. Similarly the heavy commitment to pursuit of pleasure and the characteristically unreflective character of traditional Santal culture contrast sharply with the dominant surrounding Brahmanical tradition. Their patterns of social interaction and group structure rest heavily upon a markedly patrilineal kinship system closely paralleling that of their Hindu neighbors. On the other hand, their social organization is relatively lacking in the cleavages of caste which are so prevalent all about them. Apart from the village, which is usually composed of a number of patrilineal kin-groups, territorial organization is weakly developed, and the tribe as a whole has never been politically unified so far as historical records indicate.

Even as a unique people, the Santal merit special attention. Numbering today over three million, the region they occupy contains the richest deposits of iron, coal, and other crucial industrial minerals in all of India. The chief source of their livelihood, at least since the nineteenth century, has been settled agriculture. Prior to the nineteenth century, hunting was also an important livelihood activity, but the great increase in the region's population and subsequent reduction of forest areas has made game extremely scarce; it is now difficult for the inhabitants even to obtain materials for fuel and building purposes. Though the chief source of food and cash is the annual rice crop, a few vegetables are also grown and some fruit trees maintained. Some villages are near enough to a river for fish to be readily obtained, while some villages stock their local ponds. Even prior to the introduction of modern industry in the region, the Santal had ceased to be economically independent. Although they once did some pottery making, weaving, and even iron smelting, they have come to rely almost exclusively on various local Hindu castes for these material products. Women from several low Hindu castes are used frequently as midwives. While almost no

Santal are ritually served by Brahman priests, itinerant Hindu holy men *(sadhu)* visit many Santal villages, consulting horoscopes and offering prayers for a fee or donation.

Most Santal are still agriculturists, but some have turned to less time-honored work. The chief industry in which these others are employed is as characteristically modern as the Santal themselves are not. It is the Tata Iron and Steel Company (TISCO), the largest and oldest modern iron and steel company in India. Since Santal have been employed by TISCO from its founding over fifty years ago, enough time has elapsed for change to become manifest and for the directions of change to become evident.

Jamshedpur, the industrial town which was created to house TISCO, was carved out of the jungle of Singhbhum district, Bihar, in 1908; it owes its location to the rich natural resources of the surrounding area in which the Santal and other closely related tribes form a substantial part of the population. The population of the city has been drawn from every section of India as well as a number of foreign countries and has grown to more than 218,000. Jamshedpur is extremely atypical compared to other Indian cities—the fact that it is a company town of unprecedented size accounts in large measure for its atypicality. TISCO has from the beginning completely controlled the government of the town, provided housing for most of its workers, operated a school system, and furnished other services. It is a city as much dominated by modern industry as a medieval city was by its cathedral. Traditional institutions, such as temples and sociopolitical organizations, exist but are weakly developed; the Tata Company, the Tata Workers Union, and the town's competing regional and national political parties are the dominant corporate groups. Caste persists as an institution for regulating marriage, for assigning employment in certain traditionally degrading occupations, and, to a certain extent, in structuring social interaction. But class in the traditional Western sense is increasingly displacing caste as the organizing principle of membership in social groups and in social interaction.

OUTLINE OF THE THEORY

> Solidarity which comes from likenesses is at its maximum when the collective conscience completely envelops our whole conscience and

> coincides in all points with it. But, at that moment, our individuality is nil. It can be born only if the community takes smaller toll of us. There are, here, two contrary forces, one centripetal, the other centrifugal, which cannot flourish at the same time. We cannot, at one and the same time, develop ourselves in two opposite senses.—If our ideal is to present a singular and personal appearance, we do not want to resemble everybody else [Durkheim, 1933: 130].

Santal historical experience suggests certain processes which may be applicable to all encysted societies which have conceded rank to a dominant surrounding society. It is these processes which I call the rank concession syndrome (RCS). Rank concession means the acceptance of social inferiority. Acceptance is an unfortunately ambiguous term, for it may occur in various degrees and perhaps even in a variety of forms. Nevertheless, to whatever degree and in whatever form such acceptance occurs, the syndrome becomes operative.

The evidence of Santal rank concession is manifold. The earliest historical evidence shows the Santal adopting and abandoning traits on the basis of their rank value in Hindu society. The acceptance of alien rank values can hardly ensue without rank concession. The Santal themselves say that "the *diku* are big and knowing people" and their demeanor as they interact with Hindus, particularly of high caste, confirms acceptance in fair measure of their inferior status. Even today, in a period of renewed solidarity, the Santal are striving to raise themselves to the level of the diku, as they say. The term diku is, of course, evidence enough of the ambivalence of the concession, and the history of Santal-Hindu relations has provided the Santal with sufficient reason to hate and fear as well as admire their Hindu neighbors. The earliest historical information indicates not only that rank concession must have begun considerably before British control, but that it was particularly pronounced among the traditional Santal elite. There is, for example, the myth of the Kisku king who became a Hindu and was forsaken by his fellows, and the myth of Santal society constructed as a set of functionally specialized castes with each clan occupying a caste position.

To accept inferior status is to accept the attributes of rank of the superior society, and such acceptance produces a tendency to emulation. This emulation is an effect of rank concession, and evidence for it as well. The many examples noted of Santal emulation of the dominant Hindu society are to be understood as arising from the universal tendency of societies which have conceded rank to emulate. They may also be regarded as supporting evidence, to indicate rank concession, though rank concession can be independently observed in Santal social relations with Hindus as well as inferred from Santal comments about the social superiority of the diku.

Such an emulative tendency, if put into practice, may at first lead to gross copying. Without internalization of emulative attitudes it is not likely to accomplish its objective and results in the kind of imitation epitomized by the nouveau riche. Such behavior appears gauche and thereby generates its own tendency toward internalization. Emulation may be directed toward attaining rank in interaction with social units belonging to the dominant society or within one's own society.

To concede that another society is more powerful is not tantamount to conceding it superior rank. Such a power concession generates its own borrowing tendencies, which might be termed "power-incorporative borrowing," since it involves regarding the more powerful society's traits as possessing a special efficacy, which arises by projecting onto these traits the power of the dominant society. But while power concession does not imply rank concession, rank concession is almost always accompanied by and therefore implies power concession. Rank concession will therefore always be associated with both emulation and power-incorporative borrowing. Emulation and power incorporation are often intermingled in particular acts and serve to maintain borrowed practices as well as to generate new borrowings. Power incorporation appears in its purest form in Santal curing practices, whose end is primarily curing efficacy or the belief in such efficacy, though the prestige which accrues to the practitioner who employs borrowed practices is not without significance. For example, one commonly hears Bengali and Oriya songs and Sanskritic prayers as essential elements in curing rites, though often even the practitioner does not understand the words. Such practices could hardly have been borrowed because of a peculiar "fit" of their content with traditional Santal patterns. Certainly such practices are not naturalistically more efficacious. It is rather a case of connecting status and efficacy which is expressed in the Santal reference to Hindus as "*big* and *knowing* people."

There is a third reason for the tendency to adopt practices of the dominant society, one which follows almost inevitably from the condition of being encysted. This is the external pressure brought to bear by the dominant society on the encysted society to adopt its customs. The basis for the near inevitability of this pressure is nothing less than the solidarity needs of the dominant society. Surrounding dominant societies must demand allegiance to at least the official symbols of suzerainty; acceptance of a number of other customs is likely also to be demanded as the solidarity needs of the dominant society increase. There are likely also to be a number of customs of the encysted society which the superior society finds repugnant, a reaction which invariably creates some pressure for reform of the encysted society. The pressures exerted on the encysted society may include positive rewards for compliance or coercive acts by police or even by an army.

Perhaps the oldest and most continuous of such pressures on the Santal has been governmental interference with Santal cow sacrifice and beef eating. Armed interference has been practiced everywhere and has not ceased under present democratic conditions. Cow sacrifice has continued, but in secret so as to avoid prosecution and interruption of the rites. However, some villages, fearing arrest and harassment, have abandoned such sacrifice, particularly where there are Hindus in the village who might report this activity to the authorities. There is no question that this pressure is the result of repugnance as well as solidarity needs.

The tendency to borrow which follows rank concession undermines solidarity, for all borrowing, by reducing distinctiveness, weakens solidarity, and borrowing of traits which directly represent solidarity is especially debilitating. Emulative borrowing not only eradicates boundary markers but is destructive of the internal aspect of solidarity, since it is likely to be more intense among segments of the encysted society with superior status. This differential emulation arises largely because of the following conditions:

(1) the elite tend to have more intimate contact with members of higher social units because they are often representatives of their society in intersocietal relations

(2) they may attract a coterie belonging to higher social units but personally limited in political and economic power; such a coterie derives political and economic advantage from the relationship and confers prestige on its benefactor

(3) no amount of emulation in the absence of the requisite political or economic power can achieve rank improvement; the recognition of this fact contributes to differential emulation since the wealthy and powerful are the only ones in a position to attain higher rank in exogenous relations through emulation

Such differential emulation and the type of external relations described here have been noted among the Santal as well as other tribes in the area.

The impact of differential emulation on internal solidarity is to reduce "likeness" and to provide a cultural basis for divisive status distinctions. The ubiquitousness of envy and connected witchcraft in Santal social relations, and the difficulty of relations between the educated Santal elite and the nonelite exemplify this effect. The cultural gap between the traditional, uneducated elite and the ordinary Santal was not nearly so great, nor relations between them so strained.

In addition to the borrowing tendencies which flow from rank concession and the external pressure exerted on encysted societies, there is another centrifugal tendency inherent in being an encysted society: political-judicial encroachment. The increasing jurisdiction of Hindu courts and village *panchayat* over affairs formerly handled endogenously, are an example; there are also such benefits as the issuance of scholarships, health protection, and treatment, and so forth. These encroachments on areas formerly managed by the traditional social organization diminish its authority and importance and serve to attach the Santal more firmly to the surrounding society.

All of the borrowing tendencies which flow from rank concession, plus external pressure and encroachment on encysted societies, tend to diminish solidarity. If there were no countervailing forces, the world would long ago have become more culturally homogeneous and socially unified. A highly potent force contributing to cultural distinctiveness derives from the necessity of preserving solidarity. Although the basis of solidarity has been widely explored by sociologists and structural anthropologists, there is one condition of solidarity maintenance which has not received much attention, and it is decisive within the RCS. I refer to the relation between avenues of mobility, or rank paths, as I have called them, and the maintenance of solidarity: some rank paths support solidarity and some undermine or destroy solidarity. Emulation might be assigned to the class of destructive rank paths, but since it is never a successful path in the absence of political and/or economic power, it is not by itself likely to pose a serious threat to solidarity. What is crucial, then, to the maintenance of solidarity, are the political and economic conditions of mobility.

In general, the pursuit of rank through political power by a society, whether by war or by winning elections, serves to support the society's solidarity. Increased political power promises rewards to everyone in the society, though it may be specially advantageous to an elite. To be achieved, it requires cooperation and a solid front of opposition against competing societies; efforts will be made to counter centrifugal tendencies and to increase solidarity.

Pursuit of rank through economic improvement, at least within the context of an essentially market society, has the opposite effect. It is divisive because it encourages greater emulation on the part of those who have superior economic power. Furthermore, it not only does not require societal solidarity, but is commonly inhibited by internal ties which demand redistribution of economic gain and, thereby, prevent individual accumulation and investment. Unlike success in the political rank path, which promises rewards to all, economic success is essentially individual; its ultimate achievement is "passing," i.e., social absorption in the dominant society. If rank is pursued through individual economic gain, all of the centrifugal tendencies inherent in rank concession and being encysted not only are free to run their course but are even accelerated as the rank path itself contributes to the centrifugal pressure.

The choice among rank paths obviously depends in part upon the objective conditions of the alternatives. However, whatever choice is made, including the possibility of various combined efforts, the emulation-solidarity conflict cannot be wholly evaded. This, of course, is so because emulation is inherent in rank concession and because all societies manifest some pressure on behalf of solidarity. If a society is essentially pursuing an economic rank path, solidarity will decline and the conflict will either be reduced or remain constant, depending upon the intensity of the various centrifugal tendencies which are operative. But if such a society pursues a political rank path, the pressure for solidarity will increase, and since the pressure for emulation continues, the conflict is likely to become acute. There are, of course, the other centrifugal tendencies already mentioned which also conflict with solidarity, but emulation and solidarity are especially opposed within the political rank path because both are important means to rank improvement; it is a case of not being able to have one's cake and eat it, too.

The conflict between emulation and solidarity which is engendered by rank concession has certain characteristic cultural consequences. These consequences may be understood as attempts to reduce the conflict by transforming emulated traits into somewhat distinctive ones or by claiming that they are actually indigenous. The latter practice is especially effective among tribal and peasant peoples where history-mythology is so commonly used to validate contemporary practice; all that is required is a demonstration that what is borrowed is in reality a forgotten indigenous custom. The transformation device used most often to "naturalize" borrowed practices is syncretism; occasionally there may be "innovative combination," in which the fusion of borrowed and native custom is accompanied by a change in both, and something new emerges. All of these naturalizing practices allow a society to have some solidarity and emulation, too. Since societies on an economic rank path tend to have increasingly diminished emulation-solidarity conflicts, they are likely to manifest increasingly less naturalization of borrowed customs. Instead of producing cultural movements, the culture of such societies is likely to become rapidly more like that of the dominant society. Societies on a political rank path, however, are likely to engage in considerable naturalization since the emulation-solidarity conflict will be more intense. If the conflict is intense enough, a cultural movement may ensue.

In order to check the centrifugal tendencies inherent in rank concession and being encysted, such movements seize on distinctive traits and give their practice a new emphasis; they may even invent new distinctive characteristics. Ideally, the traits emphasized will be attributes of low rank value in the dominant society since these will be most conspicuously distinctive; furthermore, these traits serve best to bind the endogenous elite to those of ordinary rank and thereby to close the gap created by differential emulation. It is, of course, difficult for the elite to carry out such customs since it often requires them to readopt what they had long ceased to practice; in some instances they may even have to overcome hard-won antipathies. This difficulty is an aspect of the emulation-solidarity conflict which may also lead to innovation and generate energy for cultural creativity.

Consider the case of a society which, after long pursuing rank on the economic rank path, switches to a political path. The leaders of such a move are likely to be the elite since they are in a position to gain office should there be a political victory and are generally most adept at political maneuvering, in both their own and the wider, dominant society; but they also have the strongest tendency to emulate, and since they have long been engaged in the effort to secure economic gains, they will have managed to internalize thoroughly the values, beliefs, and even fundamental orientations of the dominant society. The leaders of such a movement are bound to have an intense emulation-solidarity conflict. Should a cultural movement develop under these conditions, it is bound to be marked not only by a reemphasis of distinctive traits and the characteristic naturalizing tendencies which arise from the emulation-solidarity conflict, but also by strikingly innovative processes. For such an elite, the only recourse which will simultaneously preserve distinctiveness and solidarity on the one hand, and express their own deep evaluations of what is true, beautiful, and worthy of rank on the other, is to produce their own "great tradition"; or they may borrow a distinctive one from some other society than the one in which they are encysted. If they create one of their own, it is likely in fundamentals to be much like that of the dominant society but to differ conspicuously in lesser matters, retaining distinctive traditional practices. But even the fundamentals which are imported must be naturalized, both to maintain distinctiveness and to penetrate successfully the nonelite. Such cultural creativity is likely to develop a momentum of its own as personal expression, pride of creativity, and public recognition of culture-producers come into play. Thus a cultural movement which begins largely as a means of strengthening solidarity so that political gains can be made, may become genuinely creative, though it must continue to satisfy the requirements of its original impetus.

To return to Santal historical experiences, from the pre-British period in which rank concession began, to the 1930s and the advent of political democracy, excluding brief periods of nineteenth-century rebellion, solidarity was increasingly undermined. Rank concession was producing emulation, differential emulation, and power-incorporative borrowing. The evidence suggests that migration was the chief means of responding to this pressure so as to maintain the integrity of Santal society. With British control, the spreading market, growing wealth differences, and external encroachment add to the previous centrifugal tendencies. Santal solidarity continued to be sufficient to permit the rebellions of the nineteenth century, but their crushing defeat, in the absence of democratic political opportunity, meant complete closure of the political rank path. Thereafter the decline of solidarity accelerated, and Hinduized sects began to develop. An additional centrifugal force, industrial development, appeared in the early part of this century among the southern Santal. It affected all of this population by accelerating economic opportunity and increasing education, but had an additional disintegrative effect on migrant industrial workers.

Culturally, the entire period from pre-British times to political democracy, is one of increasing Hinduization. Except for the brief periods of rebellion, from the time of British control to the 1930s, the Santal pursued rank almost exclusively via economic mobility and emulation. The cultural consequence of this rank path, plus the disintegrative tendencies inherent in rank concession and being encysted, is to increase the "cultural load," i.e., to increase the number of borrowed traits which reduce distinctiveness and are internally the object of invidious comparison. This threat to internal solidarity was noted in the growing antipathy between rich and poor and the surpassing importance of envy. On behalf of traditional solidarity, efforts were made to halt emulation and participation in Hindu ceremonies, but with limited success. There is also evidence from this period of the naturalizing processes which arise from the emulation-solidarity conflict. Nevertheless, the cultural load tended to increase, particularly in unaltered Hindu attributes of rank value.

The new political rank path of the 1930s, with its need for increased solidarity, created enough pressure through the emulation-solidarity conflict to generate a cultural movement. Not only were serious efforts made to curtail borrowing and to abandon emulative practices, but various emulative characteristics previously absorbed were naturalized. While distinctive traits of low rank value received new emphasis, the great-tradition aspect of the movement imported essentially Hindu orientations into the culture, naturalizing them by means of innovative combination with traditional customs.

As an aid to thought, the major principles of the RCS as applied to encysted societies are succinctly listed below. However, this condensed statement has required some obvious oversimplification and is not meant as a substitute for the more discursive discussion which precedes it.

The RCS as Applied to Encysted Societies[3]

(1) The "cultural load" of the encysted society is inversely related to its solidarity and its ability to resist assimilation.

(2) Rank concession produces a tendency to emulate; since emulation increases the cultural load, rank concession is also destructive of solidarity.

(3) Emulation is likely to be of the differential variety.

(4) A dominantly economic rank path decreases solidarity.

(5) A dominantly political rank path increases solidarity.

(6) Differential emulation is more destructive of solidarity than random emulation; since the economic rank path is particularly conducive to differential emulation, it is particularly destructive of solidarity.

(7) Though power-incorporative borrowing is at least analytically distinguishable from emulation, rank concession is always accompanied by power concession; therefore, rank concession implies a tendency to power-incorporative borrowing as well as emulation. (Power concession does not, however, imply rank concession.) Power-incorporative borrowing also increases the cultural load.

(8) Every society has some endogenous sources of solidarity; therefore, every society that has conceded rank has an emulation-solidarity conflict of some proportion.

(9) Such a conflict manifests itself in a variety of characteristic ways, among which are the following:
   (a) indigenous claims
   (b) syncretism
   (c) innovative combination
   (d) pattern emulation—trait maintenance

(10) An intense emulation-solidarity conflict may generate a "cultural movement."

(11) A shift from a dominantly economic to a dominantly political rank path is likely to produce a severe emulation-solidarity conflict and may thereby serve to generate a cultural movement.

(12) Encysted societies are subject to the following additional threats to their solidarity:
   (a) political-judicial encroachment
   (b) pressure to accept national solidarity symbols
   (c) repugnance pressures

Some Qualifications

Before considering comparative material, I would like to make a correction or two in my formulation of the RCS, based not so much on additional comparative evidence as on more careful consideration of what I had already observed. As my colleague Paul Rosenblatt has pointed out to me, the contention that "a dominantly political rank path increases solidarity" (see statement 5) is not necessarily so, even according to my own evidence and argument. Of course, insofar as a people as a whole engage in unified support of a single political objective, solidarity is likely to be reinforced with deliberate efforts to emphasize and develop unifying cultural symbols. But it is also thinkable that such a group might be torn asunder by democratic political activity. One would think that the internal condition of the encysted society which would be most relevant would be the degree of internal differentiation with respect to wealth and source of income. For political democracy builds and solidifies economic interest-group constituencies as well as ethnic ones. Thus, statement 5 might better be rephrased in the following manner: *The less an encysted society is economically differentiated, the more likely it is that a dominantly political rank path will produce increased solidarity.* Here I have tried to correct what was simply a blunder on my part since I had noted that in certain political contexts (e.g., the Santal industrial workers living in the dominantly nontribal city), there was a tendency to rally behind militant left-wing parties. Nevertheless, on balance, Santal political efforts certainly had contributed to solidarity, not only among Santal but, as I had noted, with other "aboriginals" (*adibasi*), as well. But since the Santal were moderately differentiated internally, why was democratic political activity, on balance, unifying?

Perhaps one condition responsible for this outcome was that aboriginals seeking political leadership found the aboriginal constituency and aboriginal politics far more likely to result in success than competition based on appeal to economic interest groups. For in the former arena one's birth-ascribed status proved an invaluable asset, while in the latter it might well prove a liability. In addition the competition in the smaller pond was likely to be less rigorous; a little education, for example, goes a much longer way in aboriginal political circles. This is the seamy and self-seeking side of the answer, but there is another side, as well.

Insofar as the members of the encysted society are held to be hereditarily inferior, even the most economically successful member will come to recognize that his own ultimate success in terms of rank depends upon overcoming the racist beliefs of the dominant society. In this he is as one with the least of his brethren and this is a powerful source of solidarity. Secondly, since the dominant society is likely to justify its claims to hereditary dominance by pointing to secular inferiority, the elite of the encysted society are likely to feel that they need to help raise the secular attainments of their fellows to counter this argument. Furthermore, a modicum of secular power within the encysted society is required to increase the opportunities for success within the larger political sphere, for it requires money and influence to be a successful politician.

THE JATAVS: AN INDIAN PARIAH CASTE

Lynch's description and analysis (1969) of an untouchable caste, the Jatavs of Agra, are especially pertinent to a consideration of the

RCS. Like the Santal, the major determining external conditions (e.g., the Independence movement and political democracy), profoundly altered both their relation to Indian society at large and relation to their own culture. On the other hand, they did not enter this period as a formerly autonomous community, but rather as a despised but, nonetheless, economically and culturally integrated part of the larger society. If the Santal were a tribe in search of a great tradition, the Jatavs might well be called a caste in search of a great tradition, and, as we shall see, they seem to have found one. Before going abroad to consider widely differing situations, we have here the opportunity for a somewhat more controlled comparison. Since Lynch himself has considered the bearing of his findings on the RCS we may profit from both the data and his analysis.

The Jatavs from the city of Agra are "one of a much larger group of castes known as Camar, or leather workers (though some Jatavs would deny that they are Camars, as we shall see). In the traditional caste system of India Camars fall close to the bottom rung of the caste hierarchy because they are reputed to eat beef, which is the most polluting of foods according to orthodox Hinduism" (Lynch, 1969: 2). The author's "because" is perhaps best interpreted as meaning "according to the traditional explanation-justification"; for the evidence that Lynch himself and many others have presented makes it perfectly clear that neither a change of occupation nor change of eating habits results in a change of caste status. Caste membership in this sense is like ethnicity; e.g., being a Jew in the United States; one might even abandon the religion and still retain his ethnicity whatever the traditional definition-justification might be; to do otherwise would require something tantamount to "passing."

Agra is located about one hundred twenty miles from New Delhi, the capital of India, and has itself been a capital city several times since the sixteenth century. Apart from the Taj Mahal, the city has a number of important industries; e.g., shoes, cotton spinning, sisle mills, oil pressing mills, small-scale iron industry, and the like, but little in the way of heavy industry. The population at the turn of the century was 188,022 and has now reached 462,020 (Lynch, 1969: 22).

As untouchables, the Jatavs suffered a number of disabilities in the secular sphere and with regard to ritual interaction. In a variety of ways, depending upon region, they were held to be in a state of pollution, and therefore polluting to others. Their rank was at the bottom of the cast hierarchy and they were, in general, poor, uneducated, and powerless. However, beginning with 1935 and continuing until today, the untouchables have become the object of "protective discrimination" (Galanter, 1963: 551); i.e., subject to a variety of governmentally insured advantages aimed at reducing inequality. The grounds for treatment as a Scheduled Caste include "exclusion from entry into Hindu temples; exclusion from the services of 'clean Brahmans,' exclusion from the services of the same barbers, tailors, and so forth, used by higher castes; inability to give water to higher castes; and limited access to public facilities such as wells, schools, and roads" (Lynch, 1969: 23); where these conditions were not quite met, the criteria of illiteracy and poverty were occasionally invoked. The relevant modern legislation was provided in the constitution established in 1950 and was to remain in effect until 1959, but has since been extended for an additional ten years (to 1970).

"A certain percentage of jobs at the central and the state levels are reserved for members of the Scheduled Castes (S.C.)" (Lynch, 1969: 25). "In the sphere of political representation, there are a number of seats reserved for the S.C.s in the people's and States' Legislatures at the center and the state level respectively. These are established according to the percentages of S.C.s in the total population and filled by joint electorates in which only S.C.s may be candidates for the seat to be filled. Nevertheless both S.C.s and non-S.C.s vote for the S.C. candidate of their choice" (Lynch, 1969: 25). However, an S.C. may stand for a nonreserved general seat.

In addition the constitution of India proclaims an end to the disabilities of untouchable status, though in practice the attainment of equal rights has often waited for both militant protest and legal action, as in the case of our own civil rights legislation. A variety of special welfare programs have been developed for untouchables as well as campaigns of public education designed to help eradicate attitudes supporting untouchability.

The term "Camar" means leather worker and embraces a large number of named "castes" (*jati*). In the traditional Jajmani system linking the services and ritual interactions of various castes, the Camars had, as their preeminent occupations, tanning hides, making shoes, and removing the carcasses of dead animals which they might eat. Like other low castes, they were obligated to perform forced labor (*beggar*) for those who were dominant. In some regions they were subject to residential isolation. According to popular aphorisms, they are dark of skin in contrast with Brahmans and other high castes, though in reality, for at least some Camar castes such as the Jatavs, it is doubtful that there is even a correlation between skin color and caste rank. They are found chiefly in the North Indian states of Uttar Pradesh, Punjab, and Bihar.

The Jatavs are one of the endogamous castes among the more than one thousand castes comprising the Camars. They are found mainly in western Uttar Pradesh and the Punjab and are sometimes known as Jatiya, Jatua, or Jadav, "but the term Jatav is current in Agra City" (Lynch, 1969: 30). Of the 462,000 residents of Agra (1961 census), approximately 71,404 are Jatavs, or about fifteen percent. The combined total of Scheduled Castes of Agra comprise 97,833 individuals, so that the Jatavs constitute about seventy-three percent of that total.

Before the major development of the shoe industry of Agra, the Jatavs were mainly stone cutters, laborers, scavengers, and leather tanners (Lynch, 1969: 32). A few became contractors for the labor of their own caste, becoming wealthy enough to open up large cotton mills. From the turn of the century through World War II, the shoe industry continued to grow, establishing markets throughout India as well as the Near East and East Indies. However, these foreign markets were greatly reduced after Partition, since they were largely maintained by Muslims who fled to Pakistan. Since 1955, the slack has been taken up by the Soviet Union and other East European nations, and by Germany. In addition, this new trade has generally demanded a higher quality, more fashionable product, necessitating increased skill.

Prior to Partition, the Jatavs maintained a near monopoly of the productive aspect of the industry with only the position of "factor"; i.e., middleman between the producer and retail distributor in other hands. Factors tended to be Muslims, and whatever the facts may have been, they are now looked upon as having been much easier to get along with (i.e., less exploitative) than the Punjabis, who have increasingly come to replace them, and who are also more than successfully competing with Jatavs in developing factories, as well.

Jatavs range from unskilled laborers to "big men" who are proprietors of factories employing fifteen or more men. Many varieties of less organized production unit are also to be found. Though there have been a number of attempts to organize unions, none has really taken hold, and all work is on a piece-work basis. Presumably particularistic and paternalistic relations among the Jatavs have hindered the development of unions.

Turning now to the efforts at mobility undertaken by the Jatavs, Lynch (1969) conveniently divides them into three chronologically adjacent periods which consist of the "pre-independence period, the transitional period and the post-independence period."

The Pre-Independence Period

This period which began just prior to this century was one of gradually improving economic opportunity resulting in wealth differentiation, and increasing opportunity to observe and emulate high caste behavior. Two wealthy Jatavs came under the influence of a Brahman named Swami Atma Ram who proclaimed in a book that the Jatavs were in reality of the Warrior or Ruling Varna (*Kshatriya*) rather than lowly untouchables. The two rich Jatavs organized a Jatav Committee of Agra City whose aim was to encourage emulation of the prevailing high caste customs of the region. "A caste council of Jatavs from all of Agra City was convened, and it resolved to forbid the eating of beef and buffalo by Jatavs from that time forward" (Lynch, 1969: 68).

Mission schools and, most important, the Arya Samaj, a Hindu reformist sect seeking to associate modern humanistic goals with traditional Hindu scripture, provided contact with emergent modernizing cultural influences. The notion of mobility through achievement was given legitimacy through these contacts. The Jatavs worked hard to establish their Kshatriya genealogy and adopt Kshatriya-like ceremonies, such as shooting a cannon at weddings, and the use of bow and arrows at the birth ritual (Lynch, 1969: 71). Appropriate reasons were found for their fall to Camar status with important charter functions in the contemporary scene; it was held that they were the Kshatriya victims of Brahman enslavement, thus justifying the position held through the Independence period of opposition to the Brahmans, though the claim to simultaneous Brahmanic Kshatriya status made this opposition somewhat paradoxical. During this period they firmly rejected solidarity with other untouchable castes who they claimed, for example, were really Camars; in one instance, they were reported to have driven out of Agra the leader of a movement (The Original Hindu Movement: see Lynch, 1969: 76) who taught that the Scheduled Castes were the original inhabitants of the land and who opposed the caste system and advocated the abolition of caste endogamy.

New leaders who were literate, educated, and wealthy, began to eclipse traditional caste leaders. By the 1920s, the Jatav Men's Association succeeded in getting one of its members into the State Legislative Council. Gradually the Jatavs managed to obtain a variety of other governmental posts at the local and state level. The Jatavs used their increasing political leverage to obtain a variety of benefits for the caste and to oppose discrimination directed at its members. By the late 1930s, the Jatavs gained governmental recognition of their claim to Kshatriya status! However, World War II intervened, and subsequent events rendered this acceptance of little significance.

An indication of the degree of political awareness of the Jatav big men of the period was the letter they sent to the Round Table Conferences in England during the early thirties. These conferences were to "sound out Indian opinion on the nature of reforms for the proposed new constitution of India, subsequently known as the India Act of 1935" (Lynch, 1969: 81). The concern of the Jatavs was to let the conference know that their support was for Dr. B. R. Ambedkar, who was himself an untouchable, and not for Mahatma Gandhi, who was not. The chief issue which divided these two eminent leaders at the conferences was the question of separate elections or joint elections for the "Depressed Classes." Gandhi supported joint elections with reserved seats for the Depressed Classes, arguing that separate elections would serve to further fragment Indian society. Ambedkar, on the other hand, argued that joint elections would allow the upper castes to continue their dominance and exploitation. In 1932, the British made what was known as the Communal Award, providing special seats and a separate electorate for the Depressed Classes, though Depressed Classes were also entitled to vote for general seats. Thus Ambedkar won out for the moment. The issue was finally decided by an "epic fast" by Gandhi who refused to accept the award. With Gandhi near death, the pressure proved too much for Ambedkar, who yielded to a compromise solution (Poona Pact) in which separate electorates applied only to the primaries. Though Gandhi would, of course, have defended his fast as a form of moral suasion, and so on, it seems clear in this case, at least, that Ambedkar felt the fast to be more like blackmail; at the very least the enmity felt by Ambedkar never abated and was certainly intensified by the incident.

Transitional Period

The events of the thirties and early forties, particularly those in support of Ambedkar against Gandhi, certainly presage the volte face which characterized the post-Independence period. The Independence movement was brought home to the masses with slogans not only of independence from England and appeals to patriotism, but by the slogans of democratic freedom for the various oppressed peoples of India. Thus Ambedkar, who had received his postgraduate training in the U.S.A. and England, proclaimed the goals of "Liberty, Equality and Fraternity," while leaders of as much renown as Nehru spoke openly of their sympathies for socialism and even of a socialist India. Gandhi, whom the Jatavs, following Ambedkar, came increasingly to oppose, spoke out and acted dramatically against "untouchability," though his ideological position that the untouchables should be regarded as part of the fourth (lowest) Varna (category of castes) rather than out of the system or as untouchable was hardly well calculated to win the support of the ambitious Jatavs (Lynch, 1969: 132). Gandhi argued that upper castes have a trustee-like obligation to lower castes. Again, given the libertarian ethos of the period, this aspect of his stance smacked of a kind of paternalism ill-fitting the occasion.

By 1944-1945, the Jatavs had primarily gone over from Kshatriya claimants to identification with other untouchable castes. A Scheduled Castes Federation of Agra was formed and was linked to Ambedkar's All India Scheduled Castes Federation. In place of claims within the caste system, under the leadership of Ambedkar they came to denounce the legitimacy of the system. Direct acts of civil disobedience *(satyagraha)* were the most notable actions of this period. Agra Jatavs played a conspicuous part in the demonstrations at Lucknow, during which demands were presented to the government for reserved positions for the Scheduled Castes. A similarly aggressive action was a parade in Agra in opposition to the Poona Pact; i.e., the compromise on voting agreed to under the pressure of Gandhi's fast. The pact was symbolized in the parade by a Brahman wolf "that had run wild and was devouring the Scheduled Castes. . . . The wolf was publicly burned, and the heat of Jatav anger was not lost to the eye of the Agra upper castes who saw themselves symbolically consumed in the flames of long smoldering resentment" (Lynch, 1969: 88).

By 1946, however, a potential cleavage among the Jatavs became actualized in the campaign to fill reserved seats in the Legislative Assembly. Lynch characterizes this split as between "conservative followers of the Congress Party and the revolutionary followers of the Scheduled Castes Federation." Each put forth its own candidates in a variety of elections up to 1957. Though the Congressman won these elections, perhaps largely due to the joint election system, Lynch judges that the Scheduled Castes Federation leaders have increasingly come to command the allegiance of the Jatav caste. Since this split becomes more pronounced during the following post-Independence period, it will be examined more closely in this context. A good summary statement of the Jatavs' perception of the volte face which begins in the transition period is provided by an Agra organizer of the demonstration against the Poona Pact: "The reason for the movement was that we felt the British Raj was coming to an end and that if Congress came into control we would not have any rights because of the Congress being the party of the caste Hindus. . . . It was this movement that was the real awakening of the Scheduled Castes. From this time on they were really conscious of the battle that had to be fought" (Lynch, 1969: 87).

Post-Independence Period

The post-Independence period unfolds in the context of the universal franchise and a modern democratic constitution; indeed portions of that constitution specifically relating to untouchability were drafted by none other than Ambedkar himself. The first of the two most notable developments of this period is the conversion of most of the Jatavs to Buddhism under the leadership of Ambedkar.

The second, which followed two years later, was the development of the Republican Party, appealing for a unified vote among the Scheduled Castes. As Lynch puts it, "The Republican Party and the Buddhist movement are in a complementary relationship" (Lynch, 1969: 95); i.e., the former attends directly to secular matters and the latter "provides a psychological justification and an ideological rationalization for political militancy" (Lynch, 1969: 96).

The explicit reasons given by Ambedkar for abandoning Hinduism are that its most sacred works and tradition are based upon and provide support for the iniquitous caste system. On the other hand, Buddhism, which perhaps arose in opposition to Brahmanic influence, provided no such support, was indigenous, and provided connections with an already existent worldwide religion. According to Ambedkar, the present untouchables are the descendants of aboriginals who were conquered by sedentary Brahmans and made to do labor for them. The aboriginals were even then Buddhists and were, therefore, despised by the then cow sacrificing followers of Brahmanic religion. With the masses of non-Brahmans converting to Buddhism, the Brahmans changed their beliefs so as to more successfully compete by not only abandoning cow sacrifice, but meat eating as well, and they declared the cow to be sacred. Thereby the poor who, perforce, had to consume beef, became "untouchable." Thus, Ambedkar's historical interpretation validates the appropriateness of Buddhism as the religion of untouchables and puts the Brahmans and Brahmanism in the role of oppressors. As Gandhi intertwined Hinduism, often with his own novel interpretations, and the Independence movement, so Ambedkar linked the elimination of untouchability with conversion to Buddhism.

Apparently Buddhism has not much altered the ritual practices previously characteristic of the Jatavs, nor is it clear what Hindu beliefs in particular it has supplanted. Some effort has been made to institutionalize the learning of Buddhist doctrine, but the general impression is that the Buddhist movement is somewhat less prominant in the minds of the Jatavs than the Republican Party. As Lynch (1969: 148) puts it, "the success of the Party has somewhat quenched the thirst for the religion."

The Republican Party has succeeded in electing numerous members to the Agra Municipal Corporation which runs the city and a few members to the Legislative Assembly of the State (U.P.). Many, though by no means all, of these representatives have been Jatavs. The pressure for electoral alliance has, in various instances, led to agreements with varying constituencies having a common interest at the moment; e.g., the Muslims of the region. The party is primarily concerned with securing economic benefits for untouchables and fighting instances of discrimination.

Though the Republican Party commands the loyalty of most Jatavs, it is itself divided into at least two ideological factions, and, as indicated earlier, there is still some support among the Jatavs for the Congress Party. The Republican Party seems to be divided into a kind of left and right wing, with the former demanding, surprisingly, an end to protective discrimination; i.e., reserved seats, and so forth, and the right wing wishing to maintain these advantages. The left takes this position on the grounds that the reserved seats and other privileges cause the untouchables to continually rely on others; i.e., the Congress and the government which it dominates, rather than struggling on their own for equality. In addition, there have been frequent conflicts between party members who have obtained office and those who only hold office within the party. Since the former control patronage and have access to state power, while the latter do not, the party has occasionally found it difficult to maintain control over its elected officials. The radicals feel that protective discrimination produces "Uncle Toms"; i.e., it facilitates cooptation.

The Jatav Congress supporters are primarily led by the old big men. Most of them are not "politicians" *(neta log),* but rather wealthy entrepreneurs, and their following is small. They support Congress because Congress has the power to help or hinder their business. Their position sounds much like that advanced by Booker T. Washington in its advocacy of economic advance and steering clear of political struggle; as one of their leaders puts it, "What is necessary first is education and businessmen who have some money. When we are a minority community without any power, then it is not for us to openly criticize those in power and to revile them. . . .If I carried on as some of the members of the Republican Party do, where would I be now, as a businessman? I have made much progress by not following its way. The Hindus are in power and we are too weak to stand up to them. Only when we have educated men and powerful businessmen of our own can we stand before them as equals" (Lynch, 1969: 113).

It seems most useful to begin an analysis of the relevance of this data to the RCS by considering Lynch's own comments on this subject. As he puts it, "the analytical concepts and procedures used in the discussion of the Jatavs and of Sanskritization have an obvious relationship to what Martin Orans (1965) has called the rank concession syndrome (RCS) and the emulation-solidarity conflict (ESC)" (Lynch, 1969: 219). However, Lynch finds that in a number of ways his data do not support the claims of the RCS theory. He argues that, contrary to expectation, economic mobility has not contributed to a decline in solidarity but has, on the contrary, increased it. He claims that this is so because, in India, ranking recognizes only group achievement, not industrial achievement; therefore, he argues the big men who have prospered economically had to join their caste-mates in an effort to pull up the whole caste in their own personal interest.

Let us first consider the factual question of whether or not economic mobility has decreased solidarity. According to the theory of the RCS, it is generally divisive for the following reasons:

(1) it encourages greater emulation on the part of those who have superior economic power, thereby resulting in differential emulation and therefore in cultural and social schism

(2) such mobility is inhibited by internal ties which demand redistribution

(3) its ultimate achievement is "passing"—i.e., absorption in the dominant society—rather than improvement of the rank of the encysted society

As for differential emulation per se, Lynch does not provide systematic exploration of this question, but the data presented would certainly seem to bear it out as fact. For example, we learn that two Jatav rich men around 1900 came under the influence of a Brahman who wrote a book establishing that the Jatavs were really Kshatriya. Indeed this was the beginning of the emulative Kshatriya movement among these people. We are not able to judge from the evidence presented how differential was the effect of the movement, but we may surmise that education varied to a considerable measure with wealth so that the wealthy tended also to adhere more readily to Kshatriya ritual and other practices of rank value. However that may be, it is clear from the data at hand that the division between wealthy entrepreneurs (big men) and the rest of the society is quite pronounced. Indeed this difference now relates not only to membership and commitment to different political parties, but even to the conversion to Buddhism. It is probably somewhat of an oversimplification, but nonetheless true, on balance, that the wealthy tend to wish to remain in the Hindu fold, though there is no longer an open Kshatriya movement, while the rest of the society is making common cause with the rest of the untouchables. No doubt differential economic and related political interests account for much of this disunity, but differential emulation perhaps has also contributed.

As for our second contention that getting ahead economically as individuals, or in units of less than a societal order within the context of the market, puts pressure on traditional ties, again we lack sufficient evidence to judge. Some of the derogatory terms used for the rich like *bara petwalla* (literally "big stomach one") suggest that such strain may exist, but from Lynch's presentation it is not clear that such usage applies to rich Jatavs.

The third contention regarding passing is perhaps most important in regard to Lynch's criticism, for in effect his claim that in India individual rank is not pertinent, apparently implies, among other things, that passing is not possible. As we have indicated in discussing the Santal and other Mundari tribes, passing involving subtribal units has indeed occurred in which validation of genealogies was gradually accepted. That wealthy Jatavs might not have eventually come to the same end is not at all ruled out by previous experience. Secondly, and perhaps this is more important for the case at hand, it seems to me that the evidence is overwhelming that even in a static analysis, rank in India has many facets, of which ritual caste rank is only one. Prestige, ritual rank, political power, knowledge, wealth, and so on (i.e., any valued characteristic), may, in certain contexts, be salient and these are by no means in perfect correspondence. Every study I know, of caste ranking dealing with this question, makes this clear, and though my own evidence (Orans, 1968) indicates that this is becoming increasingly true, there is abundant evidence that these characteristics varied somewhat independently in the past (Silverberg, 1959; Hiebert, 1969). It is false that individual achievement is without significance and false that in societies like our own the rank of one's group is irrelevant. There is no doubt a difference in degree of significance of individual and group rank, which is exceedingly important, but all societies take cognizance of both. Nevertheless, Lynch has an important point to make which is well taken; I should like however to put it like this: *the more salient and all encompassing the rank of an encysted society is, the less the danger to solidarity of differential emulation, wealth differences,* and so on. Therefore, in the Indian context, wealthy Santal and wealthy Jatavs have a greater interest in the welfare of their people than they would in societies with less concern with group rank; however, concern with group rank varies widely even within a single society, not to mention a nation-state, so that it is not at all clear that being Black in the United States is less salient than being a Jatav in India! Though I find Lynch's emphasis on the importance of group versus individual rank important, I do not see that either his argument or the facts he presents indicate that the pursuit of wealth in the context of the market has not led to the kind of divisiveness I discussed. On balance, such pursuit seems to have led to the very kind of differences predicted, though the wealthy, as among the Santal, have made efforts to raise the rank of their entire group.

The second criticism Lynch offers is that with the adoption of Buddhism, rank concession has ceased to operate. The suggestion seems to be that, having opted out of the system, the ranking problem is settled. Though not much evidence on this point is offered, we do learn that the Jatavs have internalized the upper caste stereotype that they are by nature drunken, misbehaved, and the like. Perhaps Lynch thinks of rank concession differently, but as a psychological fact, I am not prepared to believe that it simply vanishes as an attitude if one loudly proclaims his equality, superiority, or simply claims to have opted out. Here one would like to see evidence that the various modes of resolving the conflict engendered by rank concession (e.g., indigenous claims, syncretism, innovative combination, and pattern emulation-trait maintenance) are no longer operative. It would appear that Ambedkar's Buddhism makes use of all these devices. What the Jatavs want, as Lynch indicates, is "equality," and I should think that at least until they get it they shall continue to entertain doubts about their own worth, however much they proclaim it.

Finally, Lynch argues that political judicial encroachment, as I called it, has not weakened the solidarity of the Jatavs; this claim seems to rest on his contention that the government is a kind of third party which mediates relations between constituents so that functions are not really usurped by a competing social unit. Secondly, he argues that insofar as the government has encroached, it has done so on all groups so that none are differentially weakened. Lynch's evidence makes it clear that traditional forms of organization and leadership have indeed declined as the result of encroachment, but he would no doubt argue that new forms of organization and leadership integrating the Jatavs with the total society, and even beyond, have arisen. Here perhaps the difficulty is simply misunderstanding of my *ceteris paribus* qualification to each separate contention regarding the RCS. I think that encroachment has operated as indicated, but as Lynch indicates, greater articulation with the state has resulted in new internally based organization and leadership. This is, of course, as true among the Santal with their Jharkhand Party as among the Jatavs. Perhaps one might profitably combine Lynch's point with my contention in the following way: *Greater articulation with the state is likely to have both positive and negative effects on the solidarity of an encysted society; the positive effects may result from the development of locally based articulating offices and organization, and the negative from the withering of previously existing offices and organization whose functions have been preempted.* The task then becomes one of evaluating which of these effects is most potent. A case like that of American Indians, in which encroachment has been nearly total without much opportunity for new locally based articulating organization and office, would then be much more destructive of solidarity than the Jatav or Santal case, where the encysted society entered enthusiastically into the democratic political arena. If the government were indeed quite distinct from competing social segments, and if it operated equally with respect to all such segments, each segment would be no worse in terms of solidarity vis-à-vis each other's segments than without encroachment, except insofar as it happened to have more autonomy beforehand to encroach upon. It would then only decline in solidarity to a point of equality. But almost no one in India (certainly not the Jatavs or Santals) believes that the government is such an impartial third party. Many groups, in fact, believe firmly that the government is dominated by something like Brahmanic Hinduism and the top, or "twice born" castes. A number of other partially independent claims of particularism having to do with wealth are, of course, also widely believed. Given the competing interests and beliefs that operate to control the Indian government, no doubt such claims are often exaggerations, though they are certainly psychologically real. However, I would argue that the claims are more than a little correct and that self-protective tendencies of untouchables-tribals, and so forth, are not without justification, however poorly such parochialism may serve the integrity of the Indian nation. Certainly the Santal and other Mundari tribals had more autonomy prior to modern times than the Jatavs. It would then follow that encroachment has proved more damaging to them than to the Jatavs. I would judge from the evidence at hand that this has probably been so.

Having attempted to incorporate some of the criticism of Lynch, it is perhaps useful finally to examine the general course of Jatav history from the point of view of the RCS to see what may be learned. The potency of birth-ascribed status has induced the Jatavs and the Mundari tribals to reform their own practices and join together politically as a unit in their own interest. Nevertheless, it seems that from the transition period to the present, the political rank path has become increasingly significant. It also seems fair to claim that the earlier Kshatriya movement, which accompanied a less political period, was predominantly emulative; furthermore, the Buddhism of the more recent political period surely qualifies as an anti-emulative separatist ideology parallel to the Mundari-Sarna movement; the former unites converted untouchables, the latter the aboriginals. In both cases, the opening of the political rank path which began in the thirties was accompanied, and, perhaps partly the cause of, an anti-emulative cultural movement. It seems that the Jatav experience is consistent with my contention that an increasing reliance on politics for mobility is likely to increase the probability of cultural separatism, which is, of course, not very surprising.

But there is another question which the Jatav material raises and which might also be asked about the Santal, viz., why did the Jatavs not become increasingly militant in their Kshatriya claims, or the Santal in their claims to acceptable Hindu status? For the Santal and various other Mundari tribals it was perhaps due to their poverty and powerlessness combined with a wide cultural gap. However, Lynch makes a point about the Jatavs in this connection which is perhaps generally relevant, so far as the present is concerned. He argues that the social system had changed sufficiently by the transition period, and certainly since Independence, so as to alter the nature of the rank game. In brief, the relevance of birth-ascribed caste rank had significantly declined in many contexts; therefore, the appeal for acceptance of high birth-ascribed status has lost much of its cogency. The "new wind" (Beals, 1962: 75) which blows in India and elsewhere carries the message of equality which means either everybody should be regarded as of equal rank and/or worth is a matter of personal attainment, possible logical inconsistency notwithstanding. A solid political constituency is, therefore, seen to be of much more worth than a plaintive cry for superior status in an ancient system of rank which is losing, but has not quite lost, its hold.

There are a few other circumstances of the Jatavs which deserve notice for our purposes. The first is that the system of protective discrimination has given rise to interminable legal and moral complications, as indicated by Galanter (1963). One of these difficulties directly relevant to the Jatavs is that under present interpretations only Hindus are entitled to the benefits of a Scheduled Caste. The claim supporting this interpretation is that the disabilities of untouchableness and presumably backwardness also disappear with conversion to another religion. Though this is largely patent nonsense, it has weakened the Buddhist movement among the Jatavs and no doubt discouraged other anti-emulative interpretation. A second point is that the Jatav occupational specialization in leather goods has always stood in their way in the traditional Hindu rank game since it is a particularly "polluting practice," according to the prevalent understandings. Therefore, the Kshatriya claim must always have suffered from this disability.

Having generally adopted the ideology of equality with all its ambiguities, the question naturally arises why the Jatavs do not lend their support to the communists. Perhaps the answer has both a particularistic and general component. Ambedkar opposed communism because of its totalitarian character; no doubt his influence has been significant. However, one might well argue that the acceptance of Ambedkar itself requires explanation. Here perhaps the class-like differences among the Jatavs facilitated a communal commitment with an interpretation of equality as the elimination of disabilities of untouchables, rather than hostility toward capitalism and the owners of capital. Though the tendencies of schism based upon these class-like differences have manifested themselves, the already developed Scheduled Caste constituency has so far led to division within the movement rather than outright disaffection. If communism were to gain a foothold, one might well imagine a Jatav chapter of the Communist Party like the Black Che-Lumumba chapter of the Communist Party in our own country.

THE BURAKUMIN: A JAPANESE PARIAH CASTE

The problem of naming the Japanese pariah caste to be discussed is like that of designating Americans of African descent. The names used in Japan range from terms like "Eta" (full of filth) or "Yotsu" (four-legged) to more positive terms reflecting various views about ethnic relations. "Burakumin" (people of special communities; see DeVos and Wagatsuma, 1966; DeReuck and Knight, 1967) is a rather neutral term, perhaps comparable to "Negro" in contemporary American usage, but, like the latter, unsatisfactory to some, precisely because it does not convey militancy and is therefore seen to signify acquiescence to objectionable conditions. Names such as "Buraku-no-hito" (people of unliberated communities) have been coined, and are preferred by certain segments of the community.

The Burakumin are "the descendants of members of a pariah or 'untouchable' caste who are still socially and economically discriminated against as the result of prejudice by members of the majority society" (Wagatsuma, 1967: 118). Like the untouchable Jatavs of India, though they are not racially different from those of higher status, they are believed by many, often including themselves, to be racially distinct. In fact, it is often only place of birth or residence which makes identification possible. The majority stereotype of the Burakumin contains the pollution notion and, as with the Jatavs, it is connected ideologically with their traditional occupation which is, again, leather working; the historical Buddhist connection seems clearly to link these instances.

Like Blacks in the United States, the Burakumin are widely regarded by the majority to be "mentally inferior, incapable of high moral behavior, aggressive, impulsive, and lacking any notion of sanitation or manners." In addition, they are often "the last hired and the first fired" (Wagatsuma, 1967: 118-119). Marriage outside the community is infrequent and often, when it does occur, results in tragedy and ostracism, though such marriages are becoming more frequent. The Burakumin perform more poorly in school than the majority population and their communities are marked by higher-than-average crime rates. Again, like Black Americans, they have suffered greatly at the hands of the majority population, and have, in modern times, developed a variety of movements aimed at alleviating their condition.

Pre-Meiji Origins

As early as the fifteenth century in Japan, there existed a category of people designated Senmin—low people. This category was further divided into artisans working in leather or bamboo or dyeing, and itinerant entertainers, prostitutes, and itinerant religious practitioners. They were forced to live in separate communities, often on dry river beds or on tax-free barren lands (Wagatsuma, 1967: 119). During the Tokugawa period in the early seventeenth century, a relatively rigid hierarchy was established with four major divisions paralleling the Hindu four-Varna system, without the priestly equivalent of Brahmans. Beneath these four levels, like the Hindu untouchables, were the low people divided into the Hinin (non-human), and the Eta (filth); the former consisted largely of those who had committed certain crimes, such as adultery, while the latter were largely the hereditary descendants of the artisans whose occupational status was polluting. The Hinin were somewhat higher in status and might regain a respectable position with an adequate sponsor and by paying a proper fee; the Eta were not permitted such mobility. Each tended to be endogamous and were required by law to wear various clothing signifying their status. "Their main occupations continued to be handling leather . . . working for police (attending, escorting and executing criminals and disposing of their corpses). Depending upon the locale, they were also engaged in dyeing cloth, making bamboo brushed (for the tea ceremony), magico-religious practices and various forms of entertainment, and very often also begging" (Wagatsuma, 1967: 120-121).

Emancipation?

In 1871, the Meiji government abolished pariah status as a legal fact. In examining the reasons for this act, one is struck by the combination of concern for appearances abroad and naked self-interest. Two years earlier, a committee of feudal clan officials met to consider the continued existence of untouchability Japanese-style which was declared to be "a national shame in foreign eyes" and a "flaw in the imperial rule" (Wagatsuma and Totten, 1966: 33). When the "shame" was eliminated in 1871, the government succeeded in eliminating the tax-free status of the land occupied by the Burakumin, denying them their trade monopolies and requiring them to enter military service in menial capacities. All feudal rights of value were thus swept away with all the social disabilities maintained, as they did not require legal reinforcement.

After emancipation came the deluge. Severe economic difficulties beset the nation. Like Jews in Europe, the Burakumin were often scapagoats during this period. This situation was made worse by the fact that the emancipation caused small farmers to fear competition from those emancipated. "Eta hunts" or "campaigns to exterminate the Eta" were carried out. In a situation closely paralleling the circumstances after United States emancipation of Blacks, those who tried to cash legal emancipation in for social equality were especially hard hit. For example, in 1872 a group of newly liberated commoners notified a neighboring village that they would no longer carry out such demeaning tasks as "driving off thieves, beggars, and disposing of dead carcasses" (Wagatsuma and Totten, 1966: 36). The villagers retaliated by denying the Burakumin fire wood which they gathered from an adjacent hillside. The struggle escalated to a quarrel and, ultimately, the murder of four Burakumin, burning down their village, and so forth. Such sporadic incidents were often directed against the government, as well, and frequently led to governmental repressive actions. It goes without saying that such flagrant attacks on Burakumin were accompanied by an endless series of more casual abuses which have been plainly shown to have scarred many, if not all, of those belonging to the category. In schools, in situations of interdining, in official documents, in epithets hurled upon chance contact, and on and on, the Burakumin have been made to feel inferior.

To the End of World War I: Emulation and Reconciliation

Organized movements among the Burakumin to improve their condition became evident around the turn of the century. "Middle class property owners" (Wagatsuma and Totten, 1966: 38) who were influenced by the imported philosophies of "liberalism, democratic rights, and socialism" formed an organization in Okayama Prefecture, the Bisaku Common People's Association, to "attempt to overcome the cold and harsh discrimination and mistreatment of Burakumin" (Wagatsuma and Totten, 1966: 38). The primary emphasis was emulative with the premise being that discrimination could be ameliorated by "improving their customs and manners, raising their moral standards, bettering themselves economically, and educating themselves so as to achieve self-reliance and independence" (Wagatsuma and Totten, 1966: 38). A similiar organization, the Greater Japan Fraternal Counciliation Society, was formed in 1903 with its inaugural meeting held in Osaka. Their declared objectives were "improvement of Burakumin moral training, customs and manners, general education, sanitary conditions, leadership, frugality and saving, and economic conditions" (Wagatsuma and Totten, 1966: 39). Cooperation with law enforcement officials and loyalty to the Emperor were further stated objectives of the organization whose leadership was predominantly that of the propertied class among the Burakumin. Whatever its local successes, these emulative movements did not much succeed in either eliminating discrimination in social interaction, decreasing poverty among the Burakumin, or even in preventing discriminatory decisions by the courts.

End of World War I to 1937 and the Japanese Attack on China

The worldwide inflation following the war hit the Japanese poor very hard. A general price rise from 1914 to 1918 of one hundred thirty percent was accompanied by a rise in income for workers of about fifty-seven percent. Meanwhile the price of rice was more than doubling. Rice riots broke out throughout Japan with raids on stores and destruction of police stations. The Burakumin who, of course, suffered even more than the general working-class population, played a conspicuous part in these uprisings as well as in various strikes.

Buraku leaders following this mass movement began to shift from emphasis on self-improvement to petitioning the government for aid and an end to the various forms of discrimination. In 1920, the government made its first financial contribution to "improve the Buraku," appropriating 50,000 yen ($25,000) (Wagatsuma and Totten, 1966: 41). Joint government and Buraku organizations were established to ameliorate the Buraku situation so as to forestall the spreading of revolutionary ideas. In spite of these efforts, the Buraku participated in strikes in a number of industries connected with the manufacture of leather goods, wooden clogs, sandals, and toothbrushes.

From the early twenties until the Chinese War, the movement was increasingly dominated by various socialist, communist, and anarchist ideological trends. Leadership passed into the hands of younger men who resented and rejected the "self-debasing" approach of the former leaders who harped continually on emulation; i.e., raising the level of Buraku culture so that the majority would accept them without prejudice; similarly they rejected the paternalism of the government. Some proclaimed that only through a socialist revolution could the Burakumin be liberated, and even some members of the Buddhist clergy developed socialist political organizations among the Burakumin. In 1922, the National Leveller's Association (Zenkoku Shiheisha) was formed by these new, more militant, leaders taking its name from the "Levellers" of seventeenth-century Britain. Their program and their flag are an interesting blend of Marxism-Leninism, Kropotkinism, Christian socialism, and Buddhism.

The following three propositions were adopted as the objectives of the Levellers (Wagatsuma and Totten, 1966: 43):

(1) We Tokushu Burakumin will attain complete emancipation by our own action.

(2) We Tokushu Burakumin demand economic and occupational freedom from the majority society and we shall obtain it.

(3) We Tokushu Burakumin will be conscious of human dignity and we shall march forward toward the full realization of human values.

The tone of the movement at its inception is nicely indicated by the following declaration which was read aloud:

> Tokushu Burakumin throughout the country, unite! Long-suffering brothers: In the past half century, the undertakings on our behalf by so many people and in such varied ways have failed to yield any favorable results. This failure was a divine punishment we incurred for permitting others as well as ourselves to debase our own human dignity. Previous movements, though seemingly motivated by compassion, actually corrupted many of our brothers. In the light of this, it is necessary for us to organize a new group movement by which we shall emancipate ourselves through promoting respect for human dignity.
>
> Brothers! Our ancestors sought after and practiced liberty and equality. But they became the victims of a base, contemptible system developed by the ruling class. They became the manly martyrs of industry. As a reward for skinning animals, they were flayed alive.... They were spat upon with the spittle of ridicule. Yet all through these cursed nightmares, their blood, still proud to be human, did not dry up. Yes! Now we have come to the age when man, pulsing with this blood, is trying to become divine. The time has come for the victims of discrimination to hurl back labels of derision. The time has come when the martyrs' crown of thorns will be blessed. The time has come when we can be proud of being Eta.

The flag adopted was one with a black background setting off a "round crown of thorns dyed blood red, an intentional symbol of the martyrdom of Christ. The flagpole was fashioned in the shape of a bamboo spear, symbolizing the militancy of the traditional peasant uprisings against injustice" (Wagatsuma and Totten, 1966: 44).

Apart from its diverse sources, this document and symbols of the Levellers are noteworthy from our point of view for their emphasis on pride, self-help, and militance. Much of the early activity consisted of militant stands against particular acts of discrimination. When an act of discrimination was uncovered demands were made for a public apology and sometimes such campaigns were successful. Frequently such opposition led to fighting. In one such instance a confrontation with an ultra-rightest group led to the formation of a "resolved to die" group. When the police intervened, those arrested turned out to be overwhelmingly Burakumin. But the Levellers continued to expand their membership and even the more conservative Burakumin organizations were forced to oppose discrimination as well as urging cultural uplift.

The remainder of the twenties was occupied with a major struggle over discrimination in the armed forces. The government reacted to the increasingly militant movement by establishing manhood suffrage on the one hand in 1928, and on the other by increasing repression particularly directed against leaders of the movement.

Within the movement there were continuous struggles between competing left-wing blocs. Particularly noteworthy was the struggle between the Bolsheviks and Anarchists over the question of centralization of control within the movement itself. Some of these internal difficulties led to the formation of offshoot organizations, though participation continued in the national organization. Continued struggle against discrimination in the armed forces led to the arrest of the leader of the National Shiheisha (Levellers), Matsumoto Jiichiro and the head of the pro-Bolshevik Proletarian Federation faction. "The police claimed to have discovered dynamite and revolvers in Matsumoto's home and hand grenades in the homes of other leaders" (Wagatsuma and Totten, 1966: 51). Ten leaders were arrested and five died in prison from "illness." It appears that the charges were, in fact, fabricated. Nevertheless, Leveller candidates participated in the elections for the House of Representatives begun in 1928, and though they were not elected, this was the start of greatly increased political participation which shortly was to prove successful.

With the depression of the 1930s which again hit the Burakumin harder than any other segment of Japanese society, the move to the left became even more pronounced, and the dominant position within the movement came increasingly to be determined by the turns of international relations and their effects on the tactics of the communist movements. These turns involving popular front cooperation or alternatively more militant activity within avowedly communist organizations, seemed to dominate politics within the Buraku. Though the connected ideology is interesting per se, it is not germane to the present explication. What is important is that the general tendency was toward the popular front and, therefore, led to widespread participation in a great variety of union, political, and antidiscrimination events, while at the same time encouraging the maintenance of various Burakumin organizations. In a word, the strategy adopted was to utilize both the Burakumin identification and to encourage identification with the proletariat as deftly as possible.

Led by Matsumoto Jiichiro, the Levellers were able to address enormous rallies involving hundreds of thousands and to gain the support of unions and left-wing parties such as the Socialist Masses Party. On this basis they succeeded in obtaining reversals of court decisions against various Burakumin, and apparently this was accomplished without "backlash." In a move reminiscent of protests within the United States, opposition was leveled against movies, newspapers, and magazines depicting the Burakumin in a demeaning manner.

Levellers began to succeed in getting elected to all levels of government, and the avowed aim of their political participation was to "strive through collective political, economic, and cultural struggles to protect and expand the rights of the oppressed Buraku masses toward their complete emancipation" (Wagatsuma and Totten, 1966: 59). By 1937, the movement was so effective that the Socialist Masses Party, with which the Burakumin associated themselves, elected thirty-seven members to the House of Representatives, thus becoming the third largest party.

In the midst of all this success it is difficult to estimate what the future might have brought, had international events taken a different turn, but however that may be, 1937 marked the onset of the Japanese invasion of China which completely interrupted the course of local events. The pressure for patriotism and support of the national effort overcame whatever little tendency to opposition that was left, and the Burakumin, as well as the rest of the movement, urged patriotic support and unity in order to avoid destruction and to be in a position to bargain for their own share. The Levellers were not quite dissolved, in spite of great governmental pressure, but they became completely inactive. The potency of national identification which was latent, prior to the war with China, is noteworthy, and somewhat surprising, given the apparent militancy and class identification which immediately preceded the war. In summarizing the Levellers' movement, it is noted that "it was almost the only organization in which some progressive Buddhist influence was found, albeit to a minor extent–a fact that may evoke comparisons with the importance of religion to the Negroes and the Jews. It was the one in which women and children played a really important part" (Wagatsuma and Totten, 1966: 63).

The Postwar Resumption of Burakumin Left-Wing Politics

The American occupation of Japan led to the establishment of a more democratic constitution which plainly forbade all forms of discrimination. Burakumin leaders were again elected at all levels of government and Matsumoto Jiichiro, who was now head of a new Buraku organization, the National Committee for Buraku Liberation, ran as a socialist for the upper house of the national legislature and received over 400,000 votes, the fourth largest total for any candidate (Wagatsuma, 1966b: 69). A variety of protests similar to those before the war characterized the movement, and in some ways it was as though there was simply a resumption of activities interrupted by the war. If there was a new emphasis in domestic politics per se, it was the tendency to recognize that the plight of the Burakumin would not be much ameliorated simply by the elimination of outright discrimination, but that equality would require positive assistance to the Burakumin by the government. The charter for such aid was the historically accurate claim that the Burakumin were suffering now from their exploitation in the past. The demand was for governmental acts of the kind which Galanter (1963) has called "protective discrimination" in reviewing practices in India.

One characteristic of the postwar movement, which is perhaps characteristically Japanese, has been the effort to reeducate and ultimately win over even those who have been overtly hostile. There apparently have been some notable successes in this endeavor.

Two historical events have combined with certain local conditions to stabilize the present shift to the left. The first of these events involved the refusal of Matsumoto Jiichiro to bow to the Emperor at the opening of the second postwar Diet. This gesture so outraged various conservatives that Premier Yoshida purged him from the Diet and all further public service on a trumped-up charge that he had cooperated with the militarists. General MacArthur acquiesced in this decision, thus turning the movement against the United States. Ultimately under extreme public pressure, Matsumota's position was restored. There was much foreign interest in the case including that from the American Civil Liberties Union, but the national government which protested the action was that of the Soviet Union, whose delegate to the allied Council for Japan demanded Matsumoto's reinstatement. The second event which intervened concerned the killing of a Burakumin woman on an American Army base in Japan. The base had been built on land formerly belonging largely to Buraku villagers, and without other means of support many of the former farmers earned some income by gathering empty shells from the base which were then sold as scrap. An American soldier at the base directed the Buraku woman to a place where there were shells and then killed her with his rifle. At first the Army claimed she was killed by a stray bullet, but eventually they admitted how she had been shot, and a formal apology was sent by our embassy to the Japanese Ministry of Foreign Affairs. A long argument arose over jurisdiction, but ultimately the American soldier was tried in Japan, and though found guilty, was sentenced to only three years on the grounds of his contrition, and so forth. In fact, he was returned to the United States without serving any time. Much additional anti-American bitterness was aroused by the event which was further compounded by the fact that a number of U.S. bases were similarly located on land formerly belonging to Burakumin; it is doubtful that U.S. authorities were aware of this fact, but the same probably cannot be said of Japanese authorities.

The domestic fact most relevant to the postwar position is that the more conservative parties of Japan have made very little effort to help the Burakumin, so that only the left offers any promise of support. It is beyond the scope of this inquiry to ask why Japanese politics is so polarized, but the fact that it is, has affected strategies adopted by the Burakumin.

By and large, religious orders have not played a conspicuous part in the various movements for improving Burakumin status, though within secular organizations, as indicated, religious symbolism has been significant. However, the nature of education in the public schools has come increasingly under the consideration of Burakumin leaders. One wing of the movement essentially advocates an education inculcating a Marxist interpretation of society accompanied by "humanistic appeals to brotherhood" (Wagatsuma, 1966a: 105). This kind of education would stress that "Japanese monopolistic capitalism subjugated to the United States of America makes use of feudalistic remnants of Japan, preserves Buraku, and creates and re-recreates discrimination" (Wagatsuma, 1966a: 105). Another point of view stresses endogenous reasons for failure in school, in addition to exogenous discrimination. Here the emphasis is directly on the pernicious effects of belief in one's own inferiority; i.e., rank concession. The argument which follows is much like that heard here among Black and Mexican-American educators; the necessity for an emphasis on pride in one's people, with developing "the feeling of resistance"; i.e. "the strength of character that would protect Buraku youth from either despair, or a defeatist desire to escape their status" (Wagatsuma, 1966a: 106). I shall have more to say about this conflict in comparing tendencies among the four ethnic groups.

That the Burakumin suffer from rank concession is obvious, and with one notable exception, from which there is much to learn, most of the characteristics of the syndrome seem operative. Differential emulation on the part of the few wealthy Burakumin was certainly the earliest characteristic tendency. Given significance of birth-ascribed status and the barriers to passing, it is not surprising that these few wealthy individuals should attempt to get their entire people to emulate; given their wealth it is also not surprising that they did not encourage opposition to the general Establishment. Again the conflicts anticipated between those emulating more and less are certainly manifest, but here we do not have the data to determine their cultural component, or, to put it differently, to determine whether cultural variation as well as difference of interest have led to the division between the wealthy and the ordinary Burakumin.

There is certainly abundant evidence also of an emulation-solidarity conflict with notable examples of syncretism and innovative combination as indicated. However, indigenous claims seem a little doubtful, unless one were to interpret the Burakumin argument that they have remained more faithful to traditional Japanese culture than the rest of the population as a form of this position. No examples are readily available of what I have called pattern emulation-trait maintenance where deep values of the outside are incorporated after being duly naturalized while distinctive traits are vigorously maintained. Certainly the record is replete with such importation of exogenous values and even of examples of their naturalization by combination with distinctive Buddhist and Christian symbolism, but there does not seem to be much attention given to trait maintenance. Indeed there is not any evidence at hand of ideological effort to raise the status of leather working, and the like. Perhaps the identification with the proletariat and the belief in its historic ennobling mission, and so forth, have somewhat mitigated the need to resolve the conflict between solidarity with their own group and emulation of the dominant surrounding society. The struggles to maintain their own association, or to simply join various working-class associations, have come to our attention, but in large measure it would appear that the total international political situation has considerably affected this kind of decision. Nevertheless, it is possible that the persistence of the commitment to Burakumin organization is connected with solidarity pressures and in complex connection with cultural maintenance and change, though the evidence is certainly not in on this question.

From what we have at hand, emulation connected with rank concession is certainly evident, but the movements which have developed seem only marginally to derive from, or be connected with, resolving an emulation-solidarity conflict. Improving social rank and economic opportunity seem to have been the name of the game. If, as suggested, Marxism is a means of escaping from the conflict by establishing an entirely different form of identification, then we have a distinctly different way of resolving the emulation-solidarity conflict than we had originally imagined. We should then add to proposition 9 earlier in this essay an (e) *redefinition of relevant group identification.* In comparing the four ethnic groups this question will be pursued further.

The question of the effects on solidarity of economic versus political rank paths does not yield such a clear answer as the RCS implies. As argued, the differential emulation characteristic of the economic path is evident even if the divisive results allegedly stemming from this variability are not. In addition, there is abundant evidence of passing, but the effects of the political rank path seem more complex than I had imagined. Certainly it is thinkable that the Burakumin have lost some cohesion by the strength of their identification with the proletariat, which identification is, of course, intimately connected with the political rank path they have chosen. *In general it would appear that as in the case of the Santal and the Jatavs, politics may suggest redefinitions of belonging so as to increase the effective constituency. It is possible of course to have both an ethnic and a special political identity, but it is also thinkable that there may be varying degrees of tension between these identities. When the tension is resolved in favor of the wider constituency some ethnic solidarity may be lost.*

Perhaps it may be concluded there is little evidence that the solidarity requirements of the Burakumin have played much part in either producing tension through conflict with emulative tendencies, or in shaping the nature of the sociocultural movements among the Burakumin. If such requirements have not been manifest, then this certainly contradicts a significant claim of the RCS. This also is a point to be considered further in developing general conclusions. Though there has certainly been a shift from a dominantly economic to a dominantly political rank path, which neatly coincides with the development of what might be termed a cultural movement (e.g., the Levellers and other Burakumin organizations), it seems that the tensions of the emulation-solidarity conflict had little to do with it. Perhaps those investigators with a deep first-hand knowledge of the Burakumin might be able to throw further light on the question.

BLACK AMERICANS

There is of course no neutral term (i.e., one without implied commitment), or one that is purely denotative for those Americans who are identified or identify themselves as of African origin. There is a long history to the naming game in this instance, and it is indicative of the course of race relations in this country. By way of example, "Afro-American" did not seem such a good idea if one thought poorly of Africa or if one wished to oppose movements to deport Blacks to the place of their ultimate origin. If I understand the present temper of the times with its preference for Black, rather than Negro or Colored, it is a preference for the plain reference to color, signifying by its directness self-acceptance and pride. It is perhaps a bit like saying one is a Jew rather than the more refined Jewish or Hebrew. It takes the very term of opprobrium used by the racist and flings it back at him as a term of pride. But this emphasis on pride has of course not always been preeminent by any means, and I propose here to examine sociocultural movements among Black Americans to see what light may be thrown on the RCS, and whether or not the RCS illuminates this history.

One suffers here from an embarrassment of riches since there already exist, of course, not only a large body of detailed studies on many particular movements and leaders, but even rather definitive works, such as John Hope Franklin's (1969) classic *From Slavery to*

Freedom, covering the entire history of the Black experience in this country. There are, in addition, large numbers of sociological or anthropological investigations covering at least a quarter of a century pioneered by such scholars as E. Franklin Frazier, Charles S. Johnson, St. Clair Drake, and others. The present demand for books and essays on Black people is of course so great, it is doubtful that even a full-time expert could consume the entire production. No doubt, as some of the old-timers, who pursued the field when it was less popular, enjoy pointing out, not so much of the over-abundant present literature deserves the attention it gets. I shall not pretend to cover all the ground available for investigation. What follows instead are carefully chosen examples meant to serve the theoretical purpose at hand; to interpret certain facts of Black American historical experience in comparative terms and in terms of the RCS. The reader of course must judge whether or not the examples provided are representative.

For the most part I shall further delimit this part of the essay by concentrating on the post-Reconstruction period because I think it most comparable in circumstance to the other comparative material.

It seems appropriate to begin a comparative consideration by asking whether or not the RCS appears at least grossly to make some sense of what has been happening to Blacks since Reconstruction. Had they in fact conceded rank, in the sense which I use this phrase? I do not imagine that today there is anyone who has followed the development of what is termed Black Nationalism who would deny that this is so. Perhaps the following testimony of Malcolm X on the cosmetic and religious manifestations of rank concession will convey something of the affect associated with rank concession:

> My first view in the mirror blotted out the hurting. I'd seen some pretty conks, but when it's the first time, on your own head, the transformation, after the lifetime of kinks, is staggering.
>
> The mirror reflected Shorty behind me. We both were grinning and sweating. And on top of my head was this thick, smooth sheen of shining red hair—real red—as straight as any white man's.
>
> How ridiculous I was! Stupid enough to stand there simply lost in admiration of my hair now looking "white" reflected in the mirror in Shorty's room. I vowed that I'd never again be without a conk, and I never was for many years.

> This was my first really big step toward self-degradation: when I endured all of that pain, literally burning my flesh to have it look like a white man's hair. I had joined that multitude of Negro men and women in America who are brainwashed into believing that the black people are "inferior"—and white people "superior"—that they will even violate and mutilate their God-created bodies to try to look "pretty" by white standards.

> This religion taught the "Negro" that black was a curse. It taught him to hate everything black, including himself. It taught him that everything white was good, to be admired, respected, and loved. It brainwashed this "Negro" to think he was superior if his complexion showed more of the white pollution of the slavemaster. This white man's Christian religion further deceived and brainwashed this "Negro" to always turn the other cheek, and grin, and scrape, and bow, and be humble, and to sing, and to pray, and to take whatever was dished out by the devilish white man; and to look for his pie in the sky, and for his heaven in the hereafter, while right here on earth the slavemaster white man enjoyed his heaven [Malcolm X, 1964: 54, 163].

To be sure one also hears in Malcolm X a retrospective abhorrence of rank concession, but Malcolm X was one of the relatively few Negroes "saved" from its debilitating effects; and not even all those who join in the rhetoric of Black Nationalism are in fact freed from the force of rank concession.

The beauty shops catering to a Black clientele, the advertisements in the Negro magazines, and even casual observation of ordinary interaction, all bear terrible testimony to rank concession. The "natural" hairdo presently in vogue, is of course the obvious counter to previous white emulation. I often noticed how in arguments between Blacks on the streets of Chicago, one or both of the parties would exclaim "Quit acting like a nigger!" I remember seeing a four-year-old middle-class Black child in a nursery school argument tell a somewhat darker child "there you go again acting like a Black nigger." This whole problem has probably been made especially acute by the fact that even among Blacks there is a correlation between color, social status, and "white ways."

But the agony of self-deprecation connected with the ambivalent belief that others are superior in some very pervasive sense, has long ago been documented by all the sociological studies previously mentioned. Indeed, what I have called rank concession in a kind of social science jargon, is much like what Kardiner and Ovesey (1951) called *The Mark of Oppression,* a designation more evocative than the one I chose much later.

In Frazier's *Negro Youth at the Crossways,* based on field work in Louisville, Kentucky, and Washington, D.C., in 1939, we find the following class distribution of rank concession: "Youth in lower-class families, which comprise about two-thirds of the Negro population, were influenced in their conceptions of themselves as Negroes by their parents' acceptance of the belief that the Negro is inferior and that his subordination to the white man is inevitable" (Frazier, 1967: 263-264). He explains the pervasiveness of this attitude among the poorer Blacks by the fact that their parents frequently grew up in such a repressive southern environment as to make such self-deprecation necessary to survival. It is thus seen as appropriate for such techniques of "getting by" as "acting like a monkey, jibing, flattery and plain lying" (Frazier, 1967: 264). The claim is made that the middle-class children, though outwardly manifesting some acceptance of inferiority, do not believe that they are inherently inferior, but rather that the white man happens to have more money and power. Thus it would be a case of what I had termed power concession as distinct from rank concession. No doubt there is something in Frazier's evaluation, though I suspect the psychology of rank concession is too subtle to be simply related to the rather intellectual avowal that one is not inherently inferior. Indeed Frazier himself proceeds to tell us that the middle-class parents, anxious for mobility, urge their children "to appear decent and respectable and thus not only to dissociate themselves from lower-class Negroes but to win the approval and respect of whites" (Frazier, 1967: 264). I imagine that such a position must be accompanied by a form of rank concession not easily separable from belief in inherent inferiority given the ubiquitous character of racism. In the case of upper-class families we learn that they do not inculcate attitudes of subordination to whites, but their tendency to identify themselves with whites and "their ambivalent attitude toward themselves as Negroes" (Frazier, 1967: 265) make it clear that even success in the economic rank game does not solve the problem.

The evidence just presented, and that of every other similar study which I have consulted, support the prevalence of rank concession, emulation, and differential emulation on a class basis.

The question as to whether or not one can detect the effects on solidarity predicted is not so easily answered. The difficulties attending this question arise partly from the fact that Blacks, unlike our three other ethnic groups, did not arrive in the New World regarding themselves as a "people" in any sense, so that they were not bound together initially in the same way as were these other groups. As for the vexing questions of how much they brought from Africa culturally, how much they retained, how much what they developed resulted from common origins and similar circumstance, and how similar was their African heritage, I shall not presume to comment; the issues are too complex, the disagreements between experts too great, and my own knowledge too fragmentary. Certainly Herskovits (1941) was correct in pointing out that the many so-called experts who had pronounced the judgment that Blacks had no relevant history, frequently based their arguments on blatant errors having no empirical basis; e.g., the claim that the variety of origin of the slaves would not have provided much shared culture among the slaves. On the other hand, the manner of disposing of slaves was not conducive to retention of traditions. There are records of African experience and ancestors and of the terrible trials of becoming enslaved preserved in the oral tradition, and some have even been in writing. Still it has been possible for Blacks as well as whites to accept the absurd folk image of Africans "swinging around in trees!" However, African culture may have shaped both Black and white American culture, it is not easy for a nonexpert and difficult even for experts to point to much that is distinctively African in the way of specific traits. As for more subtle configurations of culture, perhaps no one is quite sure. What is more important for our purposes is that considerable heat is generated by discussion of this question, and I presume that this heat comes not so much from interest in the facts per se, but in the social uses of such facts. As Herskovits argued (1941), to believe that one is completely cut off from the past or that one's ancestors were without accomplishments, is extremely damaging to one's pride and identity, given the prevailing view of things. Such a belief therefore serves as charter for rank concession, and its denial provides a basis for race pride. In addition the nonsignificance of Africa is associated with a charter for assimilation or political action on the basis of nonracial identifications; the significance of African association provides a charter for a more separatist position both socially and politically. Of course, these associations are not inevitable either logically, psychologically, or culturally, but they are socially dominant at this time. I think it is correct to claim that this concern is quite analagous to the Santal concern for history and is characteristic of the RCS.

A second possible source of distinctiveness arises from the common experience and social condition of being classified as Negro or some more pejorative label by whites. And of course the associated restrictions also serve both solidarity and cultural homogeneity. As DuBois put it: "The so-called American Negro group . . . has nevertheless a strong, hereditary cultural unity born of slavery, of common suffering, prolonged proscription, and curtailment of political and civil rights. . . . Prolonged policies of segregation and discrimination have involuntarily welded the mass almost into a nation within a nation" (DuBois, 1947: 362-363).

Thus whatever variability may have characterized initial Black life in the New World, white discrimination has resulted not only in common bonds of sympathy, but a partially distinct social and cultural world. On the other hand, even in the earliest days of this country, there were social and cultural differences between those who were slaves and those who were freemen and sometimes even slave owners; between those who were field hands and those who were "house Negroes," and between those who lived in areas with regionally different cultures. It is pertinent to consider the existence of men like the brilliant scientist Banneker, whose work received recognition by Thomas Jefferson, though obviously the combination of such genius and opportunity was a rare exception at a time when most Blacks were slaves. Nevertheless, the cultural gap between the most sophisticated Blacks and the most parochial of slaves was, without doubt, wider than any among the Santal, or the Jatavs, and perhaps as wide as any among the Burakumin.

Perhaps the most succinct pejorative description of the cultural variability arising from differential emulation is that of the Black sociologist, E. Franklin Frazier (1957) in his work *Black Bourgeoisie:*

> The Black Bourgeoisie has been uprooted from its "racial" tradition and as a consequence has no cultural roots in either the Negro or the white world. In seeking to conform to bourgeois ideals and standards of behavior, this class in the Negro community has sloughed off the

> genteel tradition of the small upper class which had its roots among the Negroes who were free before the Civil War. But more important still, the black bourgeoisie has rejected the folk culture of the Negro masses. The artistic creations of the leaders of the Negro Renaissance in the nineteen twenties, who turned to the black masses for inspiration, were regarded largely with indifference by the increasing number of Negroes who were acquiring a middle-class outlook on life. As a consequence of their isolation, the majority of the black bourgeoisie live in a cultural vacuum and their lives are devoted largely to fatuities. There have been only two really vital cultural traditions in the social history of the Negro in the United States: one being the genteel tradition of the small group of mulattoes who assimilated the morals and manners of the slaveholding aristocracy; and the other, the culture of the black folk who gave the world the Spirituals [Frazier, 1957: 98].

But as Frazier indicated in a new preface to his work, the Black bourgeoisie has undergone some important recent changes; Frazier cites the following evidence as indicative of these changes, and also of his part in effecting this change. The question was asked if his work would "contribute to modifying the mentality of this elite. . . . A partial answer was given to these questions by some of the young leaders of the sit-ins who said they did not aspire to become the middle-class Negroes described in Black Bourgeoisie" (Frazier, 1962: 14).

Emulation-Solidarity Conflict in the Post-Reconstruction Period

So far then the general formulation of the RCS seems to fit the situation of Black people well enough, but then this is perhaps the least exceptionable claim of the approach. Perhaps the most significant and least commonsense claims are those concerning the significance of the economic versus political rank paths and the alleged emulation-solidarity conflict; closely associated with these arguments is the contention that the emulation-solidarity conflict is likely to be particularly pronounced when there is a shift from a dominantly economic to a dominantly political rank path, and that this condition may give rise to a cultural movement. I propose to at least take a glance at the applicability of this part of the RCS before turning to needed corrections and additions.

It is perhaps not too much of a distortion to say that the end of Reconstruction in the late nineteenth century marked what I had called, in the case of the Santal, the closing of the political rank path. It was, of course, a period in which unbelievable violence was done to the Negro to "put him back into his place." Without examining the details of this violence one cannot claim to have any understanding of contemporary Black history or any right to a judgment of how things ought to be. Let us settle here for the terrible understatement that political participation for most Blacks who lived then in the South came to a vengeful end.

The generally acknowledged ideological and social leader of this new period was surely Booker T. Washington. The doctrine he publicly promulgated appears at first glance to do some violence to what the RCS would suggest, for, on the one hand, he championed individual economic gain in the market place even to the total exclusion of political struggle; but contrary to what might have been anticipated, he also favored a position which superficially appears as nonassimilationist. Actually, his famous pronouncement, so often cited to represent the kernel of his position, is in context of his whole program more a question of strategy than a statement of objectives. "In all things that are purely social we can be as separate as the fingers, yet one as the hand in all things essential to mutual progress" (Washington, 1902: 221-222). But Washington's objective was plainly ultimate integration, social, cultural, and economic, and even political, for that matter. For what he preached was that by hard work, frugality, making oneself economically useful, and all those well-known virtues characterized as the Protestant ethic, the Black man could raise himself individually to positions of power and respect; and he insisted that when that happened, Black men would find their place in the sun among civilized men. His point of view on this matter was exactly that proclaimed by "moderate" whites; i.e., that Blacks were despised because so many of them were despicable; that they should pull themselves up by the bootstraps. To him the ultimate proof that his strategy was a sound one was plainly his own success and acceptance by prominent white people. His strategy called for getting on with the economic task without bugging white people who had shown how hard they strike if challenged. He was willing to accept social barriers (though he often took pleasure when he was himself made an exception to such social exclusion) and political disenfranchisement if only Blacks would be allowed technical education and a chance at economic well-being. Thus there is really nothing contradictory to the processes suggested in the RCS in his position.

In cultural matters, as already indicated, he was a thorough emulation man as is appropriate to the economic rank path. Nowhere do we see him concerned with maintaining distinctive Black characteristics, or attempting to naturalize white traits by syncretism, innovative combination, or pattern emulation-trait maintenance. In the atmosphere of the post-Reconstruction South he could thoroughly rely on the white Establishment to see to it that his flock would not stray.

Of course there were militant opponents of white racism and slavery in the days before Reconstruction and notable leaders and movements even in the post-Reconstruction era. During and after World War I there was a large northern migration of Blacks. There were the terrible race riots of 1919. President Wilson, as Carmichael and Hamilton point out (1967), had proclaimed the war necessary "to make the world safe for democracy," but, as they say:

> This was the very same President who issued executive orders segregating most of the eating and rest-room facilities for federal employees. This was the same man who had written in 1901:
>
> > An extraordinary and very perilous state of affairs had been created in the South by the sudden and absolute emancipation of the Negroes, and it was not strange that the Southern legislatures should deem it necessary to take extraordinary steps to guard against the manifest and pressing dangers which it entailed. Here was a vast "laboring, landless, homeless class," once slaves; now free; unpracticed in liberty, unschooled in self-control; never sobered by the discipline of self-support; never established in any habit of prudence; excited by a freedom they did not understand, exalted by false hopes, bewildered and without leaders, and yet insolent and aggressive; sick of work, covetous of pleasure—a host of dusky children untimely put out of school. [Carmichael and Hamilton, 1967].[4]

Thus white racism sat astride the official sources of power while the KKK freely carried out lynch law where the official sources were unable or found it unpolitic to go. Under these conditions, the early Black Nationalism of Noble Drew Ali and Marcus Garvey made its mark; the former preached Islam, the Asiatic and therefore noble origin of Blacks, and advocated the development of Black-owned businesses. There is a complex genealogical connection between this early Black Islam and the contemporary movement. Marcus Garvey advocated a Pan-Negro nation arguing that the self-respect of Blacks, like other races, depended upon having a homeland. He also had a go at mobilizing Black power politically and used this power to support Coolidge in 1924.

Even in this quick overview of Black ideologists and associated movements after Reconstruction, it would be impossible to overlook the influence of W. E. B. DuBois who was a contemporary of Booker T. Washington and yet survived actively until 1963. His positions however—and one must use the plural here—are difficult to briefly summarize because they involve apparently important changes in general viewpoint as well as tactical modifications. In his earliest pronouncements he mainly takes issue with Washington on the question of what kind of education Negroes should have. He argues that the classical liberal education was as necessary for Blacks as whites; that only such an education would provide the kind of talented elite necessary to raise the general level of Blacks. Himself an intellectual of the first rank, his interest in promoting a Black intelligentsia is one continuous thread in his otherwise rather shifting position.

In general, he was more traditionally radical in every way than Washington, and his radicalism tended to increase with time. He refused to accept segregation and gradually came to argue that economic success could not be achieved without sufficient political force to guarantee free access to the market and protection of gains. After World War I, he advocated immediate civil rights for Negroes, and soon he adopted the goal of socialism as the promise of the future. But until the 1930s, he was at odds with other left-wing Blacks because of his abiding belief that the working class was the very repository of racism. Perhaps his elitest outlook also made it difficult to welcome the working class as the vanguard of a bright revolutionary future. At any rate this position made him a strange (at that time) socialist separatist, advocating social separation, the cultivation of a Black intelligentsia, race pride, and cultural pride. He was among the first to speak out for Black standards of beauty and he insisted on using darker Blacks for cover pictures on the NAACP magazine *Crisis.* In the late twenties and early thirties he, like Frazier later, proclaimed his contempt of the Black bourgeoisie who were not becoming the unique elite that he hoped for, but who were instead behaving like the prototype of the nouveau riche. DuBois is very much the hero of a wide variety of contemporary Black power advocates, and appears, retrospectively, to have been somewhat ahead of his time.

Contemporary Black Nationalism

If we were to cap this resumé with a description of contemporary Black Nationalism, I think it would be fair to say of the evidence that, in general, the developing events of this century were in accord with claims of the RCS. As more Blacks migrated to the northern cities, they provided an increasingly significant political constituency. Movements among Blacks tended to become increasingly political and, in general, more concerned with developing and maintaining Black cultural differences. DuBois and others developed and participated in literary movements which aimed at becoming broad cultural movements—without much success. Black history was studied to provide a basis for pride in the African and Afro-American experience (indigenous claims); similarly it would be easy to find many instances of syncretism and innovative combination, particularly in the world of art; however pattern emulation-trait maintenance was perhaps rather rare until the very recent period of Black Nationalism. The reason seems rather clear and is instructive. It is extremely difficult to inculcate acculturative values while simultaneously raising folk practices of generally low rank value to respectability. Under what conditions will an elite participate in such elevation of what they formerly considered the very marks of low status? The RCS model would suggest that this requires a genuine and full-scale effort to engage the poor politically, and this move is seen as a device to bind together what "class" previously put asunder. That such political engagement is increasingly a fact is,

I think, undeniable, and, in the contemporary movement, particularly as promulgated by those who march under the banner "Black Power," such pattern emulation-trait maintenance is often manifest. Thus the foods of poor Blacks, formerly a prime marker of low status, are now designated "soul foods", i.e., they symbolically represent that special quality of Blacks which whites can't quite measure up to; this same inherent defect is held to account for the relative inability of whites to be as musically expressive as Blacks. Following the great success of Malcolm X, men such as Carmichael made use proudly of certain distinctive phrases and terms of address characteristic of the formerly despised elements of the Black community. And in the case of music, such status elevation has perhaps been easiest, as sophisticated whites had long ago come to admire the superb musical creativity of Black musicians, particularly in the jazz, blues, folk, and gospel idioms. That this elevation of distinctive traits did not occur earlier seems to coincide rather well with the fact that previously the Black elite did not engage the masses in a cultural movement connected with the building of Black political power. Indeed in spite of the attention that whites give to the insurrectionist character of Black Power, it is its effort to build a political base among the Black masses which seems to me its most fundamental thrust.

Without exhaustively comparing the development of Black movements with the claims of the RCS, it is perhaps not too much to say that it has at least a gross fit with larger general trends. But if one looks at matters more closely, there seems to be at least one major flaw in the argument. In the RCS, the great driving force is plainly attainment of rank so that the determining forces appear largely to be the conditions of mobility with individual economic gain pulling largely in one direction and political power in another. Perhaps a bias in this direction is not surprising when one remembers that the idea of the RCS developed most directly from my experience in India, But the flaw in the argument which becomes so clear when one examines current Black Nationalism is simply that the most potent movement so far, in terms of genuine cultural and social separateness, plainly has been that of the so-called Black Muslims, i.e., the nation of Islam under the leadership of Elijah Muhammad and, until his death, of Malcolm X. I say the most potent because it is acknowledged by all that they succeeded in reaching Blacks of the very lowest social and economic position; the prisons have proved a particularly fertile ground for recruiting; but they not only went after the lowest of Blacks but succeeded in producing apparently profound character changes. This in spite of the fact that their moral code is not only exceptionally strict, but puritanical in its general tone. The treatment of women departs radically from any traditional Euro-American model. As is well known, they demand rather complete social separation of the races and, unlike many other Black Nationalists, do not accord whites ("white devils") the possibility of even ultimate moral equality. Plainly it is therefore possible to develop a profoundly separatist, antiacculturative movement in the absence of any political program whatsoever, for the nation of Islam has had virtually nothing to do with political power per se.

It would be easy to remedy things by simply concluding that there is obviously a variety of unforeseen ways to develop such a movement. However, I think that inspection of the most potent motivating force among the Muslims does not reveal something new that was absent among the Santal, but rather something I overlooked. I think this other driving force is the struggle for what we clumsily call identity. The constant accompaniment of rank concession is a gnawing self-hatred, and even without the exigencies of a social unity aimed at improving group status, that gnawing self-hatred cries for expression. Some men may perhaps attain surcease from this condition by identifying with all mankind, but it would appear that most men are attracted to more parochial identifications. It is this problem, which may even survive political and economic success, which probably helps to drive all nationalistic movement ostensibly aimed at quite different goals. Among contemporary Black Nationalists like Carmichael or Cleaver, the identity problem is clearly articulated and conscious efforts are made to deal with it.

To put these final conclusions back into the frame of the RCS, one must add that *besides the emulation-solidarity conflict there is the rank concession-identity conflict. This latter conflict seems capable of combining with or independently developing a social movement with characteristics much like that already described for the former.* It may however be an unfortunate truth that the rank concession-identity conflict is especially likely to be mitigated by morally justified hatred. One can only hope that men will prove artful enough to find a less self-defeating path to self-acceptance.

Currently all Black political organizations and movement, following the leads provided by the Muslims and Malcolm X, attend to the problem of racial pride. Among these organizations the most newsworthy has of course been the Panthers. The close connections of some of its founders with the Muslims has profoundly influenced the movement. Unlike the Muslims however, and more like Malcolm X toward the end of his life, the Panthers are overwhelmingly political. Though relying strongly on Black solidarity, they do not disdain alliances with white groups when it is deemed to be of political advantage. Avowedly Marxist, but having learned their lesson from the Muslims, they particularly strive to recruit that element of the population disdained by even the Marxists, i.e., the "lumpen proletariat" which in Panther literature is frequently referred to simply as "lumpen." Acts defined as criminal by society at large and performed by the lumpen are regarded as prerevolutionary, perhaps after the suggestion of Fanon (1966). One who has performed such acts and comes to see them in the light of this ideology may regard his former life not as one filled with unredeemed sin, but, on the contrary, as that of a kind of unconscious revolutionary and as an exploited victim of an evil social system, *viz.*, capitalism.

The Panthers, with their modified Marxism replacing the tight racism of the Islamic Black movement, seem to be having some success particularly among the lumpen in and out of prisons. Obviously the Panthers' revolutionary stance and support for the lumpen will not be conducive to recruiting nonlumpen Blacks. In terms of the concerns of the RCS, the Panthers organization is an exemplification of the modifications I have offered earlier to statement 5 regarding the unifying effects of a political rank path. Here economic differentiation and its accompanying social and cultural facts result in political opposition and schism within the Black community. In addition, the partial redefinition of relevant identity from Black to Black proletariat (including lumpen) results in complex cross-pressures. The particular hostility of the white Establishment to the Panthers and other similar groups, which seems to be caused more by its Marxism than its armed militance (there are other armed militant nationalistic groups which are non-Marxist but who receive only sporadic attention), makes commitment to this movement especially costly. Nevertheless, the appeal of this movement and the Muslims to the lumpen is a new thing in the world, and it may be widely emulated.

It is interesting, in considering the future of Black-white relations, to consider contingently what might happen if poverty were to be eliminated, and civil rights were really provided on an equal basis. That, of course, was the first objective of Martin Luther King. Through the process of nonviolent struggle for these objectives, King hoped to redeem the racism of whites and its complement of self-hate among Blacks. He, like others, maintained that political, social, and economic objectives were intimately linked with the elimination of racism. No doubt racism, like all hereditary status, would fare more poorly under circumstances of relative equality of secular power. All of this seems reasonable to me, and yet perhaps we should not write off racism so easily as though it were simply a dependent variable. Since it seems capable of playing such a powerful identity role, it may well be cause as well as result of secular inequality; it may even be capable of upsetting the applecart once it has been set aright. It was certainly not secular inequality which unleashed the torrent of antisemitism which swept Nazi Germany! I say all this not as a charter to lessen the significance of eliminating secular inequalities, but simply to lay matters on the table as they seem to be.

CONCLUSIONS

The empirical material suggests the possibility of redefinition of identity for political purposes; i.e., specifically adopting the identity of proletarian. Such identification exists on some scale among each encysted society under consideration, but is most pronounced among the Burakumin. Clearly the nature of ethnic relations among the proletariat must condition the opportunities for such identification. It would therefore seem that in the near future this would be extremely difficult to accomplish among Blacks, since, as indicated, the evidence is abundant that the non-Black proletariat is an especially prejudiced segment of the population. No doubt competition among working people has powerfully conditioned such attitudes, but this is a fact of life in all the societies under consideration, i.e., the labor market. Perhaps the differentiating factor is the prevalence or opposition to socialistic ideologies characteristic of the general working-class population. It would appear that the Japanese trade unions have the strongest socialist tendencies and the American unions the least. If this is a critical variable, one would expect the Burakumin to be most subject to redefinition of identity and Black Americans least with the Jatavs and Santal in between. For the present this seems to be the case, though I may of course be right for the wrong reasons.

Perhaps the most thoroughgoing revision in the RCS suggested by the comparative evidence, is in connection with the contention that "(5) A dominantly political rank path increases solidarity." As noted earlier, and repeatedly, it has been necessary to qualify this contention in a number of ways. It would be more accurate to say that the effect of political efforts at mobility is indeterminate with respect to solidarity; perhaps I am correct in suggesting that internal economic differentiation is a major determinant of whether or not such a rank path is solidifying or divisive, but the evidence is not in on that question. Certainly there are also instances known to us, and even among the people discussed in this paper, in which joint efforts in the economic sphere have been attempted. No doubt such efforts also contribute to solidarity so that the RCS contention that "(4) A dominantly economic rank path decreases solidarity" is, at the very least, an overstatement and should probably be qualified in a manner similar to what has been suggested about the political path; particular market conditions might provide the necessary qualifications. Still the evidence does seem to support RCS claims regarding the particular strains on solidarity produced by market activities. As my colleague in sociology, Edna Bonacich has suggested to me, perhaps what I have really done is to suggest mobility paths be searched for their individuating and solidifying effects. Some of these have been specified, but doubtless there are many other possible effects not envisaged in my cursory treatment.

In the body of the text, I refer to such developments as the all-Black Che-Lumumba chapter of the Communist Party, the Panthers who stand as examples of an effort to combine proletarian identification with a strong measure of separatism. There are, no doubt, characteristic tensions in such a position, though, as indicated, the Burakumin experience with the Levellers suggests that it is by no means an impossible position to maintain. But it is distinctly different in a number of ways from exclusively separatist positions. Thus one hears some denunciation by Black leftists of such trait-maintance practices as the appreciation of soul food. Such talk is dubbed " pork chop nationalism." It may also be true that hatred of "whitey" is psychologically more effective than opposition to the "system"; in addition race hatred has the advantage of binding together those who profit from the system as well as those who are exploited. That, of course, is not an advantage from the point of view of left-wing objectives since it requires muting system opposition, but it does make for a large constituency.

The evidence seems to be that antiassimilationist solidarity efforts are much more pronounced among the Santal and currently among Black Americans than among the Burakumin or the Jatavs. One might imagine either that the former are more concerned with solidarity than the latter, or in greater danger. Perhaps a greater proletarian identification among the two pariah leather working groups has made ethnic identity less relevant. In the case of Blacks the solidarity emphasis may be due to the fact that they are more heterogeneous and less socially bound together than the other groups, though certainly the Burakumin also embrace rather diverse communities. Surely the Burakumin have suffered as much as any people from the mark of oppression, so that one might imagine a stronger anti-emulative tendency arising from identity needs, but it does not seem to be so. It appears that the Santal and Blacks have produced movements with greater emphasis on identity and the Jatavs and Burakumin with more emphasis on political power. It is thinkable that on occasion the psychological and political emphasis might be in conflict, even when one or the other is ostensibly in the foreground.

When Mrs. Rosa Parks in 1955 refused to give up her seat to a white man, the civil rights movement led by Martin Luther King had its beginning. Of course it seems highly probable that even if Mrs. Parks had acted otherwise the movement would have unfolded, but why did she make this protest at the time and place that it occurred? To suggest that she was just very tired that evening would hardly account for such an exceptional event. Besides we know that she had been an official in the NAACP. The answer that Martin Luther King gave to this question was that "She had been tracked down by the Zeitgeist–the spirit of the times" (Lomax, 1962: 100). While the language employed is uncongenial to contemporary social science, my own reading concerning the four ethnic groups frequently led me to consider the significance of what we anthropologists might speak of as historical factors. One learns that various Japanese leaders were profoundly embarrassed about the continued existence of slavery in Japan in 1870 when so many other nations had abolished it. Though such embarrassment is hardly sufficient explanation for abolition, it seems not at all improbable that it played its part. If one further examines the Burakumin case, it is obvious that the spread of Marxist ideology has profoundly affected the development of that movement. Largely exogenous historical currents are also evident in the Jatav movement, dominated as it has been by the ideologies which sparked the Indian Independence movement. The Santal and other Mundari tribals, though somewhat less well connected, still show the influences of the Independence movement in general. Indeed Indian villagers of all kinds speak of "the new wind" (Beals, 1962: 75), and usually mean the spirit of progress, equality, democracy, and the like. I do not mean to suggest that the new wind is itself unexplainable or that where it blows is simply a matter of chance; but I think it is probable that none of the movements under discussion is entirely to be understood by examination of purely endogenous conditions. Since all the cases under consideration have occurred in a world whose ideologies are now profoundly interconnected, one must be cautious about interpreting even the most striking parallels as occurring solely because of similar conditions.

For example, each of our societies has begun on a largely economic rank path and become increasingly political, a fact connected with increasing democratization throughout the world. This shift has been accompanied with varying stress by cultural efforts to maintain distinctiveness. Indeed I might even claim with somewhat less assurance that in each case the shift from the dominantly economic to the political rank path was accompanied by something that might be termed a sociocultural movement, as the RCS predicts. But if it were not for the new wind, would any of these events have transpired? Consider the fact that the gypsies of Europe began a civil rights movement in 1967!

The RCS Reconsidered

After examining a preliminary version of a paper similar to this one, Vera John, a colleague in psychology, made some criticism of the term "rank concession"; if I understood her correctly, she felt quite perceptively that the term suggested a kind of voluntarism which was both misleading and objectionable. Concession does convey an act which might or might not have occurred. In fact when I coined the phrase rank concession, I thought that perhaps not all conquered or exploited peoples did feel that their rulers were superior. Indeed the Chinese, both at home and abroad, seemed to me a notable counterexample and certain other more esoteric cases were also known to me. I think also that in varying degrees, the new ideologies developed by each society have succeeded in reducing the feeling of inferiority. Since the mere fact of less power does not seem to result in rank concession, I think it useful to consider endogenous as well as exogenous factors. If my terminology sounds as if it were purely an endogenous question, it is indeed misleading. As the question now stands, I think the causes of rank concession need further study. Because the terms suggest the significance of endogenous factors as well as exogenous, the RCS might be objectionable as a charter since it deflects blame from the system or exogenous individuals. On the other hand, the RCS might serve as a charter for self-help toward liberation since it does not suggest that only exogenous conditions are relevant! Be that as it may, such questions ought not to be confused with its empirical validity which is, of course, independent of its charter uses.

Finally, turning attention to the works of Lewis (1966) and Berreman (1966a, 1966b) on the "culture of poverty" and caste cross-culturally, it is possible to join at least one emphasis common to these various approaches. I refer to their emphasis on the significance and manifestations of self-deprecation. Consider the fact that the stereotypes held by the dominant society in all four cases before us, diverse as they are culturally, are virtually identical. "Shiftless, lazy, amoral, unable to postpone pleasure, impulsive," are absolutely universal in their application. In fact, they are nothing more than negative forms of virtue which are held to explain and in fact justify positions of wealth and power. Given a society with rich and poor it would seem that such a stereotype is necessary to the rich in order to justify themselves and to resist the blandishments of the poor. Given rank concession, in addition, the stigmatized poor must come to accept the stereotype and thus become the victims of a self-fulfilling hypothesis. They come to despise themselves and to behave in ways consistent with the stereotype, thus reinforcing the stereotyped view for both themselves and the rich. Viewed from this perspective, the self-deprecation which may be the psychological heart of what Lewis (1966) terms the culture of poverty, is not only an endogenous belief perpetuated from generation to generation, but a belief fundamentally rooted in the continuing relations between poor and rich; this would, of course, be especially true where poverty coincides with ethnicity and where mobility is low. If this perspective has some merit, the question is not so much one of the relative contribution of exogenous and endogenous conditions to the culture of poverty (Current Anthropology, 1969), but rather an interactive process in which exogenous and endogenous conditions mutually reinforce one another.

> The laws of changeless justice bind
> Oppressor with oppressed;
> And close as sin and suffering joined
> We march to fate abreast.
>
> [Washington, 1902: 222]

A Revision of the RCS as Applied to Encysted Societies[5]

(1) The "cultural load" of the encysted society is inversely related to its solidarity and its ability to resist assimilation.

(2) Rank concession produces a tendency to emulate; since emulation increases the cultural load, rank concession is also destructive of solidarity.

 (a) *The more salient and all encompassing the inherited rank stemming from membership in an encysted society is, the less the danger to solidarity from emulation.*

(3) Emulation is likely to be of the differential variety.

(4) *Insofar as an economic rank path results in differential emulation, is destructive of internal ties by limiting redistribution, and results in "passing," it decreases solidarity.*

(5) *The less an encysted society is economically differentiated, the more likely that a dominantly political rank path will produce increased solidarity.*
   (a) *Redefinition of identification to expand a constituency may result in cross-pressures which weaken the solidarity of the encysted society.*

(6) Differential emulation is more destructive of solidarity than random emulation; since the economic rank path is particularly conducive to differential emulation, it is particularly destructive of solidarity.

(7) Though power-incorporative borrowing is at least analytically distinguishable from emulation, rank concession is always accompanied by power concession; therefore, rank concession implies a tendency to power-incorporative borrowing as well as emulation. (Power concession does not, however, imply rank concession.) Power-incorporative borrowing also increases the cultural load.

(8) Every society has some endogenous sources of solidarity; therefore, every society that has conceded rank has an emulation-solidarity conflict of some proportion.
   (a) *Any person who has conceded rank will have an emulation-identity conflict which will operate much like the societal emulation-solidarity conflict.*

(9) Such a conflict manifests itself in a variety of characteristic ways, among which are the following:
   (a) indigenous claims
   (b) syncretism
   (c) innovative combination
   (d) pattern emulation-trait maintenance
   (e) redefinition of relevant group identification, e.g., proletariat.

(10) An intense emulation-solidarity *and/or emulation-identity* conflict may generate a "cultural movement."

(11) *A shift from a dominantly individuating avenue of mobility to a dominantly solidifying one is likely to produce a severe emulation-solidarity conflict and/or a severe emulation-identity conflict, and may thereby serve to generate a cultural movement.*

(12) Encysted societies are subject to the following additional threats to their solidarity:
   (a) political-judicial encroachment
   (b) pressure to accept national solidarity symbols
   (c) repugnance pressures

(13) *Positive effects on solidarity arising from increased connection with the state may result from the development of locally based articulating offices and organization.*

NOTES

1. Léopold Sédar Senghor is President of Senegal and also an eminent poet (Moore, 1962: 8-9).

2. This English translation by George DeVos and Hiroshi Wagatsuma appears in DeReuck and Knight, 1967: 135-136.

3. All of the listed propositions should be qualified by *ceteris paribus.*

4. The quotation from Woodrow Wilson was originally taken from his article, "Reconstruction in the Southern States," which appeared in the January 1901 issue of the Atlantic Monthly.

5. As indicated in my first statement of the propositions in this essay, they should be qualified by *ceteris paribus.* New ideas suggested by the comparative evidence are in italics.

REFERENCES

BEALS, A. (1962) Gopalpur, a South Indian Village: Case Studies in Cultural Anthropology. New York: Holt, Rinehart.

BERREMAN, G. D. (1966a) "Concomitance of caste organization," pp. 308-324 in G. DeVos and H. Wagatsuma (eds.) Japan's Invisible Race. Berkeley: Univ. of California Press.

--- (1966b) "Structure and function of caste systems," pp. 277-307 in G. DeVos and H. Wagatsuma (eds.) Japan's Invisible Race. Berkeley: Univ. of California Press.

CARMICHAEL, S. and C. V. HAMILTON (1967) Black Power. New York: Vintage.

Current Anthropology (1969) "Book review of culture and poverty: critique and counter proposals." (April-June).

DeREUCK, A. and J. KNIGHT [eds.] (1967) Caste and Race: Comparative Approaches. London: Churchill.

DeVOS, G. and H. WAGATSUMA [eds.] (1966) Japan's Invisible Race. Berkeley: Univ. of California Press.

DuBOIS, W. E. B. (1947) "Three centuries of discrimination." Crisis 54 (December): 362-363.

DURKHEIM, E. (1933) On the Division of Labor in Society (G. Simpson, trans.) New York: Macmillan.

FANON, F. (1966) The Wretched of the Earth. New York: Grove Press.

FRANKLIN, J. H. (1969) From Slavery to Freedom: A History of Negro Americans. New York: Random House.

FRAZIER, E. F. (1967) Negro Youth at the Crossways. New York: Schocken.

--- (1962) Black Bourgeoisie. New York: Collier.

--- (1957) Black Bourgeoisie. New York: Macmillan.

GALANTER, M. (1963) "Law and caste in modern India." Asian Survey 3: 544-559.

--- (1961) "Equality and 'protective discrimination' in India." Rutgers Law Rev. 16.

HERSKOVITS, M. J. (1941) The Myth of the Negro Past. New York: Harper & Brothers.

HIEBERT, P. (1969) "Caste and personal rank in an Indian village; an extension in techniques." American Anthropologist 71, 69: 434-453.

KARDINER, A. and L. OVESEY (1951) Mark of Oppression: A Psychosocial Study of the American Negro. New York: W. W. Norton.

LEWIS, O. (1966) "The culture of poverty." Scientific American 215, 4 (October): 19-25.

LOMAX, L. E. (1962) The Negro Revolt. New York: Signet.

LYNCH, O. (1969) The Politics of Untouchability. New York: Columbia Univ. Press.

MALCOLM X (1964) The Autobiography of Malcolm X. New York: Grove Press.

MOORE, G. [ed.] (1962) Seven African Writers. England: Northumberland Press.

ORANS, M. (1968) "Maximizing in Jajmaniland." American Anthropologist 70, 5 (October): 875-897.

--- (1965) The Santal. Detroit: Wayne State Univ. Press.

REDFIELD, R. and M. SINGER (1954) "The cultural role of cities." Economic Development and Cultural Change 3, 1: 53073.

SILVERBERG, J. (1959) "Caste-ascribed 'status' versus caste-irrelevant roles." Man in India (April-June): 148-162.

VALENTINE, C. A. (1968) Culture and Poverty: Critique and Counterproposals. Chicago: Univ. of Chicago.

WAGATSUMA, H. (1967) "The pariah caste in Japan: history and present self-image," pp. 118-140 in A. DeReuck and J. Knight (eds.) Caste and Race: Comparative Approaches. London: Churchill.

--- (1966a) "Non-political approaches—the influences of religion and education," pp. 88-109 in G. DeVos and H. Wagatsuma (eds.) Japan's Invisible Race. Berkeley: Univ. of California Press.

--- (1966b) "Post-war political militants," pp. 68-87 in G. DeVos and H. Wagatsuma (eds.) Japan's Invisible Race. Berkeley: Univ. of California Press.

WAGATSUMA, H. and G. TOTTEN (1966) "Emancipation: growth and transformation of a political movement," pp. 33-67 in G. DeVos and H. Wagatsuma (eds.) Japan's Invisible Race. Berkeley: Univ. of California Press.

WASHINGTON, B. T. (1902) Up From Slavery. New York: Doubleday, Page.

Part II

RACE, CULTURE, AND URBANIZATION: CONTEXTUAL ASSESSMENTS

Attempts to analyse the coherence and functioning of populations and subgroupings within them, especially those experiencing the impact of externally induced change, are often subject to limitations imposed by research designs, resource constraints, parochial assumptions, and the peculiar character of the group being studied. The three chapters included in this section are notable for the ways in which each circumvents these problems. They are distinguished both by the singular perspectives of the authors and the distinctive methodologies they employ. Through the use of methods and indices particularly appropriate to the circumstances in which the peoples studied are found, each of the investigators renders his subject matter visible in an unusually cogent fashion.

In Chapter 5, Helen Safa casts clear retrospective light on the responses of the Puerto Rican urban underclass to enforced urbanization. She adduces compelling data (gathered through community ethnographic methods and buttressed with more specific and detailed followup survey material) showing the ways in which the strengths of community, mutual aid, and kinship networks were destroyed with the enforced removal of the Puerto Rican shantytowns. The longitudinal character of her study focuses attention on and anticipates the problems encountered in the public housing with which the shantytowns were ultimately replaced.

Whereas Safa's assessment of adaptation to urbanization involved comparisons over time, Martin Diskin, using survey data in an anthropological field situation, assesses the impact of urbanization (the diffusion of economy-connected life styles in a regional market system) on the villages, towns, and the city in the valley of Oaxaca, Mexico. He directs his attention to the distribution and relative retention of traditional attitudes and cultural practices in the face of increasingly rational marketing practices of new capitalists in the town plazas. Variations in outdoor sleeping preferences and estimations of the appropriate length of ritual celebrations are among the unusual data Diskin uses to demonstrate the peculiar character of urbanizing Oaxaca.

Safa laments the loss of certain traditional community strengths as a result of the Puerto Rican government's modern housing policy. In anticipation of formal government policy, Diskin suggests that the loss of traditional cultural forms need not be as great as it has been in other circumstances where the confrontation between tradition and modernization has taken place.

TenHouten, Stern and TenHouten are moved to investigate novel approaches to the collection of data in a context in which more standard approaches fail to accurately estimate the community's character. In Chapter 7, these authors examine, on a more microscopic level than did either Safa or Diskin, patterns of leadership in four urban communities. Their research demonstrates the importance of time and place as critical determinants of access to indigenous leadership in poor communities. They emphasize the importance of obtaining samples in a manner which does not assume that all inhabitants of poor urban areas are attached to dwelling units. The significance of their effort is indicated in their finding that in each of the four communities studied leaders were concentrated in commercial, as opposed to residential, areas.

TenHouten, Stern and TenHouten find discrepancies in the political values and attitudes among residents in the four communities studied (although, notably, the sense of political inefficacy seemed to be nearly universal). And, although their methodology enabled them to locate previously inaccessible leaders, the variations they found in the structure of informal leadership leads them to the conclusion that no universal generalizations can be made regarding leadership in poor urban communities. Of course, it remains to be seen whether the differences they find are systematic and applicable to other communities similar to the types considered in their research.

5

Puerto Rican Adaptations to the Urban Milieu

HELEN ICKEN SAFA

□ PUERTO RICO IS A PRIME EXAMPLE among developing nations of rapid economic growth and social change. As a result of Operation Bootstrap, instituted by the Commonwealth Government, in the postwar period, life expectancy increased to age 70 and over, per capita income tripled (Planning Grant Application, 1967: 19), the labor force has moved out of agriculture into urban employment, and illiteracy dropped from thirty-two percent in 1940 to eleven percent in 1965 (Lewis, 1966: xiv).

Most of this economic growth has centered in the San Juan Metropolitan Area.[1] In 1960, this area accounted for almost half of the island's net income and nearly three of every four nonagricultural jobs created in Puerto Rico between 1950 and 1960 (Planning Grant Application 1967: 19). Within San Juan are located most of the island's manufacturing, financial trade, and tourism activities, including a jet airport and major port; tall glass and steel office and apartment buildings; luxury hotels; and suburban Levittowns with large, modern shopping centers. All of this has been built within the past two decades.

The heavy concentration of activities in the San Juan Metropolitan Area acts as a powerful magnet for rural migrants looking for a better job and other urban amenities. Migration began during the depression of the 1930s and has continued to the present, as agricultural development in the rural areas failed to keep pace with urban industrial growth. Between 1940 and 1960, the population of the San Juan Metropolitan Area more than doubled (from 300,000 to 648,000). It is estimated that today the area contains about 750,000 people, or thirty percent of the island's population (Planning Grant Application, 1967: 20). Though the majority came from adjacent regions, particularly in the early period, migrants to San Juan have been drawn from all parts of the island. Thus, in the "shantytown" studied here, the birthplace of rural migrants represents nearly every municipality on the island. There is no apparent tendency for families from the same region to settle in one neighborhood, as is the case in some cities in Africa and Asia. This may be true of a recently settled shantytown, but where a neighborhood has been established over one generation, as in this case, any initial concentration from one area has been dispersed and replaced by more recent migrants.

How have these migrants fared in the city? Where do they settle and what kind of jobs and other opportunities do they find? In short, what is the process of adaptation by the Puerto Rican poor to the urban milieu? These are some of the questions I shall attempt to answer by examining three different patterns of urban adaptation found in the San Juan Metropolitan Area: a shantytown, a public housing project, and the movement out of the shantytown to an *urbanización* or private subdivision. These neighborhoods will be examined as parts of an overall process by which poor rural migrants gradually adapt to the urban milieu. Families move back and forth between one neighborhood and another, and maintain contact with friends and kin left in their old residence. Kinship and friendship patterns, as well as jobs, public services, and other institutions also provide links into the larger urban society of which these neighborhoods form a part. Thus, none of these neighborhoods can be viewed in isolation. In fact, the emergence of the shantytown, public housing project, and urbanización as distinct neighborhoods can be explained largely as the result of the need for an urban labor force to fill the demands of San Juan as the ever-expanding economic, political, and cultural center of Puerto Rico. Yet each neighborhood is characterized by a different type of population and life style which distinguishes it from other neighborhoods in the metropolitan area.

The first place of residence for most poor rural migrants is the shantytown. Unable to find housing within their very low income range, migrants during the thirties and forties illegally occupied marginal public lands located chiefly along the vacant banks of the Martin Peña Channel and on the periphery of Old San Juan, property which at the time no one saw fit to develop for residential or commercial use. In Puerto Rico, residents of shantytowns never acquire legal title to the land they occupy, which remain public property, but houses are bought and sold in perfectly legal transactions. The government even compensates shantytown residents for the loss of their homes in slum clearance programs.

This location along marginal, public lands distinguishes shantytowns from what may properly be called a slum. Slums are usually located in the center of the city, close to the central business district, and consist of once adequate structures which have been converted from their original use to tenements housing many times their anticipated occupancy. Shanties, on the other hand, are, from the outset, inadequate structures; most are rather flimsily constructed with makeshift materials and unskilled labor provided by the owner and his friends. Thus, while slums are a form of blight or deterioration characteristic of highly industrialized cities, shantytowns are commonly found in the preindustrial cities of underdeveloped areas where marginal, public lands still exist on the urban fringe.

Approximately 86,000 people now reside in the five-mile belt of shantytowns extending along the Martin Peña Channel, with an average density of nearly one hundred persons per acre (Caplow et al., 1964: 41). Though efforts have been made to reduce this figure through slum clearance and public housing programs, new shantytowns are constantly springing up as the stream of migrants continues to pour into the metropolis. Although the growth of shantytowns along the channel has been checked, the decline of population in this area recently averaged less than one-half of one percent per year, while new shantytowns are beginning to emerge in more outlying areas (Caplow et al., 1964: 228).

Shantytowns thus represent one pattern of adaptation by the Puerto Rican poor to the urban milieu. Measured in terms of time and the number of families involved, shantytowns are still the most important mode of urban adaptation in Puerto Rico and closely resemble other Latin American squatter settlements described in the literature (e.g., Mangin, 1967a; Turner, 1969). However, Latin American squatter settlements often develop into substantial working-class neighborhoods, as residents improve their homes and pressure the government for public services such as paved roads, water, electricity, and the like (Mangin, 1967b). In this way, residents of Latin American squatter settlements are absorbed into the city as part of an overall process of neighborhood change, without necessarily having to change their place of residence. Puerto Rican shantytowns also show improvements over time, but they have little chance of developing into normal working-class neighborhoods. Most shantytowns are eradicated by an ambitious urban renewal program which clears these areas for other residential, commercial, or public purposes and relocates a large percentage of shantytown families in public housing.[2] The shantytown studied here has been completely cleared to make room for a major highway and new housing under the Model Cities Program.

Public housing represents yet another pattern of adaptation by the Puerto Rican poor to the urban milieu. It differs radically from shantytowns both in physical design and resident composition. Public housing is a totally planned community, with rows of mass-constructed apartment buildings separated by carefully laid-out streets and lawn areas. Each project houses a minimum of several hundred families. As originally conceived, public housing was designed as a temporary, stopgap measure for low-income families unable to meet their housing needs. In reality, however, it often tends to become a dead-end street, as many of the residents become part of a permanent public housing population. This is due both to the nature of the population recruited into public housing, drawn largely from the poorest segment of the shantytown, and to the nature of the regulations imposed on the project population, which effectively curb strong incentives for upward mobility. Thus, as compared to the shantytown, the range of incomes in public housing is generally lower and narrower, and the project population contains a high percentage of households (in some projects) headed by women, unemployed, aged, and others who are at a decided disadvantage in competing in a limited low-income job market. Thus, public housing provides a refuge for the poor who have little hope of upward mobility.

However, not all families who leave the shantytown, either through forced relocation or personal choice, end up in public housing. Many do not qualify for public housing, chiefly because their incomes are too high. The demand for public housing is much greater than the supply, and first preference is given to families living in urban renewal areas whose homes are destroyed by clearance programs. In addition, many poor families prefer to remain in shantytowns, because they like the security and independence of owning their own home, however squalid, and may not be able to afford even the minimal rents in public housing. The eradication of a shantytown in one area may therefore lead to the creation of new shantytowns in other parts of the city, or to increasing population in existing shantytowns.

Another alternative open to migrant families who have been able to improve their socioeconomic status through years of effort is to move into a private subdivision or urbanización, usually located in the suburbs of the metropolitan area. An increasing number of shantytown families have chosen this alternative in recent years, moving in particular to the lower-priced urbanizaciones designed for families of moderate income. The urbanización thus represents yet a third mode of adaptation for the Puerto Rican poor to the urban milieu.

While they are part of the same overall process, each pattern of adaptation described here performs a different function for the urban poor of San Juan and serves a somewhat different population. The shantytown is a rather heterogeneous community, composed of recent rural migrants and a core of old-time residents, where the poor may gradually learn to build up their skills to deal with the complexities of the urban milieu. Public housing, on the other hand, provides paternalistic care for the poor who are most lacking in resources and often creates a permanently dependent housing population. Those families who move out of the shantytown or public housing into an urbanización have generally broken out of the cycle of poverty and made the first step toward entering the middle class. This is one reason why a home of one's own in an urbanización has such great symbolic value for the Puerto Rican poor. It represents security, independence, and status. It means that the migrant's hopes for finding a better life in the city have been realized.

METHODOLOGY

The data presented here were gathered at various times during a decade of tremendous change in the structure of Puerto Rican society. The original study of the shantytown and the public housing project was conducted in 1959-1960 under the auspices of the Urban Renewal and Housing Administration of the Commonwealth of Puerto Rico, which was concerned with determining the changes resulting from the relocation of shantytown families in public housing. The author spent several months as a participant observer in a specific shantytown and public housing project chosen at random within the San Juan Metropolitan Area. On the basis of the data gathered through participant observation, we drew up a standardized interview schedule which was then administered to a sample of 100 families in each of these neighborhoods. In all, a total of 200 families, or 474 individuals, were interviewed during the original study, since both husband and wife, and, where possible, an adolescent child, were interviewed in each family.

The shantytown studied in 1959-1960 was completely cleared to make room for a major highway transecting the San Juan Metropolitan Area. All the residents have moved, some to public housing, some to other shantytowns, some to urbanizaciones. During the summer of 1969, I attempted to trace all of the shantytown families whom I had known personally as a participant observer ten years earlier. Fortunately, I was able to locate every one of the families, all of whom now live in various parts of the San Juan Metropolitan Area. In this restudy, I conducted lengthy taped open-ended interviews with seven families, or fifteen individuals, several of whom had been children ten years before. The perspectives of this younger generation turned out to be one of the most interesting aspects of the restudy.

The data presented here cannot claim to be a representative sample of all the Puerto Rican urban poor, nor even of the particular neighborhoods in which these families live. Time and money did not permit a complete restudy of the survey sample drawn during the original study in 1959-1960 and the sample is too small to be representative. However, my previous acquaintance with these families, plus the rapport I have established with them, permits me to draw many inferences which would not be possible with a more impersonal survey sample. The restudy is, in essence, a view of the changes that have taken place in Puerto Rican society during the past decade from the perspective of a few families who have lived through them.

I have chosen to emphasize the qualitative data obtained through participant observation rather than the quantitative data obtained through the survey sample. The sample was designed to verify the range of certain patterns and trends observed in the field, and to compare the shantytown and housing project in terms of certain demographic and socioeconomic characteristics. Since the results of the survey sample are limited to data collected in 1959-1960, they have been reported only where they prove useful in distinguishing the different life styles in each neighborhood, and none of the original tables have been reproduced.

The variety of life styles described here belies the notion held by many social scientists and the public at large that the poor are a homogeneous, downtrodden, disorganized, and apathetic people. On the contrary, this study emphasizes the resourcefulness and optimism of the urban poor of Puerto Rico and their ability to respond to changing socioeconomic conditions. Is their optimism warranted? Have they obtained their just share of the benefits of the fantastic economic growth that has occurred in Puerto Rico in the postwar period? We shall conclude this paper with a reexamination of these questions in the light of the data presented.

THE PUERTO RICAN SHANTYTOWN[3]

The Setting

To the outsider, the area along the northern border of the Martin Peña Channel appears to be one unbroken mass of tiny shacks packed one upon the other in hopeless confusion. There is no apparent order to the maze of narrow, twisting alleyways, few of which deserve to be called streets. Nothing appears to distinguish one section from another, and so outsiders often lump the whole area together under the infamous name of El Fanguito (Little Mud Hole).

For people living in the area, however, El Fanguito is only one of the shantytowns, perhaps the best known. Each shantytown has a special name—Tras Talleres, Shanghai, Los Bravos de Boston—and with the name goes a distinct personality. The boundaries are not geographic, though a street leading in from the main thoroughfare may come to serve as a dividing line. Nor do the boundaries always coincide with the political division into *barrios* set up for administrative purposes. The primary factor distinguishing one shantytown from the other is a group sense of belonging together.

Each shantytown has been built up gradually over many years through the joint efforts of the families settling there. Twenty-five years ago, for example, the first families arriving in Los Peloteros (the shantytown described here) found only *mangle* (mangrove) growing in the swampy land destined to become their homes. They cleared and filled in a few plots to construct their houses. They built plank sidewalks connecting different parts of the shantytown. Today street names such as O'Higgins, Colón, and Pablo Nuñez commemorate the struggle of these first families in the places where they established their homes.

For many years Los Peloteros was without water and electricity. The entire barrio depended upon one single public faucet. Now running water and electricity have been installed in almost every house, thanks to the work of barrio committees. Committees have also succeeded in having a few of the principal streets filled in with gravel, but still none are paved or provided with lighting. Because of the low level of the land during heavy rains, water drains into the shantytowns from higher ground, turning the streets to mud. Pools of stagnant water are left behind for days, creating additional health hazards. No sewage system was ever installed in Los Peloteros, and most families have latrines in the back of the house. For many years the channel served as a main sewage outlet for the city as a whole, and the polluted waters produce a foul, nauseating stench.

Los Peloteros is essentially a residential neighborhood. It provides no jobs and only limited shopping or other commercial facilities. For daily purchases, Pelotereños rely largely on local stores selling a limited variety of canned goods and other nonperishable items. But usually the big weekly purchase of food *(la compra)* is made at a larger *colmado* or grocery store in Barrio Obrero, an old working-class district located near the shantytown. Clothing and household goods are also bought in Barrio Obrero, where they are colorfully displayed on racks outside the store. This is where Pelotereños go to movies or dances on Saturday nights or where men meet at a favorite *cafetín* (cafe). In a sense, Los Peloteros is considered merely a poorer offshoot of el barrio, as it is popularly called.

Earning a Living

The largest percentage of Pelotereños born in the rural area started out as landless agricultural laborers, earning less than $500 annually. In San Juan, migrants have a greater variety of job opportunities open to them than in the stagnant rural area, but most are still employed as unskilled laborers. Some men in Los Peloteros work at the docks, or in one of the new factories in the metropolitan area, while the more skilled may be employed as artisans in construction projects. Most men in the shantytown are concentrated in these blue-collar occupations. Another group is employed as service workers in restaurants, hotels, and other institutions catering to the thriving tourist trade.

None of these jobs offers real economic security. The employment of longshoremen depends on the volume of the shipping trade and they may go for weeks without working. Construction workers may stay at one job for six months or as long as two years, but unless they are permanently employed by a construction company, they must constantly look for new projects. Factory and service jobs are also subject to layoffs and slow periods. Thus the rural migrant has not really escaped the insecurity of agricultural wage labor. Unless he commands some marketable skill, he may find himself in as precarious a position as he was in the rural area.

Nevertheless, most migrants feel they have improved their life chances considerably through migration to the metropolis. Their salaries, for example, are much higher than what they could have earned in the rural area. In 1959-1960, skilled artisans, experienced longshoremen, or factory workers in Los Peloteros earned $2,000 and $3,000 annually, which is relatively high for the Puerto Rican poor. Annual incomes in Los Peloteros among the families interviewed ranged from under $500 to over $5,000 annually. Since income depends almost entirely on wages, the number of persons working in the family is of crucial importance in determining annual income. The lowest incomes are found in homes where no one is working and where families depend largely on public welfare and subsidiary economic activities. On the other hand, where more than one person is working, total annual income often exceeds $3,000 and may go as high as $5,000.

The poorest elements in the shantytown are found among the aged and among fatherless families where the husband has died or left his family, through separation or divorce. These families depend largely on the minimal allotments provided by public welfare or social security supplemented by subsidiary economic activities. For example, Carmen, a young widow with five small children, received $40 a month from public welfare and also worked three days a week as a laundress. She later took up with an older man who, as she pointed out, could afford to support her because his children were already grown. This man no longer allowed her to work and provided her with most basic necessities. Most Pelotereños would not frown on extramarital relationships of this type because they are fairly stable and fulfill an obvious economic need.

Illegal activities such as prostitution, the selling of illegal lottery tickets or *bolita* and the manufacture of illegal rum or *cañita,* do not appear to be as widespread in the shantytown as popular opinion would assume. Of course there are differences between shantytowns in this respect, and Pelotereños appear to pride themselves on having a very small percentage of prostitutes, drug addicts, or other social deviants. Pelotereños told me of police raids where convicts were brought into the shantytown to cut the mangle or growth along the channel where the stills for making cañita were hidden. Some men escaped by hiding in the water of the channel for hours. However, these appeared to be isolated incidents and illegal activities did not seem to constitute a major source of income for most shantytown residents.

Some Pelotereños, especially the young, place their hopes for the future on migration to the mainland. Recognizing the limited possibilities of upward mobility on a wage basis in Puerto Rico, they hope to earn higher salaries for the same type of work in New York. One young couple, for example, planned to go to New York to save enough money to return to Puerto Rico and buy their own home in a nice urbanización. Many are actually returning[4] as the economy of Puerto Rico provides more jobs for the low-income group.

Variations in Life Style

It should be clear from the above account that the shantytown, though obviously limited to the poor, is by no means a homogeneous community. Incomes vary considerably, depending on the occupational skill of the breadwinner, the number of persons employed in the family, marital status, stage of the life cycle, and other factors. All of these differences contribute to important status distinctions within the shantytown, which might appear small to the outsider. But to the Pelotereño, for whom the shantytown is the primary reference group, these differences are very important.

Socioeconomic differences are expressed in widely divergent life styles within the shantytown community. Houses, for example, vary considerably in size, condition, and value. Some have as many as three bedrooms while the majority have only one and a few are just one-room shacks. Some are well kept and painted, while others are in need of major repairs. The poorest houses are built over the banks of the channel, and supported by wooden piles built into the water. Plank sidewalks just wide enough to allow two people to pass each other provide the only means of access. Further inland, conditions improve somewhat and houses are higher priced. Thus, residence within the shantytown itself is ranked differentially, depending upon distance from the channel.

Differences in life style are also manifest in the quantity and quality of household furnishings. Better homes may be furnished with a living room set consisting of matching sofa and chairs and perhaps a coffee table, while the very poor may not even have a chair to sit on. A decade ago, most shantytown families still used a kerosene stove, a great fire hazard in so congested a neighborhood.[5] Now more families have gas stoves and most own a refrigerator, sometimes secondhand, which is usually prominently displayed in the living room. The most common article in any household is a radio. All day long it is turned on full volume, blaring out mournful love songs, and reporting the latest news and baseball scores, or winners at the race track. In recent years, television has grown increasingly popular, and families without one often gather together in a neighbor's house to see the latest *novela* (soap opera) or variety show.

Diet exhibits less variation, and in 1959-1960, most families spent \$15 to \$20 a week on food. Meat is reserved for special occasions, and some families supplement their income by raising chickens or pigs. The common staple of the poor is rice and beans, and as one man exclaimed bitterly, "Nosotros los pobres, a mediodia comemos arroz con habichuelas y a la tarde habichuelas con arroz" ("At noon we poor eat rice and beans, and at night beans and rice"—meaning warmed-over leftovers).

Variations in life style are not confined to material items such as houses, furnishings, and diet. There is considerable variation among both adults and adolescents in educational levels, though most of the younger generation in the shantytown have had far more schooling than their parents, few of whom went beyond eighth grade. Education is recognized as a primary avenue of upward mobility for the Puerto Rican poor, and is emphasized for girls as well as boys, reflecting the new occupational opportunities open to women. Tumin has pointed out the importance of a high school education in terms of occupational mobility in Puerto Rico (Tumin and Feldman 1961: 301). A high school education should qualify a person to become a white-collar worker, in clerical or sales jobs, or in the government bureaucracy—jobs which provide a degree of security and prestige which most workers in the shantytown have never known.

While there is considerable drive for prestige and status among the more mobility-oriented families in the shantytown, it is checked by a strong feeling of community solidarity, reflected in extensive patterns of mutual aid. Very poor families may store milk and other perishables in their neighbor's refrigerator, or tap their neighbor's electricity or water supply (in cases where their own service has been curtailed, usually for failure to pay their bills). Small articles like tools or electric irons are constantly being borrowed, and women often look after each other's children. Men cooperate in the repair and improvement of their homes (and cars, when they have them), though skilled workmen are paid for some of the more difficult tasks. Even food is shared. Some old men living alone in the shantytown depend almost completely on neighboring families for their meals, for which they contribute little or nothing. As one woman remarked when a neighbor passed by with a plate of rice and beans, "Nadie aquí pasa hambre" ("No one goes hungry here").

The pattern of sharing and mutual aid acts as a leveling device in the shantytown community, much as the sponsoring of religious fiestas redistributes wealth in the peasant community (Wolf 1957: 4-5). The person who shares his possessions with others is rewarded with greater prestige than the one who withholds them exclusively for his own use. By his ability to help others, he demonstrates his own relative prosperity. Thus, sharing supports status distinctions based on differences in living standards, at the same time as it prevents these distinctions from becoming too great. By distributing benefits which might otherwise be confined to a few families, sharing tends to equalize some of the differences in socioeconomic standing in the shantytown. Too high a degree of internal differentiation would weaken neighborhood solidarity by destroying its basic homogeneity. The pattern of sharing, on the other hand, ties Pelotereños together in a system of mutual aid; community cohesion is reinforced through the interdependence of its members.

Relatives, Neighbors, and Friends

The solidarity of the shantytown community serves a very important function for the newly arrived urban migrant and even older low-income residents of the modern metropolis. As the first place of settlement for most rural migrants, it provides a familiar setting within which migrants may gradually adapt to the new way of life in the city, and from which they are gradually absorbed into other parts of the metropolitan area. Thus, there is a constant flow of people into the shantytown from the rural area, and from the shantytown to other urban neighborhoods.

Despite this constant turnover of personnel, however, a core of old-timers, made up of some of the original settlers of the shantytown, contributes greatly to neighborhood stability. They form a stable nucleus to which new migrants can attach themselves and they provide an important source of leadership and continuity to the community. Don Andrés, who was one of the leading officers of the Popular Party in Los Peloteros, has lived in the neighborhood for twenty years and his wife for thirty. He works as an electrician at the docks, where he was secretary of his labor union. He also has his own electrical business on the side. The family now lives in an urbanización created as a result of a housing cooperative formed in Los Peloteros, of which Don Andrés was president (see section on the urbanización).

Kinship provides important links among residents in the shanty-town. Over half the Pelotereños in our survey sample have relatives living in the neighborhood and they generally see each other daily. Though they may not occupy the same dwelling unit, such a kin group is often a tightly knit functional unit, cooperating extensively in the care of children and other household tasks. Newcomers to the shantytown often settle near a relative who helps them find a house, a job, and generally orients them to the urban milieu. The house of one old couple in Los Peloteros was surrounded by the houses of their children and grandchildren, copying the same settlement pattern to which they had been accustomed in the rural area.

Relatives are particularly important to women for bonds of kinship are emphasized in the maternal line. Though they associate extensively with their neighbors, women often remark that their only real friends are relatives—usually other females such as a mother, sister, or daughter. Children come to know their mother's relatives far better than those of their father, simply because they see more of them. The bond with the maternal grandmother may be particularly strong, reflecting her dominant position in the kin group.

Marriage among residents of the shantytown seems to be fairly frequent as a natural consequence of limited contact with the outside world. In addition, many *compadres* are chosen from among immediate neighbors. *Compadrazgo* is a ritual kin tie important in Latin America because it establishes a bond between the baptismal godparents of the child and his real parents; thus compadrazgo serves very often to incorporate neighbors and close friends into the kinship system. The choice of neighbors as compadres in Los Peloteros hence reinforces already existing neighborhood ties.

Ties of kinship, marriage, and compadrazgo integrate a neighborhood indirectly since they provide an additional point of contact among unrelated people as well. For example, Pedro does not only know Uncle Juan in the next block, but also Luis, who is Juan's next-door neighbor. Thus, Pelotereños live in what Bott has termed a "highly connected network" of relatives, neighbors and friends (Bott, 1957). Almost everyone in the shantytown knows everyone else and the outsider is spotted immediately. Men, too, tend to find most of their friends in the immediate neighborhood and spend much of their leisure time in a local cafetín, or bar. The same crowd of men commonly congregates nightly in a favorite locale to drink, gab, listen to the jukebox, or play a game of dominoes. The proprietor often becomes one of the gang and it pays to be his friend, since he may be called upon to extend credit when cash is low.

Social control in the shantytown rests not so much on outside authority as with neighbors themselves. The highly connected network of relatives, neighbors, and friends permits no deed to go unnoticed, and deviant behavior reflects not only on the person himself, but on his family and friends as well. Responsibility for the regulation of neighborhood affairs rests largely with the men. For example, men may attempt to end a fight between neighbors or to tell a drunkard to do his drinking elsewhere, while women are hesitant to intervene in nonfamily affairs. The previously mentioned barrio committees to improve conditions in the shantytown also are largely made up of men.

However, the shantytown is characterized by a dispersed and fluid form of leadership rather than a rigid hierarchy of authority. There is no central authority in the shantytown community, or even a central point at which people may gather, such as the plaza in the Puerto Rican rural town. The integration of the community, instead of being based on a hierarchy of established authority, is built up through a series of small, overlapping segmentary groups, each of which is composed of perhaps a half-dozen neighbors. Contact between the groups is maintained by persons who are members of more than one group, because of close friends, compadres, or relatives living in other parts of the neighborhood. Such persons serve as connecting links through which the more extensive associations of kinship and friendship operate to build up an overall neighborhood unity.

Those primary group ties which extend beyond the boundaries of the shantytown to the wider metropolitian community tend to bind Pelotereños more closely to members of their own class subculture than to other class groups in the city. Ties of friendship, marriage, and compadrazgo rarely cross class lines. The bars and stores which Pelotereños frequent, cater almost exclusively to a lower-class clientele. The more important life crisis ceremonies, such as weddings and baptisms, are confined largely to the immediate network of relatives, neighbors, and friends and seldom include persons from other socioeconomic levels. Thus, shantytown families are drawn together by their common position at the bottom rung of the social ladder and by their physical and social isolation from other classes in the metropolis.

The Shantytown and the Outside World

The highly personal nature of relationships within the shantytown contrasts sharply with the impersonal relationships between Pelotereños and other classes in the metropolis. Most transactions with persons of higher status are segmentary and specialized, and leave the poor in a subservient position, as a customer in a store, a worker in a factory, or a patient in a clinic, dependent upon others "who own the instruments of production, provide the work opportunities and sell the commodities to be bought" (Mintz, 1956: 141).

Formal institutions maintain the subordinate-superordinate relationship of the poor to the rest of urban society. Certainly the poor belong to political parties and labor unions, and attend churches and schools in the wider metropolitan community. But participation by the poor in these citywide institutions is highly selective. Because of the large number of parochial and other private schools on the island, even attendance at public school is generally an indicator of lower-class status in Puerto Rican society.

Though they are nominally Roman Catholic, few Pelotereños ever attend Mass, and even when they do, they are unlikely to meet people from other class levels in the formal setting of public worship. The poor seldom participate in the social or educational activities sponsored by the church, nor are they likely to attain positions of prestige and authority in the Catholic hierarchy.

The weakness of religious faith among the Puerto Rican poor and the tolerance of different denominations prevent religion from acting either as a unifying or a divisive force in the community. The tremendous growth of the Pentecostal movement in Puerto Rico in recent years as well as other sect movements catering to the poor (compare Cook, 1965) could serve as a catalyst for some form of class solidarity, but tends instead to divide the poor into a number of competing religious groups. These fundamentalist sects usually concentrate on bible reading, hymn singing, and other religious activities, which offer the poor an escape from this world rather than a means of improving their chances here on earth.

Politics is a far more explosive and emotional issue in shantytown life than religion, though active participation is generally limited to a few local residents, headed by the *comisario,* or ward boss. Doña Felisa, the popular former Mayoress of San Juan, for many years commanded a considerable following among the urban poor of the metropolitan area, and the comisario served mainly as her agent. He was not a leader in his own right, nor did the local branch of the Popular Party[6] which he headed have a permanent, functioning membership. When called upon to do so by the Mayoress, the more enthusiastic Populares would board a caravan of cars headed for a show of party strength. Such demonstrations were particularly frequent in election years, when political sentiments ran strongest. But active support of the party on the part of the rank and file was largely limited to these momentary outbursts of allegiance.

As in the rural area (Mintz, 1956: 394-399), politics is seen primarily as a set of reciprocal obligations between the party and the voter. Shantytown families justify their invasion of public land by displaying a Popular banner outside their house. The main job of the comisario is to provide services for the party faithful–gravel for roads, lumber for the repair of homes, bail for the jailed–in exchange for votes on election day. Doña Felisa used to hold weekly "open houses" at which people would present their particular problem or request–ranging from a faulty sewer to a need for public assistance because of a wayward husband. Though open to everyone, these sessions were naturally most popular among the poor, who utilized them to circumvent the red tape of government bureaucracy and to secure preferential treatment. Thus, a person in need of medical services would be directed to the doctor in charge rather than being admitted through regular clinic channels.

The labor union movement in Puerto Rico is still relatively weak and fragmented, and many men in the shantytown reported no unions in their place of employment. Despite the long-standing tradition of informal cooperation among neighbors and relatives, Pelotereños have little confidence in setups requiring a formal commitment on the part of a membership bound only by secondary association. Even the housing cooperative formed in Los Peloteros lost many members as families tired of bureaucratic delays and red tape. It is assumed that outside the primary group (that is, outside the small circle of relatives, neighbors, and friends), everyone is out for himself.

The weakness of the labor union movement among workers in the shantytown is indicative of their lack of class consciousness. Though Pelotereños commonly refer to themselves as *los pobres* (the poor) and are conscious of similarities in socioeconomic status, they do not see themselves as a separate social segment cut off from other class groups in the society. *Their sense of group consciousness and cohesion does not extend beyond the boundaries of the immediate community.* Aspirations for the future are always voiced in terms of individual or familial upward mobility rather than in terms of class solidarity. They are more concerned about the status differences among themselves and with improving their position in the shantytown community than with the status gaps between themselves and other class groups in the metropolis.

Other than the primary group ties mentioned earlier, there are no mechanisms to extend the sense of solidarity beyond the shantytown to the larger class collectivity. Politics, religion, and labor unions are all directed and controlled from above, with the poor acting as passive followers rather than active participants. Thus these institutions become mechanisms by which the poor may be kept in their subordinate position, rather than a means of assuring them participation in the larger society. A few individuals in the shantytown do use politics, religion, or labor union activity as avenues of social mobility, like the comisario or Don Andrés, who was president of the housing cooperative in Los Peloteros. These persons serve as brokers (compare Wolf, 1956) between the poor and these formal institutions, and while they may advance their own status aspirations, they do little to alter the subordinate position of the poor in the larger society.

Nevertheless, in terms of their own individual life histories, most migrants feel that they have greatly improved their own life chances and those of their children by moving to the city. They have a wider choice of jobs, they earn more, and they have better educational opportunities for their children. They may not have improved their socioeconomic status relative to the larger society, but nonetheless, for them, their gains are real. They do not question the legitimacy of the existing social order, they merely want to improve their own position within it.

PUBLIC HOUSING

Relocation in public housing does not foster upward mobility among the Puerto Rican poor. It succeeds in dislodging them from the shantytown, and in breaking old kin and neighborhood ties, but it does not integrate them into the larger urban community. Public housing is just as isolated as the shantytown from other class groups in the metropolis–perhaps more so. The physical design of the projects, the rules and regulations governing selection of tenants and continued occupancy, and the paternalistic policy of project management, all contribute to the residents' alienation from the larger society and to their continued subordination within that society. Despite its original intent, public housing has become a trap for a totally dependent, frustrated, and often resentful lower-class population.

The Design of a Planned Community

El Capitán the project, and Los Peloteros, the shantytown, are totally different types of neighborhoods. Narrow twisting alleyways have been replaced by broad, paved streets. Flimsily constructed shacks have been replaced by three- and four-story-long concrete buildings, separated by stretches of green lawn traversed by pedestrian walks. The lawns provide additional play areas for the children; concrete benches allow people to sit and talk; and clothes drying and garbage disposal are communally enclosed.

The size of the housing project and its architectural design help to isolate it from the surrounding community. El Capitán consists of 140 buildings housing 2,610 families and covering several city blocks. The neighborhood surrounding the project is largely upper middle class, including many "continentals." or people from the mainland. Their life styles are so different that relationships between the two groups are usually impersonal, if not hostile. Contact between the project and the older working-class neighborhood on its southern border is effectively cut off by a major six-lane highway dividing the two neighborhoods.

Most public housing in Puerto Rico is built with the aid of federal funds and follows mainland standards of design and layout. All of the buildings in El Capitán are of the same rectangular design and are laid out in linear fashion following a carefully conceived, orderly plan. Communication among neighbors in the project is difficult because the building design fails to recreate the small neighborhood groupings found in the shantytown. There tends to be more conversation across the lawn between neighbors in buildings facing each other than between neighbors at opposite ends of the same building. The emphasis is on maximum use of space with large communal green areas, which fail to give project residents the sense of privacy or independence they found even in the densely populated shantytown. As a result, tenants take no pride in their community and leave the maintenance of buildings and grounds entirely up to project management.

Architects have tried to make up for the sterile, monotonous layout in building design through the provision of community facilities. While Pelotereños must leave the shantytown to seek most essential services, project residents can meet many of their needs within the local community. The commercial center located at the fringe of the project includes food stores, variety shops, a pharmacy, a barbershop, a bar, a dairy bar, a shoe repair shop, a dry cleaner, and a laundromat. A modern supermarket and large public park are located immediately across from the project and are used chiefly by project residents. El Capitán is also equipped with a health center and medical dispensary, two nurseries, milk station, a breakfast center for preschool children, a library, an elementary school, and a junior high school. Adults may attend night classes at the school in literacy and primary grades. A home economics class for women meets in one of the apartments. For recreation, the project is equipped with baseball, softball, basketball, and volleyball courts, and the beach is just a few blocks away. The community center offers numerous club activities every night of the week and church services of different denominations are held every Sunday.

None of these facilities is shared by residents of the surrounding neighborhood, since most of them cater especially to the needs of the project population. Even the public beach and park located outside the project are shunned by neighborhood residents who resent and fear this lower-class element in their midst. Likewise the supermarket across the street from the project caters to the lower-class tastes and buying patterns of project residents and is seldom frequented by middle-class patrons from the surrounding area.

Community facilities in the project fail to integrate families with residents of the surrounding area and also provide few points of contact between project residents. For example, stores are all located at one end of the project instead of being dispersed throughout the neighborhood as in the shantytown. Tenants who shop there are not necessarily known to each other. Illegal businesses selling a variety of basic food products in trucks called *ambulantes* have sprung up throughout the project community. In a sense, they fill a gap left by the lack of the shantytown *tienda* or cafetín. Like the latter, ambulantes are scattered throughout the neighborhood and serve a steady clientele. Here people may buy small quantities and on credit, which is not permitted at the supermarket. However, since these ambulantes are located outdoors, and offer no shade or seating place, they do not induce customers to linger as do the stores in the shantytown.

However, the marked improvement in physical conditions has had a noticeable impact on the health of the project population. Compared to Pelotereño children, project children show a much lower incidence of illnesses such as diarrhea, pneumonia, anemia, and internal parasites, afflictions which are normally associated with poor sanitation and other bad living conditions. The outdoor latrine in the shantytown has been replaced by a modern bathroom with flush toilet and shower. The kitchen is equipped with a sink, cabinets, and a two-burner electric hot plate in place of the kerosene stove used by most Pelotereños. Eighty percent of the project residents interviewed own a refrigerator, which must be bought by tenants themselves.

In general, project apartments have more appliances and better furnishings than all but the best houses in the shantytown. Almost twice as many of the project residents interviewed had a radio, refrigerator, and television set compared to the shantytown sample. Almost ninety percent of the families in the project sample had complete living room sets, while one-quarter of the families in the shantytown sample had no living room furniture at all. Even if we hold income constant, the number of items owned is still higher in El Capitán than in Los Peloteros. Since project residents no longer have such a fear of hurricanes, fires, and other adverse physical conditions, they are more willing to invest in household furnishings and appliances than shantytown families.

Socioeconomic Status

While most project residents agree that public housing represents a distinct physical improvement over the squalor of the shantytown, they consider a move to public housing a step backward in the social scale. It means that they have abandoned—at least for the time being—all hope of owning their own home or of being their own boss. They are forced to abide by the rules and regulations of project management, which stifle individual initiative and incentives for upward mobility. They are living among the poorest elements of the shantytown community, since project regulations exclude the more prosperous, upwardly mobile families from admission into public housing.

Since El Capitán is a federal project, requirements for admission are fixed by the federal government, and are based chiefly on the need for housing and on income. Because of the tremendous shortage of public housing in Puerto Rico, top priority is given those families living in older, cleared shantytowns, the inhabitants of which were among the first migrants to the metropolitan area. As a result, project residents on the average have lived in the metropolitan area a longer time than shantytown families, and a greater percentage of project residents were born in the city. In contrast to the shantytown, the project population is not a predominantly migrant population; it is a more highly urbanized group that has cut most of its ties with the rural area.

The wide variations in income found in Los Peloteros are sharply reduced in public housing. Thus, in 1959-1960, only five families in the project sample had incomes over $3,000 yearly, compared to three times this number in the shantytown sample. At the same time, there was a higher concentration of project families in the lowest income group; thirty-five families in the sample from El Capitán had annual incomes under $1,000 compared to twenty-four in the sample from Los Peloteros. The concentration of families from the lower socioeconomic segment of the shantytown also can be seen in the high proportion of renters who enter public housing. While most homes in the shantytown are owner-occupied, only forty-four percent of the project sample owned a house prior to moving into public housing.

As in the shantytown, income is intimately related to the number of persons working in the household. In all but one case, project families with annual incomes over $3,000 had more than one person working in the household. However, only twelve families from the project sample could count on more than one salary compared to thirty-four families in the shantytown sample. This may be due partly to differences in household composition. Project regulations regarding household composition discourage the addition of working adults to the household, whereas these persons contribute significantly to family income in the shantytown. Thus, there are fewer project households in which relatives are residents and contribute to the family income. In addition, the number of women reporting employment is lower in the project than in the shantytown. (Project regulations also discourage women from reporting employment, as we shall note later.) As a result, compared to Pelotereño households, the number of persons with earning capacity in the project household decreases, while the number of dependents remains the same or may even increase. Thus, the project family's earning power is reduced as the burden of support falls in most cases on a single breadwinner.

The rate of unemployment among male heads of households is higher in public housing than in the shantytown. Some of these men are permanently disabled or too old to work. One man with ulcers had been employed by the government on road construction for 38 years, but received no pension because he was not in civil service. He and his family of four children relied primarily on a monthly welfare payment of $32. His wife made a little extra cash as an Espiritista (spritualist medium), but her reputation was too dubious to draw much of a clientele. Understandably they complained even of a minimal rent of $4.50 monthly.

Many fatherless families are recruited for public housing because their choice of alternative private housing is very limited. The number of households headed by females increases from twenty in Los Peloteros to thirty-five in El Capitán. Since there usually is no male breadwinner, these families are often totally dependent on public welfare and annual incomes frequently fall under $1,000. For example, Luis Lopez left for the mainland six years ago and his family has not heard from him since. The youngest of nine children was then only an infant and Luis' wife, Esmeralda, has struggled desperately to keep her large family together. She was very happy when the family was moved into El Capitán four years ago, because the housing was nicer than they could possibly have afforded outside. The family depends largely on public welfare and occasionally Esmeralda takes in some laundry. But she is unable to do much because of a bad case of ulcerated varicose veins, which has left her almost crippled.

The reasons for the increase in the number of fatherless families in public housing were studied in a brief analysis of Housing Authority records conducted in 1963 (Safa, 1965). It seems that because of their low incomes, fatherless families are more likely to apply for public housing and they are also much more likely to be accepted. Once in public housing, they are less likely to leave. It could not be determined on the basis of the data available whether residence in public housing actually leads to a weakening of the conjugal relationship and marital separation, although our data strongly suggest that the man's authority is severely threatened in public housing (see section following). It does seem that public housing recruits unstable families with a history of marital separation, so this would also contribute to the higher number of fatherless families.

Project regulations discourage tenants from increasing their earning capacity. Increases in income are penalized by raises in rent, and if families pass beyond a certain income level, they are evicted. Many families prefer to deliberately restrict their earnings in order to stay within the income limits. Women are less likely to have a steady job than women in Los Peloteros and many prefer temporary employment or subsidiary economic activities which can be hidden from project management—and public welfare officials. For example, one woman collects garbage from the neighbors and sells it to a man who raises pigs. Another prepares *fiambres,* or hot lunches served in metal containers, to sell to workers in the vicinity. However, the number and type of subsidiary economic activities in public housing are restricted by project regulations which forbid the use of the apartment for commercial purposes. Thus, the only bar in El Capitán was run by one of the local comisarios or ward bosses of the Popular Party, who had undoubtedly used his political influence to secure this lucrative trade.

There are some upwardly mobile families in El Capitán. A good number of project adults have gone to high school, especially among the men, reflecting generally the longer urban residence of most project residents. (Educational opportunities are generally more available in the city than in the rural area, where most shantytown adults grew up.) A higher percentage of project men are working at clerical jobs, although the bulk are still employed as artisans or service workers. But these upwardly mobile families fail to identify with the project community, and therefore cannot exert the leadership which the core of old-timers provided in the shantytown. They consider themselves above their neighbors and prefer to associate with people of higher status outside of the project community. They leave as soon as they can, and leadership is never given a chance to form. Thus, the stimulus provided by the range of socioeconomic levels found in the shantytown is lacking in public housing, and project management is left with a dependent population of fatherless families, aged, and other disadvantaged households, who remain in public housing because they have nowhere else to go.

Interpersonal Relations

Project families tend to look toward housing management to provide the leadership and direction lacking in their own community. Social control, for example, has passed out of the hands of the residents themselves—as in the shantytown—into those of project management. Since public housing belongs to the government, most tenants feel it is up to management or the police to settle disputes between neighbors, or to punish a delinquent youth. As one man in the project commented, "Aqui todo es publico y uno no puede mandar en nada" ("Here everything is public and one has no authority in anything"). Bitter disputes between neighbors in the project are common, and often involve a question of jurisdiction or responsibility. Thus, arguments frequently occur over the cleaning of the stairways, which serve six to eight apartments. Each resident feels it is his neighbor's responsibility and most would prefer to leave it to project management.

Men feel particularly threatened in the project community, because many of the traditional bases of male authority in the shantytown have been taken from them. A man no longer owns his own home, he can no longer repair and maintain it as he wishes, and he must report every change of income or occupation to project management. The leadership functions exercised by barrio committees in the shantytown have been usurped by project management, which sets and enforces all rules and regulations governing the project community. Men do not even have a place to get together in the project. The local cafetín and pool parlor have been eliminated, and the community center does not attract them because it is not conducive to informal meetings among friends who just want to chat awhile.

Many men claim not to have a single friend living in the project. Complaining of the isolation, one man exclaimed; "Aqui se muere uno y no hay quien le haga un favor. No hay hermandad, ni buenos vecinos" ("One can die here and no one would do you a favor. There's no brotherhood or good neighbors"). Another man, referring to the difficulty in making friends in the project, remarked; "Como no abren la puerta uno no puede ni hablar" ("Since they don't open the door, one can't even talk"). Men seem to suffer most from the loss of the highly localized network of social relations in the shantytown.

Women in El Capitán, on the other hand, still find most of their friends in the immediate community. Women more easily reestablish close relationships with neighbors since they are home most of the day and continue to engage in extensive patterns of mutual aid. They meet other women as they shop, hang out the laundry, or attend a baby contest in the community center. Many of the activities in public housing are directed at women or children: child care services, recreational programs, public health lectures on prenatal care, and so forth. This adds to the man's sense of isolation and lack of recognition in the project community.

One might expect the oppressive nature of life in public housing to be reflected in a greater sense of class consciousness. Project residents come much closer than shantytown residents to being a true, full urban proletariat; they are an older, urban population, a lower and more homogeneous socioeconomic group, and they are very resentful of the paternalistic control exercised by project management. Resentment has been reflected in increasing dissatisfaction with the Popular Party, the government then in power, which was closely identified with project management. Compared to Los Peloteros, there was a marked increase in the number of *Estadistas* or statehood sympathizers in El Capitán particularly among the men. Estadistas complained of raises in rent, in water bills, and in bus fares, all of which were blamed on the government. One man denounced public housing as merely an exploitation of the poor by the government, and many tenants would like the opportunity to buy their apartments rather than renting.[7] Estadistas felt that if Puerto Rico became a state, then Puerto Ricans would be regarded as first-class (American) citizens, and welfare payments would equal those paid in New York.

By voting for the Statehood Party (a faction of which is now in power under the label of the New Progressive Party), project residents hope to extend and increase the benefits of the welfare state to equal those paid in the United States. They feel that legal incorporation of Puerto Rico into the United States will bring an automatic increase in standards of living to the island, especially for the poor. The United States is equated with progress, material wealth, and greater benefits for the poor. Thus, far from asserting any radicalism in terms of greater class consciousness and class solidarity, Estadista supporters merely reaffirm dependency by seeking greater aid from the United States.

Project residents have no greater desire to oppose the existing order than shantytown families. Some did oppose the Popular Party, but their vote for statehood is basically conservative and expresses a desire to extend the benefits of paternalism rather than end it. This paternalism has fostered a feeling among the poor (and other class elements in the society) that Puerto Rico cannot progress on its own, and needs the economic and political protection of the United States (compare Seda Bonilla, 1969: 304-308). Like the shantytown residents, they accept the social hierarchy of Puerto Rican society as given, with Americans at the top of that hierarchy, and seek only to increase the benefits to themselves within the existing order.

THE MOVEMENT OUT OF THE SHANTYTOWN: THE URBANIZACION[8]

In 1959-1960, we studied two basic patterns of adaptation by the Puerto Rican poor to the urban milieu; the shantytown and public housing project. We have shown that each caters to a different type of urban poor population and performs a different function for these families. The shantytown acts as a filter, through which migrants are sorted out and launched on their respective urban careers. Public housing takes those families with the least resources and provides them with paternalistic care which often hinders further their chances for upward mobility.

Principal Sources of Change

All of the shantytown families studied in 1959 have now moved. Los Peloteros no longer exists. It has been totally cleared to make room for a new highway along the borders of the Martin Peña Channel and new housing under the Model Cities Program.[9] The families interviewed in this restudy have been living in their present residences for periods ranging from three to seven years, most moving directly from the shantytown. In the summer of 1969, I returned to Puerto Rico to see what changes had taken place in the lives of these families in this ten-year period, and in particular what effect the move from the shantytown had on their life styles and life chances.

Three of the seven families interviewed are living in public housing, three are living in a variety of private housing developments or urbanizaciones, scattered throughout the San Juan Metropolitan

Area, and one family is renting a home in a "pocket" slum in the center of the city. The move itself is indicative of the mobility patterns in each family, since the more upwardly mobile families have gone to urbanizaciones while the less mobile have gone to public and rental housing. However, all the families feel they are better off now than ten years ago, and cite at least three main sources of improvement in their lives:

(1) The move itself. Even the families living in public housing feel that this is a distinct improvement over their former residence in the shantytown. Like the public housing residents studied earlier, they cite the improvement in physical conditions, and the loss of fear of hurricanes, fires, and so forth. Those families living in urbanizaciones are very proud of their accomplishment and frequently point to additions and improvements they have made in their home in the short time they have lived there. What they miss most about the shantytown is its central location in the metropolitan area and the friends they knew there.

(2) Changes in the life cycle. It became apparent that the problems of poverty are most critical for young families with small children. Many of my principal informants are now older and no longer need worry about young children to support. In come cases, these children are now an additional source of family income. However, as they marry and have children of their own, their children are facing many of the same problems which their parents faced ten years ago.

(3) Changes in the nature of Puerto Rican society as a whole. The overall increase in income, education, industrialization, suburbanization, and other indices of modernization, are all reflected in the lives of these families. Thus, educational levels have increased noticeably, many more of the poor are finding factory jobs, and all but two families are now living in the suburbs of the metropolitan area rather than in the center of the city where they lived previously.

Attitudes and Values

Though progress has not been uniform for all families (or even within families), all of them share a firm belief in the possibility of upward mobility and in the value of individual initiative. They sincerely believe that anyone who wants to can get ahead. If he does not, it is his own fault; i.e., he is lazy, ignorant, or has too many vices. For example, Doña Ana now lives in an urbanización built as a housing cooperative with government support. Her niece, whom she raised, has married and lives nearby in a private middle-class urbanización. The niece now has two small girls, whom Doña Ana cares for so her niece can continue to work to help maintain their middle-class life style. Doña Ana notes that it is difficult to improve, but one must try:

> One has to try to improve. If we hadn't tried to improve, we would never have improved, never. We would have always gone backward, because we aren't rich and we have no income except what my husband earns. One has to try to improve, to do his part. If one has vice, and leaves it, and saves for something else which one needs, right? Well that's what we have done and save. What's more, there are times we haven't gone to the movies so as not to spend.

Education continues to be emphasized as a primary avenue of social mobility. Flor, the oldest of ten children from a very poor shantytown family, is now twenty-one, has married and has two young boys; she managed to complete high school despite her father's fierce opposition and states:

> Today anyone, no matter what age or what class, can study all he wants to. If he doesn't study it's because he doesn't want to, he's lazy and doesn't want to progress. Because there are so many kinds of work. Every day they invent more occupations, more careers, much training.

However, as in the United States, even high school graduates are having difficulty securing a good job. And, even factory jobs, or trades like electricians or beauticians, now require a high school diploma. For example, Don Andrés' son cannot even be licensed as an electrician without a high school diploma. Don Andrés wanted his son to be an engineer, and was prepared to put him through college. But Junior dropped out of school at the age of fourteen, went to New York, has been married twice, and is now back in Puerto Rico with his second wife and newborn baby, living with his family and looking for a job. Thus, progress is not uniform, even within the same family. In terms of the position which Don Andrés has attained, as a well-paid electrician, secretary of his labor union, and active in the Popular Party, his son appears to be distinctly downwardly mobile.

Don Francisco has been more successful. All eight children have finished high school, and a few have gone on to college (though only the oldest daughter seems destined to complete her college education). Don Francisco is a self-employed builder, and owns a home in a middle-class urbanización, and has just bought a small farm in the mountains to which he hopes to retire. Don Francisco is a firm Seventh Day Adventist and credits his religion with his success because it taught him how to live and not waste his money on vices. He is a firm adherent of the Protestant ethic, and believes that life is a never-ending ladder of social mobility, ordained by God, which individuals ascend step by step with His help:

> Lean on a good tree and good shade will cover you. That is, if I stick by someone worse off than I, in what conditions shall I put myself? In the same conditions as he, right? But if I try to improve I don't try to look back. . . . In such a scale lives humanity. Thus we should all be looking up. He who wants to remain below, it's because he wants to. . . . Rising and falling, rising and falling. That no one will protect us from because that is the word of God.

Even among those not so religious, there is a strong belief that the division of society into rich and poor is part of the natural order of things, and that the poor must accept their subordinate position rather than protest against it. Most feel that there is nothing to be gained from collective action, that the only escape from poverty lies in individual initiative and personal gain. Labor unions are regarded as "an organization of lazy people who live off others." The Popular Party is accused of apathy, inefficiency, and domination by Muñoz, its long-time leader. Radical political ideologies are regarded with great suspicion; the poor fear the examples of the flight of exiles from Cuba (many of whom came to Puerto Rico), the bloody violence in the Dominican Republic, and the riots among Blacks in the United States, and feel that all these groups are inspired by

Communists, who are merely out for their own gain. Tito, the same young man who complained about unions and who has lived for some time in the United States, explained:

> Communism is a method invented to oppress the poor. Live off the poor like the unions. That the poor person can't express himself nor do anything. Only live under the boot, under the yoke of the government.

These families are afraid that Communism will take away the small gains for which they have struggled so hard, because "under Communism all the money stays in one pocket." They measure their gains primarily in terms of material acquisitions–a home of their own, nice furniture, a color television, perhaps a car, or even a telephone (still a luxury in this class).

Those living in public housing find other sources of satisfaction, like Rosa, mother of ten, married to an alcoholic. A monthly social security check of $150 gives greater financial security than she has ever known. Similarly, Carmen, the widow with five young children mentioned previously, now lives comfortably on the $82 a month she receives from the government for herself and her three grown children now living at home. Her oldest daughter works and also gives her an additional $15 a week. Compared to ten years ago, when she tried as a widow to raise four small children on a welfare check of $50 a month, her life now is less of a burden. Here we see the importance which changes in the life cycle (such as the age of children, the age of the household head) have on the family's sense of well-being, despite little change in their economic status.

There is little recognition of conditions in the larger society which contribute to poverty in Puerto Rico, such as colonialism, a monopolistic and stagnant agriculture, and continued economic and political dominance by the elite (compare Pabón, 1967: 182). On the contrary, most of these families favor continued dependence upon the United States and feel Puerto Rico is too small and poor to stand alone. In all the families, at least one member or a close relative has been to the United States, some several times, and all comment favorably on their experience: it is easier to find work, salaries are higher, and they do not feel discriminated against, though most lived in all-Puerto Rican neighborhoods. They see no conflict in being both Puerto Ricans and American citizens and feel that Puerto Rico can best assure itself of continued progress by preserving its alliance with the United States, "the most powerful nation in the world." The possibility of migration to the mainland offers another outlet to frustrated Puerto Ricans who see no hope of progressing on the island, and this tends to further reduce discontent among the poor.

Thus, as Lewis (1966: 24) found, most of the Puerto Rican poor are politically conservative and committed to the status quo. Lewis is correct in his assessment that the Puerto Rican poor are not class conscious and have "little revolutionary spirit or radical ideology" (Lewis 1966: 24). However, this is not, as he suggests, because the poor are apathetic or alienated from the rest of the society. On the contrary, the poor identify strongly with the progress which Puerto Rico has made in the postwar era, even though this progress may be of little direct benefit to them. They cite the new expressways, office buildings, and luxury apartment houses "as big as those in New York" as evidence of Puerto Rico's prosperity, and they are proud of it. Thus, the expansion of the economy and the overt signs of prosperity have created the illusion that anyone can get ahead if he really wants to.

CONCLUSIONS

Despite the fantastic economic growth of the postwar period, it seems clear that not all segments of society have shared equally in Puerto Rico's progress. The benefits of modernization have trickled down to the poor in terms of better health, better education, and better housing, but it has not noticeably altered their subordinate position in the larger society. In fact, it would appear that while the income of all classes has increased, the gap between the rich and the poor is growing wider (compare Feldman and Kendrick, 1968: 40).

However, we have seen no sign of protest among the urban poor families described in this paper. Rather than emphasizing class solidarity or radical ideologies aimed at the overthrow of the existing order, the poor seem intent on promoting their own individual mobility within that order. They are remarkably optimistic and profess a strong belief in the possibility of upward mobility based upon individual initiative, thrift, education, and so forth, and measured by property and other material acquisitions.

Several factors account for their optimism. First, we have seen that they measure success in terms of very modest gains—a home of their own, a stable income, a better education for their children. Many of the urban poor, particularly in the shantytown, are rural migrants for whom there was no possibility of escape from grinding poverty in agricultural labor. They feel they have improved their own life chances, and those of their children, by moving to the city where there is a wider choice of jobs (though still insecure and underpaid), where there are more educational opportunities (evidenced by the dramatic rise in educational levels between the adult and adolescent generation), and which offers a greater variety of urban amenities like stores, movies, and especially public services such as hospitals, public transportation, and water and electricity. Migrants have improved their own standard of living, even if they have not raised their socioeconomic status in the larger society (compare Nelson, 1969: 15).

Secondly, the economic and social structure of Puerto Rican society is open enough, at the present time, to absorb those who are upwardly mobile. Class lines are not so rigid that they constitute absolute impediments to social mobility, as in some Latin American and other developing countries. The middle class has grown rapidly, as evidenced by the proliferation of white-collar and service jobs and the spread of middle-income urbanizaciones. The gap between families with the lowest and highest incomes may be growing, but the number of families in the middle-income range has also increased (Andic, 1963: 21).[10] This increase appears due largely to the growth of urban industrial employment and the decline of poorly paid agricultural labor (Andic, 1963: 24-25).

The families studied here no longer subscribe to the fatalistic attitudes of a static, agrarian society, in which peasants jealously guarded their few possessions and felt that no one could improve his own position except at the expense of others. This "image of limited good," as Foster (1965) has termed it, coincides with an economy of scarcity in which there are no new opportunities or possibilities for upward mobility. However, it would seem that as societies develop and the economy expands, this fatalistic ethos is replaced by a new value system emphasizing competition, hard work, thrift, and other attitudes commonly associated with the Protestant ethic.

On the other hand, economic development alone will not eradicate poverty and social inequality. The persistence of poverty in the most developed nations of the world, including the United States, should have long since destroyed that myth, yet it continues to be applied to developing nations with the same blind faith. It is not recognized that in a class-stratified, capitalistic, colonial society like Puerto Rico, the benefits of modernization accrue mainly to the elite who already own property and other economic interests. Development by itself does not destroy their differential access to sources of power, prestige, and wealth. This can be accomplished only be a redistribution of the resources of the society.

A major redistribution in the resources of Puerto Rican society, however, is not likely to be accomplished as long as the poor support the legitimacy of the existing order. The evidence presented here suggests no sign of protest or strong opposition to the present political and economic system; on the contrary, there is a strong commitment to the status quo and to continued dependence upon the United States. In the words of Milton Pabón (1967: 182), a Puerto Rican political scientist:

> The Puerto Ricans, in the great majority, tend to identify the limited electoral participation which they enjoy within an incomplete political system with the existence of full democracy. Similarly, they tend to identify the palpable benefits of economic development with the efficiency of the majority party of the government, or with the political protection or economic help of the United States [author's translation].

One can argue that the optimism of the Puerto Rican poor is unwarranted, and that the persistence of mass poverty and competition for low-level jobs may breed future discontent, particularly among the younger generation. There is some evidence in our data to support this argument. The political conservatism of the older generation stems from their belief that they have something to gain from the existing order, and that if they have failed, it is because they have not met the requirements of that order: i.e., they have too little education, too few skills, or they have wasted their money on gambling, drink, or other vices.

However, many of the younger generation of urban poor, who were born and raised in the metropolitan area, have met the requirements of the system. They have worked hard, saved their money, and obtained a high school education. Yet as the number of high school graduates increases, many are having difficulty finding jobs,[11] or, like Don Andrés' son, they discover that even a high school diploma is now required for the same jobs their parents filled with far less education. The resentment and frustration experienced by this group are likely to be much greater than those of migrants or even of the dependent poor who have remained in public housing.

If the poor meet the requirements of the system, and are still denied an opportunity, or if the system keeps raising requirements in order to continue excluding a significant portion of the population from its just share of material prosperity, then the conditions would appear to be ripe for radicalization and even revolution. At present, however, Puerto Rican society is far from this stage. Prosperity has been of sufficient benefit to the poor to reduce the likelihood of any strong opposition to the existing order. The poor have shown themselves remarkably responsive to changing socio-economic conditions and look forward to the future with great optimism.

NOTES

1. The San Juan Metropolitan Area includes the urban core of Old San Juan, Santurce and Río Piedras plus the five municipalities of Cataño, Bayamón, Carolina, Guaynabo, and Truijillo Alto.

2. By mid-1965, a total of 16,600 housing facilities had been provided by the governmental housing agencies, through their various types of programs (not just public housing) benefiting not less than 78,000 persons within the municipality of San Juan alone (Planning Grant Application, 1967, Part III: 82).

3. Much of the material in this section has appeared previously in my article, "The Social Isolation of the Urban Poor," in AMONG THE PEOPLE: Encounters with the Poor, edited by Irwin Deutscher and Elizabeth Thompson, ©1968 by Mrs. Meiling Weiss, Basic Books, Inc., Publishers, New York. Reprinted by permission of the publisher. The shantytown is described in the present tense, except for the data collected ten years ago through participant observation and the survey sample.

4. In the past few years the return of migrants to Puerto Rico has matched, and, in some cases even surpassed, the number leaving. In 1961, for example, there was a net in-migration from the mainland to the island of 1,750 persons. (Figures supplied from New York City Migration Division of the Department of Labor, Commonwealth of Puerto Rico.)

5. In the Model Cities area along the Martin Peña Channel, fires in 1953, 1958, and 1966 destroyed a total of 200, 175, and 66 homes respectively (Planning Grant Application, Part III: 65).

6. The Popular Party is the political party headed by Luis Muñoz Marín which governed Puerto Rico from 1940 to 1968.

7. The Commonwealth Government has instituted a plan in some of the older projects whereby families can buy their apartments and take over their own maintenance.

8. Parts of the material in this section are reprinted from my article in Psychology Today Magazine, September 1970, by permission of Communications/Research/Machines, Inc.

9. For a description of the housing and other facilities planned for this area under the Model Cities Program, see Planning Grant Application, 1967, especially Part III.

10. Andic's analysis of income distribution in Puerto Rico has been questioned by other economists who argue that there remains far more inequality than he suggests (e.g., Jose Herrero and Rolando Castaneda, Revista de Ciencias Sociales). In particular, they feel he has underrepresented the number of low-income families because his data deal only with the employed. Since I am not an economist, I do not feel qualified to evaluate these arguments, but it would seem that there is a greater persistence of poverty than Andic suggests. The objective of this study was not to measure the actual changes in the structure of Puerto Rican society, but the perception of these changes by low-income families.

11. Feldman reports that in his restudy of a sample of Puerto Rican heads of households originally interviewed in 1954, most respondents are less optimistic about getting a better job and believe education to be less accessible to their children in 1965 than in 1954 (see Feldman and Hurn, 1966: 391-392).

REFERENCES

ANDIC, F. (1963) "Income of wage earner families and economic development of Puerto Rico, 1941-1953." Caribbean Studies 2, 4: 14-27.

BOTT, E. (1957) Family and Social Network. London: Tavistock.

CAPLOW, T., S. STRYKER, and S. WALLACE (1964) The Urban Ambiance: A Study of San Juan, Puerto Rico. New York: Bedminster Press.

COOK, S. (1965) "The prophets: a revivalistic folk religious movement in Puerto Rico." Caribbean Studies 4, 4: 20-35.

FELDMAN, A. and J. M. KENDRICK (1968) "The experience of change in Puerto Rico." Howard Law J. 15, 1: 28-46.

FELDMAN, A. and C. HURN (1966) "The experience of modernization." Sociometry 29, 4: 378-395.

FOSTER, G. (1965) "Peasant society and the image of limited good." Amer. Anthropologist 67, 2: 293-315.

LEWIS, O. (1966) La Vida: A Puerto Rican Family in the Culture of Poverty–San Juan and New York. New York: Random House.

MANGIN, W. (1967a) "Latin American squatter settlements: a problem and a solution." Latin American Research Rev. 2, 3: 65-98.

--- (1967b) "Squatter settlements." Scientific American, 217, 4: 21-29.

MINTZ, S. (1956) "Cañamelar: the subculture of workers on a government owned sugar plantation," pp. 314-417 in J. Steward (ed.) The People of Puerto Rico. Urbana: Univ. of Illinois Press.

NELSON, J. M. (1969) Migrants, Urban Poverty and Instability in Developing Nations. Cambridge, Mass.: Center for International Affairs, Harvard University.

PABON, M. (1967) "La integración política en Puerto Rico," pp. 174-186 in S. Lewis and T. Mathews (eds.) Caribbean Integration. Rio Piedras, Puerto Rico: Institute of Caribbean Studies, University of Puerto Rico.

Planning Grant Application for a Model Neighborhood in Demonstration Cities (1967) San Juan, Puerto Rico: Urban Renewal and Housing Corporation, Commonwealth of Puerto Rico.

SAFA, H. (1970) "The poor are like everyone else, Oscar." Psychology Today (September).

--- (1968) "The social isolation of the urban poor: life in a Puerto Rican shantytown," pp. 335-352 in I. Deutscher and E. Thompson (eds.) Among the People: Encounters with the Poor. New York: Basic Books.

--- (1965) "The female-based household in public housing: a case study in Puerto Rico." Human Organization, 24, 2: 135-139.

SEDA BONILLA, E. (1969) "La función de la cultura en los procesos politicos." Anuario Indigenista 29: 295-316. Mexico: Instituto Indigenista Interamericano.

TUMIN, M. and A. FELDMAN (1961) Social Class and Social Change in Puerto Rico. Princeton: Princeton Univ. Press.

TURNER, J. (1969) "Uncontrolled urban settlement: problems and policies," pp. 507-534 in G. Breese (ed.) The City in Newly Developing Countries. Englewood Cliffs, N.J.: Prentice-Hall.

WOLF, E. (1957) "Closed corporate peasant communities in Mesoamerica and Central Java." Southwestern J. of Anthropology 13: 1-8.

--- (1956) "Aspects of group relations in a complex society: Mexico." Amer. Anthropologist 58: 1065-1078.

6

Persistence of Tradition in the Urbanization of the Oaxaca Valley

MARTIN DISKIN

□ ECONOMIC DEVELOPMENT STRATEGIES are often aimed at the alteration of the structure of a society so that imported infusions of capital may be better employed. This usually means changing work habits and notions of time in order to ensure predictability of the work force as well as the dissemination of skills that facilitate the mechanization of production. Thinking along these lines tends to favor either the actual movement of people toward the city and the industrial machine or the absorption of ideas and techniques that emanate from the city. The naive assumption that urbanism is required for economic development is no longer tenable. In fact, anthropologists are increasingly turning their attention toward the identification of elements that help societies modernize in ways most consistent with their own historical background and cultural integrity (Belshaw, 1964; Geertz, 1963; Salisbury, 1962, 1970). This chapter will consider a regional system in southern Mexico in an effort to describe its potentiality for autochthonous change.

This regional marketing system, the remnants of an ancient urban civilization, embodies some well-developed techniques for the movement of goods and ideas, and, I believe, for the communication of beneficial changes. Although the region depends on subsistence agriculture of rather low productivity, it has good communications, a stable market network, and high economic incentive of its inhabitants. So it seems to possess many of the necessary attributes for growth commonly cited (compare with Mosher, in Wharton, 1969: 6-11) and its "backwardness" is probably due to the capital scarcity and limited technology. This system will be described through an overview of its operation at the widest level as well as a more careful examination of a key segment, a market town.[1]

THE REGIONAL MARKETING SYSTEM

A regional marketing system provides an excellent opportunity to study the movement of goods, people, and ideas. Information can be collected from as many points of view and from as many types of actors as exist within the context of culturally standardized modes of participation.

For Nash (1966: 58-80), a regional marketing system, likened to a solar system, exhibits many qualities of the modern social and economic world—cultural differentiation between villages, production specialization based on ecological differences, extensive use of money, rationality of choice, pursuit of self-interest and profit. In counterposition to these "modern" and efficient qualities are what Nash and other writers call "leveling" mechanisms; i.e., forces militating against economic development. They include technological limitations on agricultural production, the fracturing of estates through partible inheritance, as well as a variety of cultural mechanisms whose impact is to discourage the individualistic amassing of wealth. Included here, as well as in many places in Mesoamerica, are the forced participation in the governance of the community—the civil-religious hierarchy, the *cargo* system—and forced expenditure in ritual—the mayordomía system.

It should be made clear that two different kinds of social careers are being described here. The first takes place in the modern or "open" community (Wolf, 1956), permits interaction with other communities, enables the movement of people and things across boundaries, and permits the display of symbols of wealth and mobility as well as the conscious jockeying for higher status through manipulation of friendships, fictive kinship arrangements, and strategic marriages. The other career centers around membership in a community, and participation in the ritual and ceremony that define the significant social group. Here ties of kinship are important as well as the donation of services and wealth to the community. This describes a "traditional" mode of activity that usually flourishes within the closed corporate peasant community (Mintz, 1959; Wolf, 1957).

From the perspective of economic development (i.e., capital formation and the growth of firms), the traditional pattern is stagnant. Community service is seen as creating leveling rather than accumulation. On the other hand, the traditional viewpoint sees the developmental perspective as being risky, deracinated, and not entirely honorable. Both perspectives contain some truth. The level of welfare in the most remote villages is lower than in the more urbanized places. Public health workers find a good deal of resistance to programs of innoculation and sanitation, although the village diet may be basically good. But full-time entry into the market world jeopardizes subsistence activity, forces behavior patterns that may be felt to belong to a different moral sphere, and causes loss of community membership in favor of an ill-defined group–modern Mexican society.

Seen from the distance of the national economy, there are series of regional systems whose histories have left them with specific constellations of cultural traits and which act as centers of reception of goods, people, and ideas from without, as well as serving as suppliers of the same for the nation. The aggregate impact of this is what Mintz (1959) calls an internal market system.

These two modes have been characterized as a national orientation and community orientation (Wolf, 1956). For others, these two models evidence a dual society, or plural society (Smith, 1960; Furnivall, 1948). From this national perspective the policy goal is usually one of "incorporation" (sometimes called "castillianization" in Mexico), assimilation, and other euphemisms expressing the general lack of acceptability of the subcultural level. Here I propose to examine this situation from the local level. Economic activity is easily evaluated, both by the observer and the actors themselves, in the terms discussed above. Reference to economic and marketing activities provide a convenient frame through which to elicit attitudes and preferences. For any given individual in this system, an important arena for action within either a traditional or modern mode is the market place.

One goal of this chapter, then, is to show how the social system is regularly recreated in the market places and how, in a real sense, people may "shop" for ways of life and careers. While the outlines of action are provided by the system of markets in the valley of Oaxaca, the subject matter is the constitution of a region according to the various places in which people live and which live in their minds. The task is to elucidate the region as a whole so that movement or stasis become intelligible.

CITY AND TOWN IN THE VALLEY OF OAXACA

The valley of Oaxaca is actually a highland plateau divided by several ranges of hills into several smaller valleys. Its northernmost point is about twenty kilometers above the city of Oaxaca. Extending to the southeast is the valley of Tlacolula formed by the sierra de Juarez on the northeast and a small range of hills (cerro Guibini) on its southwest. The *valle grande* region extending to the south is bounded on its west by more hills and the valley of Zimatlan. It is defined on its western side by the sierra de Zimatlan which effectively marks a cultural as well as a physical boundary.

These natural features accommodate a population distribution whose centers of greatest density and activity are in the small valleys where communication is easy. They bind together groups of communities that make extensive use of environmental difference so as to yield a large number of agricultural and production specialties.[2] A glance at the map confirms that this arrangement pertains today. It is also clear that the city of Oaxaca occupies a focal position by virtue of its lying at the hub of the wheel formed by the radiating spokes of the smaller valleys.

Finally, in purely physical terms, we see that the valley of Oaxaca has three types of communities: the city, those centers of activity and movement in the small valleys; the towns; and the smaller, somewhat isolated places that are often located in irregular settings, the villages.

The valley of Oaxaca is an old region. Human occupation probably dates back several millenia before the Christian era. Recent work indicates that this may have been the site for the domestication of several food staples (maguey, maize) that are still of considerable importance today (Flannery, 1968). The urban settlement of Monte Alban, built at the beginning of the Christian era, gave evidence of regional coherence through pottery styles and other artifactual materials, that included most of the present limits of the valley of Oaxaca region. The region under consideration here is generally the same as the Valley Zapotec Zone (Spores, 1965: 963). Archival evidence at the time of the Spanish conquest strengthens the view that the region maintained its integrity and presents evidence as well for existence of a marketing system (Diskin, 1967: ch. 11). This system presumably served the Zapotec inhabitants of the zone, but was also used by the Aztecs during their phase of imperialist expansion, and was taken over by the Spaniards during the Colonial period. In fact, the most impressive evidence for the sustained, systemic nature of the region is seen in the multiplicity of functions that are performed in the same basic pattern. Even today, services and transactions of a political, religious, legal, economic, and medical nature all move within the avenues of communication that have existed for centuries.

Market day *(día de plaza)* in the towns is the day when cases are heard, taxes collected, and the doctor is in the clinic. Market places–called plazas (Diskin, 1969)–are located in the center of the communities. The sites of the major plazas are all towns, places that are of central importance in their particular zones. There are eight plazas in the Oaxaca system, each one convening once a week. In addition there are plazas that convene in smaller villages, most of which last for one or two hours each day. They usually occur in the morning and are devoted to the sale of food. There are also plazas that occur according to the annual ritual calendar and are located in the town whose patron saint is being celebrated. These occasions, particularly the large pilgrimages, since they often extend beyond the limits of the system, provide opportunities for people to learn about different geographical settings, cultural configurations, and life styles.

The town plazas are the most important sites for this discussion. It is in the town plaza that people regularly come from every sector of the Oaxaca regional system. The typical allocation of time for villagers includes a visit to the nearest town plaza, not once a week, but over the course of a year. This trip is made more often than a trip to any other setting. The town dweller naturally has weekly experience with the plaza. In addition, merchants who are reasonably well capitalized and have sizable inventories of goods use the town plaza as the focus of most importance for buying and selling. Furthermore, there is a balance struck between the income-generating potentiality of the countryside and the appearance of new goods, as well as the number of large merchants that regularly appear in any given plaza. I have shown elsewhere the actual proportion, for all town plazas, of those who offer for sale products they produced in small quantity: usually village-based peasant agriculturalists or artisans, those who enter the market having adapted to the scale of the town; larger-scale producers and middle-men based in the towns, and those who are specialists in deriving a living by traveling between production sources and buying in quantity and reselling in the plaza; and city-based full-time traders and businessmen (Diskin, 1969: Tables 2, 3, 4).

The systemic quality of the region is of utmost importance in understanding the meaning of place and the probable paths people take. The unifying factor that connects all parts of this region is the cyclical market system. For every person here, irrespective of the community in which he lives, the weekly occurrence of a market provides a link with communities of different shapes. This fact of life provides the rationale for the study and analysis of a region through its market places.

The plaza is the arena where the entire system is represented every week. It is not only convenient but necessary to use the plaza as a base for the study of the system. However, to fail to include significant aspects of life that are located outside of the plazas, activities that are not primarily economic but political, social, and religious, would seriously distort an appreciation of this region.[3] These features are briefly sketched below.

MAJOR SOCIAL AND CULTURAL DIMENSIONS OF THE REGION[4]

The elements of this system have become differentiated under the impact of contact with external agencies and their internal economic specialization. For analytic purposes it is possible to give broad profiles of the major characteristics of villages, towns, and the city. The categories used to describe these elements are:

(a) *cultural boundaries*—the extent to which community affiliation is buttressed by specific definitions and symbolized by repetitive ceremony

(b) *sociopolitical structures*—the standardized mechanisms for the regulation of public affairs

(c) *economic organization*—the nature of economic activity, whether subsistence or specialized production or another special part in the chain of distribution

(d) *ideology*—the intellectual framework that informs and evaluates the disposition of effort, time, and wealth

The Villages

Cultural boundaries. The communities are villages whose inhabitants are distinguishable from each other through commonly recognized clusters of customary behavior (*costumbre*). Dialect differences in Zapotec (in the Tlacolula valley, and, to some extent, elsewhere) are noticeable from village to village. In addition, informants recognize distinction between community styles of dress, community ceremonials such as marriage (*fandangos*), the sponsorship of saints' celebrations (mayordomías), specific religious village observances such as patron saint feast days, lenten celebrations, peacemaking rituals (*contentadas*), and the functions of ritual specialists (*huehuetes*). The validity of these cultural boundaries is generally agreed upon by the local people.

Sociopolitical structures. To a large degree these villages regulate their public affairs through locally chosen officials by means of customary techniques. Some form of the civil-religious hierarchy survives in an attenuated form throughout the whole region. While the villages are no longer landholding corporations, they do exercise considerable pressure on individual landowners and thereby qualify private ownership in the direction of community responsibility. Community service in the form of the *tequio* is one manner of satisfying community obligation.

Production organization. While all villages in the area are predominantly agricultural, because of environmental variation some have specialized in the production of particular crops. In addition, virtually every village engages almost exclusively in certain crafts, such as sandal-making, weaving, rope-making, basketry, pottery, wood-working, and stonework (*metates*). The full list of these craft and agricultural specialties constitutes the culturally standardized inventory of necessary goods.

Ideology. Most activities on this level—social, political, and economic—are seen as reinforcing community boundaries. Production and distribution have as their goal the maintenance of an approved life style. People try to make enough *por el gasto* (to meet expenses); this includes the "replacement fund" (agricultural implements, animals, and house-building) and the "ceremonial fund" (hospitality and gift-giving at mayordomías) (Wolf, 1966: 6-7). This may be characterized as a consumption ideology since internal consumption is the goal for any goods produced.

The Towns

Cultural boundaries. These are the market towns—the fixed points of the weekly round of economic activities. These towns are not clear entities in any cultural sense, exhibiting instead some measure of internal differentiation into modern and traditional sectors. The criteria that differentiate the sectors are roughly the same as for the villages; i.e., language (whether Spanish or Zapotec as a main language), dress, and ceremonial participation. There are also residential qualifications. The traditionalists tend to live farther from the commercial center of the town than those attracted by a modern life style and economic activities (Diskin, 1967: ch. 5).

Sociopolitical structures. Here most public affairs are controlled by officials whose offices are in accordance with state dictates. The presence of a central administrative authority is felt more keenly than in the villages. While there are ceremonial offices (*mayordomo*) and ritual specialists (*huehuete*), their function is not so much that of social regulation as of the maintenance of the cultural integrity of the traditional sector of the town. Political activity is mainly framed in terms of state and national organization. The community exercises no jurisdiction over land. National politics is active here.

Production organization. The number of production specialties increases on this level and there is a considerable amount of occupational diversity. Some activities, such as baking and butchering, flourish because of steadier demand and the larger buying public. More services may be sold here because of the weekly markets and their need for part-time labor. There are also a large number of nontraditional employment opportunities for townsmen: jobs connected with the government offices located in town; jobs relating to the transportation of goods and people (bus drivers and helpers, auto mechanics, taxi drivers, truckers); and special services dealing with maintenance and repair of modern goods (watch repair, radio repair, electricians). In addition to this expanded list of occupations, there is greater opportunity for job mobility here than in the villages.

Ideology. The ideology of consumption is represented by those people who engage in agricultural and traditional occupations (rope-making basketry, metate production, baking), and also by those who engage in some forms of nonagricultural wage labor (mason, brick maker, water carrier, porter). However, the notion is widely held that economic activity can be cumulative and expanding. What might be called an ideology of capital formation is expressed by many people and is openly contrasted by these people with the consumption ideology of the villagers (Diskin, 1967: ch. 5).

The City

Cultural boundaries. The central Oaxaca area is felt by Mexicans from other parts of the country to be unique with respect to the physical type of its inhabitants, certain typical foods, and, possibly, musical styles. Participants in the plaza system who come from the city identify themselves more with the nation than with local communities. Although the merchants often make conscious efforts to adopt local cultural styles, this is in order to sell more effectively in the plazas.

Sociopolitical structures. The legal system, administrative procedures, and political processes that affect life on this level are virtually identical with those that pertain throughout the nation.

Production organization. Economic activity is geared toward profit. Cash is universal. Within the limits set by the general level of industrialization, there are a great number of occupations, and an accompanying increase in occupational mobility. Economic activity is responsive to the changes of the national and international market.

Ideology. Those whose activities are based in Oaxaca are concerned primarily with capital formation. They organize their activities in line with capital accounting, and rational procedures involving the measurement of units of labor, costs of transportation, and margins of profit.

The repetitive meeting in the market place of these three social layers of the region accomplishes several tasks. First, it provides the opportunity for peasant villagers to provision themselves with the full list of culturally approved goods deemed basic to their physical and sociocultural well-being (see Appendix A). Second, it enables peasant producers to encounter the aggregate demand for their production on a regular basis. Reasonable adjustment can be made between the particularistic demands of a given community and the impersonal quality of the market place. The market place makes it possible for a person to be traditionally upright, embedded in a round of ceremonial and ritual obligations at home, and at the same time to act in an impersonal setting that is characterized by profit maximization behavior in the market place.

The peasant's presence in the plaza provides the audience for urban economic specialists whose life experiences are very different. These "outsiders," well aware of the integrity of the plaza system and the regional coherence it embodies, try to adapt to local patterns of culture and etiquette. Their efforts to speak a few words of Zapotec, usually concerning number and price, as well as information about the cultural requirements underlying demand (e.g., when certain pottery is needed for a given celebration) while done consciously in the name of salesmanship, is a further measure of the internal integration of the system. All the plazas in the Oaxaca valley have a consistent flavor, one that facilitates movement for the most isolated villager and sets rules for the outsider.

The stability of this flow is demonstrated by an examination of town receipts for the use of market space over the period May 1960-October 1965 (Table 1). The table disguises weekly fluctuations but points out that these differences are evened out within a short time. The large increase in October is due to the imminence of the celebration of the Day of the Dead, the most important celebration of the year, which occurs at a time when the harvest is in and there is maximum cash available.

Thus, in any given plaza, all the parts of the system are visible. The plaza may be conceived of as a series of successive layers whose totality comprises the structure of the region.

THE ROLE OF TOWNS IN REGIONAL CHANGE: THE CASE OF TLACOLULA

The town is the location in the Oaxaca region where most social and economic movement takes place. It is an especially good context in which to assess the region's capacity to accommodate change; for example, the role it plays in the processes underlying social mobility.

The information used in the analysis of a town comes from the writer's residence in Tlacolula from October 1964 to November 1965. One body of information was collected by direct observation and participation in the life of the town. It reflects the personal

TABLE 1
MEAN MONTHLY INCOME FROM PLAZA FOR TLACOLULA 1960-1965

	1960	1961	1962	1963	1964	1965
January	–	602	569	577	584	715
February	–	606	588	514	679	654
March	–	652	579	587	695	693
April	–	538	573	525	770	693
May	604	611	541	508	667	686
June	535	575	514	543	667	687
July	519	553	487	531	704	655
August	587	567	517	531	686	653
September	575	632	464	528	733	764
October	1107	1043	1133	1227	1444	1235
November	557	473	483	446	584	–
December	605	559	522	547	700	–
Monthly mean for year	835	621	581	593	739	743

NOTE: All sums in pesos. 12.50 pesos = $1 U.S.

relations in which my wife and I became involved, attendance at all the significant religious and festive observances, a great deal of informal gossip in the context of visits, work, eating and drinking with Tlacolulenses. The second source of data is a sample survey done on the whole town. Since Tlacolula is rather large (about 7,500 population), it became clear that impressionistic treatment would be insufficient; after nine months of residence I designed a questionnaire and administered it with the help of local people. To complete the picture, I obtained statistical information from the town census of 1963 (see Appendix B). The statistical breakdown enabled me to determine exactly what population to sample in terms of residence in the town and occupation.

The town of Tlacolula has the status of seat of a *distrito* of the same name. Administratively it contains the apparatus for tax collection, a judge, a jail, and an elementary and secondary school. For those areas not included here one must go to Oaxaca. The idea one gets from historical sources confirms the central political and administrative role of Tlacolula for the district for the past two centuries at least.[5] Internally, the town is divided into sections (*secciones*). In addition to these formal divisions, there are also named neighborhoods that refer to features of interest for the inhabitants such as *el rastro,* the slaughterhouse (the neighborhood near the old slaughterhouse); *el Calvario,* the neighborhood that regularly uses the church of the Calvary, built on a hill; *el centro,* that part of town that houses the administrative offices, some larger stores, the main church, the market place and the schools; and *el barrio,* the most remote neighborhood, farthest from the highway. There are annual celebrations during weeks leading up to Holy Week when various neighborhoods sponsor feasts, games, and music. At such time, there is a rather intense atmosphere of competition. Occasionally, membership in one of these locations has definite social value. Those of el centro describe themselves as more cosmopolitan than those of el barrio. This judgment seems to be agreed upon by the residents of both places.

The town of Tlacolula impresses the observer with the juxtaposition of modern and traditional elements where both are enacted and accepted as legitimate. For example, in the Sunday plaza when festivities are in progress at the church (which adjoins the plaza), a striking contrast is seen between the ancient and formal behavior of the mayordomo and members of his party, to the accompaniment of the tom-tom and the *chirimia* (a flute), and the hubbub of the plaza with evidence of its modern technology, general impersonality, and devotion to business. One sees newly married couples going from the church ceremony to the ceremonial fandango that confers community legitimacy on the union, in a new taxi festooned with decorative hubcaps, ornate paint jobs, and "o-you-kid"-type slogans stenciled on the side fenders.

General demographic and descriptive data concerning the population of Tlacolula are summarized in Appendix B. It is noteworthy that ninety-three percent of the sample were born in Tlacolula and continue to reside there. Of those who were born elsewhere, they maintain the identity of their birthplace and see themselves as residing in Tlacolula but not "from" Tlacolula (in Spanish the distinction is between "de dónde es Ud" (where are you from), and "dónde radica Ud" (where do you reside). While that points to a strong sense of regionalism that is widespread in Mexico, it also points to the fact that participation in this regional system does not require mobility in the usual sense but enables a firm base for identity and location in one part while one's activities may diffuse throughout the region. It is common to hear of people living in different towns, villages, or in Oaxaca for a number of years, or even permanently, while maintaining allegiance and affection for the customs of one's home community. I suspect that the above also explains why there is very little out-migration (Martinez Rios and De Luna Mendez, 1960: 123). It is quite possible and convenient for a person here to act within the context of several identities; e.g., villager, farmer, and marketer. It also suggests that there are significant benefits in each community type of a nutritional, emotional economic, and recreational nature. Given a fundamental reason to stay put, it is possible to enjoy the rest of the system by attendance at the plazas and occasional visits to Oaxaca.

The distribution of occupations among the sample indicates that agriculture is the most important activity compared to craft production and commerce (Table 2). While the sample does not include certain occupations found in Tlacolula (barbers), direct observation indicates that there are more occupational possibilities in town because of the larger public to service in the form of work attendant on the plaza (helping set up merchants' stalls), the large number of people that buy ready-made food served on market day, as well as transportation specialists (taxi drivers) and others.

TABLE 2

JOB CLASS IN A TLACOLULA SAMPLE

Artisan	Business	Agriculture	Service	Professional	Total
79	32	179	82	6	378
20.9%	8.5%	47.4%	21.7%	1.6%	100%

The increased list of possibilities for making a living, the town's capacity to absorb new roles, is in itself no proof of Tlacolula's greater urbanism. In order to make such a judgment, it is necessary to learn something about the meaning of such activity and, perhaps more important, to get a sense of the alternatives present for different life styles. Is it possible to continue a traditional style of life in Tlacolula in a satisfying manner? Is it possible to see the town as a step up on the way to the city? For information on these points, a series of questionnaire items was constructed that compiles data on specifics of language use, dress, cooking arrangements, and sleeping arrangements for males in the sample (Appendix C: items 12-17).

The questionnaire also included a series of attitudinal items concerning commitment and participation in two significant ceremonial observances–the mayordomía (celebrations of saints) and fandangos (wedding celebrations). These two events are often discussed by people in this region as being diagnostic of the particular flavor of a given community. They both require the *guelagueza* for their support and maintenance. This is a ritualized, reciprocal, gift-giving custom that effectively locks an adult into a pattern of rights and obligations to give and receive for a lifetime, and is therefore a sensitive measure of attachment to a mode of living that is recognized as traditional. Much conversation is devoted to evaluations of these events and there seemed to be a spectrum of opinion concerning the optimal number of days for their adequate enactment. Two parallel questions were asked concerning the mayordomía and the fandango[6] (Appendix C: items 18,19). Since the responses were parallel for these questions, I reproduce the frequency distribution for the question on number of days that a mayordomía should last (Appendix C: item 18e; see also Table 3). The clearly bimodal nature of this distribution suggests that, clustered around three and eight days respectively, there is a constellation of attitudes and behaviors that we might designate as traditional and modern.

Commonsense suggests that consistency be sought between the traditional-modern dichotomy with respect to the life style dimension: one would hypothesize that a composite picture could be derived of mode of dress, language, and attitudes concerning ceremonial. Traditional thinking concerning modernization or urbanization suggests that, in a more or less continuous fashion, each step toward urbanization can be characterized by a discrete cluster of traits that imply all clusters beneath it. An example is Kay's study of Tahitian consumer purchases (Kay, 1964). The data in fact do not fit the requirements of scalogram analysis and indicate that either the process being studied is not continuous or there is no compulsion to be mobile in the manner usually conceived; i.e., movement from particularism, physical movement to a larger community, more education, and different dress.

Two aspects of the population helped characterize a modern and a traditional orientation. They have to do with the use of the local Indian language, Zapotec, and the specific manner in which people arrange their beds. The use of Zapotec helps us understand the direction of common activities an individual chooses. If Zapotec is the language for ordinary occasions, the person is well rooted not only in town activities as opposed to city affairs but also is probably involved heavily with the round of ceremonials that define the more

TABLE 3

HOW MANY DAYS SHOULD A GOOD MAYORDOMIA LAST?

Number of Days	Responses	Percentage
1	19	5
2	60	16
3	91	24
4	53	14
5	28	7
6	23	6
7	5	1
8	79	21
9	9	2
No opinion	6	2
No answer	8	2
	381	100

TABLE 4
ZAPOTEC SPEAKERS AND LIFE STYLE

	Mean mydm. days	% Empl.[a]	% Shoes	% Pants	% Shirts	% Raised Hearth	% Charcoal	Artisan	Business	Service	Professional	Agriculture	Totals
Suelo si	5.4	63	2	52	35	2	2	17	0	0	0	85	102
Tablas si	5.1	77	10	82	76	17	6	19	2	2	0	40	63
Colchon si	4.0	84	36	94	75	72	14	14	1	1	0	17	33

NOTE: Suelo, tablas, and colchon refer to sleeping arrangements. The si means that the respondents speak Zapotec. Other categories are self-explanatory.

a. Refers to choice of occupation between store clerk and seller in plaza. The former is considered a more modern choice.

traditional sector of the town. In Tables 4 and 5, use of Zapotec is indicated by the response "si" (yes), or no.

The way a person sleeps, on a straw pallet on the floor (*petate*), on a bed made of planks laid across two sawhorses (*cama de tablas*), or on a mattress with springs (*cama de colchon*), says a good deal about that person's income. People are conscious that the sleeping arrangement must be consistent with other expressions of wealth or poverty. Thus we have six possible arrangements of these two attributes and the tables are arranged to demonstrate that the responses to these two questions help organize the results of other questions about life style (compare with Appendix C: items 14-17).

TABLE 5
MONOLINGUAL SPANISH SPEAKERS AND LIFE STYLE

	Mean mydm. days	% Empl.[a]	% Shoes	% Pants	% Shirts	% Raised Hearth	% Charcoal	Artisan	Business	Service	Professional	Agriculture	Totals
Suelo no	2.6	92	8	100	92	15	0	5	0	1	0	7	13
Tablas no	4.0	95	5	100	100	24	24	4	1	3	0	13	21
Colchon no	2.5	82	83	100	96	70	74	7	3	4	1	4	19

NOTE: Suelo, tablas, and colchon refer to sleeping arrangements. The no means that the respondents do not speak Zapotec.

a. Refers to choice of occupation between store clerk and seller in plaza. The former is considered a more modern choice.

The general correspondence between cultural values, life styles, and occupations shows that Tlacolula can accommodate a wide range of possibilities. This range includes all the observances and types of social relations found in the villages. In fact, Tlacolula has the reputation of having a vigorous ceremonial life.[6] Its mayordomías are said to be more sumptuous and longer lasting than those of any other community in the vicinity. Also, the overwhelming majority of people who were asked, claimed participation in the guelagueza, the mechanism through which these feasts are provisioned. The sponsors of these feasts (mayordomos) are people who practice traditional occupations, combinations of agriculture and craft production, and are apparently quite successful. This points out that, within the institutional matrix sketched here, especially the traditional sector, there are behaviors that must be considered "rational" and "developmental" (Cook, 1970).

CONCLUSION

I have tried to show here that the usual notion of progress from the village setting toward the city is not a necessary one. A coherent region such as this one is able to make itself available to most of its inhabitants without the violation of their self or community identities. This seems to be the reverse of most ideas of economic development and urbanization. Instead of promoting social and cultural change in a direction which renders the recipient population more able to use outside capital, it is suggested here that the institutional mechanisms necessary for beneficial development of the region are already present. They are the simultaneous participation in both culturally distinct community affairs that presuppose long residence, kinship connections, acceptance of local definitions and usages, as well as information and experience about a much wider and complex area of behavior.

The town population leans more heavily toward a modern orientation than would be expected for the villages (about three-quarters of the response to the mayordomía question favor a shorter celebration). But at present there is no sign that these traditional observances are diminishing in strength or in degree of commitment by their practitioners. The town, for some people, is simply a certain environment in which to practice traditional modes of behavior.

The region as a whole does need to increase its agricultural production. For this it needs knowledge of improved techniques and the tools to implement them. While this would only be the beginning of the solution to a serious problem I hope that the material presented here makes it clearer that any policy designed to help this region, however well meaning, *must* take serious account of the capacity of the market system to reach people. Within the known set-up of the market places, more people comfortably travel through their social system now and are able to evaluate the different arguments for change, than through any external scheme. Therefore, the introduction of forces or agents for change should be made in the plazas of Oaxaca.

By planning for change in this manner, the coherence of the region could be preserved and used. Thus, the economic shrewdness that exists and is seen in the use of new production opportunities, economies of scale, and the filling of new roles as they become feasible, could be built upon. By trying to work out of the present structure (i.e., different levels of organization tied together into a culturally intelligible system), individual participants will become able to choose more freely the direction in which to harness their energies. The moral tone of social action with its tradition of decentralized local politics and respect for community autonomy seems a strikingly modern aspect of this region, one that should not be swept away in the name of progress. It is to be hoped, then, that those who allocate the national and state budgets focus on making the financial and technological tools available for growth, while being respectful of ways of life that have been worked out over many centuries.

Economic development in underdeveloped regions may not involve as much housecleaning as is usually supposed. Serious thought must be given to the capacity for change in places like the valley of Oaxaca. This would mean that the process of development (i.e., improvement of the level of welfare of the inhabitants) should be largely undertaken by those whose lives are most affected, even though certain key resources may come from the outside.

NOTES

1. The fieldwork done for this paper was funded by the Oaxaca Markets Study Project, Department of Anthropology, UCLA, Ralph L. Beals, principal investigator. My participation in the project was from October 1964 until November 1965.

2. The specialization of production may not be as firmly related to ecological diversity as implied here (Flannery, et al., 1967). Nevertheless, whether it represents a cultural superposition on the landscape or not, it is, as practiced today, generally consistent with environmental differences such as altitude, quantity, and distribution of rainfall, availability of raw materials (wood, clay, lime, and stone).

3. The Geographic Relations of 1580 mention Tlacolula, and its position between Mitla and Oaxaca implies that it was of considerable significance. In the eighteenth century Teotitlan del Valle was the administrative seat of the district before it was moved by Royal order to Tlacolula.

4. The town was also polarized between those whose religious expression adhered to Catholic orthodoxy both in worship and through church sodalities, and those for whom the mayordomía was more significant. The latter group was constantly urged through sermons from the pulpit to give up its "uncivilized ways." In its view, the services of a priest were strictly secondary. The matter reached an unfortunate head in an assault on a priest leading to the interdiction of the town by the bishopric.

5. This seems to be true of many of the market towns of the region. Tlacolula's fame has spread in part because of a book written by a local person (Barriga-Rivas, 1952) that was made into a movie.

6. Recent studies have shown that rational economic behavior and economic development are not incompatible with traditional cultures and societies. That development does not occur seems to reflect scale of operation, available credit, and technological limitations (compare Tax, 1963; Belshaw, 1955, 1964; Salisbury, 1962, 1970).

APPENDIX A
ORIGIN OF PRODUCTS SOLD IN THE TLACOLULA PLAZA

Products	Number of Market Stalls	Villages in Tlacolula Subsystem	Tlacolula	Tlacolula and Villages	Other Plazas	Other Traditional Zones	Oaxaca	Other Regions in Mexico
Raw wool	4	X						
Ocote, papas, tejocote	27	X						
Wood for construction	1	X						
Wood products (furniture)	2	X						
Black beans	13	X						
Charcoal	7	X						
Red pottery	15	X						
Tomatoes	18	X						
Penca de maguey	2	X						
Patio brooms	1	X						
Serapes	4	X						
Herbs, garlic, spices	25	X						
Chile verde	4	X						
Baskets	5	X						
Limes, chilacayote, chico zapote	3	X						
Tejate, agua fresca	14		X					
Nieve	6		X					
Small fondas	9		X					
Large fondas	8		X					
Barbeque	13		X					
Bread	57		X					
Beef	30		X					
Pork, pork products	42		X					
Live turkeys, chickens, pigs	93			X				
Tortillas (clayudas, blanditas)	20			X				
Maiz criollo	4			X				
Alfalfa	8			X				
Eggs, chicks	15			X				

APPENDIX A (continued)

Products	Number of Puestos	Villages in Tlaco-lula Subsystem	Tlacolula	Tlacolula and Villages	Other Plazas	Other Traditional Zones	Oaxaca	Other Regions in Mexico
Green pottery	10				X			
Cal	12				X			
Black pottery	7				X			
Miltomate, garlic (vegetables)	28				X			
Tenates, petates	9				X			
Peanuts	2				X			
Cheese	5				X			
Tinware	5				X			
Candles	4				X			
Candy	3				X			
Huaraches and parts	6				X	X		
Sugar cane	10				X			
Wooden toys	1					X		
Medicinal herbs	2					X		
Dried chiles	16					X		
Gourds	1					X		
Dried fish, shrimp	10					X		
Fiber products (ixtle)	13					X		
Copal, pataste, ojo de venado	4					X		
Cacao, coffee	28					X		
Pottery (white)	3						X	
Manteles, napkins	2						X	
Games of chance	1						X	
Popcorn	1						X	
Used clothes	1						X	
Knives, machetes	2						X	
Comic books	2							X
Shoes	1							X
Govt. corn	11							X
Bolt cloth	6							X
Factory-made blankets	3							X
Bananas	12							X
Chile	4							X
Fancy fruits	2							X
Sacks for corn, rice	2							X
Powdered soap	2							X
Solder, hardware	1							X
Oranges	9							X
Cantaloupe, watermelon	12							X
Shawls	3							X
Notions	22							X
Tomatoes	3							X
Ready-made clothes	40				X		X	X
Straw hats	5					X		

APPENDIX B

SELECTED CENSUS DATA FOR THE TOWN OF TLACOLULA (1963)

Male	Female	Total	Family Heads Male	Family Heads Female	Total
3622	3794	7416	1305 81%	314 19%	1619

Average Family Size – 4.2

Birthplace

Tlacolula	Tlacolula District	Valley of Oaxaca	Other	Total
355 93%	3 1%	15 4%	8 2%	381 100%

Education by Sex

Sex	Education in Years: 0	1-4	5-6	7-12	No Answer	Total
Male	131 46%	116 41%	33 12%	6 2%	0	286 (75%)
Female	52 55%	36 38%	5 5%	2 2%	0	95 (25%)
	183 48%	152 40%	38 10%	8 2%	0	381 100%

Age and Education

Age	Education in Years: 0	1-3	4 and Over	Total
0-30	10	20	40	70 18.4%
31-40	28	37	22	87 22.8%
41-50	26	39	16	81 21.3%
51-60	51	14	7	72 18.9%
61-70	29	13	8	50 13.1%
Over 70	14	5	2	21 5.5%
	158 41.5%	128 33.6%	95 24.9%	381 100%

APPENDIX C
TLACOLULA QUESTIONNAIRE

______
Sec. ______
Date ______

1. Name ______
2. Age ______
3. Where were you born? ______
4. How long have you been living in Tlacolula? ______
5. How many years of school have you had? ______
6. Civil status ______
   (a) How many children do you have? ______
7. What is your main occupation? ______
8. What other occupations do you have? ______
9. *AGRICULTURE*
   (a) Are you a landowner?
      Yes______ No ______
   (i) Who works your land?
      (a) you ______
      (b) on shares ______
      (c) day laborers ______
      (d) other ______
   (ii) Under what arrangement is it worked?
      (a) ejido ______
      (b) on shares ______
      (c) day laborers ______
      (d) other ______
   (b) What do you plant? ______
   (c) Does your work allow you time for other occupations? ______
      (i) Which ones? ______
10. *CRAFT*
   (a) Are you a maestro?
      Yes ______
      (1) Where is your shop? ______
      (2) Are you the owner of the shop? ______
      (3) How much do you earn a month? ______
      Apprentice? ______
      No ______
      (1) Do you work with a maestro? ______
      (2) Are you an operaio ______ oficial ______
      (3) How are you paid? ______ salary? ______
11. *BUSINESS*
   (a) What do you sell? ______
   (b) Where?: in this plaza? ______, in other plazas? ______, which ones? ______, at home? ______
   (c) Do you have a stall? ______
   (d) Are you ambulante (sell while walking around) ______
   (e) Do you deliver? ______.
      To private parties ______
      To middle-men ______
      To stores ______
      Foreign shipments ______
   (f) If you sell to a middle-man, where does he sell? ______
   (g) How much do you earn a month? ______
   (h) Where do you buy your merchandise? ______
   (i) From whom? the producer? ______ middle-man? ______
   (j) Where is the merchandise made that you sell? ______

APPENDIX C (Continued)

12. Do you speak Zapotec? ______
13. Do you speak it outside of your house? ______; in your business? ______
14. Do you wear (men); factory-made pants ______, homemade-style pants ______, factory-made shirt ______, homemade-style shirt ______, shoes ______, sandals ______, barefoot ______, native-style sash ______.
15. Do you wear (women); dress (modern style) ______, blusa (traditional blouse) ______, traditional skirt ______, shoes ______, sandals ______, barefoot ______.
16. Do you cook; charcoal brazier on the floor ______, brazier-elevated ______, with firewood ______, with charcoal ______, oil or kerosene stove ______, gas stove ______,
17. Do you sleep; on the floor on a straw pallet ______, on a bed of boards ______, on a bed with a mattress ______.
18. Have you attended many mayordomías? ______
    (a) Have you played any part in a mayordomía? ______
    (b) Which part? ______
    (c) Did you enjoy it? ______, why? ______
    (e) In your opinion how many days should a good mayordomía last? ______ why? ______
19. (a) Have you attended a fandango? ______, how many days did it last? ______
    (b) Have you attended many fandangos? ______
    (c) In your opinion, how many days should a good fandango last? ______ why? ______
20. Have you brought guelagueza to fandangos or mayordomías? ______
21. What is your opinion of the custom of guelagueza? ______
22. (a) Have you served on any post in city hall? ______
    (b) Which one? ______
    (c) Have you served as: block chief? ______, section chief? ______
    (d) Would you like to have a town post? ______ why? ______
23. (a) Do you buy ready-made tortillas? ______
    (b) Do you buy corn for tortillas? ______
    (c) What kind of corn do you prefer? ______ why? ______
    (d) Who grinds it? ______
    (e) Where do you get it? ______
24. In your opinion, what do the following terms mean?
    (a) Catrin ______
    (b) Gente de razon ______
    (c) Llope ______
25. Which of these terms describes you? ______ why? ______
26. In your opinion, which of the following occupations is better?
    (a) day laborer ______, water carrier ______, why? ______
    (b) baker ______, peasant ______, why? ______
    (c) tortilla maker ______, servant ______, why? ______
    (d) sandal maker ______, business man ______, why? ______
    (e) seller in the plaza ______, clerk ______, why? ______

i– Attitude of interviewee: receptive ______, neutral ______, suspicious (desconfiado) ______
ii– House type: adobe ______, wattle and daub ______, brick ______, with yard ______
iii– General observations and more details of the interview ______

REFERENCES

BARRIGA-RIVAS, R. (1952) La Mayordomía. Mexico.

BELSHAW, C. S. (1964) Under the Ivi Tree: Society and Economic Growth in Rural Fiji. Berkeley: Univ. of California Press.

--- (1955) In Search of Wealth. American Anthropological Association Memoir 80.

COOK, S. (1970) "Price and output variability in a peasant-artisan stoneworking industry in Oaxaca, Mexico: an analytic essay in economic anthropology." American Anthropologist 72: 776-801.

DISKIN, M. (1969) "Estudio estructural del sistema de plaza en el valle de Oaxaca." America Indigena 29, 4: 1076-1099.

--- (1967) Economics and Society in Tlacolula. Oaxaca, Mexico. Ph.D. dissertation. UCLA. Unpublished.

FLANNERY, K. V. (1968) "Archeological systems theory and early Mesoamerica," pp. 67-87 in Anthropological Archeology in the Americas. The Anthropological Society of Washington. Washington, D.C.

--- , A. V. T. KIRKBY, M. J. KIRKBY, and A. WILLIAMS, Jr. (1967) "Farming systems and political growth in ancient Oaxaca." Science 158: 445-454.

FURNIVALL, J. S. (1948) Colonial Policy and Practice. Cambridge: Cambridge Univ. Press.

GEERTZ, C. (1963) Peddlers and Princes. Chicago: Univ. of Chicago Press.

KAY, P. (1964) "Guttman scale of Tahitian consumer behavior." Southwestern J. of Anthropology 20, 2.

MALINOWSKI, B. and J. DE LA FUENTE (1957) "La economia de un sistema de marcados en Mexico." Acta Antropologica 1, 2. Mexico, D.F.

MARTINEZ RIOS, J. and G. M. DE LUNA MENDEZ (1960) Efectos Sociales de la Reforma Agraria en al Ejido de Guelavia, Estado de Oaxaca. Universidad Autonoma Nacional de Mexico. Instituto de Investigaciones Sociales. Mexico, D.F.

MINTZ, S. (1959) "Internal market systems as mechanisms of social articulation," in V. F. Ray (ed.) Intermediate Societies, Social Mobility and Communication. Proceedings of the 1959 Annual Spring Meeting of the AES, Seattle.

NASH, M. (1966) Primitive and Peasant Economic Systems. San Francisco: Chandler.

PARSONS, E. C. (1936) Mitla, Town of the Souls and Other Zapateco-Speaking Pueblos of Oaxaco, Mexico. Chicago: Univ. of Chicago Press.

SALISBURY, R. F. (1970) Vunamami. Berkeley: Univ. of California Press.

--- (1962) From Stone to Steel. Melbourne.

SMITH, M. G. (1960) "Social and cultural pluralism." Annals of the New York Academy of Sciences 83: 763-777.

SPORES, R. (1965) "The Zapotec and Mixtec at Spanish contact," in G. Willey (vol. ed.) Handbook of Middle American Indians, Vol. 3, Part 2. Austin: Univ. of Texas Press.

TAX, S. (1963) Penny Capitalism. Chicago: Univ. of Chicago Press.

WHARTON, C. R., Jr. [ed.] (1969) Subsistence Agriculture and Economic Development. Chicago: Aldine.

WOLF, E. R. (1966) Peasants. Englewood Cliffs, N.J.: Prentice-Hall.

--- (1957) "Closed corporate peasant communities in Mesoamerica and Central Java." Southwestern J. of Anthropology 13, 1.

--- (1956) "Aspects of group relations in a complex society." American Anthropologist 58: 1065-1078.

7

Political Leadership in Poor Communities: Applications of Two Sampling Methodologies

WARREN D. TenHOUTEN
JOHN STERN
DIANA TenHOUTEN

□ PRESENTED IN THIS PAPER are two sampling methods for the study of informal political leadership. These methods are utilized in a study of four urban areas, including two ghettos, a barrio, and an Anglo slum. The two methods developed for this study—site sampling and snowball sampling—are used to analyze relationships between informal leadership structures, and political values and attitudes in a Community Action Program evaluation.

SAMPLING METHODOLOGY FOR THE STUDY OF A COMMUNITY'S INFORMAL LEADERSHIP

Leadership structure can be studied either from the top or the bottom of a community's social hierarchy. The method of studying leadership from the top involves a selection of informants to identify

---

AUTHORS' NOTE: *This research was supported by a contract, B00-5122, with the U.S. Office of Economic Opportunity. Many helpful suggestions were provided by Laurence Barss and Stanley F. Greenberg of Barss-Reitzel & Associates, Robert Burns of TransCentury Corporation, Joan Laurence of*

leaders, presuming the designated persons are leaders, and then studying them through interviews or other techniques. This, the "reputational technique," has been used by Hunter (1953), Barth and Abu-Laban (1959), and Jennings (1964). Hunter had the heads of civic organizations provide lists of top leaders and then had a panel of fourteen judges select the leaders from the resulting corpus of names. From the standpoint of statistical inference, this method provides no rationale for selecting organization heads or judges. Further, there is no method for deciding if the resulting top leaders lead, or that there exists any structure of leadership. Barth and Abu-Laban's study of a black community followed much the same procedure, as they employed persons in organizations, whom the researchers knew personally, to identify judges, who, in turn, identified leaders. Jennings (1964) used a variant of the reputational technique, with results no more replicable than those of Hunter or Barth and Abu-Laban.

The shortcomings in the reputational technique are not easily overcome, but it is clear that an adequate and replicable methodology must involve (1) random selection of judges or informants, and (2) a model for reaching a statistical decision about the existence of leadership in the community.

If informants are selected at random, it is probable that the leadership structure will not be accurately delineated. Persons choosing leaders may be unaware of formal or informal political processes and leadership. Respondents are likely to designate persons visible in the media, or those involved in community service organizations, or who hold appointive or elective office. Such designations are thus influenced by the media's perceptions of what constitutes the "respectable" leadership of a community. In the black communities, such designations are likely to focus on middle-aged ministers, officials in service organizations, and so forth—persons who, in other words, are likely to be politically conservative, nonmilitant, and nonradical. In sum, the lack of information about leadership processes from the bottom of the social

---

Applied Research Enterprises, and Gerald McWorter of Spelman College. The study is especially indebted to Deanna Beeler and Diane Dimperio of UCLA, who were responsible for implementing the sampling design in the field. Data analysis was carried out by Lillie Durr, Donald Long, and Kenneth Ong of UCLA, and editorial assistance was provided by Francoise Kendall and Karen Mitchell of UCLA.

hierarchy limits the value of such randomly selected persons as informants.

The following example of a community leadership structure suggested an alternative method. In an informal study of migrant farm laborers, workers were observed to have been radicalized to some extent, and demands were being made on the growers. It was hypothesized that a nationally known organizer who had recently visited the area had radicalized these workers. Interviews with the workers indicated, however, that they did not, in general, identify this person as the source of personal leadership. Instead, they named a local grocer who, in turn, identified the known leader. Thus, the leader led the workers, not directly, but in a two-stage process of which workers were not cognizant. Obviously, reputational techniques of studying leadership would not have uncovered this leadership structure.

SNOWBALL SAMPLING IN BIASED NETS

It is possible to employ methodologies for the study of community leadership which do not require selection of informants, which make no assumption concerning the existence or identity of any leaders, and which are sensitive to detecting a multiple-stage structure of the type illustrated in the farm worker study.

The statistical method consists of two discrete stages. First, an initial sample of community residents is chosen through some random process (here, site sampling). Second, persons initially sampled are asked to identify others who speak for them. The persons thus identified as leaders are interviewed, in turn, and asked to identify those who speak for them. These persons are then interviewed, the process continuing until an adequate view of the community's leadership structure is generated. This methodology is called *snowball sampling.* It was developed by Goodman (1961) and has been used to study political radicalism (Trow, 1957), diffusion of medical innovation (Menzel and Katz, 1955-1956; Coleman et al., 1966), attributes of arrestees in the Watts riot (Tomlinson and TenHouten, 1970), and informal leadership among naval aviation cadets (Hollander and Webb, 1955).

An important feature of snowball sampling is that it provides a basis for making an inference to a network of interpersonal relations. The persons selected for the zero-stage snowball sample (a subset of the initial random sample), plus all the persons designated as leaders by these persons, who constitute the one-stage sample, plus the persons designated as leaders by the one-stage sample, and so on, constitute the snowball sample. If these n persons are listed in the same order to make an n x n matrix, showing all choices, this matrix can be used to estimate properties of the N x N matrix of the N persons in the total population. Thus, statistical inferences are made about the population matrix from the sample matrix. This is an unusual type of inference in social research, as it assumes the universe of content is not a sample of N individuals, but a matrix of relationships between these individuals. This differs from the usual procedure of sampling individuals (or other elements) as the unit of analysis, which is based on a politically conservative assumption, namely, that social reality is comprised of an aggregate of individuals, as opposed to a social system. This methodology-in-use contributes to a view of society consisting of relatively autonomous individuals.

A number of questions were pretested to elicit leadership designations. The form of the questions, of course, operationally define informal[1] political leadership. The following question was used:

> Do you know anyone like yourself who cares about what's happening in this community? I already know most of the community leaders, but I need your help in getting the opinions of people *you* feel know what's happening.

If this did not elicit a name, the interviewer added:

> Can you think of anyone around here who seems to know a lot of people in the area, someone who talks to a lot of people and gets in on things that are going on?

If this was not understood, the interviewer added:

> Someone who usually knows things like who is moving in and out of the neighborhood, who has jobs and who needs jobs, and how people can get help for their problems.

Almost all responses produced names of persons who were known personally rather than by reputation. When asked how they knew the designated persons, typical answers were: "Personal friend"; "You know. You hear about these things in the neighborhood"; "I have seen him a couple of times."

When asked about their leaders' activities, respondents indicated these persons are involved in political activities, rather than just being friends. Greenberg (1969: 29) cites some examples:

> We had a rent strike and [he] was trying to get the rent changed.
> She talked to me about coming to some community meeting.
> Sold garbage cans below the market price.
> Bringing information back to the residents.
> Working for us on a highway dispute.

The data for the study indicate that leaders are more active in talking about community issues than are nonleaders. All respondents were asked: "Do you talk about community issues with other people around here?" For the site samples (comprised largely of nonleaders), it is estimated that fifty-two percent of residents in the target areas would answer "often" or "sometimes" to this question, as opposed to seventy-nine percent of persons in the snowball samples who are named at least once as leaders.[2]

STATISTICAL INFERENCE

Two basic types of statistical inference can be made from a snowball sample:

(1) A statistical inference can be made about the zero-stage probability sample to the entire population or stratum from which persons are drawn. This sort of inference can be made without snowballing, as only persons in the zero-stage are used in inference. Any such random sample can be thought of as a special case of a snowball sample with zero stages and one name, so that only individuals in the random sample are interviewed (Goodman, 1961: 150). It cannot be expected that persons in higher-stage samples are representative of the entire population. They are different in the trivial sense that they are identified as leaders. They should be

different than nonleaders, substantively, in socioeconomic status, political values, and so forth.

(2) A statistical inference can be made to the population from which stages zero through s are drawn. This is essentially an inference to the matrix of all relationships for the population. Any entry φ_{ij} of this matrix is "1" if person i names person j among his k choices, and "0" otherwise. If i = j, the entry is set at "0"; i.e., it is illegal for persons to choose themselves. The entire matrix can be known only through interviewing the entire population. But a number of properties of the structure can be estimated without bias through the theory of snowball sampling. It is possible to use a sample a-matrix to estimate the chance expectancy for parameters such as:

(a) the total sample size, given k and s;

(b) the distribution of sociometric status (choices received) in the matrix;

(c) the distribution of personal influence in the matrix (an individual's personal influence is defined as the total number of individuals who name him, or who name individuals who, in turn, name him, and so on, until all steps are considered);

(d) the number of reciprocated choices;

(e) the number of "cycles" of a given length;

(f) the number of s + k person, s-step, k-direction relations.

Such measures of structure lead to a decision about the existence of a friendship structure or a leadership structure. If it is decided that no leadership structure exists, then no further analysis is justified. In such a case it would be concluded that: (1) there is no leadership structure, or (2) there is a leadership structure, but the field operations and methodologies employed did not measure it.[3]

If it is decided that there is a structure, then the structure can be described in terms of its gross statistical properties, its more detailed properties (by the theory of linear graphs), and by its organizational connections and prevailing attitudes. It can then be related to other sets of variables, processes, and social structures. The methodology used for testing the leadership hypothesis derives from the theory of biased structures (nets). It is described, along with data from the four-city study, in Appendix B. It is sufficient here to note that evidence for the existence of leadership was found in all four cities.

SAMPLING RESIDENTS IN POOR URBAN AREAS

The communities for this study were selected from the National Opinion Research Center's Permanent Community Sample of urban areas (Rossi and Crain, 1968). From a sample of fifty cities, each having Community Action Programs in poor target areas, a sample of five were selected with probability of selection proportional to city size. Within each selected city, one neighborhood center jurisdiction from an OEO Community Action Program (CAP) was chosen by simple random sampling. These areas are called target areas by the CAP and will be called that here. The data pertaining to leadership from one community were not usable, so four target areas in four cities are used. The method for selecting cities provides a limited basis for inference to poor urban areas within cities in the United States. Of course, any such generalizations are highly tentative, as they employ a sample size of only four. It will be shown that these cities exhibit considerable variations in informal leadership structure and political values.

The selected target areas are named Ghetto A, Ghetto B, Barrio, and Slum. Ghettos A's and B's populations are predominantly black, Barrio is predominantly Mexican-American, and Slum is predominantly Anglo. Because these areas, as well as the samples in this study, are nearly uniform in ethnic composition, explicit control of the ethnicity of sampled individuals is redundant and not included. The black communities were larger in population than nonblack communities (the excluded city was a smaller black area). Size of the target areas of these cities as defined by the CAPs are: Ghetto A, 86,200; Ghetto B, 100,300; Barrio, 12,800; Slum, 10,000 (see Table 12). The names and population sizes of the cities are not given, to respect the confidence of the respondents.

Within the selected target areas, there was considerable variability in the incomes of individuals and in home values. To reduce this variability, and produce samples that consist mainly of poor persons, subsets of all city blocks containing fifty percent or more deteriorating or dilapidated houses were designated as the sampling frames for the target areas. Since the census data used to determine housing conditions were gathered eight years prior to the 1968 surveys, some coding of blocks was carried out in the field.

In community surveys, individual persons or social units such as families are the usual units of analysis. If a list of all persons in a target area is available, a simple random sample of persons can be drawn. In such situations it may also be possible to use control beyond or through stratification. But in situations where such lists are not available, persons or groups are ordinarily located through dwelling units (Ross, 1963: 27) or clusters of dwelling units (DUs), which can be constructed from city blocks, census tracts, or other primary sampling units. City block cluster sampling is a widely used method in community surveys.

Such methods assume that for each person or other unit there corresponds exactly one DU. Such an assumption is probably justified in a middle-class white suburb. In Calcutta, however, this procedure would produce enormous biases, for the higher classes are apt to live in DUs, while the lower classes are more likely to inhabit the streets. In poor urban areas of the United States, a situation intermediate between these exists. As a result, sampling of residential areas produces bias in that persons not attached to DUs have a reduced chance of being sampled. Further, there are individuals or groups who do not have, or do not wish to reveal, connection to particular families or DUs. The U.S. Census Bureau hypothesizes the urban black population to have been underestimated by about twenty percent in the 1960 census. Black leaders have suggested this underenumeration to be closer to forty percent. It can be argued that much of this underestimated population consists of blacks who are not locatable through DUs—the conventional sampling method used by the Census Bureau.

Further, sampling methods which result in systematic under-sampling or underweighting of any group yield data which may be distorted. Individuals least associated with housing units may include those persons most involved in informal leadership—persons who are militant, or who are alienated from the political system. A sampling design inclusive of such groups has political consequences. Definitions of nonresidential groups as a special stratum, and pursuit of an inclusive sampling design are our objective. An appropriate sampling methodology is outlined in the following section.

SITE SAMPLING

If it is not feasible to sample persons without large biases through selection of DUs, it is necessary to look for persons where they are, when they are there. This can be done by dividing the target population with regard to variables pertaining to place and time. The sites were selected according to three criteria.

First, as discussed above, if persons are not associated with residences or residential areas, then they may be associated with commercial areas.

Second, persons may be either inside of structures or outside, on the streets. Whyte describes a class of men who frequented "outside commercial" sites, and defines "corner boys" as those "who center their social activities upon particular street corners, with their adjoining barber shops, lunchrooms, pool rooms, or clubs" (Whyte, 1955: xviii). Research by Liebow (1967) and Clark (1967: 27) suggest that in black ghettos residents are concentrated in commercial areas and on streets.

Third, persons may be sampled during both the day and night, so that persons available at only one time would not be omitted. The following sites were chosen by the first author and by Professor Gerald McWorter of Spelman University:

(a) commercial inside day (CID)
(b) commercial inside night (CIN)
(c) commercial outside day (COD)
(d) commercial outside night (CON)
(e) residential inside day (RID)
(f) residential inside night (RIN)
(g) residential outside day (ROD)
(h) residential outside night (RON)

To estimate population characteristics from such a sample, it is necessary to estimate: (1) the number of persons at each site; (2) the total population size; and (3) the between-site mobility of the persons who ordinarily frequent each site; i.e., the way the persons whose interviews were initiated at each site normally distribute themselves over the eight sites.

This study produced some empirical outcomes suggesting that biases resulting from block cluster sampling may derive from problems other than those occurring in the field. Persons located in commercial areas, who are outside of residences at night, reported that they were only infrequently available in the RID and RIN sites: no amount of call-backs would locate these persons. The RID and

RIN interviews tended to generate respondents who—relative to persons located at other sites—were not involved in informal community leadership. The overall evaluation of the data from the site samples produces a picture of black and poor communities being more militant, alienated, and organized, than do block cluster samples.

Site sampling also deals directly with sampling biases created by movement within the community, and movement in and out of the community. Conventional residential sampling methods assume not only that individuals are associated with housing units, but also that persons ultimately interviewed in this manner do not differ significantly from those never located or interviewed. Consequently, those individuals who spend much time in the community but almost no time in housing units are undersampled, as are those persons who carry out many activities beyond the boundaries of a community. Membership in a single community is, after all, a function of the activity that defines community.

The methodology for estimating parameters from a site sample involves a system of weighting the means and variances for each site. This methodology is described in Appendix A. In addition, operational definitions for the commercial-residential, inside-outside, and day-night criteria are given.

One consequence of site sampling can be seen in a comparison of the distribution of community persons and community leaders across sites. Since the estimates of the number of persons at each site determines the number of persons interviewed at the site, the percentage distributions of the site sample across the eight sites estimates the distribution of the community across the sites (the site samples contain the zero-stage snowball samples). The leaders within each snowball sample are defined as the persons who are named as leaders by one or more persons. The percentages of the site sample members and leaders whose interviews are initiated at each site are shown in Table 1.

The data show that in all four target areas, leaders are interviewed at different sites than are members of the site samples. In each target area the site sample members are concentrated in the RID and RIN sites (the CID and CIN sites contain thirty-three percent of the site samples in Ghetto A). These residential sites contain few leaders: in every target area, the CIN site contains the greatest percentage of leaders. The strongest variable differentiating the site sample from the leader sample is commercial-residential. The percentages of the site and leader sample members in commercial areas are Ghetto A 39 and 80, Ghetto B 11 and 90, Barrio 14 and 95, and Slum 6 and 91. These large differences suggest residential sampling may contribute to biases through the underrepresentation of persons who are leaders and are more apt to be sampled in commercial areas.

COMMUNITY LEADERSHIP STRUCTURES

There are a number of variables useful for describing sociometric structures. Frequently used variables are sociometric status, the number of reciprocated choices, and the number of cycles of a given length. The relevance of these variables depends on the relationship in question. If designations of friends are asked, certainly the number of reciprocations and of cycles are important. Friendship is ordinarily a symmetric relationship, thus symmetric configurations can be expected. But leadership is in general asymmetric and at least partially transitive. If A chooses B as his leader, and B chooses C, B would not be expected to choose A, and C would not be expected to choose either A or B. Thus, neither reciprocations nor cycles are expected to occur.

TABLE 1
Site Sample at Each Site and Persons With Leadership Status at Each Site by Target Area (in percentages)

	Ghetto A		Ghetto B		Barrio		Slum	
Site	Site Sample	Leaders	Site Sample	Leaders	Site Sample	Leaders	Site Sample	Leaders
CID	20	17	5	1	3	17	3	6
CIN	13	36	6	46	6	38	2	55
COD	4	17	0	24	4	24	1	18
CON	2	10	0	19	1	16	0	12
RID	40	16	48	9	43	5	54	3
RIN	21	4	41	1	40	0	40	3
ROD	0	0	0	0	2	0	0	3
RON	0	0	0	0	1	0	0	0
Total	100 (n=270)	100 (n=70)	100 (n=252)	100 (n=75)	100 (n=156)	100 (n=58)	100 (n=184)	100 (n=35)

Sociometric status (SS) is certainly an important dimension of leadership, and is one measure of leadership status. But it is limited in its usefulness by the fact that it does not take into account the status of the person(s) choosing. For example, it should contribute more to a person's leadership status to be chosen by three persons who themselves have leadership status than by three persons lacking such status.

A measure of leadership status which takes into account the status of those persons who are choosing is given by Goodman (1961: 158-159). This variable is called personal influence (PI) and is defined as follows:

> A person's personal influence score is the total number of individuals who name him, or who name individuals who name him . . . until all steps are considered.

Given these two measures of leadership, a typology of leadership statuses can be constructed. Nonleaders lack both SS and PI; social leaders are accessible to persons who themselves lack status, so they are high on SS and low on PI; influential leaders are less accessible to persons without status, but are chosen by persons with status, so they are low on SS and high on PI; elite leaders are high both on SS and PI. Appropriate operational definitions for these four statuses depend on the number of designations made by each person and the distributions of SS and PI in the matrix. For the one-name snowball samples used in the present study, the following definitions are used:

(1) nonleaders receive no choices (SS = 0 and PI = 0);

(2) social leaders receive one or more direct choices, but no indirect choices (SS $\geqslant$ 1 and SS = PI);

(3) influential leaders receive one direct choice and more than one indirect choice (SS = 1 and PI $\geqslant$ 2);

(4) elite leaders receive two or more direct choices and have indirect choices as well (SS $\geqslant$ 2, PI > 2, and PI > SS).

Apart from the substantive advantages of this typology, it is a statistical necessity given the high level of dependence between sociometric status and personal influence. For the four target areas combined the level of ordinal association (gamma) between the variables SS and PI is .96 ($z = 30.1$, $p < .001$, $n = 516$). In addition, there are dependencies between SS and leadership type, and between PI and leadership type.

All sociometric substructures which contain persons with PI scores of six or more have been diagrammed. The most complex structure is found in Ghetto A, where the statistical analysis (Table 14) indicates the highest level of community leadership structure. There are two such substructures in Ghetto A, Ghetto B, and Barrio, and there is one in Slum.

One of the large substructures, from Barrio, is shown in Figure 1. In this figure, persons A through L are zero-stage nonleaders. The three persons receiving more than one choice tend to be chosen by

Sociometric Status, Personal Influence Score, and Leadership Types (Identification: SS, PI, LT*) for Members of a Snowball Leadership Sociogram: Barrio Snowball Sample

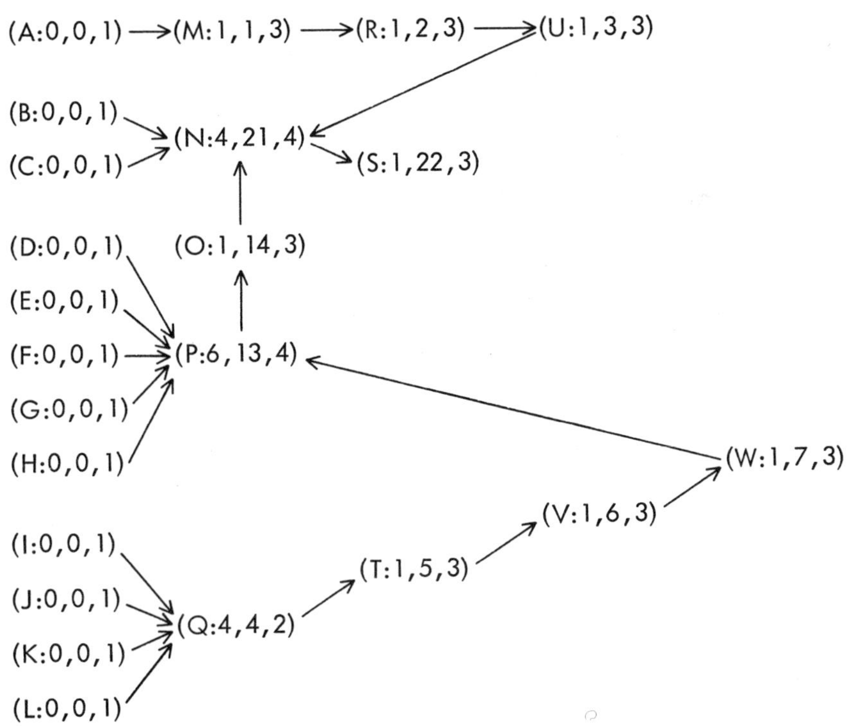

* Non-leaders 1, social leaders 2, influential leaders 3, elite leaders 4.

Figure 1.

TABLE 2
Distribution of Leadership Statuses by Target Area: Snowball Samples (in percentages)

Status	Ghetto A	Ghetto B	Barrio	Slum
Nonleaders	45	30	43	47
Social leaders	25	26	21	32
Influential leaders	25	38	30	17
Elite leaders	5	6	6	4
Total	100 (n=163)	100 (n=110)	100 (n=112)	100 (n=110)

TABLE 3
Nonleaders (NL), Social Leaders (SL), Influential Leaders (IL), and Elite Leaders (EL) Interviewed at Each of Eight Sites by Target Area: Snowball Samples (in percentages)

	Ghetto A				Ghetto B			
Site	NL	SL	IL	EL	NL	SL	IL	EL
CID	1	0	32	33	3	0	45	14
CIN	1	0	23	0	0	0	26	57
COD	8	31	3	17	3	0	3	0
CON	7	56	16	33	8	100	10	0
RID	51	10	26	0	37	0	13	29
RIN	32	3	0	17	49	0	3	0
ROD	0	0	0	0	0	0	0	0
RON	0	0	0	0	0	0	0	0
Total	100 (n=73)	100 (n=32)	100 (n=31)	100 (n=6)	100 (n=37)	100 (n=30)	100 (n=38)	100 (n=7)
	$\chi^2(15)=120, p<.001$				$\chi^2(15)=156, p<.001$			

	Barrio				Slum			
	NL	SL	IL	EL	NL	SL	IL	EL
CID	7	9	42	0	4	0	46	0
CIN	2	5	28	0	0	0	23	50
COD	0	18	17	14	4	6	8	0
CON	18	68	3	86	9	89	15	0
RID	28	0	10	0	48	0	8	0
RIN	39	0	0	0	34	0	0	50
ROD	4	0	0	0	0	6	0	0
RON	2	0	0	0	0	0	0	0
Total	100 (n=46)	100 (n=22)	100 (n=29)	100 (n=7)	99 (n=44)	101 (n=18)	100 (n=13)	100 (n=6)
	$\chi^2(21)=102, p<.001$				$\chi^2(18)=97, p<.001$			

nonleaders: eleven of their fourteen choices come from the zero-stage. Person Q is a "social" leader, receiving his four choices from nonleaders; N and P are "elite" leaders, being chosen by both nonleaders and leaders, and having personal influence in excess of their sociometric status. The other eight leaders are "influentials," receiving all their choices from persons who are themselves chosen.

In the large sociograms there are altogether five persons receiving four or more choices: of the thirty-nine designations of these leaders, thirty-four come from zero-stage nonleaders. Also, respondents with high PI scores tend to be chosen by persons in the higher stages who themselves have status: these persons have high PI scores and low SS scores. Table 2 shows that leaders are comprised mostly of two types—social and influential. The number of elite leaders is small in every target area. The leadership structure in Slum differs from the other three areas, as there are more social than influential leaders.

SITE AND LEADERSHIP STATUS

Table 3 shows that leaders are concentrated in commercial sites, and that distributions differ across sites for site sample members and leaders. Within the snowball samples, site is related to sociometric status, to personal influence, and to leadership type. The leadership typology provides the clearest view of the relationship between site and status. The data show a strong relationship between site and leadership type in every target area. Further, there is a cross-target-area consistency in the sites occupied by nonleaders, social leaders, and influentials.

(1) In all four target areas, nonleaders are concentrated in the residential inside sites (RID and RIN): the percentages of nonleaders in these sites are Ghetto A 83, Ghetto B 86, Barrio 67, and Slum 82.

(2) In all four target areas, social leaders are most apt to be found and interviewed in the commercial outside sites (COD and CON): the percentages of social leaders in these two sites are Ghetto A 87, Ghetto B 100, Barrio 86, and Slum 95.

(3) In all four target areas, influentials are most apt to be found and interviewed in the commercial inside sites (CID and CIN): the percentages of influential leaders in these sites are Ghetto A 55 (with twenty-six percent in the RID site), Ghetto B 71, Barrio 70, and Slum 69.

(4) The elite leaders represent only a few persons, and their distributions could have arisen by chance alone; no generalization about these leaders can be made from these data.

The sites at which members are found in stage one and above in the snowball samples in part reflect field procedures used by interviewers to locate designated leaders. It is of interest that the interviewers (who employed a wide range of tactics) were able to find social leaders in commercial outside sites and influential leaders in commercial inside sites. But this does not imply that the leaders spend most of their time in these sites. Independent information was obtained on the availability of these leaders, as all respondents were asked where they would have been during the day and at night on the day preceding their interview. The following questions were asked:

(1) Suppose I had tried to interview you yesterday at about this time; where could I have found you? (Probe for inside-outside, residential-commercial.)

(2) Suppose I had tried to find you yesterday (in the afternoon/at night, where were you then? (Probe for inside-outside, residential-commercial.)

Each person in the sample is thus asked to provide one location for day and one for night on two different occasions.

Since the results in Table 3, and in response to these two questions, do not vary much by target area, the data in Table 4 are shown for the four areas combined. The outcomes are generally

TABLE 4

Nonleaders, Social Leaders, and Influential Leaders Reporting That They were Available at Various Sites on the Day Preceding Their Interview: Snowball Samples (in percentages)

Sites	Nonleaders	Social Leaders	Influential Leaders
Commercial inside	11	48	80
Commercial outside	2	32	10
Residential inside	87	20	10
Total	100 (n=194)[a]	100 (n=69)[a]	100 (n=28)[a]

a. Since each respondent can provide one response for day, and one response for night, the number of responses is greater than the number of respondents.

coordinate with those in Table 3. In the nonleader group, eighty-seven percent of the responses indicate availability in the residential inside sites. The modal response of both commercial and influential leaders are commercial inside, but the concentration in the commercial inside sites is substantially greater among influential leaders.

The concentration of social leaders in commercial outside sites, and of influential leaders in commercial inside sites, is consistent with substantive interpretations of social and influential leadership. Social leaders, chosen by nonleaders, are the visible opinion leaders on the streets of poor urban communities. Their turf is the street, and they are not as apt to be found in business establishments and organizations. Influential leaders, on the other hand, are not visible to nonleaders, and are chosen by social leaders. They are less apt to frequent the streets, and are more likely to be involved with business enterprises and organizations.

The two types of leaders were not found to differ by age and sex characteristics. When the leaders were categorized by age and sex, the distributions differed little from area to area: the percentages for the four areas are younger males 16, younger females 10, older males 42, and older females 32.

CHARACTERISTICS OF LEADERS AND NONLEADERS

Table 5 shows that leaders have somewhat higher socioeconomic status than do nonleaders. In each area, leaders rather than nonleaders are more apt to report they have white-collar occupations: however, the differences are significant only in Ghetto B and Barrio. The outcomes for education show a significant difference in Barrio, where sixty-four percent of leaders and thirty percent of nonleaders are high school graduates.

TABLE 5

Leaders and Nonleaders with White-Collar Occupations, High School Educations, and Organizational Membership by Target Area: Snowball Samples (in percentages)

	Ghetto A		Ghetto B		Barrio		Slum	
Characteristic	**Lea.**	**Nonlea.**	**Lea.**	**Nonlea.**	**Lea.**	**Nonlea.**	**Lea.**	**Nonlea.**
White-collar	44 (n=77)	36 (n=56)	47 (n=89)	16 (n=38)	55 (n=58)	26 (n=42)	30 (n=79)	22 (n=54)
High school graduate	82 (n=86)	86 (n=65)	81 (n=55)	80 (n=40)	64 (n=53)	30 (n=43)	39 (n=80)	42 (n=53)
Belong to organization	30 (n=93)	20 (n=73)	36 (n=98)	26 (n=43)	45 (n=64)	34 (n=47)	35 (n=86)	19 (n=57)

Respondents were asked to name important groups or organizations comprised primarily of community residents. Persons able to identify one or more organizations were asked if they were organization members. Table 5 shows the percentages of those who report membership. In all four areas, leaders rather than nonleaders are more apt to belong: however, the difference is significant only in Slum.

As suggested by the above data on the age-sex composition of social and influential leaders, males are more apt to be leaders than are females. This difference is significant only in Barrio, where eighty percent of the leaders (n = 75) are male, as compared to fifty-one percent of the nonleaders.

COMMUNITY LEADERSHIP STRUCTURES AND POLITICAL VALUES

In this section, informal leadership status is related to sampling site and to militancy and radical ideology. The analysis is carried out for the snowball samples. In addition, responses of (site sample) residents are compared to (snowball sample) leaders for selected questions pertaining to political values and attitudes.

Sociometric configurations generated through snowball sampling provide a vehicle for the analysis of the diffusion of political beliefs, attitudes, and behaviors. Leadership choices are directed upward from low- to high-status persons. In general, political leadership should work in reverse; i.e., downward from persons with higher-leadership status to those having lower status. In black and Chicano communities in the United States that are becoming politically mobilized, it might be expected that leaders will be more militant and radical than nonleaders. In Anglo slum areas, political ideologies are currently more conservative, and residents of these areas regard cooperation with existing allocators of socioeconomic and political resources as a viable strategy.[4] It is probable that radical and militant leadership prevails in a higher proportion of poor ethnic communities than in poor nonethnic communities.

MILITANCY, RADICAL IDEOLOGY, AND LEADERSHIP STATUS

This section describes measures of militancy and radicalism, and relates these variables to sampling site and to leadership type.

Intercity comparisons are made between leaders and nonleaders; within the leadership group, social and influential leaders are compared.

The militancy index is constructed from responses of snowball sample members to the following questions:

(1) Some people say the police rough up people unnecessarily when they are arresting them or afterwards. Do you think this happens to people in this neighborhood?

(2) If a disturbance like the one that occurred after the assassination of Martin Luther King broke out here, do you think you would join in, would you try to stop it, or would you stay away from it?

(a) Even if you didn't join, would you feel in sympathy with people who did join?

Respondents were asked to give a yes or no response to this item, the yes being scored high on militancy. Respondents who would join a disturbance, or who would feel in sympathy with a disturbance, are scored high for their responses to 2 and 2a. Inclusion of the second part of this question makes the item meaningful for persons who ordinarily would not participate in a disturbance due to their age or sex: urban riots are dominated by young men, but women and older men could be highly militant and support such activities without participating themselves.

The final measure scores persons giving the militant response to both items as high on militancy, and low to persons giving the militant response to one or no items. Thus, operationally, respondents are defined as militant if they perceive police as brutal, or would join in or sympathize with a riot. This variable may have more salience in the black and Chicano areas than in Slum, but all sampled groups were able to interpret the questions. Responses to these items did not form a Guttman scale.

Radical ideology is measured by responses to three questions:

(1) Do you think all people have the right to have enough money to live on, whether they are working or not?

(2) How would you feel if everyone got paid the same amount of money, no matter what his job was? Do you think it would be good or bad?

(3) Do you think that everyone in this country has a right to have a nice home even if he doesn't earn much money?

TABLE 6

Respondents High on Militancy and Radicalism by Sites: Snowball Samples (in percentages)

Sites	Militancy		Radicalism	
Commercial inside	47	(n=217)	51	(n=238)
Commercial outside	45	(n= 97)	56	(n=104)
Residential inside	38	(n=554)	52	(n=627)
	$\chi^2(6)=6.76$, p .05		$\chi^2(2)=.77$, n.s.	

Respondents are given high scores for a yes to 1, good to 2, and yes to 3. The respondents who scored high on 2 or 3 items are scored high on radicalism, and other are scored low. The items form a Guttman scale with a coefficient of reproducibility of .96, a minimum marginal reproducibility of .70, and a percent improvement of 26.[5]

Table 6 shows the percentages of snowball sample members interviewed in commercial inside, commercial outside, and residential inside sites who are scored high on militancy and radicalism. Militants are more apt to be associated with the commercial than the residential inside sites; however, these differences are small and barely significant. Persons with radical ideology are slightly more concentrated in the commercial outside sites, but the differences are not statistically significant. Thus, the data show weak dependencies between site of interview, militancy, and radical ideology. These differences are in the direction suggested in the discussion of sampling biases resulting from disproportionate sampling of persons associated with residential sites.

Table 7 shows the data relating leadership statuses to militancy and radical ideology. Since the sample sizes for elite leaders are small (ranging from four to eight) their percentages are not listed separately. There are systematic variations among the target areas. In Ghetto A, leaders are higher on militancy and radical ideology than are nonleaders; in Ghetto B, however, the obverse is the case: nonleaders are more militant than leaders. In Barrio, as in Ghetto A, leaders are more militant and higher on radical ideology than nonleaders. In Slum, small percentages of leaders and nonleaders are militant. As suggested, riots and confrontations with police are foreign to this community's experience, and such involvements are not criteria for leadership.

Table 7 also shows that within the leadership group, influentials are more apt to be militant and have radical ideology than are social leaders. The difference is significant only in Barrio, where influentials are twenty percent more apt to be militant, and twenty-one percent more apt to be radical. This differentiation within the leadership types of Barrio extends to a number of other variables to be discussed, but these differences will not be shown.

In each city, there are variations in the political values and behaviors of members of differing substructures, which, to some degree, obscure between-area differences. These within-leadership variations are not shown here (for an analysis of the substructures, see Greenberg, 1970).

In addition to the questions pertaining to militancy and radical ideology, information was obtained regarding a wide variety of political values and attitudes. Responses to these questions by members of the site samples and leaders in the snowball samples (receiving one or more choices operationally defines a leader) makes it possible to compare probability samples of residents and their leaders. The percentages shown for residents are derived from the site samples. Percentages for each of the eight sites are first computed, and estimates of the target area parameters then constructed with the weights shown in Table 11.

A small proportion of target-area residents are leaders, so the distributions of residents closely resemble those of nonleaders in the

TABLE 7
Nonleaders and Leaders High in Militancy and Radicalism by Target Area and Social or Influential Leadership Type: Snowball Sample

	Ghetto A		Ghetto B		Barrio		Slum		Total	
	%	(n)	%	(n)	%	(n)	%	(n)	%	(n)
Percentage Militant										
Nonleaders	45	(53)	53	(32)	29	(38)	11	(37)	35	(160)
Leaders	62	(73)	37	(78)	55	(55)	15	(41)	45	(247)
Social	62	(34)	39	(28)	42	(19)	9	(23)	40	(104)
Influential	58	(33)	35	(43)	62	(29)	28	(14)	47	(119)
Percentage Radical										
Nonleaders	50	(56)	73	(37)	28	(43)	37	(46)	47	(182)
Leaders	65	(83)	71	(88)	53	(58)	42	(53)	63	(282)
Social	73	(38)	70	(33)	45	(20)	43	(30)	60	(121)
Influential	78	(37)	67	(48)	66	(32)	47	(19)	67	(136)

TABLE 8

Residents and Leaders Perceiving Domestic and Foreign Repression by Target Area

	Ghetto A		Ghetto B		Barrio		Slum		Total	
	Res.	Lea.	Res.	Lea.	Res.	Lea.	Res.	Lea.	Res.	Lea.
Items	% (n)	% (n)	% (n)	% (n)	% (n)	% (n)	% (n)	% (n)	% (n)	% (n)
Concentrations camps[1] (yes)	44 (151)	60 (42)	25 (189)	21 (61)	12 (111)	36 (44)	31 (105)	17 (23)	29 (722)	34 (170)
Assassination plots[2] (yes)	88 (188)	93 (58)	78 (208)	86 (57)	79 (124)	74 (49)	77 (122)	77 (26)	81 (820)	84 (190)
Police brutality[3] (yes)	68 (245)	66 (59)	69 (218)	63 (68)	44 (112)	80 (51)	36 (156)	37 (27)	57 (925)	65 (205)
Vietnam a mistake[4] (yes)	85 (191)	88 (51)	74 (216)	71 (63)	62 (128)	66 (53)	58 (139)	60 (20)	72 (858)	73 (187)

1. "There have been some rumors about concentration camps being set up in this country. Do you think there's any truth in that?"
2. "Some people say that one or more of the recent assassinations—of the Kennedys and Dr. King—were part of a plot. Do you think there is any truth in that?"
3. "Some people say the police rough up people unnecessarily when they are arresting them or afterwards. Do you think this happens to people in this neighborhood?"
4. "Do you think the United States was right or wrong to send troops to fight in Vietnam?"

snowball samples. The percentages of militant and radical responses shown in Table 7, for instance, do not differ significantly from the percentage estimates based on the site samples. This could be expected, as nonleaders in the snowball samples were highly concentrated in the zero-stage snowball samples, which were selected from the site samples by a random process, in which the number of persons selected from each site was directly proportional to the weights assigned to each site.

The data presented are categorized by three dimensions of political values and attitudes. Table 8 describes perceived political repression; Table 9, perceived political participation and powerlessness; and Table 10, attitudes to militant and radical strategies.

POLITICAL REPRESSION

Table 8 shows that about one-third of the residents and leaders (who give definite responses) believe concentration camps are being set up in the United States. This perception is most widespread in Ghetto A, where forty-four percent of residents and sixty percent of leaders agree with the statement. In Ghetto B these percentages—twenty-five and twenty-one—are far lower. In Barrio, as in Ghetto A, the data indicate residents are less apt to believe in the rumors than are leaders: the percentages—twelve and thirty-six—however, are much lower. In Slum, the reverse outcome is obtained: thirty-one percent of the leaders as compared to seventeen percent of the residents agree with the statement.

More than eighty percent of residents and leaders in the four cities believe that one or more of the assassinations (of the Kennedys or Martin Luther King) were part of a plot. These percentages are high in every city: the highest level of belief in assassination plots is found in the black ghetto leadership (ninety-six percent in A, eighty-eight in B). The percentages are lower in the resident and leader samples of Barrio and Slum. Of the respondents who believed there existed a plot, forty-eight percent indicated the murders were intended to repress blacks, thirty percent indicated the intent was to repress the poor, and thirteen percent believed the murders were political. Responses to these two items do not show that blacks, Chicanos, or poor persons have developed elaborate ideologies in opposition to the political system. But they may imply a high level of alienation from, and distrust of, that system.

There exists widespread belief that police rough up people unnecessarily in their own neighborhoods: fifty-seven percent of residents and sixty-five percent of leaders for the four areas. In Ghettos A and B, about two-thirds of residents and leaders perceive police brutality. In Slum—where the actual level of police brutality is undoubtedly lower—about one-third of the residents and leaders believe that police rough up people unnecessarily. For these three target areas, residents differ little from leaders. But in Barrio, there is a pronounced difference, the percentages being residents, forty-four and leaders, eighty.

Nearly three-fourths of both samples believe the United States government was wrong to send troops to Vietnam. This perception is endorsed most in Ghetto A and is least agreed with in Slum. There are no significant differences between residents and leaders, which suggests that leaders identified for their involvements in local issues may not function also as leaders for noncommunity issues. Of course, Vietnam is only one such issue, and there may be other issues external to the community for which the leaders identified here would again be designated. Also, the Vietnam War does, to some extent, involve community organizations and political processes.

POLITICAL PARTICIPATION AND POWERLESSNESS

Residents and leaders were asked about their views on the need for "major changes in the way the country is run," and whether any government officials care about them.

Table 9 shows that major changes are viewed as necessary by a majority of the responding persons in each area. In Ghetto A and Barrio, leaders are somewhat more apt to respond that major changes are necessary than are residents; in Ghetto B and Slum, the reverse outcomes are obtained, in a weaker form.

Responses to the question about city government indicates high levels of perceived powerlessness. Nearly two-thirds of residents and leaders respond that they have little or no say in these decisions: this outcome varies neither by target area nor by leadership status.

A majority of residents and leaders in each area perceive that there are no government officials in their city who care about what they want. Again, this shows a widely shared feeling of political powerlessness among urban poor. For the total sample, the percentages are residents, sixty-one and leaders, fifty. In the three ethnic areas, residents are more apt to make this response; in Slum, the leaders are more apt to make this response.

These data thus show high levels of perceived powerlessness vis-à-vis national and local political entities. There are some differences between residents and leaders, but for these items the differences are small, and indicate the perception that change is necessary or that the poor lack power, is widely shared in poor areas, and does not constitute criteria for leadership status.

DIRECT ACTION POLITICAL STRATEGIES

The relatively conservative Slum community, as shown by the above data, defines political strategies in terms of cooperation with, and participation in, local political processes. In the three ethnic areas, however, direct-action strategies are often perceived as vehicles for bringing about change—both locally and nationally. Data on such strategies for the three ethnic areas are shown in Table 10.

The percentages of respondents disagreeing with the statements that blacks (Mexican-Americans) cannot win their rights without violence is shared by a majority of Ghetto A respondents, by about one-fourth of Ghetto B respondents, and by less than one-fourth of the Barrio respondents. Resident-leader differences are consistently small.

TABLE 9
Residents and Leaders Perceiving Lack of Political Participation and Powerlessness by Target Area

	Ghetto A		Ghetto B		Barrio		Slum		Total	
	Res.	Lea.	Res.	Lea.	Res.	Lea.	Res.	Lea.	Res.	Lea.
Items	% (n)	% (n)	% (n)	% (n)	% (n)	% (n)	% (n)	% (n)	% (n)	% (n)
Changes needed[1] (yes)	59 (238)	78 (60)	67 (219)	60 (67)	64 (135)	82 (55)	62 (140)	59 (27)	63 (886)	71 (209)
Decisions[2] (none or very little)	64 (224)	60 (57)	68 (219)	64 (72)	66 (123)	70 (57)	71 (138)	64 (31)	67 (898)	64 (217)
Officials concerned[3] (no)	64 (191)	40 (52)	66 (176)	57 (53)	60 (85)	45 (38)	58 (100)	67 (24)	61 (707)	50 (167)

1. "Some people say that things will never really get better for poor people without major changes in the way the country is run. Do you think this is true?"
2. "How much say do you think people in this neighborhood have about the decisions made by the city government?"
3. "Are there any government officials in this city who really care a lot about what people in this neighborhood want?"

TABLE 10
Residents and Leaders Endorsing Various Direct Action Political Strategies by Target Area

	Ghetto A		Ghetto B		Barrio		Total	
	Res.	Lea.	Res.	Lea.	Res.	Lea.	Res.	Lea.
Items	% (n)	% (n)	% (n)	% (n)	% (n)	% (n)	% (n)	% (n)
Minority rights[1] (no)	59 (222)	56 (57)	29 (235)	24 (70)	13 (149)	20 (54)	36 (606)	33 (181)
Demonstrations[2] (yes)								
Join in?	26	33	10	3	4	9	14	14
Try to stop it?	17	14	14	22	27	54	18	29
Stay away from it?	57 (206)	53 (55)	76 (231)	75 (68)	69 (141)	37 (54)	68 (578)	57 (177)
Sympathy for demonstrators[3] (yes)	70 (181)	82 (60)	63 (216)	48 (67)	43 (127)	60 (52)	01 (524)	63 (179)
Would you boycott supermarket[4] (yes)	78 (247)	86 (66)	94 (251)	97 (75)	62 (149)	82 (57)	01 (647)	89 (198)
Would threatening manager help[5] (yes)	21 (229)	18 (65)	17 (244)	11 (74)	7 (150)	9 (54)	16 (623)	13 (193)
Would you agree to do it[5a] (yes)	15 (239)	14 (62)	7 (247)	1 (73)	6 (150)	6 (54)	08 (636)	07 (189)

1. "Do you think that black (Mexican-American) people today can win their rights without using violence?"
2. "If a disturbance like the ones that occurred after the assassination of Martin Luther King broke out here, do you think you would: (a) join in? (b) try to stop it? (c) stay away from it?"
3. "Even if you didn't join in, would you feel in sympathy with people who did join?"
4. "If a big supermarket in the neighborhood wouldn't hire blacks (Mexican-Americans), do you think it would help if you stopped shopping there and told all your friends not to shop there—would you do it?"
5. "How about threatening the manager with damage to his store—do you think it would help? (a) would you agree to do it?"

The question about disturbances similar to those occurring in the aftermath of the assassination of Dr. King indicates that only fourteen percent of residents and leaders would join in, that about one-fourth would try to stop such disturbances, and that about sixty percent would stay away. The residents of Ghetto A would turn out in substantial numbers: thirty-seven percent as compared to their leaders' twenty-four percent. In Ghetto B, however, interest in riots is virtually nonexistent. In Barrio, where leaders are consistently more militant than residents, nineteen percent of leaders, but almost no nonleaders, would participate. In this target area, however, there would be active resistance to such an eventuality, and a small majority of both residents and leaders would try to impede it, more than in any other area.

In these three areas the percentages of residents and leaders who would *sympathize* with a riot are far higher—ranging from eighty-two for Ghetto A leaders to forty-three for Barrio residents.

The strategy of boycotting a supermarket is endorsed in all three areas. Support for this activity—in contrast to more violent alternatives—is higher in Ghetto B than in Ghetto A. The percentage who believe that threatening a manager or carrying out the threat would be effective, is far lower—but higher in Ghetto A than in Ghetto B. Again, this suggests Ghetto A's population is more predisposed to militant or violent strategy than that in Ghetto B.

In Barrio, as in Ghetto B, the percentage is small of respondents who perceive that threats to a manager or carrying out those threats would be helpful. The resident-leader difference for these items follows the general pattern—with leaders more apt to give militant responses in Ghetto A and in Barrio, but not in Ghetto B. These differences are not statistically significant, however.

AN OVERVIEW OF THE DATA

Although the data in Tables 8, 9, and 10 do not provide a comprehensive view of community politics, they do describe the general political climate in the areas, and differences between residents and leaders. Ghetto A residents are militant, alienated from the political system, and predisposed to violent strategies. The leadership in this area appears to be more militant than are residents. Ghetto B residents are as militant and nearly as alienated as those of

Ghetto A, but favor somewhat more moderate strategies. Leaders are less militant than residents. Barrio residents are less militant than leaders, and less apt to endorse radical ideology and militant strategies. In this area, influential leaders are more radical, militant, and predisposed to direct action than are social leaders. This difference between leadership type does not occur in the other three areas. In all four areas, residents and leaders share the view that major changes are needed in the way the country is run, and share a perception that they cannot influence decisions, nor are they cared about by politicians.

Thus, there are differences among target areas in political values and attitudes. There are differences between residents and leaders that vary from area to area. Within the leadership, there are differences between social and influential leaders in one area. These outcomes suggest that no universal generalization about either ethnic or nonethnic communities can be made, and that there may be wide variations in the structure of informal political leadership in poor urban communities. Certainly the empirical outcomes underlie the necessity for sampling more than one urban area (for any ethnic or racial group) before generalizing as to urban areas in the United States. For example, if the study had been confined to Ghetto A, it would have been tempting to hypothesize that black leaders are more militant and predisposed to violent strategies than are the people they lead throughout the United States. But the inclusion of Ghetto B indicates such a generalization may be invalid. In an important sense, a study of a community—even with large samples of residents and leaders—still constitutes a sample of one and does not provide a basis for statistical inference.

SUMMARY AND CONCLUSIONS

There are two basic types of samples which can be drawn from communities and organizations. First, samples of elements that make up target populations—such as work groups, families, individual persons, or other social units. Second, samples of the social organization of a target population, from which inferences may be made about networks of relations among component elements. Such networks might consist of informal communication among work groups in a firm; friendship choices in a school social system; leadership relations among persons in a community. Sampling networks may on occasion be more relevant than sampling of the elements themselves. This task, however, receives minimal attention in sampling theory and social research.

The combination of site sampling and snowball sampling provides a vehicle for studying community leadership. Dependencies between the two methods are reflected in substantive outcomes: in all target areas, leaders are concentrated in commercial as opposed to residential areas; among leaders, social leaders are concentrated in commercial outside areas, and influential leaders in commercial inside areas. In addition, there are between-area variations in political values of residents (primarily nonleaders) and leaders.

The snowball sampling methodology can measure leadership structure effectively if the zero-stage respondents are representative of the target population. Here, where site sampling was used, evidence for leadership structure was found in all four target areas, and its internal structure related to other variables. Yet, in a comparative study of three urban areas in which systematic sampling of residents was used, the resulting zero-stage snowball sample did not lead to the statistical decision that leadership even existed (see 3 ff). This illustrates the importance of obtaining samples in a manner which does not assume that all inhabitants of poor urban areas are attached to dwelling units.

Snowball sampling could be used for evaluating the effectiveness of social programs. As mentioned, this was the purpose for which the present study and its methodology were developed. A social program may directly involve only a small percentage of a community's population, and yet have profound consequences in socioeconomic change and political mobilization, via a process of indirect leadership. That is, if some set of residents who are involved in or knowledgeable about a program undergoes changes in political values, attitudes, and activities, and communicates these alternatives to persons it leads, and so forth, eventually the nonleaders may undergo changes in their political values and behavior. This can occur even if *no* resident, whether or not a leader, perceives the informal structure of leadership or his position in it.

This methodology might be used also for organizing communities toward political or economic action. Snowball sampling could be used to generate a set of persons with leadership status who might be approached for possible inputs into a political process. Such sampling may be carried out for organizing residents of a community, or workers in a firm or industry. On the other hand, the methodology could be used by political incumbents to obtain information about informal leadership in order to repress or inhibit such organizing.

APPENDIX A: SITE SAMPLING

ESTIMATES OF PARAMETERS

Consider a parameter such as a population mean μ, that for simple random sampling can be estimated without bias by a sample mean $\bar{X}$. The problem is to develop a sampling scheme so that μ can be estimated without bias from the sample mean for the sites; i.e., a weighting scheme such that the statistic $\bar{X} = W_1 \bar{X}_1 + \ldots + W_K \bar{X}_K$ estimates the population mean for a K-site sample.

It would not be appropriate to weight the mean $\bar{X}_k$ by the proportion of persons who are at site k, n_k/n. For this gives extra weight to persons who are active in "occupying" sites and are consequently more apt to be sampled. If the entire population were stationary—so that each person is in one site with a probability of 1 and elsewhere with a probability of 0—then the sites would be strata and the scheme to be presented would result in unbiased estimation through stratified random sampling.

In order to determine the weights it is necessary to establish estimates for: (1) the number of persons ordinarily available at each site, n_k; (2) the total population size N; and (3) the distribution of the n_k persons at site k over the K sites.

The distribution of persons over sites is estimated, for this survey, by asking all respondents who are sampled at site k where they could have been interviewed the day before. Respondents were asked: "Suppose I had tried to interview you yesterday at about this time; where could I have found you?" and "Suppose I had tried to find you yesterday afternoon (at night); where were you then?" The answers for all persons interviewed at each site are averaged to create an estimate of how the n_k at site k distribute themselves. Let p_{kj} represent the proportion of the sample at site k reporting they were available yesterday at site j. Each of these proportions can vary from 0 to 1.

An overall view of how persons who occupy many sites is generated as follows. For the persons interviewed at site 1, $p_{11} n_1$ is allocated to site 1, ... $p_{1K} n_K$ to site K. These values measure activity in appearing at the sites. In all, the allocation for persons at site k is

$$m_k = \sum_{k=1}^{K} p_{kj} n_k . \qquad [1]$$

The m_k are then divided by their sum m. to produce the site weights $W_k = m_k / m_{\cdot\cdot}$ Given this, an estimator of the population mean is given by

$$\bar{X} = W_1 X_1 + \ldots + W_K \bar{X}_K . \qquad [2]$$

The estimated error variance about this mean at site k is then given by

$$s^2_{\bar{x}_k} = (1 - n_k / N)(s^2_k / n_k). \qquad [3]$$

The estimated error variance about the mean is then (Kish, 1965: 79)

$$s^2_{\bar{x}} = (1 - W^{-2}) \sum_{k=1}^{K} W^2_k s^2_{\bar{x}_k} . \qquad [4]$$

It is not proven here that the method of weighting means and error variances results in entirely unbiased estimates of parameters. A limited amount of simulation of the method indicates that any sampling biases inherent in the method are not large.

A further source of possible bias derives from estimates of the number of persons at each site, the probabilities that persons at one site are available at other sites, and the population size. The biases resulting from these sources of error could conceivably outweigh reduction in bias through the inclusion of groups hard to locate through other methods. Since even crude estimates of the sizes of strata in stratified random sampling generally reduce error variances, it should not be expected that this source of error would create large biases. The self-reported locations provided by respondents regarding their locations the previous day is a more difficult problem. These questions turned out to be sensitive questions to ask, so it was not feasible to obtain more elaborate descriptions, or pursue independent validity checks. The data showed, however, that there was considerable similarity between the persons interviewed at the various sites.

The weights, W_k, for sites in the four target areas are shown in Table 11.

TABLE 11
Weights Assigned to Each of Eight Sites by Target Area: Site Samples

	Site							
Target Area	**CID**	**CIN**	**COD**	**CON**	**RID**	**RIN**	**ROD**	**RON**
Ghetto A	.08	.06	.00	.01	.45	.38	.01	.01
Ghetto B	.06	.02	.01	.00	.39	.52	.00	.00
Barrio	.01	.01	.01	.00	.47	.49	.00	.01
Slum	.08	.06	.00	.01	.45	.38	.01	.01

SITE SAMPLING AND SNOWBALL SAMPLING

In choosing subsets (not necessarily proper) of the probability samples of the M persons for the zero-stage snowball samples, sampling in sites was carried out with probability of selection proportional to the W_k. If the desired zero-stage snowball sample size is n_o, then $n_o W_k$ persons were interviewed at site k, through simple random sampling.

OPERATIONAL DEFINITIONS OF SITES

A *commercial area* is dominated by structures that are used for business or organizational purposes. Examples are: bars, markets, service centers, doctors' offices. Commercial areas can intersect with residential areas; e.g., where apartments are rented above business establishments. *Residential areas* refer to DUs, streets, and street corners in noncommercial areas, and places such as parks, church grounds, and so forth.

The *inside-outside* distinction refers to the respondents' relations to structures mentioned above, and whether or not they are inside such structures.

Day is defined as 9 a.m. to 4 p.m.; *night,* 7:30 p.m. to midnight. No interviews were initiated at other times.

Interviewer instructions were generated for making random selection of locations within each site, and for choosing persons at random in these sites.[6]

APPENDIX B: SNOWBALL SAMPLING IN BIASED NETS

THE EXISTENCE OF COMMUNITY LEADERSHIP: A MODEL FOR NO STRUCTURE AS THE NULL HYPOTHESIS

In the theory of snowball sampling, the statistical modes assumes there is no structure in the population α-matrix. That is, it is assumed that all choices by all persons are made independently and at random. Such a network of random connections between persons in a structure is called a *random net.* In the terminology of the theory of random nets, from each individual there issues k choices, so that the particular persons upon whom a choice settles is determined by a chance device selecting from the entire population with equal probability. The resultant net (matrix of relations) is a random net with "axone density" of k (Rapoport and Horvath, 1961: 280). In the target areas analyzed here, one name is followed for each person in the sample, and $k = 1$.

The assumption that all choices are made randomly will not, in general, be true in human populations where choices are leadership designations. That is, most target populations have some sort of leadership structure. Since this statistical assumption is empirically the most dubious assumption, it is designated the null hypothesis.

It is difficult to describe the biases that can exist in a net. If the persons chosen are not chosen by chances with equal probability, they can be chosen in innumerable ways.

This problem is approached by considering a single gross parameter that provides a crude measure of the existence of structure. This is defined by Rapoport (1953) as the degree of "overlap" between the acquaintance circles of acquainted individuals. If we denote this "bias parameter" by the symbol θ, then $\theta = 0$ means that there is no structure and that the population matrix of relations is a random net. But if there is a maximum overlap, so that $\theta = 1$, then the net is not random. A net that is not random is defined as a "biased net." *With this terminology, we can regard* $\theta = 0$ *as the null hypothesis for the theory of snowball sampling.* If θ is positive, then some alternative hypothesis must be true. In statistical analysis, the consequences of a range of true alternative hypotheses can be used to generate a power curve. But the exact consequence of an alternative hypothesis ordinarily requires an assignment of an exact numerical value to the parameter in question.

Consider a large population for which $\theta = 0$. In this case, if there are s stages and every individual names k persons, the expected number of persons interviewed is

$$N\,[1 - \Sigma_i\,(1 - p_o)^{i+1} b_{sk}(i)]. \qquad [5]$$

where $b_{sk}(i)$ is the proportion of the population which is named either directly (in one step) or indirectly in s steps by i different persons, and where p_o is the proportion of persons in the population in the zero-stage snowball sample (Goodman, 1961: 161). Note that the $b_{sk}(i)$ are scores of *personal influence.* Thus, the distribution of personal influence scores (PIs) determines the expected size and shape of the snowball sample.

This theory can be used to generate the expected number of persons entering the snowball sample for stages 1 through s. This distribution can be compared to a modification of the *observed* distribution of persons entering the snowball sample for sample at each stage. This modification, to be dealt with presently, deals with the problem of missing data caused by persons making no choices who are, in turn, interviewed.

First, data are needed on the cities for constructing the expected distribution of new persons entering the snowball sample at each stage, $n_1^* \ldots n_s^*$. If the target area is, with respect to leadership, a random net, then this distribution should differ only by chance from the modified observed distribution $n'_1 \ldots n'_s$. But should there be influence in the community, so some $b_{sk}(i)$ are greater than chance expectancy, and θ is positive, then the distribution of n' should be stochastically smaller than the expected distribution.

The expected number of persons that enter each stage of a snowball sample depends on three factors:

(1) p_o, the proportion of persons in the community interviewed in the zero-stage snowball sample. This is the rate of the number of persons in the zero-stage snowball sample to the total population size N: $p_o = n_o/N$.

(2) k, the number of persons identified as leaders to be followed up. In all four cities, k is set at one, so one-name snowball samples are obtained. The name followed up for each person is the first name given for

which an interview could be obtained. Thus, if a person's first two choices refer to persons who do not live in the community, cannot be found, or for other reasons cannot be interviewed, the person's third choice was used as his first named.

From the zero-stage sample of $p_o N = n_o$ persons, kn_o choices issue to some set of persons. Some of the chosen persons will be in the initial sample, and others will be newly contacted. We will call the fraction of the N persons that are newly contacted p_1. These n_1 people constitute the one-stage sample. This process continues through s stages. We denote the proportion of the population in the entire sample as

$$X_s = \sum_{i=o}^{s} p_i. \qquad [6]$$

For a random net, Rapoport (1951) shows that the expected proportions of persons in the s = 1 stage is a result of two independent events: (1) That the person chosen does not belong to an earlier stage. This probability is $1 - X_s$. (2) That one of the choices from a person in the s stage does connect with a person in the s = 1 stage. This probability is $(1 - (1 - /n)^{knp_s})$. Since events (1) and (2) are statistically independent, the probability that both occur is the product of their separate probabilities.

Therefore (Rapoport, 1953: 514),

$$p_{s+1} = (1 - X_s) \; [1 - (1 - 1/N)^{knp_s}] \qquad [7]$$

For large n the bracketed term is approximately $(1 - e^{-kp_s})$, so that

$$p_{s+1} = (1 - X_s) \; (1 - e^{-kp_s}). \qquad [8]$$

Equation (8) indicates that the total sample size will increase most rapidly when p_o and k are large, a result that is of course consistent with Equation (5) from the theory of snowball sampling. Since the sample size is incremented with each stage, the total size of the snowball sample will vary directly with s.

(3) s, the number of stages to be followed. In all cities but Slum, five stages are followed. In Slum, s = 6 with one person interviewed in stage six.

Table 12 shows the data used in testing the null hypothesis of no leadership structure for the four target areas.

The observed number of persons at each stage of the snowball sample can differ from the expected distribution for two reasons.

TABLE 12

Distribution of Snowball Samples by Target Area

Target Area	Population Site (n)	Zero Stage Site (n)	P_o	Expected Distribution n^*_o	n^*_1	n^*_2	n^*_3	n^*_4	n^*_5	n^*_6
Ghetto A	86,200	78	.00086	78	78	78	78	77	77	–
Ghetto B	100,300	47	.00047	47	47	47	47	47	47	–
Barrio	12,800	40	.00313	40	40	40	39	38	38	–
Slum	10,000	53	.00530	55	53	52	51	50	48	47

First, structural bias in the population (the existence of leadership) will cause the snowball sample to funnel, with the number of persons entering each new stage of the sample smaller than expected by chance. Second, persons chosen into the snowball sample may themselves provide no (less than k) names for which interviews can be obtained. This, of course, also reduces the number of new persons entering the sample at each stage. Since there is a substantial amount of such missing data, it is necessary to modify the observed distributions so they reflect what *would* have happened if the persons who did not choose new persons had chosen persons.

An acceptable way to deal with this problem is to assume that persons who provide no new names are choosing at random.

Since there is little funneling in the above expected distributions, a slightly more conservative procedure is used. It is assumed that missing data can be replaced by *new* persons. In Table 13 the n represents the actual number of persons interviewed at stage j, m_j of "missing data" cases that name no one who is interviewed, and n' the "adjusted observed distributions" to be compared to the "expected distribution" in the statistical tests. In this table $n_i' = n_i + \sum_{j=1}^{i-1} m_j$. That is, the adjusted observed number of cases at stage i is the number of persons interviewed at stage i, plus the number of persons in the previous stages who made no choices.

In all four cities, the distributions observed (after correction for missing data) are more concentrated in the lower stages of the snowball sample than would be expected in random nets. Chi-square test comparing the observed and expected distributions can be carried out to test the null hypothesis of a random net.

TABLE 13

Adjusted Numbers of Observed Persons Entering Snowball Samples at Each Stage by Target Area

Target Area	Snowball Stage	Number of Cases, n_i	Number not Making Choice m_i	Adjusted Observed Number, n_i^*
Ghetto A	0	78	–	78
	1	51	20	51
	2	16	5	36
	3	10	0	35
	4	7	0	32
	5	6	–	31
		168		
Ghetto B	0	47	–	47
	1	36	39	36
	2	27	11	36
	3	13	0	33
	4	13	4	33
	5	7	–	31
		143		
Barrio	0	40	–	40
	1	36	12	36
	2	19	6	31
	3	13	7	31
	4	5	1	30
	5	2	–	28
		115		
Slum	0	55	–	55
	1	42	17	42
	2	25	13	42
	3	9	2	39
	4	7	1	39
	5	3	1	36
	6	2	–	35
		143		

Table 14 shows that in each target area, it is improbable that there is *no* leadership structure. The chi-square for Ghettos A and B are significant beyond the .001 level, Slum is significant beyond .01, and Barrio beyond .05. It appears that the level of community organization is greater in Ghetto A than in the other three areas. In all four cases there is adequate justification for analysis of substructures and of the leadership properties of individual persons. The table also shows contingency coefficients (C) which control for the differing sample sizes, to provide crude measures of dependence (departure from the null hypothesis). A second measure is provided by estimates of the bias parameters, θ. The product moment correlation between these two measures is $r_{C\theta} = .99$.

Ghetto A, by both indicators, has a higher level of organization than do the other three areas, which do not differ significantly from each other.

SNOWBALL SAMPLING: DESIGN DETAILS

The number of persons in the zero-stage snowball sample of each city is a subset of the persons in the site sample who could have been selected. All persons in the zero-stage snowball sample, however, are in the site sample.

Persons in the site sample but not in the snowball sample are coded "no information" for leadership status in the analysis. It is not proper to assume that persons in the site samples are low on any measure of leadership, in the same way that such an inference would not be justified for persons in neither the site nor the snowball samples.

In every city, a one-name sample was obtained. The person's one choice was defined as the first name he gave of another person who was located and interviewed. If a person selected for inclusion in the zero-stage sample did not produce the name of a person who was

TABLE 14
Chi-Square Significance Tests for the Existence of Informal Political Leadership, Contingency Coefficients, and Estimates of Bias Parameters by Target Areas

Target Area	Chi-Square	df	Contingency Coefficient	Probability	Bias Parameter Estimate, $\hat{\theta}$
Ghetto A	113.5	3	.63	$p < .001$	.71
Ghetto B	18.9	3	.34	$p < .001$	.48
Barrio	9.1	3	.28	$.02 < p < .05$	.41
Slum	16.2	4	.32	$.01 < p < .001$	.45

actually found and interviewed, he was retained in the site sample but not in the zero-stage snowball sample. But if persons chosen in the snowball samples (at any stage) gave no names of persons who were then interviewed, they were deleted from the snowball sample.

NOTES

1. Formal leadership in a community would refer to public officials and high-status persons in private or public organizations.

2. The site sample size of these estimates is based on 1,097 site sample respondents and 232 snowball sample leaders. The percentages are weighted by the criterion described in Appendix A (Table 11).

3. In a similar study of two rural and three urban target areas, evidence for the existence of community leadership was found in the rural areas but not in the urban areas. From independent information about these three urban areas, it seemed most unlikely that there was no leadership structure. (One area, in fact, was so highly organized that the evaluation research was partially controlled by, and channeled through, a local organization originally funded as an OEO Special Impact Program.) Thus, it appeared leadership had not been measured. This could have resulted from the snowball sampling methodology or from differentiation in the "initial" samples from which the zero-stage snowball samples were chosen at random. The snowball sampling methods differed little for the two studies, being carried out by the same subcontractor with much of the same personnel. Consequently, it is likely that the initial sample design was responsible for the failure to detect leadership. The major sampling technique was residential sampling. In one community (a black ghetto), a systematic probability sample of every fourth household was drawn; in the other two areas (a black ghetto and a Chicano barrio), systematic sampling of residences was supplemented by quota sampling in commercial areas. These methods differ from the site sampling method used in the present study. It may be the case that the site samples contained higher proportions of persons connected to leadership substructures than did the residential and quota samples. If so, this project constitutes a control group for an inadvertent quasi-experimental study of sampling methodology in poor urban areas.

The study of three urban areas was carried out by Westinghouse Learning Corporation under a contract with OEO, to evaluate the impact of Special Impact Programs. The snowball sampling analysis is described in TenHouten (1969b).

4. There are exceptions; e.g., the Isla Vista area of Santa Barbara, California, is dominated by university students whose leaders are militant and radical: residents of this community have engaged in a series of confrontations with the forces of "law and order."

5. A Scale Gradient Analysis (TenHouten, 1969c) indicates ninety-eight percent of the variance in response patterns is accounted for by the model of a Guttman scale; i.e., the Determinant of Scalability is .98 ($t = 21.2$, $p < .001$).

6. These are described in a manual available on request.

REFERENCES

BARTH, E. A. and B. A. ABU-LABAN (1959) "Power structures and the Negro sub-community." Amer. Soc. Rev. 24 (February): 69-76.

CLARK, K. B. (1967) Dark Ghetto. New York: Harper & Row.

COLEMAN, J., E. KATZ, and H. MENZEL (1966) Medical Innovation: A Diffusion Study. Indianapolis: Bobbs-Merrill.

GOODMAN, L. A. (1961) "Snowball sampling." Annals of Mathematical Statistics 32 (March): 148-170.

GREENBERG, S. F. (1970) Mobilizing Poor Communities. Report to U.S. Office of Economic Opportunity. Cambridge, Mass.

--- (1969) A Story and a Study. Cambridge, Mass. Unpublished.

HOLLANDER, E. P., and W. B. WEBB (1955) "Leadership, followership, and friendship: an analysis of peer nominations." J. of Abnormal and Social Psychology 50 (March): 163-167.

HUNTER, F. (1953) Community Power Structure. Chapel Hill: Univ. of North Carolina Press.

JENNINGS, M. K. (1964) Community Influence. New York: Free Press.

KISH, L. (1965) Survey Sampling. New York: John Wiley.

LIEBOW, E. (1967) Tally's Corner. Boston: Little, Brown.

MENZEL, H. and E. KATZ (1955-1956) "Social relations and innovations in the medical profession: the epidemiology of a new drug." Public Opinion Q. 19 (Winter): 337-352.

RAPOPORT, A. (1953) "Spread of information through a population with socio-structural bias: I. the assumption of transitivity." Bull. of Mathematical Biophysics 15 (December): 523-533.

--- (1951) "Nets with distance bias." Bull. of Mathematical Biophysics 13 (June): 85-91.

--- and J. HORVATH (1961) "A study of a large sociogram." Behavioral Science 6 (October): 279-291.

ROSS, H. L. (1963) "The inaccessible respondent." Public Opinion Q. 27 (Spring): 269-275.

ROSSI, P. H. and R. CRAIN (1968) "The NORC permanent community sample." Public Opinion Q. 32 (Summer): 261-272.

TenHOUTEN, W. D. (1969a) Site Sampling and Snowball Sampling in Biased Nets: Methodology for the Study of Community Organization. Report to U.S. Office of Economic Opportunity. Los Angeles.

--- (1969b) Community Leadership Structure in Special Impact Areas: Snowball Sample Analysis. Report to Westinghouse Learning Corporation. Los Angeles.

--- (1969c) "Scale gradient analysis: a statistical method for constructing and evaluating Guttman scales." Sociometry (March): 80-99.

---, F. TZUEN-JEN LEI, F. KENDALL, and C. W. GORDON (1970) "Ethnic composition, social contexts, and educational plans of Mexican-American and Anglo high school students." Amer. J. of Sociology.

TOMLINSON, T. M., and D. TenHOUTEN (1970) "Methodology: Negro reaction survey," pp. 127-137 in N. Cohen (ed.) The Los Angeles Riots. New York: Praeger.

TROW, M. A. (1957) "Right wing radicalism and political intolerance: a study of support for McCarthy in a New England town." Ph.D. dissertation. New York: Columbia University.

WHYTE, W. F. (1955) Street Corner Society. Chicago: Univ. of Chicago Press.

Part III

URBAN INSTITUTIONS: PROBLEMS AND PROSPECTS

In the political writings of the American left, minorities have increasingly been treated on the analogy of an internal colony. While none of the chapters in this section adopts that stance they do deal with various aspects of the custodial relationship between minorities and the American majority or the agencies responsible for managing one or another aspect of that increasingly uneasy relationship. It is possible to view each of these chapters as descriptive of strategies of minority management, although this is not the intent of the authors. Roy Simon Bryce-LaPorte comes closest to adopting this position the opening chapter of this section. He argues that along with the abolition of the slave plantation the society has created special institutions (or altered traditional institutional functions) to quarantine and service stigmatized black people. It is because the custodial and exploitative aspects of many present-day urban institutions are consistent with the defining characteristics of slave plantations that an analysis of the latter is instructive to an appreciation of these qualities of the former.

Each of the next four chapters touches on problems created by the custodial character of present-day urban institutions. Lisa Peattie is most explicit in recognizing the problems created by the custodial character of public housing as an institution. Not the least of these she contends is the stigmatization of the residents. Using one public housing site, she discusses, in a detailed and sometimes personal way, the interlocking administrative, class, demographic, and other elements which generate and confirm the spiral of decay and degradation which has been the fate of so many public housing projects.

Jane Mercer is concerned with the phenomenon OF labeling—how the stigma of retardation is given official sanction—and ultimately the institutional consequences of that process for the differential treatment of the inmates. Her discussion broadens the views developed in the first two chapters of this section. It indicates that the analysis is applicable to persons who are not the descendants of plantation slaves. Further, it suggests that the public schools, too, ultimately relate to minority students, to their detriment, in a custodial capacity.

Gary Goodpaster, in Chapter 11, is concerned with the institutional deprivation of, in this instance, Mexican-Americans. The economic exploitation of the migrant farm worker is akin to the experience of the plantation slave. Here the analogy to Bryce-LaPorte's thesis is, perhaps, most direct. But, Goodpaster's concern is with the paradox of reform, how partial amelioration of abhorrent conditions merely serves to perpetuate the institutional conditions which sustain the deprivation. If abolition of the institution is not to produce a functional equivalent, then, to paraphrase Bryce-LaPorte, the larger society's attitudes toward the afflicted also must change.

In the concluding chapter in this section, Harlan Hahn considers that community agency which is most clearly involved in a custodial function. Hahn discusses the conflicting pressures upon law enforcement agencies to professionalize and to decentralize police practices. In the end he urges modification of law enforcement activities so as to accentuate service, as opposed to the caretaker functions.

8

The Slave Plantation: Background to Present Conditions of Urban Blacks

ROY SIMON BRYCE-LaPORTE

THE BLACK PRESENT AND THE SLAVE PAST

□ IT IS GENERALLY agreed upon that the largest number of blacks to come to the United States since 1619 came from Africa directly or by way of the Caribbean. They came as slaves and the most common setting in which they lived, worked, and died was the slave plantation. It can be generally agreed upon, therefore, that the slave plantation constitutes the contextual base-line necessary for arriving at a *fuller understanding*[1] of the behavioral, institutional, and communal development problems of black people in the United States as well as other parts of the New World. This recognition led Wagley to juxtapose against the terms Euro- and Indo-America the term Plantation America (Rubin, 1960: 3-5). And, it was none other than E. Franklin Frazier, known for his vehement attacks on the Herskovitzian contention of African survivals, who argued that Plantation America was indeed interchangeable as a term with Negro America, inasmuch as the principal historical influence on New World black culture was indeed the plantation (Rubin, 1960: v).

The reference to the slave plantation past as genesis of black problems, black perspectives, and black behavioral patterns is not limited to scholars. Activists, black and white, have made similar claims. Malcolm X, the celebrated black nationalist martyr, and one of the most successful mobilizers of urban black masses, made repeated pleas for an understanding of slavery, if the present-day plight of black people was to be understood. The people in the street in time of greatest crisis, anxiety, and disorganization (i.e., riots) have been known to offer as the on-the-scene reason for their actions "to get Mister Charlie [Whitey] off our backs." The language may be contemporary ghetto talk or grass roots rhetoric, but the sentiments and consciousness behind it are historical, sociological, and point to a persistent problem for black people in America since the beginning of black slavery.

While the organizational-residential context under which blacks were found yesterday may have been the slave plantation, today it is the urban ghetto. Whether the shift from past to present, rural to urban, or plantation to ghetto had serious consequences for the life chances, the institutional and political-economic development, and the life style and culture of black people, is a question deserving serious sociological study. Whether the contentions of scholars and the charges of activists are valid or correct demand empirical examination. As such, these mandates fall within the scope of what C. Wright Mills (1959) called "sociological imagination"–a history-conscious and problem-derived sociology. This approach requires a sociological understanding of the earliest conditions and problems under which blacks lived before any sociological, historical, psychological, or humanistic sense can be made of their present situation.

It is indeed questionable whether presently available data and empirical technique of the social sciences and history could yield with any precision new and significant information on the nature of relationship between the slave plantation background and present-day conditions and behavior of black Americans. To do so requires information on millions of black, often alien, oppressed, and illiterate souls, spread over one-third of the United States, over the last three hundred fifty years. Merton (1957: 8-9) has warned against premature reaction of sociologists to the demands of their problem-oriented colleagues and problem-ridden public for sociological study and action on such overwhelming problems. His warning was not against ethical or political action per se, but rather against engaging in premature, futile action in light of the stage of development of sociology itself. In other words, Merton advocated a scientific priority of "appropriate" method and knowledge accumulation over problem resolution. This is in line with his premise that the former will determine, in the final analysis, how much the latter can be accomplished. His position here differs fundamentally from that of Mills (1959: 125-126), who gives priority to public issues and advocates that methods should not limit the choice of problem but largely evolve from the effort to solve such problems. Yet it is by the sociologically imaginative use of what Merton calls "middle-range theorizing"—the effort to pull together theory and fact by way of intermediate levels of formalized derivations and conceptualizations—that the slave plantation, a phenomenon of the past, can be described in sociologically useful terms (Marsh, 1964).

American slavery and the slave plantation have been studied for a long time by historians, most of whom pursued their study of the peculiar institution in the ideographic tradition of their field. Stanley Elkin's *Slavery* (1963) represented an important departure from this tradition, in some ways a controversial breakthrough, when he pursued the analogy between American slave system and the Nazi concentration camp as institutionalized systems of closure and totalitarianism which had similar effects on the personality, family, organization, leadership, forms of associational or collective behavior of their inmates—slaves and prisoners-of-war. Sociologically speaking, various questions can be raised about the assumptions and implication of Elkins' work (see Bryce-LaPorte, 1968; Lane, 1971). Notwithstanding, his conceptualization of the slave plantation as a closed and total system is of extreme sociological and political importance. Sociologists have been giving increasing attention to such closed, all-embracing organizations. They refer to them as total institutions. However, there have been few indications that sociologists recognized the slave plantation as a total institution[2] (Bryce-LaPorte, 1969: 2-8; Yetman, 1970).

Sociologically speaking, the visualization of the slave plantation as a total institution permits a shift in its study from that of a peculiar institution to that of a peculiar type of institution. Further, it permits the slave plantation to be studied as a level of social organization about which there is a growing bibliography (Goffman, 1961: 123-124). Such a visualization opens the door toward the scholarly discussion of the validity of the historical depth of black problems, black perspectives, and black patterns of behavior which have come to characterize the rhetoric of black nationalism. This research could begin to uncover new dimensions of the condition and quality of urban life for the majority of black people in the United States today, and to evaluate and give meaning to the various ways in which blacks have tried to change those conditions. But, before these potentials can be pursued, the plantation must be treated in terms which will overcome its datedness and peculiarity, as a subject matter, and which will reveal its relevance to present-day concerns.

THE SLAVE PLANTATION AS A TOTAL INSTITUTION

The term total institution is associated principally with Erving Goffman (1961: xiii) who defined it as "a place of residence and work where a large number of like-situated individuals cut off from the wider society for an appreciable period of time, together lead an enclosed, formally administered round of life." Scott Greer (1955) makes the distinction between social organization, the regulatory form of human social behavior, and the social organizational unit, a collective recognized for its specific combination of human social roles and relation. The American slave plantation was both. While it had features of both community and instrumental-formal organization, it was neither. Rather, it shared most of the peculiarities that distinguish other total institutions from these two forms. At the same time it could have been differentiated from other total institutions in terms of its uniqueness or specificity. As a dynamic structure these bases of differentiation and uniqueness were not static, but underwent changes and thus differed over time (Bryce-LaPorte, 1968; Edgerton and Sabagh, 1962: 263-270; Etzioni, 1961; Hillery, 1963; Thompson, 1959: 26-36).

On the level of purposes and ends, the slave plantation enjoyed similar, though not necessarily the same, specificities as other total institutions. The primary extrinsic goal of the slave planters was economic profit, but it was economic for prestigious rather than capitalistic interests (Genovese, 1966a: 28-31). Thus, prestige, aggrandizement, and power over other men was the primary intrinsic goal of the slave plantation, as is common among total institutions. Ownership and management of the slave plantation and immunity from slavery were bases of prestige. Akinness to those who owned was a basis of prestige and aggrandizement. Such prestige and aggrandizement were gained at the expense of others who were owned or managed, were not immune to slavery, or were more akin to slaves.

The specialized instrumental responsibility of the slave plantation was commercial agriculture. Its other instrumental responsibility was the custody of a segment of the population perceived of as unusually different. This latter responsibility is a universal service-function among total institutions, even though all "unusually different" segments are not viewed as negative. They may be unusually "good" (i.e., religious men), or they may be unusually "bad" (i.e., dangerous criminals). They are in some commonsense terms, however, not conceived of or treated as "equals" with the rest of the "normal" world.

The condition in the United States was one of great access to land and limited access to free, exploitable labor. The slave population allowed, on the one hand, the maximum exploitation of slaves as producers of profit, or, perhaps more accurately, income derived from commercial agriculture, and later, commercial breeding. At the same time it allowed for maximum confinement, containment, and control of these slaves who were so "different" (unequal) and socially undesired by the larger society. It must be recalled that the early non-Hispanic colonizers of the United States were not motivated by the zeal of extending empire and converting or colonizing people. Rather, they came as minorities to America in flight from what they perceived as corrupted or constraining conditions, in hope of establishing religious and political utopias based on homogeneity (Handlin, 1958; Morgan, 1958). Thus, not the Indians but their lands were seen as frontiers, and Africans were seen as human tools rather than human targets of frontier development. The physical and cultural differences and distance between Africans and Anglos gave cause and were conducive to the establishment of the slave plantation as a custodial institution (Jordon, 1969; Szasz, 1970: 123-139). The rural setting and economic orientation of the colonizers were conducive to the establishment of the slave plantation as a commercial, agricultural organization; in other words, a formal-instrumental institution.

Black slavery was supported by advocates and arguments. The proslavery arguments which covered a wide range were neither inconsistent collections of thoughts nor a mere repertoire of rationalizations. They were assertions by the society and its planter class of the various purposes or ends of the slave system (McKitrick, 1963; Wish, 1961). Even though the assertions were inaccurate and injurious to the Africans and Negroes, these arguments were components of the ideology of the South and its conception of Africans and Negroes qua laborers, citizens, and human beings. These arguments were aspects of a theory of human nature, in this case a psychology and ideology which justified and stipulated the treatment of slaves (Goffman, 1961: 129-180; Kardiner and Ovesey, 1962: 42-47). They fit almost perfectly Goffman's (1961: 5) typology of total institutions by which he categorizes their specialized service-function to the larger society:

> (1) those established to care for persons felt to be both incapable and harmless; (2) those to care for persons felt to be incapable but unintended threats to the community; (3) those established to protect the community against those who are intended threats to its welfare; (4) those established to pursue some worklike and instrumental task. . . .

Another level of comparable specificities between the slave plantation and other total institutions involves external relations, especially with the larger society. The history of the slave plantation in the United States indicates that it endured for almost three hundred years and as it did, it evolved internally as well as in its external relations. Limitations on its endurance and involvement were imposed by the peculiar historical, physical, geographical, and social constraints in which it was located. Hence, not only was the slave plantation never totally static, it was never totally isolated or totally independent of its larger habitat. Its history, then, is one of variation in interdependence, and consonance with the larger southern society (Thompson, 1959; Genovese, 1966a).

The American plantation was at the start a rather dependent, circumscribed, and noninfluential organization. It was formed and charged in the beginning as an instrument of the colonial, economic, and governmental bureaucracy. Its hold on its inmates was temporary and limited. In time it developed power, potentiality for exchange, scope, and pervasiveness. It highest level of development was reached when it became a producer of cash crops, having great land resources which its instrumental specialization–extensive commercial agriculture–also demanded.

During the period of American political independence and democratization and the discontinuance of regular market relations with Europe, the importance of the custodial and specialized service-functions of the slave plantation were underplayed. Manumission increased and the slave trade was declared illegal. At that time the slave plantation was losing its custodial and specialized service-functions; it was approaching a more rehabilitative and communal model. These developments were short-lived and finally reversed with new demands for American cotton now being refined with the recently invented cotton gin. Then, external recruitment of slaves was resumed, and antislave trading laws were ignored. When external recruitment proved inadequate and insufficient to meet the demands, internal recruitment, eternal retention, and status ascription were reemphasized. Thus, the custodial services were reinstated anew and an ideology of difference and therefore need for control were revived.

The period of reassertion or reemergence of the slave plantation was one which witnessed the broadening of the scope of control to embrace the slave's entire life (not just all activities at a given time in life). It was a period marked by the increase in power and pervasiveness of the slave plantation over the slave society and the men who populated it. Planters dominated the politics and economics of the South, directed policies in their own self-interest, and defended that interest to the point of war. The pervasiveness of the slave plantation extended to freeman, runaway slave, and paroled slave even in location as far as the West and the North, as well as cities of the South. It affected the treatment they received, the laws legislated against them, and threatened their security and freedom.

To the extent that the above plantation was impermeable, it gave to its slaves more material security and benign or paternal treatment than the freemen and runaways of the rural South experienced. This increased the pervasiveness and saliency of the plantation over its inmates, inasmuch as it may have led to their cherishing physical comfort or security over freedom. To the extent that the plantation was permeable, it reflected the basic status differences and theory or psychology of human nature held by the larger society about slaves, and treated them as inferior beings who could be exploited, contained, and suppressed. The pervasiveness of the slave plantation regimen was quite evident in the frontier and antebellum cities where the restriction of slaves and even of free blacks by symbolic barriers, such as Jim Crow laws, very much resembled plantation standards.

Another level of comparable specificities which the slave plantation shared with total institutions includes contextual features represented by its structure, processes, and inherent paradoxes and dilemmas. By and large, the slave plantation was characterized by a coercive-alienative control-compliance structure. Given its profit-making goal, such a control-compliance structure was incongruent. From Weber through Etzioni, sociologists have recognized that the most effective employer-employee relationship in profit-making organizations is one of manipulation and calculation with interjection of human relationships. The slave plantation (like most total institutions) was likely incongruent for another reason: its dual goal or orientation (Etzioni, 1961). Prestige, the universal intrinsic goal, rivaled and undercut profit-making (Genovese, 1966a). In addition, the slave plantation shared other structural features or concomitants of total institutions. These included a single authority structure headed by the master or his surrogate, often an overseer, and the rigid stratification and segregation of slaves and management. Also included were various institutionalized structural modes of impositions and deprivations which differentiated the two segments and discriminated against the slaves—imposition of residence and occupational roles, generalization of status, regimented schedule of activities, deprivation of privacy, power, property, free movement, indigenous primary organization, cultural development, and role repertoire.

The context represented by the slave plantation also included a set of interrelated operations and practices which were part of the overall rationale for the successful and smooth running of the estate. Among these were recruitment, socialization and mortification, a privilege-punishment system, tension-release, rehabilitation and aggrandizement, discharge, stigmatization, and so on. Many of these processes were mutually exclusive or inherently incompatible and thus, together with the other levels of complexities, created various dilemmas and paradoxes for the slave plantations, as is true of other total institutions.

The dilemmas of custody versus rehabilitation and of mortification versus aggrandizement (Edgerton and Sabagh, 1962: 263-272; Etzioni, 1961; Tittle and Tittle, 1965: 216-221) were very closely related in the case of the slave plantation. Comparable to other total institutions, the conflict experienced by the slave plantation in regard to these two dilemmas was light. In establishments where discharge is common and thus confinement is not indefinite, the issue of how much to stress custodial stringency over rehabilitation becomes a very serious issue. Similarly, in establishments where inmates are derived basically from external sources and are likely to be returned to such sources, the judicious balancing between mortification and aggrandizement becomes a problematic matter. These conditions were true for the plantation only in colonial days.

In the heyday of the slave plantation, however, custody was generally perpetual, inmate status was ascriptive, recruitment and selection criteria were internal, and physical discharge took the form of death or escape. Moreover, the tasks and roles required of slaves by the plantation regimen were generally rudimentary and routinized. Thus, there was no great urge for intensive socialization or rehabilitation programs outside of the case of newly arriving Africans or former free Negroes or city slaves. Likewise, the issue of mortification versus aggrandizement was academic for slaves other than the cases just mentioned. That is not to say that the dilemmas did not exist or that problems did not emerge. The fact that slaves had to be punished for and prevented from escaping or doing harm to their managers or owners would indicate that the mortification process was not always thorough. In fact, it could only be so if slaves were killed psychologically or physically. In addition, even though slave aggrandizement tended to be pathologically limited, and false-based, it created problems of maintaining a totally conforming slave population.

A third dilemma, the humanistic-bureaucratic dilemma, is not unknown to most formal-instrumental organizations or to communities or societies with bureaucratic orientations. However, it perhaps was more strongly felt on the slave plantation than most other total institutions (Moore, 1941: 191-201). The slave plantation attempted more than any other total institution to view and treat its inmates as subhuman property. On the other hand, the slave plantation came closest perhaps, among many though not all total institutions, to acknowledging the equality in essence (not in status) between inmates and staff, particularly by its sponsorship and participation in family formation among slaves, and between them and managers or owners. The moral, ethical, and emotional sensitivities, and the feelings of attachment and commitment associated with the family, particularly with women and children, were more than the closedness and cruelty of the slave plantation could contain. It is true that the sponsoring of family formation and limited recognition of slave families were largely efforts to perpetuate the slave plantation and assure an internal and reliable source of recruitment. But it is also true that these practices brought out of slaves their human qualities and drew out of other men a degree of action, attention, and attentiveness that men reserve only for other men. Such practices represent the kinds of concerns and involvement which beset institutions whose work is people (Goffman, 1961: 74-92). Thus the statement by U. B. Phillips (1963: 217) that: "All in all, the slave regime was a curious blend of . . . tyranny and benevolence, of antipathy and affection."

A last level of comparable specificities between the slave plantation and other total institutions is the behavior of inmates. From the distance of the outside observer, the context presented by the slave plantation as the slave appears as one of total and eternal closedness and control, one of exploitation, imposition, and deprivation, in which the slaves never could have a chance to consciously and deliberately act without reckoning with the staff and standards of the slave plantation. Even if this picture has been exaggerated, the presence of the slave plantation did have a great impact on and saliency for the slaves' behavior. While it should not be doubted that the behavior of slaves may have had some impact and saliency for the staff and standards of the plantation in return, it seems futile to deny that the behavior of slaves, especially the sustained, overt, collective-organized behavior, had to be very adjustive and truncated. Slave behavior was comprised of varying efforts, some in submission to the dictates of the plantation and others in defense against it. The circumstances of the slaves, especially those on large plantations, were not only different from the circumstances of the masters and free citizens, but so, too, their culture. In other words, the culture of the plantation was dual or plural, inasmuch as there were pronounced qualitative and quantitative differences in role repertoire and organization between the slave and the master classes.

The plantation staff did not promote the emergence or reemergence of roles of leaders or of organizations among the slaves which were property, protest, or power-oriented. To have permitted this would have been irrational and a direct contradiction of all the purposes or ends of the slave plantation, its ideology, and psychology of human nature, and would have threatened its entire structure as a total institution. Ownership of physical property among plantation slaves was almost nonexistent and limited to a few household supplies, extra clothing, and the crops or stocks that they raised on land assigned to them. They were neither free to purchase nor work toward purchase of durable, basic property; e.g., land, home, and the like. They were most often not able to purchase their own freedom. They could not own other persons as property or slaves and whatever claims they had over the spouses, children, and parents were largely offset, inasmuch as in legal terms they were all mere properties of the master. The pretense of ownership also reached overdetermined dimensions with some slaves, who at one and the same time acted as property and as part owner of plantation and planter (in much the same way children "own" their homes and parents).

The conditions of control, totalness, and remoteness which characterized the slave plantation of the United States made open, sustained, collective resistance among the slaves virtually impossible. Most successful resistance was individual, isolated, and rapid. Collective resistance was generally sporadic, if open, and necessarily subtle, if sustained. In other words, the very conditions which made open, sustained, collective resistance infeasible made more subtle, covert, elusive forms of group action sustainable and prevalent. The "underlife" (Goffman, 1961: 199) or "covert culture" (Rose, 1969: 623-630) of the slaves consisted largely of secondary adjustments and these adjustments were, in fact, subtle, elusive roles and sociocultural relations which were prevalent and sustained among them. The underlife of the slaves included, among other things, prayer meetings, spirituals, prayers, folk tales, an invisible church, and an underground escape organization. Passing as white or as free; slipping away from plantations through slave country and to free society; and the protecting of other slaves engaged in these maneuvers, represented the most daring forms of successful resistance that American slaves were able to sustain. Thus, psychological and physical flight (rather than direct fighting) became the modus operandi of American slave resistance (Bryce-LaPorte, 1969: 6).

Flight or escape required a different form of organization and role repertoire than would be required for open defiance. It demanded more invisible and covert bonds and called for cunning and elusive skills rather than confrontation and power. By the same token, however, such escapist actions were not sufficient to accommodate the large numbers desiring freedom. Thus, many slaves found relief in removal activities; e.g., folklore, spirituals, prayers, work songs, and in institutional ceremonies; e.g., prayer meetings, festive celebrations, and so forth. Such removal activities and institutional ceremonies had tension-releasing rather than freedom-rendering functions. They represented means of expression and rebellion rather than means of strategic revolution. Given the limitations of these expressive and escapist actions, there were occasional sporadic riots and rebellions among the slaves. However, these more aggressive forms of defiance were generally suppressed in a vicious public manner and were often followed by the tightening of custodial regulations. The failure of resistance was in part an indication of the lack of necessary organization and power for successful paramilitary rebellion. On the other hand, the failure indirectly boosted the invisible underground approach or led to sublimation of tensions in some other ways.

Few rehabilitative activities of the kind that could precipitate, or stem from, an anticipation of freedom, were permitted or promoted among slaves. Education and religion for slaves were discouraged in the earlier periods. They were encouraged as the institution began its decline around the period of the Declaration of Independence. Fear of rebellion and the reemergence of the importance and predominance of the slave plantation were accompanied by reimposition of bans on both education and religion. The rationale for discouraging these practices was neatly tied to the purpose and nature of the slave plantation. It made no sense to allow education as long as slaves were to be seen as subhumans, condemned to perpetual confinement and exploitation, and replaceable by local and foreign recruits. The nature of work on the plantation required no sophisticated training and time for such superfluous activities could be used in maintenance and production activities (Woodson, 1919: 153-154).

Once banned, open education of slaves was not permitted or practiced by most planters or managers of plantations. Religion was once more permitted among the slaves only when planters were convinced that it would make slaves more placid rather than provoke unrest and instill a human awareness among them. Planters took their house slaves to services with them and often-times permitted field slaves to attend their own prayer meetings alone. Both religious practices were largely viewed as removal activities which modulated the distance, severity, tension, and formality of the slave plantation regimen with highly charged emotionalism, and, at the same time, promoted conformity and compassion.

The slave plantation regimen discouraged the founding of indigenous or autonomous primary bonds among slaves. It finally conceded some such privileges in the form of the family only when it seemed necessary for the perpetuation of the plantation and not threatening to its custodial or specialized responsibilities. In the early stages of slavery the sex ratio was imbalanced. Slaves were mostly men, separated from their traditional family, and given the sex imbalance and the barriers to interracial sex; the presence of a stable union between slaves on the level of family may have been uncommon. Over time, family or quasi-family relations among the slaves were largely of the nature of secondary adjustments—unsponsored, extralegitimate, and somewhat covert relations. As secondary adjustments, awareness and acknowledgment of them by the staff were less likely, and accommodations to allow for these unions were at first unusual and gradual.

It was, however, during the days of the larger plantation and of the abolition of the slave trade that the slave family was promoted with vigor: a behavioral pattern among inmates that is sponsored or demanded by the staff. It was now of the nature of a primary adjustment. The fact that awareness and acknowledgment of family formation among the slaves led to certain allowances and accommodations by masters and overseers, indicates that the plantation as well as slaves underwent categorical redefinition. The direction of this change, inasmuch as it involved the total institution adjusting to its inmates, illustrated the point mentioned but left undeveloped by Goffman (1961: 188-193). The fact that the change involved promoting a family among inmates of a total institution suggests that it is as important to understand the conditions leading up to the emergence of the slave family, as it is to understand the consequences of the plantation context on such a family. The latter is too often stressed at the expense of the former.

The "sponsored" slave family was a means of providing and rearing internal recruits (Frazier, 1939). It was presumed that since every member of the family was a slave he was controllable and thus the unit would not be disruptive of the larger establishment (Elkins, 1963: 53-55). Consequently, the slave family was most likely truncated as a primary group, both in terms of internal roles and of institutional functions and form. This holds true using either the African and Anglo-American models of family during that time. Slave narratives suggest that this did not mean that strong bonds were not sustained among members of families separated forcibly or never permitted to relate freely. Furthermore, the strength of such bonds was reaffirmed by the risks slaves took and the routes they tramped to be near or visit their loved ones.[3] The presence of the family unearthed the self-awareness and primary group loyalties among slave families and also elicited emotional involvement and sympathetic concern for the slaves from the staff and larger society. Such expressions of humanity confounded in direct and indirect ways the impersonal, subhuman standards normally imposed on the slaves and subhuman limitations associated with them.

Over and beyond the various latent forms of behavior of the slaves, the slave plantation regimen organized and scheduled most of the activities of the slaves. A substantial part of the activities of the slaves was regimented, routinized, and closely supervised. They were the behaviors of the slaves which were most directly reflective of and determined by the context represented by the slave plantation. Such behaviors were also of the nature of what Goffman calls primary adjustments. They consisted of obedience and conformity of the slaves to the clear, premeditated, fixed mandates of the plantation regimen. It is largely in reference to these levels of behavior that Elkins (1963) seemed to conclude that the slave plantation was responsible for the predominance of the "Sambo personality"—dependence, childishness, and obedience—among American slaves. In fact, Elkins' analysis falters because he inadvertently accepts the prophecy or stereotype of the clown or Sambo as given (Allport, 1958: 27). He then proceeds to explain it and to use it to chastise the system, as well as to demand a change. Elkins equivocates between self and the presentation of self, between being and acting. In so doing he contributed to the apparent verification of a stereotype which certainly is not completely accurate about the past and that has proven more harmful than useful as a ploy for correcting the present situation (see Bryce-LaPorte, 1971).

Goffman clearly does not subscribe to the notion of a totally subdued or zombified self as being characteristic of inmates. Rather, he views them (inmates) as merely powerless to successfully destroy or radically transform the total institution. This powerlessness lies in the success of the total institution in containing and controlling the inmates as well as in the frailty of any collective disruptive efforts which they may be able to try against the total institution. Thus, Goffman portrays a picture of the inmates as a powerless and trapped aggregation of human beings, whose only rational recourse other than submission is escape. Other than physical escape which could mean death or suicide, Goffman (1961: 319) accentuates various modes of subtle social acts, involved psychological mechanisms of escape, expression, and self-management which occur because of the nature of the human self. "Recalcitrance is not an incidental mechanism of defense but rather an essential constituent of the self" (see Lasch and Frederickson, 1967: 315-329). In addition, he suggests various organization inconsistencies and external contingencies which create conflict in and affect the control and efficiency of the establishment. Hence, it may be argued that there are those levels on which the totality of the total institution is not thoroughly effective; e.g., in confrontation with the selves of its inmates, in its own organizational operation with people as work, and its dependency on and permeability to the larger society (with all the latter's dynamics, conflicts, and complexities).

In Goffman's own analysis, he makes some reference to slavery and estates, but at no time concentrates on the slave plantation as a specific referent. The consequence is that his treatment of institutional-societal relationship is less developed than the first two considerations. Somehow, there is seemingly no explicit beginning and no conceivable end to a specific total institution or type, and very little discussion takes place on the stages of relations which developed between institution and society. So that, where Goffman speaks about the moral career of the inmate (i.e., preclinical, clinical, and postclinical), he treats change on the level of the individual, less on the level of the institution, and certainly not on the level of the larger social order. Even though in another context Goffman (1963) speaks about the roles of watchdog agencies and the "wise" friends and ex-inmates who protect the welfare of inmates, he does not deal with the possibility of a successful revolution by inmates; that is to say, the complete destruction of the total institution and the total elimination of structural and stigmatic legacies it often leaves behind. To leave this sociological task undone would be to fail to derive added insight from the slave plantation experience, with all its historical and dynamic uniqueness as a type of institution whose beginning and end is recorded. Such lessons for the cause of black men still affected by its normative pervasiveness should not remain unlearned.

The question can be asked, in what total institution have inmates been able to accomplish by their own actions a lasting or substantially effective revolution? That is, in what cases have inmates been able as a collective to overthrow their oppressors and destroy the custodial system, the ideology of inequality which led to its establishment, and the stigma and social restrictions which follow membership in it? Even though studies of inmate culture have shown differences in the disposition of different types of inmates toward their stay and status, none shows inmates as ordinarily defined carrying out such a successful revolution. Riots or massive demonstrations by prisoners in penitentiaries and concentration camps tend to be futile or limited beyond the expressive or spectacular level. Such actions may call attention to the adversity of their conditions, but have not directly or immediately ended in their release or the destruction of laws or penal institution. Perhaps there are cases of greater success among the religious bodies or communities which become secular as a group, but it is not known to begin with how much they had already undergone some transformation from total institutions to community in the eyes of the larger world and certainly where the label and ideology associated with their confinement are positive, their chances of losing identity are less problematic.

With reference to slavery, it is generally believed that it was the slave revolts in the Caribbean and Latin America that reached the heights worthy of being considered successful revolution. They were successful inasmuch as they represented massive, persistent, organized actions by slaves that led to various impressive arrays of autonomous states—from sovereign nations to almost closed communities. Such feats were impressive because they overcame the power and pervasiveness of the old order and supplanted it with a new order. The existence of such autonomous forms of social organization negated the myths of inequality entertained about black men and used to rationalize slavery: black slaves proved that they were able to defeat white men by force and wit; establish and defend the frontier, property, and community they gained in battle or in pursuit; and shelter and protect those persons dear to them, as well as those practices that they valued culturally. Those black men had to be accepted as men, and equals, even if reluctantly and only temporarily.

There have been many efforts made to explain the differences between North American slave resistance and those of the Caribbean and Latin America. Often these explanations beclouded the fact that there were also unsuccessful revolts, unorganized and expressive protests in the other Americas, as well; that slaves betrayed other slaves there, too; and that ex-slaves and half-breeds were often used to suppress slave revolts and runaways in proportions unknown to North America. Given the condition of slavery, especially wherever that meant being inmates, it could not be expected that slaves could succeed without a conducive outer setting—both social and environmental. Hence, it is beyond the individual or institutional level that one must turn for an adequate understanding of the form, persistence, magnitude, and impact of resistance. One must turn to the larger society, to specific times in the history of that society for insights into the relationship between the total institution and the society and the definition entertained by that society about the inmate population. In other words, the synchretic, micro-sociological approach must be combined with a historical (perhaps comparative) macro-sociological approach to yield that fuller understanding.

In the case of the United States, slaves generally lacked both the internal and external prerequisites for successful massive defiance. As we have argued in opposition to Elkins, this did not lead them (at least not all of them) to be subdued or zombified. Rather, as Goffman's approach would suggest, they were powerless and trapped, and thus more likely to be elusive than either aggressively resistant or internally submissive. In the absence of elaborate internal or external prerequisites, escape ranged from subtle social and personal modes of eluding and expression to physical escape by way of running away, suicide, or infanticide. The conditions of control in the American slave plantations, the presence of a small minority of free blacks or friendly whites, the strangeness and severity of the physical environment, and the absence of extra-institutional rivalry, prohibited massive action. Given the conditions which runaway and rebel slaves encountered in the antebellum United States, many were forced to live by concealing their real identity (passing), or by residing beyond the boundaries of the societal-legal setting in which slave planters held hegemony. Thus, in the United States much external assistance took the form of the Underground Railroad and various abolitionist and colonization movements which organized and assisted escapees and agitated for freedom. However, for the most part, outside support for inmates was weak in political power, small in number, and generally not aggressively or violently defiant in action (John Brown and his like being exceptions).

Elkins sought to demonstrate that in the case of the American slaves, not only did they not engage in massive revolutionary efforts, but that whenever they did, it was usually under the leadership of ex-slaves or freemen (in Goffman's term, wise ex-inmates). Moreover, these and other efforts by abolitionists were ineffectual in the South because of the nature of the southern antebellum society. It lacked institutional arrangements for channeling rivalry, reform, or radicalism; it was pervaded by complete consensus in favor of slavery and in opposition to any intermediate arrangements for gradually rehabilitating the slaves; and it treated all advocates of abolition and equality as subversives or outside agitators (Elkins, 1963: 192-222). Moreover, planters were not merely head staff of their estates, but they formed the dominant class in their society. Proslavery interests thus permeated slave society (Genovese, 1966a).

The slave plantation was, in fact, simply a regional and historical adaptation of the institutionalized embodiment of white racism which pervaded the larger society. Slavery took other forms in the urban areas and in the North, and where slavery did not exist discrimination, segregation, and exploitation did. Perhaps the most telling point was the fact that freedom, especially in the antebellum South and the "open" West, were often deprived of a property base and frontier land on which to build communities, economic-political power, and to gain self-respect (Bittle and Geiss, 1957: 147-160; Berwanger, 1967; Litwack, 1961; Pease and Pease, 1963; Hill, 1946).

Even after Emancipation and the Civil War, they were denied the acre and the mule, and, later gerrymandered from local control. They were denied land and frontier—the basic prerequisites of honor, mobility, and power in a traditional, agrarian, feudal state. Inasmuch as the country was developing a capitalistic structure, this disenfranchisement proved to be crucial in crystallizing their alienated relationship to property—the principal source of power in a capitalistic society. They were condemned as a people to be nonpropertied producers operating under tenuous imaginings of political freedom and power.

Denial of the basic material resources to which they were always attached also meant denial of the symbols and statutes through which they would have been redefined and recognized as free, equal, and competitive men by their fellow men. In the words of Frank Tannenbaum (1946) they were denied a "moral personality"; i.e., social recognition as equal men, free and equal citizens, and therefore the opportunity to prove their manhood and citizenry.

Slavery, as Tannenbaum (1946: vii) argued, is a matter of definitions of men by other men, which come to give reason and rationalization for certain institutional arrangements by which some men are differentiated, segregated, and exploited by their fellow men. Thus, the appearance and disappearance of a specific institutional context (i.e., plantation), does not necessarily lead to a change in the basic definition, but may reflect efforts to resolve conflicting pressures or regulate coalescent interests. Hence, it is as important to study the developmental history of the society, the institutional forms in which its ideologies regarding man and race are expressed at different times, the degree to which these changes in form reflect the impact of modernization rather than ideological revolution, as it is to study the nature and consequences of these contextual changes on the condition and quality of life of the "pariah group" and on their chances for real rehabilitation (full acceptance as equals).

When seen from a comparative perspective, slavery and plantation are not coterminous, each can exist independently of the other. In the United States not only was there urban slavery and other nonplantation forms of slavery, but it was noticeable particularly in the South that as certain locations become more urbanized, plantations were replaced by smaller compounds (with slave quarters and fences, nevertheless). Plantation codes were replaced by Jim Crow municipal laws, while patrollers continued to exist. In line with one notion of urbanization, city slaves had more choices, greater freedom of movement, and a more diffused authority structure, but as slaves they still had to deal with the reality of confinement. Physical closure was merely replaced by legal closure (Cash, 1941: 103; Franklyn, 1956: 1-9; Wade, 1964: 266-270; Meier and Rudwick, 1969: 14-19). What is more, the official cessation of slavery did not lead to the total redefinition of black men as equals. Van Woodward (1968) correctly referred to equality as the "deferred commitment." In Tannenbaum's view, American blacks were not permitted to shed their slavery "stigma"; what Goffman (1963) calls "spoiled identity."

The Ghetto as Legacy

The abolition of slavery and the disappearance of the slave plantation can be likened to the sudden destruction of any total institution by external sources. The afflicted are usually transferred to another total institution of the same type. That is, until their social definition or the larger society's attitudes toward the afflicted have changed, special institutions will be developed to quarantine and service them. In fact, whatever the emergency structure that is used to hold and treat them, it will then take on the identity of a specific total institutional type by virtue of its association with the afflicted; e.g., prisons for the bad, asylums for the mad, reservations for the useless, concentration camps for the dangerous and disdained.

Accordingly we should not be surprised that blacks have increasingly been quarantined into urban ghettos, disproportionately to more conventional but negatively labeled total institutions, and continue to be deprived of frontier, property, power, or community. Nor should we be surprised at the structure of these new locales. These present forms of institutionalized racism are simply indications that blacks have not undergone, in the eyes of their former masters, what Peter Berger and Thomas Luckmann (1966: 157-167) call "successful alteration"; that is, sufficient rehabilitation to be accepted fully and treated as equals. But, as pointed out, the denial of rehabilitative opportunities is perhaps as old and institutionalized as the denial of a rehabilitated status. Slavery was intended as a self-sustaining system—a vicious circle of injustice. Military and moral pressures, perhaps economic ones, mitigated against the persistence of slavery, and urbanization mitigated against the prevalence of plantations today, but underlying those new urban forms is that lingering definition of blacks as contemptuous unequals. And, built into their operations and structures, are mechanisms for maintaining and reinforcing that inequality.

The ghetto then is not simply an expression of unequal urban development or black social disorganization. It is an expression of the racial legacy of slavery, an urban adaptation of a historically racist society. The phrase "to get Mr. Charlie [Whitey, the Man] off our backs" is at minimum a recognition of that legacy. It is obviously derived from the "ole Massa" tradition of slave plantation days. And while from an objective analytic stance the ghetto as an urban condition is much more complex and diffused than the plantation, it is sociologically fruitful to speculate on the similarities and differences between the two social forms. In fact, the concern should go further to inquire as to which blacks utter such phrases and when. It is already interesting to hear them uttered at times of anxiety and desperation, for it implies a visceral, deeply felt sentiment. To the extent that under such circumstances the range extends beyond just ghetto dwellers to the wider black population, it is suggested that as an utterance of purpose and a definition of a situation, this sentiment is widely shared as well.

Compared to a plantation, the proximity of the black ghetto to the model of a total institution is less obvious but, nevertheless, it shares a number of the anticommunal characteristics of the old estate, even though slavery as such is nonexistent in the United States today. The black ghetto is not owned or managed by a single individual, but from the eyes of the black residents it is often perceived to be owned and managed by one single, exclusive group–whites. "Mr. Charlie" or "the Man" is thus a label for a collective referent with little regard for sex, age, ethnicity, or number. It is a recognition that on certain structural levels the discrete stratification of people into white dominant and black subordinate within a given urban residential context is not a crucial break from the usually smaller and more simplified plantation structure but rather an elaboration. The extrinsic objectives of the ghetto may be multiple, diffused, and less clear, but its intrinsic function is quite similar to the plantation–to legitimize differences among men in terms of the superiority and acceptability of those who *can* live outside from those who *must* live within. Its specialized functions are not always as utilitarian as the plantation, but it shares the same universal functions as the plantation–to exercise custody over the different or unequal. Even though the physical boundaries are often less static than the plantation, they are always made salient in the minds of the populace. Police patrols, uncooperative taxi drivers and garbage collectors, unscrupulous real estate agents and merchants, and uninhibited black pedestrians by their activities give clues to the shift in boundary from time to time. The scope of confinement in the ghetto is felt particularly by those residents who are unemployed or for other reasons are forced to perform most of their daily activities within the local neighborhood. This saliency is even more sharply felt by those whose imaginative or intellectual faculties are not sufficiently developed to provide psychological escape from their physical confinement.

Even though urban power structures are generally viewed as more diffused and democratic than older agrarian forms, in the case of the black ghetto the power structure tends to be in fact unilaterally white and remote at its highest levels. This is still generally true notwithstanding the growing number of black mayors in American cities, particularly as it becomes clear that white suburban and rural populations, through state and county governments, maintain a stranglehold on urban development. The problems and priorities of the ghetto dwellers are often ignored or relegated to a secondary concern unless by implication the calamities they represent are thought to have far-reaching effects. Property and space are often owned by white absentee proprietors. And whether the loci of employment are local or not, unemployment tends to be high, various forms of social ills tend to prevail, and there is a proliferation of welfare and other forms of corrective or paternalistic agencies. In addition to many such agencies being nonrepresentative and substandard, they tend to be noneffective in that they neither improve the rehabilitative potential of their wards nor change the local conditions. They also tend to be ritualistic, repressive, or both. In fact, they often come to recruit, mortify, and stigmatize ghetto dwellers and, therefore, add to the justification of their continued segregation. With reference to their repressiveness, it takes various forms from outright police brutality to more subtle means of social control and cooptation. The institutional network of the ghetto thus represents a system of social control and sanction which is executed by way of coercion and control of crucial services, and is designed to stifle creative individuality, daring leadership, or viable local cooperative groupings which in any way seem likely to undermine the present relationship of the ghetto to the rest of the society, and by extension, blacks to whites.

The ghetto then represents many things not normally found in standard white or establishment-sociological literature. It is an arena of uneven confrontation of will, sentiments, and interests between two traditionally antagonistic forces—those who have achieved status, power, frontier, and community, and those whom they seek to prevent from acquiring such bases of equality (Clark, 1965). Surely it is also the arena of contest and simultaneously of communion (Kaplan, 1969: 164-176; Neuwirth, 1969: 148-165; Pitt, 1970) among blacks themselves, who, beyond the monolithic category imposed upon them by their white antagonists, comprise an array of different statuses, types, and stages of social development. But despite these internal differentiations there is a shared feeling among blacks of not having undergone a complete change of status in the eyes of white America, that their conditions have not completely changed *as a people,* even if as individuals. Thus, as *black* individuals, their acquired statuses are threatened and their individualities overlooked, especially in time of crisis.

It is this frustration, ending sometimes in futility, sometimes in a fighting mood, that such utterances express. It is a feeling of unfulfillment. For, while the ghetto has become the modern birthplace of some of the most ambitious and creative schemes for self-development and self-determination in black America, it has also been the scene of some of the most telling catastrophes and fiascos. The situation is complicated further by accumulated tensions. Tension converts into distress as the immediate causes seem not solely externally derived but also stem from internal frictions and disabilities within the black population (Comer, 1967: 21-27; Kramer, 1970: 213-253). Thus the utterances and actions of ghetto dwellers reflect not simply a style of life nor preoccupation with the present problems, but rather the burdensome baggage of bondage and oppression which antedates the contemporary and the urban.

The history of the United States has been described as that of the acquisition and development of the frontiers. Some historians have declared the frontier closed since the end of the nineteenth century (Turner, 1893). Elkins and McKitrick (1954: 321-353) have ably demonstrated that an important concomitant of frontier development has been local community development and participatory democracy. The history of the United States has also been described as a movement from rural to urban society. Urbanization, of which all of this is a part, is then a historical process, one that has been defined as an increase in the complexity or scale of social organization (Greer, 1962: 33-66). However, urbanization is not equally experienced or shared by all people or regions of a society at the same given time (Myrdal, 1957). That is just part of the complexity. Another aspect of that complexity has to do with a *masking* process which accompanies urbanization, so that old issues are constantly being replaced by new issues whether they have been solved or not. Given the inequality in the society, the problems of the powerful and articulate in-groups are often attended to and solutions sanctioned over those of less powerful, inarticulate groups. In fact, given the power and interest of those who constitute the "establishment" (and I use this term broadly to include even radical white groups in the United States, who are not in power or in conformity with government but nevertheless control resources and hold privileged positions, in terms of being heard, over nonwhites), the problems and priorities of black people are often submerged or redefined away from the uncomfortable and the threatening basic truths of American racism that predate and underlie urban arrangements.

It is a basic truth of American racism that blacks as a group in the United States have been deprived of frontier and its concomitants of community and participatory democracy in respectable proportions. The plantation may have had, as one of its original functions, the opening of the frontier (Gray, 1933; Lynd, 1967: 139), but while the slaves labored within the plantation to open that frontier, they were deprived access to that frontier and of the opportunity to establish communities in the Plantation South. On the other hand, the notion of frontier, a more symbolic phenomenon, a sense of having to achieve, acquire, and exercise domain over something of value, must have been cherished by them in their quest to be equals in the American context. Given the racialistically instigated lag between black and white political-economic developments, it is not difficult to understand that the notion of frontier persists among blacks when it may have died or even been transformed in meaning for whites. Given the power differentiation between whites and blacks today, it is not difficult to appreciate either, that the present urban institutions are not oriented to articulate or reinforce the notions of black people over those of whites. Hence the very notions of ethnic frontier and local neighborhood which are essential for providing impetus to the movement of black people at this particular moment in their history, are issued negative values or secondary priorities by urban agencies, in contrast to the thrust of the Black Muslims, CORE, Republic of New Africa, black studies, the Black Academy of Arts and Letters, the Institute of the Black World, and other current expressions of nationalist domain.

The history of the United States has been described as that of waves of immigrants, each motivated by notions of frontier, which they eventually came to enjoy either as plural communities or as parts of an integrated political-economic arrangement or both (Gordon, 1964; Lewis, 1969). Blacks, one of the oldest immigrant groups, despite the coercive nature of their arrival to America and their brutalizing slave experience, rejected colonization as a scheme and instead chose the way of other Americans–integrated communities or homogeneous ones within a presumably larger integrated total society. Both efforts were frustrated by the two evil white twins of power and paternalism. They have been refused the alternative of equal participation in American democracy or that of establishing control over their own communities. Thus the frustration continues, and the fundamental fight for black frontier persists.

The scene of the struggle today is the ghetto—"getting Mr. Charlie off our back," "getting ourselves together," and "getting our own thing going." The very ordering of these phrases is suggestive of the mood, momentum, and direction of current black movement—from the earlier unorganized, tension-releasing rebellions of Watts, Newark, and Detroit, to the later organized community- and institution-building struggles at Ocean Hill-Brownsville, Newark, and Atlanta. The present can be history and black ghettos are the scene of the making of history by blacks in the interests of blacks.

The ghetto must be studied as frontierland which black people are trying to conquer, to develop as *their* community, to engage in participatory democracy as they seek to build and develop that community. The ghetto is not closed frontier, as establishment-historical literature would suggest. It is merely urban. It is claimed developed territory, defended by white interest groups, much more powerful and sophisticated than the red Indians of old. The drive for domain and the quest for community control by the "newly arrived" black people are repeatedly frustrated by the self-proclaimed new "natives." The contest of urban confrontation is so complex and dynamic that the underlying issues are constantly being masked by new overlays of cumulating epiphenomenal ones. The urban conditions continue to be an overwhelmingly complex institutional arrangement which reflect interest intrinsically at odds with the desire of black people, but interests often distant and beyond "where black people are at."

The rhetoric of the ghetto must not be dismissed any more than the actions of its speakers. Cool talk, angry talk, crazy talk, jive. Whatever it may be called it must be looked upon with historical consciousness as existential and perceptive cues for new sociological insights on the meaning and nature of black urban behavior. The urban condition of most black people, the ghetto, must be studied for its retentions, parallels, and departures from the slave plantation. The particular behavioral and organizational nuances and limitations of ghetto people must be studied in comparison with slaves, as Malcolm X suggested. The complexity of ghetto culture must be studied in terms of a "bicultural underlife" of a people of whom most have been segregated and suppressed ever since their arrival in the continental USA (Blauner, 1970: 347-366; Valentine and Valentine, 1970). As such it is a set of behaviors and expectations too often unnoticed, misunderstood, and disregarded. It is pregnant with secondary adjustments and esoteric symbolism, which confound white citizens and scholars alike. The conception that white men hold today of black men and the institutionalized system by which it is expressed must be studied against the plantation past. It is only then we can begin to answer how far we have come, how much further we must go, and what means are necessary to get there. *Dig it? Then let's get with it.*

NOTES

1. We use the term "fuller understanding" to indicate knowledge which transcends the questions of persistence and functions of social behavior to include process, meaning, and origin as well. Such knowledge is not derivable from traditional synchronic and rigid studies, but must result from historical, comparative, and intuitively creative humanism as well (see Evans-Pritchard, 1951: 58-62; Smith, 1962).

2. This was evident in the enthusiastic manner in which the late Horace Cayton responded to my own decision to pursue such a study. It would, he felt, create a new vista for viewing the slave plantation.

3. Historians are producing increasing evidence of a high percentage of two-parent households and marriages among ex-slaves during Reconstruction. While we must be careful not to generalize too extensively from these findings, they do make plausible the argument that an ethic and the bonds of family organization were already strong among all classes of blacks, even during slavery (Blassingame, 1970; Guttman, 1971; Robinson, 1969).

REFERENCES

ALLPORT, G. (1958) The Nature of Prejudice. New York: Doubleday.

BERGER, P. and T. LUCKMANN (1966) The Social Construction of Reality. Garden City, N.Y.: Doubleday.

BERWANGER, E. H. (1967) The Frontier Against Slavery. Urbana: Univ. of Illinois Press.

BITTLE, W. E. and G. L. GEISS (1957) "Racial self-fulfillment and the rise of an all-Negro community in Oklahoma." Phylon 18: 247-260.

BLASSINGAME, J. (1970) "A social and economic study of the Negro in New Orleans." Ph.D. dissertation, Yale University. Unpublished.

BLAUNER, R. (1970) "Black culture: myth or reality?" in N. Whitten and J. Szwed (eds.) Afro-American Anthropology. New York: Free Press.

BRYCE-LaPORTE, R. S. (1971) "Slaves as inmates; slaves as men: a sociological discussion of Elkins' 'Sambo' thesis," in A. Lane (ed.) The Debate on Slavery: Stanley Elkins and His Critics. Urbana: Univ. of Illinois Press.

--- (1969) "The American slave plantation and our heritage of communal deprivation." American Behavior Scientist (March-April): 2-8.

--- (1968) "The conceptualization of the American slave plantation as a total institution." Ph.D. dissertation. UCLA. Unpublished.

CASH,M. V. (1941) The Mind of the South. New York: Random House.

CLARK, K. (1965) The Dark Ghetto. New York: Harper & Row.

COMER, J. (1967) "The social power of the Negro." Scientific American 216 (April): 21-27.

EDGERTON, R. and G. SABAGH (1962) "From mortification to aggrandizement: changing self-concepts in the careers of the mentally retarded." Psychiatry 35, 3.

ELKINS, S. (1963) Slavery: A Problem in American Institutional Intellectual Life. New York: Grosset & Dunlap.

--- and E. McKITRICK, Jr. (1954) "A meaning for Turner's frontier." Parts I and II. Pol. Sci. Q. 69: 3 & 4.

ETZIONI A. (1961) A Comparative Analysis of Complex Organizations. New York: Free Press.

EVANS-PRITCHARD, E. E. (1951) Social Anthropology. London: Cohen & West.

FRANKLYN, J. H. (1956) "History of racial segregation in the United States," in T. Reid (ed.) Racial Desegregation and Integration. Annals of the Amer. Academy of Pol. and Social Sci. 34: 1-9).

FRAZIER, E. F. (1960) "Introduction," in V. Rubin (ed.) Caribbean Studies: A Symposium. Seattle: Univ. of Washington Press.

--- (1939) The Negro Family in the United States. Chicago: Chicago Univ. Press.

GENOVESE, E. (1966a) The Political Economy of Slavery. New York: Random House.

--- (1966b) "The legacy of slavery and the roots of black nationalism." Studies on the Left 6, 6: 2-36.

GOFFMAN, E. (1963) Stigma. Englewood Cliffs, N.J.: Prentice-Hall.

--- (1961) Asylums. Garden City, N.Y.: Doubleday.

GORDON, M. (1964) Assimilation in American Life. New York: Oxford Univ. Press.

GRAY, L. C. (1933) History of Agriculture in the Southern United States to 1860, Vol. 1. Washington, D.C.: Carnegie Institute.

GREER, S. (1962) The Emerging City. New York: Free Press.

--- (1955) Social Organization. New York: Random House.

GUTTMAN H. (Forthcoming) The Invisible Fact. New York: Pantheon.

HANDLIN, O. (1958) "Editor's preface" in E. Morgan. The Puritan Dilemma: The Story of John Winthrop. Canada; Little, Brown.

HILL, M. (1946) "The all-Negro society in Oklahoma." Ph.D. dissertation. Univ. of Chicago. Unpublished.

HILLERY, G. (1963) "Villages, cities, and total institutions." Amer. Soc. Rev. 28.

JORDAN, W. (1969) White Over Black: American Attitudes Toward the Negro, 1510-1812. Baltimore: Penguin.

KAPLAN, H. M. (1969) "The Black Muslims and the Negro American quest for communion." British J. of Sociology 20 (June): 164-176.

KARDINER, A. and L. OVESEY (1962) The Mark of Oppression. New York: World.

KRAMER, J. (1970) The American Minority Community. New York: Thomas Y. Crowell.

LANE, A. [ed.] (1971) The Debate on Slavery: Stanley Elkins and His Critics. Urbana: Univ. of Illinois Press.

LASCH, C. and G. FREDERICKSON (1967) "Resistance to slavery." Civil War History 13.

LEWIS, Sir A. (1969) "Black power and the American university." Princeton University Magazine (Spring): 8-12.

LITWACK, E. (1961) North of Slavery. Chicago: Univ. of Chicago Press.

LYND, S., Jr. (1967) Class Conflict, Slavery, and the United States Constitution. New York: Bobbs-Merrill.

MARSH, R. (1964) "The bearing of comparative analysis on sociological theory." Social Forces 43, 2.

McKITRICK, E. [ed.] (1963) Slavery Defended. Englewood Cliffs, N.J.: Prentice-Hall.

MEIER, A. and E. RUDWICK (1969) "A strange chapter in the career of 'Jim Crow.' " The Making of Black America, Vol. II. New York: Atheneum.

MERTON, R. (1957) Social Theory and Social Structure. Glencoe: Free Press.

MILLS, C. W. (1959) The Sociological Imagination. New York: Grove Press.

MOORE W. (1941) "Slave law and the social structure." J. of Negro History 26, 2.

MORGAN, E. S. (1958) The Puritan Dilemma: The Story of John Winthrop. Canada: Little, Brown.

MYRDAL, G. (1957) Rich Lands and Poor Lands. New York: Harper & Brothers.

NEUWIRTH, G. (1969) "A Weberian outline of a theory of community: its application to the 'dark ghetto'." British J. of Sociology 20 (June): 148-163.

PEASE, W. H. and J. H. PEASE (1963) Black Utopia. Madison: The State Historical Society of Wisconsin.

PHILLIPS, U. B. (1963) Life and Labor in the Old South. Boston: Little Brown.

PITT, J. P. (1970) "The meaning of black consciousness: community or communion?" Paper given at 55th Annual Meeting of the Assn. for the Study of Negro Life and History. Philadelphia.

ROBINSON, A. (1969) "The reconstruction in Memphis." Senior essay. Yale. Unpublished.

ROSE, A. (1969) "Varieties in sociological imagination." Amer. Soc. Rev. 34.

RUBIN, V. [ed.] (1960) Caribbean Studies: A Symposium. Seattle: Univ. of Washington Press.

SMITH, M. G. (1962) "History and social anthropology." J. of the Royal Anthropological Institute 67, 5.

SZASZ, T. S. (1970) Ideology and Insanity. New York: Doubleday.

TANNENBAUM, F. (1946) Slave and Citizen. New York: Knopf.

THOMPSON, E., Jr. (1959) 'The plantation as a social system." Plantations of the New World. Pan American Union, Washington, D.C.

TITTLE, C. and D. TITTLE (1965) "Social organization of prisoners: an empirical test." Social Forces 43, 2.

TURNER, F. J. (1893) "The significance of the frontier in American history," in E. E. and F. M. (eds.) The Early Writings of Frederick Jackson Turner. Madison: Univ. of Wisconsin Press.

VALENTINE, C. and M. L. VALENTINE (1970) "Blackston." Progress report of a community study on Urban Afro-America. Unpublished manuscript.

WADE, R. C. (1964) Slavery in the Cities–The South, 1820-1860. London: Oxford Univ. Press.

WAGLEY, C. (1960) "Plantation America: a culture sphere," in V. Rubin (ed.) Carribean Studies: A Symposium. Seattle: Univ. of Washington Press.

WISH, H. (1961) Antebellum. New York: Capricorn Books.

WOODSON, C. (1919) The Education of the Negro Prior to 1861. Washington, D.C.: Assn. for the Study of Negro Life and History.

WOODWARD, V. (1968) The Burden of Southern History. Baton Rouge: Louisiana State Univ. Press.

YETMAN, N. (1970) "Testing Elkins' 'Sambo' thesis: some preliminary problems." Unpublished manuscript.

9

Public Housing: Urban Slums Under Public Management

LISA REDFIELD PEATTIE

□ PUBLIC HOUSING is an institution which is arresting not only for its particular and, in some cases, unique characteristics, but also because, seen as a system, it presents in microcosm many of the basic issues with respect to policy for American cities. That a look at public housing tends to focus on problems is not surprising; we find it possible nowadays to fill whole volumes with essays on "urban problems" ranging from air pollution to zoning issues, and the fact that a number of the characteristic urban problems are represented in that particular part of the urban environment we call public housing is surely not remarkable. What I would like to suggest is that the problems of public housing are not simply the result of their (so to speak) drifting in and lodging there, but rather that they are created by the way the institution functions.

It is proposed here that it may be useful to think of "public housing" in any city as a complex system, the output of which is, as a whole, system-determined–that is, not truly planned or intended, much less controlled, by any particular individual or group, but rather the outcome of the regular interrelations of a number of components or subsystems. These would include the politics of housing at the federal and state level; both formal and informal bureaucratic processes shaping the housing bureaucracy; social and economic forces, especially those governing social mobility trends and patterns; community-based politics within the projects and their surrounding communities; and group processes within the projects. These are not best thought of as discrete factors affecting life in public housing, for each of these components of the situation is responsive to other components in such a way that they may be said to form a system. In understanding how the systematic interrelations of these components of the public housing system produce the galloping decay which characterizes public housing generally, we may possibly arrive at some useful insights relevant to other problems of our cities.

Racial problems are notable in public housing; some of the most conspicuous and troubled ghettos we have are public housing projects. Racial issues easily come up in discussions of public housing. A good deal of tenant organization draws on feelings of solidarity among blacks, as Chavez has drawn on Spanish-American ethnic feeling to organize the farm workers. Yet the point of view taken here will be that the racial issue in public housing only complicates and deepens issues which arise out of the way this urban institution, like the rest of our cities as institutional systems, sort, deal with, and shape the lives of the poor, regardless of race. Here again, a look at this single institution may shed some light on the problem more generally.

Public housing as an important institution is closely associated with the big cities. Although the 2,200 local public housing authorities include many with very few units, twenty cities with populations of over half a million contain one-third of all federally aided public housing. New York City, with 150,000 units, has the largest single aggregation of units.

It has come to seem normal for big city governments and big city services to be in financial trouble, but public housing has more than its share of such troubles. The system involves a combination of government subsidies and rents. Federal or state subsidies in general cover only debt service; operating expenses are to come out of rents. Rents in a public program for low-income people cannot rise to keep up with the increasing costs of operations and maintenance, due to inflation and project aging. The result is a growing financial squeeze.

Statistics from the Department of Housing and Urban Development (U.S. News and World Report, 1969: 66) showed sixteen housing agencies nearly bankrupt, and another nineteen on their way to bankruptcy; fifty-one of the eighty-two largest authorities, accounting for nearly two-thirds of all public housing, were in critical financial shape.

Associated with the financial problems, both as cause and as result, are a wide range of other difficulties. Aging boilers break down in midwinter. Empty apartments are occupied and vandalized by groups of young people looking for a place to gather. Windows are broken by stonethrowing. Halls are littered with garbage which never made it to the incinerator. Playgrounds glitter with broken glass. Mailboxes are so often broken into that they are no safe receptacle for a welfare check. Women are afraid to go out of doors at night for fear of being mugged and robbed.

All this has a peculiar poignancy when one considers that the public housing program had its origins in the great social reform movements of the thirties as a program to solve the "slum problem." Public housing has not solved the slum problem. Instead it has succeeded in giving it new packaging under public management. Certain large high-rise housing projects—St. Louis' Pruitt-Igoe, Boston's Columbia Point, New York's Bedford-Stuyvesant—have developed national reputations as conspicuous, high-density, publicly managed slums, and stand as monumental reminders of the failures of our management of urban life (Salisbury, 1958; Moore, 1969; Rainwater, 1970).

A CASE IN POINT: LIFE IN A "BAD PROJECT"

To take a concrete example, consider the situation in one of the "bad" project areas in Boston where I have spent a little time.[1]

Included in the "management area" I will be describing are two contiguous projects. They are immediately recognizable as public housing. Perhaps in Boston the project format—brick apartment buildings, sited over a several-block area rather than arranged along the street grid—would be enough to so identify them, for Boston generally lacks the high-rent version of this kind of building which is found in New York. But in any city *these* buildings would be immediately recognizable as institutional housing for the poor by their extraordinary drabness. The grounds outside the dirty red brick buildings are landscaped mainly in chain-link fence demarcating asphalt play space, rows of clothes lines, and, in some cases, traces of discouraged grass. The management office (or, as the tenants call it, the rent office) guards itself behind heavy window screens: there is reason for defense, as typewriters have been stolen from the office overnight. Nearby is the maintenance shop—tools and parts, more chain link, a couple of desks, and workmen coming and going.

In this pair of projects live 980 families or single individuals, 712 of the units being occupied by blacks. Each project has a small group of old people who pay their rent promptly, struggle to make ends meet out of their tiny checks, worry about their door locks and about going out at night. Most of the apartments are occupied by families. Two-thirds of the apartments are occupied by families with no working member, and supported by some sort of public assistance. The largest number, over half, received aid from the Federal Aid to Dependent Children program. There are in the projects over twice as many children under sixteen as adults twenty-one to fifty-nine. There are ten times as many children as adult males under sixty, compared to a ratio of 1.2 citywide.

The day world of the project also includes an office staff of a manager, a couple of assistant managers, a cashier and three clerical workers and a maintenance staff of a superintendent, about fifteen skilled workers, and an equal number of laborers. Although in this project the manager and a couple of maintenance men are black, the staff is, as in the Boston system generally, almost entirely white. They are also generally middle-aged or aging, with an ethnic tinge of Irish. In the Boston system, a carpenter or plumber in public housing makes eighty percent of the union scale for outside, but the work is more secure, less arduous, and indoors. It is the sort of job which appeals to a man who is getting too old to shove heavy pipes around or whose arthritis bothers him when he works outside in cold weather. Neither maintenance nor the office provide much likelihood or scope for upward mobility, but both maintenance and office staff have almost absolute security. That security is prized; most staff got their jobs through some sort of political connection. Finally, there are social workers and organizers attached to the health center, the antipoverty group, the community center—people both black and white, generally young, with some sort of commitment to social change.

When I first came to this project I spent the morning in the maintenance shop talking to the maintenance men and watching the flow of events. The general atmosphere reminded me of nothing so much as an old-time movie of life in the trenches in World War 1. Here comes a group of five old ladies, indignantly complaining about a number of small plumbing repairs on which they have been awaiting service. Next comes a policeman, asking that the men remove a large, dead German shepherd dog; this generates a long discussion of means, since no scoop shovel is available. (Although dogs are prohibited in the leases, the project is full of large dogs, seen as one means of protection against being robbed.) An angry male tenant demands a new refrigerator; the maintenance superintendent tells him that his one-bedroom apartment is not entitled to a large refrigerator, all there is on hand. Another old lady comes in to complain about her leaking ceiling; the clerk tells me that roof has been awaiting repair for two years. Laborers come looking for implements to clean out the bottom of one stairwell; by the end of my visit, this stairwell, last cleaned perhaps five months before, yields three truckloads of black, wet refuse. The carpenter is asked to board up a window, since while they are engaged in sweeping up a cellar, boys keep throwing stones. One of the laborers asks me to request "central" to provide bolts on the *inside* of the maintenance office so that the staff can lock themselves in when men are trying to steal their pay checks. The superintendent shows me a set of yellowing snapshots showing shrubs and flowers in the project in its earlier days: "It used to be lovely here; we had a different class of people then," he says.

Here, already, are a number of the basic themes: the interaction between deteriorating physical environment and social behavior and social selection; the prevailing sense of difficulty, decay, and distrust; the conflicts between housing authority staff and tenants; the sense of things having once been better.

The superintendent was speaking to a sense of social change within his lifetime—the processes which have tended to replace the original tenants of public housing with a class quite different from that which occupied the projects when they were first built. When "the projects" first opened their doors during the latter part of the Depression, they housed mainly working-class and lower-middle-class families temporarily dropped from the currents of economic success by the vicissitude of the business cycle. Public housing was not seen as stigmatizing; it was something for which to exert oneself. In the 1930s some Public Housing Authorities limited the number of public assistance recipients they would accept, and others would not admit any (Schorr, 1966: 234). In Boston, at least, families tell me you really had to "know someone" to get an apartment in a project. As prosperity came back to the United States, these first families, or many of them, moved out of the projects. Their places were taken by a sort of underclass, in high proportion black, very different from the original residents. They were the displacees of urban renewal or highway building with nowhere else to go, the families on welfare, the people with no real prospects of movement out.

The very demographic characteristics of this new population spell trouble. In Boston, and I suspect elsewhere in the country, the projects with the high incidence of vandalism and of juvenile delinquency, with all the attendant consequences of conflict and general insecurity, are projects with a big load of children. When, as in the project described earlier, kids under sixteen outnumber adult males by ten to one, it does not take a gift of prescience to know that there are problems ahead. Results: urine in the elevators, marking on the walls, broken windows and broken lights, stolen checks, and vandalized empty apartments.

Other problems, less easily measured, result directly from the shape of the new-tenant population pyramid. A student who had spent some time working with a tenant group in the project described, commented feelingly to me on the sense of a social world dominated by women—women hostile to men, needing men, and trying to keep from being overwhelmed by the problems of a woman managing alone, bringing up children poor. A tenant leader in another city said to me, in a discussion of militance and fear in tenant organization, "If you're bringing up kids in poverty, I don't know what you've got left to be scared of."

The rent structure of the projects contributes to maintaining a largely dependent tenant population. Families are supposed to move when their income gets over a certain level. Income of young people in the family is usually counted. This policy is not illogical if public housing is viewed as a scarce social welfare resource. But if consistently carried out, it ensures that families who are making a success of it in the American system will leave, and be lost to the projects as potential leaders. Meanwhile it reduces the incentive of remaining tenants to get employment—and leads to cheating both by tenants who fail to report their income in order to stay, and project staff who let some families remain even though over-income. Even so, and while rents fall further and further behind operating expenses, there are people too poor for public housing.

It is not true that public housing tenants are the poorest of the poor. It has been calculated that families with less than $1,500 income have half the chance of getting into public housing as those with $1,500 to $4,000 (Schorr, 1966: 234). Moreover, as one writer puts it: "Compared with an equivalent number of dwellings in slum or blighted areas, public housing developments undoubtedly include fewer aged indigents, transients, single men and women, hoboes, crackpots, criminals, panhandlers, prostitutes, alcoholics, bohemians, taxi-dancers and other social misfits" (Dean, 1966: 257). But the proportion of welfare recipients has risen, the class level has fallen, and the poor who are in public housing have special visibility.

Past history, and the present situation, color the relationships between the tenants of public housing and the sorts of people who staff them—the people who collect the rents, sweep the grounds, clear the clogged drains. Looking at Boston as an example, at least, it would appear that the staffs of housing projects are people much like the first round of tenants, and, indeed, got their positions often in the same way—i.e., through knowing some local politician. Some of them have lived in public housing in the past. Or they used to visit their granny in public housing. So when they see the deterioration of the bad projects in conjunction with the appearance of a "different class of people," their reaction is that of people seeing their old neighborhood run down as new people have moved in. "They've run out the good tenants by accepting without screening, by not forcing people to abide by the rules. This place was beautiful. I have pictures of how it used to be in the old days . . . flowers . . . window boxes."

The age and social background of the project staff are such as to make for strong disapproval of "that element" among tenants characterized by sloppy housekeeping and failure to exert strong discipline over children and teenagers. "I lived when law was law. . . . I never got into trouble with the law. . . . If somebody gave me a good boot, I respected him for it. . . . I grew up in a house with a wooden floor. My mother scrubbed it clean every Saturday. My father used to say, 'You don't have to be rich to have soap and water' . . . . There's no respect for old people . . . . They aren't looking after the kids . . . . They want to put the responsibility on society; where's the husbands?" People like this are likely to react to poor families, especially to the FADC mother swamped under a bunch of children and minimal income, with downright hostility or with that sort of patronizing attitude which is almost worse–like the lady clerk in one project office who was explaining to me that "welfare is just like a disease, you know."

These attitudes are strongly reinforced by the project situation. For the project is undergoing conspicuous physical deterioration, easily laid at the door of the tenants. The maintenance system becomes continually more backlogged. The two working painters at the project I am describing are far behind with possible work; many apartments have not been painted for years. The two carpenters cannot keep up with mailboxes to be replaced (broken on the first and fifteenth when the welfare checks come in) and doors and locks to be replaced (for similar reasons). One old man who had been a general butt for aggression once had his lock ($18 each) replaced five times in one day. The plumbers are keeping up with requests for servicing drain stoppages, but problems with the boilers cannot be unconnected with the many badly leaking faucets which permit the hot water to keep running. Worst of all is the situation with respect to replacing broken windows. There are two glaziers on the project, and last time a count was taken there had been 7,720 broken window panes awaiting replacement.

The project has a staff of laborers which has the task of cleaning the grounds and lower stair halls. They are short-handed, owing apparently to difficulty in getting central to provide replacements, and absenteeism in the functioning staff is high. The men who are on duty perform their work with a certain lackluster air; indeed, as they will frankly tell you, they are discouraged, and do not feel like trying very hard. "They dirty as fast as you clean." The project grounds have a sort of dusting of bits of broken glass and trash. Within the buildings, the situation in the public halls is often considerably more forbidding. There is supposed to be a system by which various tenants, one each landing, take turns sweeping down the halls. This system is difficult to make work; if a tenant moves out (transiency in the project is high) it destroys the scheduling, and all it takes is for one assigned tenant to fail to sweep, or to sweep late down a hall already cleaned below, for those who are trying to comply to become exasperated with the whole system and with their fellow tenants. In any case, it would be hard to mop the halls in the winter because of the broken windows which convert scrub water on the floors into a bone-breaking sheet of ice.

At the time I visited this project, there were approximately two hundred vacant apartments. Some had been vacant for as long as seven months—in the face of a stringent shortage of low-rent housing in the city, and a considerable waiting list for public housing. One reason they are vacant is that while there is a waiting list for public housing, prospective tenants are by no means anxious to move into *this* public housing. Another reason is the problem of arranging for the already overloaded maintenance staff to do the work required to put empty apartments into rentable condition. Again, windows are a substantial bottleneck, especially since empty apartments almost immediately have their windows broken; one apartment alone, for example, had 48 panes to be reglazed. In each case, all work must be completed in coordination and the apartment rerented at once, or vandalism will undo the labor. A sheet of vandalism reports I saw included a record of $700 worth of damage to a vacant apartment which had just been readied over the weekend for occupancy.

Since any vacant and, therefore, undefended apartment is likely at once to be invaded by the teenagers (for whom the project environment offers too little otherwise in the way of space out of the way and the surveillance of adults), the problems of vandalism and of empties tend to grow in tandem. An empty apartment invites vandalism. The prevalence of vandalism discourages other tenants from trying to control the young people, and to turn instead to focusing on their own defended family territory; vandalism and empty apartments contribute to a general sense of insecurity on the part of tenants, and motivate families who have some other housing option to move out. Every moveout is an additional job for maintenance, in disposing of abandoned furniture and clearing up the vacant apartment. Vacant apartments and trash in the halls in turn support vandalism and the creation of more empty apartments to be vandalized.

It is hard to run such a project as a "tight ship," and in such a setting administrative control over tenancy struggles and may break down. Families move out without notifying the manager, and neighboring families do not trouble to report their departure. Low-income families live at the edge of financial crisis and it is easy to get behind in the rent. Low-income families, living in a setting they hardly regard as desirable, and may see even as hostile, may indeed not feel a strong sense of obligation to pay their rent on time. The management staff, living in its own way from crisis to crisis, finds it hard to mobilize the time and energy to meet with tenants in arrears and try to arrange for payment. Tenants find there are no strong sanctions on nonpayment. Other tenants, seeing arrearages mount up among some project families, feel in their turn less obligation to pay. At the time of writing, over forty percent of the tenants in the project here described were behind in rent payments; arrearages amounted to over ninety percent of the monthly rent roll. Rent arrearages contribute, with the vacancy rate and the vandalism, to draining the system of the financial resources to put things in order.

This process may, in some circumstances, reach epidemic proportions unknown in Boston (so far). It is likely that the record up to now has been set by St. Louis. In 1969, a tenants' rent strike in effect brought their system to collapse, and a reorganization under a coalition of tenants, local business, civic, and labor interest, found it necessary to put up a kitty to finance reorganization. St. Louis' Pruitt-Igoe project recently had sixty percent of the apartments vacant. Whole buildings were in effect abandoned with broken windows, unlivable, and unlived in.

In such a project, the maintenance of security for person and property becomes a continual preoccupation. There are demands for more police protection; there are attempts at volunteer patrols. There is demand for better lighting, for better locks, for peepholes at the doors. There is a recurring theme of the breakdown of social controls; people hesitate to try to prevent delinquent behavior, or to testify against persons they have observed in some delinquent act for fear that "they'll break my windows."

There are several major social control problems which overload the maintenance system and the tenants alike. The problem of vandalism, affecting mainly vacant apartments, is thought to be one of "kids," generally boys under sixteen. Breaking and entering, which affects occupied apartments, and at times the offices of the project staff or of social service groups, is thought to be carried out by older youths. Probably the same is true of breaking open mailboxes. Littering halls and throwing trash out of windows is probably done largely by housewives, although a contributory force may be children deputized by some overworked mother to carry down the trash, who get just so far with the load. The sense of personal insecurity contributes to trash littering, as people hesitate to carry their trash into dark halls or cellars. Muggings and purse-snatchings are evidently by older youths, perhaps not necessarily residents of the project. In any case, although people—both staff and residents—refer continually to "problem families," there seems to be no firm evidence to prove that problematic behaviors are generated by a limited number of problem families. It may be that these behaviors are, instead, common aspects of the functioning of rather ordinary people overloaded with problems, undersupplied with personal and financial resources, and living densely crowded in a difficult environment which lacks enough usable space and public amenities.

In any case, in such a situation, efforts to organize the tenants to take action about their common problems have continually been thwarted by the tendencies of distrust, mutual recrimination, and pervasive sense of "it's-no-use," generated by a social situation that does not work. The difficulties of life do not create a sense of common cause among the tenants; rather, they instill mutual suspicion and an inclination to blame the unpleasantness of the project environment on other tenants.

There are, nevertheless, in our sample project several such efforts, and a rather active schedule of meetings, although the same few faces seem to attend each one. The most active group at the time of writing was a tenant committee which dealt with the HUD "modernization program." This effort, supported by top staff of the authority anxious to develop more active and responsible tenant organizations, has labored under considerable difficulties, even though its access to a sum of money specifically earmarked for project improvements (which the local group is to arrange to spend) gives it a degree of visibility and credibility. The election for committee members, under HUD directives, held rather hastily and on a Saturday, which was also a welfare check day, did not elicit a great deal of participation from project residents. The resulting task force has found the process of getting visible results worked through a complex system involving citizen participation, the local and federal bureaucracy, and technically difficult considerations of cost and feasibility, an extremely frustrating one.

There is also a project funded by OEO to explore the feasibility and desirability of having the tenants take over part or all of the management of the project. This group has had the contracted services of a black consulting group from New York, got from them a favorable report on feasibility, *if* the project could be brought physically up to snuff, *if* adequate tenant organization could be created. It is, at this writing, proceeding to explore the matter further with the tenants. To an outsider, these discussions sometimes appear a particularly poignant example of the recurrent process by which blacks move into neighborhoods sufficiently decayed to admit them, or become the mayors of bankrupt cities.

HOW PROBLEMS REINFORCE EACH OTHER IN PUBLIC HOUSING

This description of the state of things at one public housing project is, of course, focused on troubles. This no doubt reflects in part the writer's preoccupation. But it is more than that, for a preoccupation with troubles is indeed characteristic of the participants in the setting described whether tenant, manager, maintenance man, or social worker. Troubles are characteristic of the setting.

There are the physical troubles of drab esthetics, physical inconvenience, deteriorating buildings and equipment; the problems of psychic tone—the sense of sliding down from the "good old days," the sense of its not being worth trying; there are emotions of bitterness, of anger, of distrust, of sense of stigma; the problems of population structure, of the basic difficulties experienced by a population so heavily dominated by women with children; the problems of delinquency, of crime and vandalism, of drug addiction; the problems of reconditioning and renting vacancies, of rent collection, of financial balance.

Yet enumerating the difficulties in one troubled project is hardly the beginning of understanding. The troubles of this project are interrelated. They have a systematic character. The accelerating character of public housing problems may be understood as an aspect of the functioning of a system which an engineer might describe as containing a large number of mutually reinforcing feedback loops—vicious-circle processes in which the institution responds to its external problems by changing in ways which make the problems worse.

The different class of people who have been coming into the projects with their difficult-to-manage population structure and their general lack of commitment to the social settings, makes for increased wear-and-tear on projects. The interaction between project staff culture and the new tenant culture makes for conflict, which probably increases project deterioration, which, in turn, greatly increases the sense of conflict. As the projects become less desirable environments, turnover and vacancy rates rise; the rise in turnover and vacancy rates increases the burden on maintenance, which in turn accelerates deterioration of the environment and reinforces the sense of an adversary relationship between tenants and the authority staff. This process tends to reduce the work motivation and efficiency of the staff, which in turn accelerates the process of environmental deterioration. As the environment deteriorates, the project becomes more and more a housing option of last resort, which means that the more stable and stabilizing families will tend to move out, and those who move in will tend to be those with fewer resources, both personal and financial.

The tenants who take the housing of last resort are at the bottom of the income scale. With rents tied to income, revenues sink. Meanwhile vacancy rates and rent arrearages rise; wear and tear on the projects increase and expenses go up. The result is a slide toward financial collapse—which, in turn, lowers the motivations of staff to try to make it work successfully, and increases the tendency to develop an adversary relation with tenants. The whole thing could, no doubt, with patience, be elegantly diagrammed.

Is this project typical? No, it is not; it is one of the worst in the Boston system, indeed, the one I myself find most unpleasant. But there is more to this question. It turns out that public housing systems are extraordinarily stratified for systems which present uniform rent schedules and supposedly, at least, uniform entry procedures and criteria. There are bad projects and there are good projects. A good project has a low transiency rate; people rarely move out; vacancies are rare and are immediately filled when they occur. Their maintenance tends to be good; one even sees vines and flowers. These projects have few children and fewer families on welfare. They tend to house a better class of people, as the maintenance men would say. There are certainly substantial differences in median income between families in the good projects and those in the undesirable ones.[2]

In comparing one project with another in a given city, since rents are uniform, the vacancy rate may reasonably be considered a fairly good index of overall social desirability. The possibilities for social stratification within one city's public housing system may be dramatically represented by the situation in St. Louis; in the nine projects of that city, vacancy rates recently ranged from .03 to over 60%. In other words, these publicly managed housing systems succeed in developing the sorts of social stratification between better and worse which are characteristic of urban neighborhoods in the private market.

One may well become curious as to the mechanisms through which this kind of stratification takes place. One is definitely aware of a sense on the part of management and tenants that social segregation is natural and desirable. The reformers in Boston went through a prolonged struggle, not wholly successful, to break down what at one time was an almost complete racial segregation in the projects. But there is segregation by class, also, in the same spirit of social management. In a neighboring city I was told that of two contiguous public housing projects one was for the good families and one for the bad families, and that when a family made application, the manager made an administrative judgment as to the proper placement. In another outlying system in the Boston area, the managers of the most desirable project took pains not to report vacancies to the city urban renewal commission, but to reserve these spots instead for families in which they had some particular interest. Other families without such resources of influence were relegated to the more isolated and more undesirable projects. Blacks were not only concentrated in the project with the highest total vacancy rate, but in the buildings at one end of that project.

The Department of Housing and Urban Redevelopment promulgated regulations, which became operative early in 1969, intended to work against this process. A new applicant, once deemed acceptable, was to be offered first a place in the project with the highest number of vacancies, then one in that with the second highest, and so on; if he declined all high-vacancy projects, he would go to the bottom of the list again.

In practice, these procedures seem to have failed in their intended effect. In Boston, a review of the first year's experience showed that approximately two-thirds of all applicants offered the three high-vacancy (or "bad project") locations rejected the housing. Only fifteen percent of applicants offered housing accepted tenancy in one of the three locations. Vacancies in the projects continued to increase. Whites continued to flee the developments with heavy concentrations of black tenants, and housing segregation by race (probably also by income) within the system accelerated. Interview and observation would suggest that there is a selection of tenants who because of age, or medical needs, can develop cases for exceptions which tend to abet the process of social stratification. One factor in Boston involves relocation workers who, feeling that the renewal agency has in the past made commitments to families in renewal areas that they might remain in their own neighborhoods, take some pains to help their clients develop such rationales for exception. No doubt in any city, persons who really know their way around inside the urban political and administrative structure can find a way to get at the more desirable places. Thus it turns out that it is only the families without resources of expertise and influence, the recently arrived Puerto Ricans, the blacks displaced by urban renewal, the welfare mother at her wit's end to keep her family together, who get routed into those projects which become, in effect, the housing of last resort for those with no other option.

What is it that determines which projects become public slums, and which remain desirable projects? I know of no definitive study, and I myself cannot weigh the various factors or be sure I have them all. I suspect that, in general, high-rise projects more easily become slums—but I have seen a system in which rather dirty high-rise brick buildings were the "desirable" projects, and row housing, one story high, was the bottom of the system. Physical isolation from the rest of the city seems to contribute to making a project a slum. I have seen some evidence that a particularly gifted and energetic manager may keep a project "good" when other forces might be tending to push it into the spiral of decline. The surrounding neighborhood must be relevant. A project under the control of some local politician who lets the agency admit only the right sort of families will tend to remain fairly well kept up. One might agree with the maintenance men I interviewed that the central factor in the character of a project is the class of people who live there—but the class of people is not an independent variable. It varies with the other components of the public housing system, and in its turn affects the way the project is managed and maintained.

HISTORY OF THE PUBLIC HOUSING SYSTEM

This system as it now functions must be understood in terms of its history. For certainly in the functioning of the system as it now stands, a central element is the project format, the fact that "public housing" mainly consists of and is socially identified with large, high-density, physically distinctive publicly managed rental units for low-income people. And this project format can only be understood historically, in the way in which economic forces (the Depression) interacted with intellectual forces (social reform movements) to create the projects.

Public housing was born within the time, and out of the turmoil, of the national crisis of the Depression. Out of the sense of crisis and problem came diverse currents which oddly intersected in a characteristically American combination of social idealism and commitment to private profit. There was a strongly felt need to stimulate the economy. There was a strongly felt need to do something about the "third of a nation" suffering most severely from the Depression. These currents of aid to private enterprise and drive for social reform joined in the context of some characteristically American ideas about the nature of social reality: (1) public expenditure should support, in case of necessity supplement, but should certainly not compete with or substitute for private enterprise; (2) allocation of goods and services to the needy is generally to be preferred to income transfer; and (3) the slum, seen basically as a physical environment, is the breeder of social pathology (this principle is probably a subinstance of the American faith in material progress as the key to social betterment).

The result was a program:

> To provide financial assistance to the states and political subdivisions thereof for the elimination of unsafe and unsanitary housing conditions, for the eradication of slums, for the provision of decent, safe and sanitary dwellings for families of low income, and for the reduction of unemployment and the stimulation of business activity [U.S. Housing Act of 1937, as amended].

and the operationalizing of this program through the building of "housing projects."

The carrying out of the program was entrusted to local housing authorities, quasi-autonomous local bodies responsible not only for the decisions on the operational management of projects once built, but also on the amount, character, and location of additional public housing. Each local housing authority is usually governed by a five-man appointed board. These boards have generally been highly conservative. A recent survey showed that a prevailing lack of aggressive action on the part of local authorities to increase the existing public housing stock is a reflection not only of the difficulties of and the political pressures against such aggressive programs but also of the preconceptions of a sizable proportion of the commissioners themselves (Hartman and Carr, 1969). In part, this conservatism is no doubt a function of the processes by which appointments are made, and the general insulation of the local housing authorities from the processes of issue politics within the municipalities. However, it seems reasonable to attribute some of this character to the preoccupation with the problems of operational management of the housing projects themselves. The focus of attention has been the management of going real estate. This situation has, in turn, helped to develop the kind of stable bureaucratic structures within public housing administrations which make focus on institutional stability and the maintenance of job security for the bureaucrats a dominant response to any attempts at innovation or flexible response to situations.

The project format, inherently stigmatizing in a society in which poverty is a source of stigma, then became a basis for additional problem-enhancing feedback as the advent of economic prosperity permitted many of the stable white working-class families who were the early tenants to get on their feet and move out, and replaced them with families of lower social status, many of them on welfare or for other reasons perceived as problems, and many of them black.

The project format has helped in isolating residents from the surrounding community both by supporting in the surrounding communities a sense of distrust and derogation of "those people" in the projects, and among the project residents a sense of being unwanted. It has helped to develop in the people who work on the staffs of public housing agencies a sense of being the custodians of a category of people damaged and perhaps somehow inherently defective. Finally, the project format as a conspicuous manifestation of social problem has helped to make public housing a rather unpopular program politically. Public housing became identified with the bad projects. Neighborhood residents tend to regard the prospect of public housing in their midst much as they might the suggestion that they come to shelter a colony of lepers, and, in recent years, the only sort of public housing which seems to have any considerable political negotiability is that for the elderly.

Meanwhile, this theme intersects with another of the political themes of our times—the growing militancy and political muscle of blacks in the central cities. This development manifests itself at the citywide level in demands by tenant groups for better service, in a push for greater degree of tenant participation and tenant power in decision-making, and in the pressure for the public housing bureaucracy to hire blacks. One system, that of St. Louis, is currently being run in part by a militant (mainly black) group of tenants who organized a massive rent strike in the projects.

At the project level, management and maintenance staff see that while the social class level of the tenantry is sinking, the tenants' political position is rising vis-à-vis the project bureaucracy: "You can't tell the tenants anything now." Some adapt themselves to this situation with grace; many with a sense of difficulty and grudge. All "middle America's" sense of being squeezed between black militance, upper-middle-class liberalism, and the unruliness of youth, is evoked by interviews with some of the persons in this position.

Public housing comes to present an interesting case study of the interplay between racial and class definitions of a problem situation. The replacement of the first families of working-class and lower-middle-class style by the "welfare mothers" and the rest of "that element" has also tended to be in large part a substitution of black families for white. Racial conflict comes to play an important role in the projects, both among tenants and between tenant and management. There is the brick in the window of the isolated black or white family, the occasional youth gang conflict which has racial overtones. A black woman who had lived for many years in the bad project described earlier told me that when the first black families came they did not dare sit outside in the evening for fear of being fire-bombed. Nowadays the attempt to organize tenants tends to borrow from the whole rhetoric and symbolism of black militance in general. In view of the tendency for racial and class categories to coincide, it is not surprising that some of the feeling of resentment on the part of housing authority staff or on the part of some tenants against others should generalize as racial feeling. But the issues involved seem to be basically ones of social class and social structure, and to a degree which surprises me, are so defined by persons in the system. Interviewing project staff in Boston, I thought it remarkable in view of the circumstances, that the low-level staff were not more racist than they are. In general, the staff tended in interviews to define the issues in terms of social class. "You can find the same problem in ______,–a large project housing predominantly white and very poor tenants. A black woman tenant in a predominantly white project spoke as bitterly about the other tenants as could ever happen if the racial roles were reversed: "My husband's in the service. We'll get out of here, not like those others. They're ignorant. They're sick–You try to organize anything . . . nothing work, everything dies."

There are issues here which are specifically issues of public policy, issues which have to do with the public management, the public responsibility, and the public visibility of the system as it now stands. And there are also issues which have to do with the problems of the poor and their relations to their neighbors in any setting. The focus here on the project format should not obscure the fact that such problems as high vandalism, rent arrearages, high transiency, delinquency, personal insecurity, a tendency for tenant and manager to develop an adversary relationship, and for maintenance to be inadequate, are just as characteristic of the housing situation of the poor in the private market. They simply develop a particular salience here because they are public problems and public responsibility.

STRATEGIES FOR REMEDIATION

There is a whole bundle of strategies for doing something about public housing which revolve around ideas of the social composition of the tenant body. The tone of voice may be different, but the root of the problem is nevertheless identified much as it is by the angry maintenance worker who sees "that element" as responsible for tearing down what had been a functioning system. Thinking here may be centered around the concept of problem families, in which case what is proposed is more rigorous screening of prospective tenants, home visits by the management staff, a more punitive policy of more frequent evictions for nuisance behavior.[3]

Or thinking may revolve around ideas of a more balanced social composition of the tenant body; i.e., a lesser dominance by really low-income people. There might be policies which would encourage tenants to remain in the projects as their income rises, rather than following a policy of over-income evictions. Various formulae are proposed under which families within a single dwelling environment might be on various plans for payment, and might shift from one plan to another as their circumstances changed. Any scheme which permits the public housing authority to own or lease certain units, rather than whole buildings, would, of course, have this outcome.

Some such strategies do indeed seem appropriate ones for improving the conditions within housing projects. But are they strategies for improving the conditions of life for poor people? It is no doubt unfair to reduce the argument to the level of absurdity by saying that if there were *no* poor people in the projects (the rent collections of course reflecting this), most of the problems would disappear. But even an intermediate strategy of scattering the very poor among the more well-to-do raises some of the issues implied in such a reductio ad absurdum.

At the other extreme, there have been discussions of the problems of public housing which have implicitly defined the problems as architectural—those of the high-rise project format. Favorite material for this point of view is the Pruitt-Igoe project in St. Louis which delighted cynics of smart-aleck architecture by being a huge architect-professional success, and then failing conspicuously in practice. The leading problem here seems to have been some wide interior galleries, designed with the best of intentions as a street and play space accessible to mothers who would have trouble accompanying their kids to ground level; these turned out to be areas which being unpoliceable and, in the social structure of the project, undefended by informal social organization, were in fact wholly unsafe. Yet Pruitt-Igoe's sixty percent vacancy rate (whole buildings are abandoned), while extreme, is by no means unique in the St. Louis system which, when the big rent strike began, had an average vacancy rate of forty percent for the whole system, so attribution of a strong causal role to architecture would require a look at all projects. It has become clear that the projects in general are ill-designed, lacking especially in those intermediate sorts of spaces which, while outside the tight-locked family apartment door, can still be defended; so no doubt better design could improve the situation.

Going further, it is not hard to conclude that high-density high-rise housing is inherently an unsuitable environment for families, and more unsuitable still for poor families. One of my students recently undertook a study of the social and political organization of tenants in a large project (Lipschutz, 1967). Although she began with a strong prejudice against attributing any great importance to the physical environment, she came to see the project environment as critical to the functioning of the system. She saw the large project as one which made social life difficult by continually bringing tenants into abrasive contact with each other through public shared spaces and shared facilities doomed to malfunction. Furthermore, one can see that the projects have done much to give the whole idea of public housing a bad public image, which has in, turn, contributed to the difficulties of the system and its tenants. These considerations lead to various strategies for dispersing the public housing stock and rendering it both less visible and, hopefully, less special, either by small "scattered-site" units in public ownership, or through the leased housing program in which a local authority can lease units in privately owned housing, to be occupied by tenants eligible by income and housing need for public units.[4]

Looking at the classically impacted character of public housing bureaucracies and the expensively legalistic specifications in the building of public housing units, other strategies go toward reducing the dominance of the public sector in the system. The leased housing program leaves ownership and most of the management function in private hands. (It turns out that the public authority still in practice at least does some coping with tenants.) The Turnkey I program leaves the development function in private hands, but has the public authority carrying out management as before. The Turnkey II program would let housing authorities turn over the operating management of public housing projects to private groups, including tenant groups. Meanwhile, the federal government is beginning to talk hopefully about selling the public housing units to the tenants.

It seems appropriate to say at this point that reducing the dominance of the public sector, while perhaps having some limited utility, as a general approach to the problem seems to attribute entirely too much causal importance to the public character of public housing as the source of its problems. There seems no reason to believe that the management and maintenance problems of poor families in high-density setting would disappear once the buildings passed to tenant ownership. Furthermore there seems no reason to believe that the tenants would, in fact, be anxious to own property in a project environment like that described earlier. Indeed, one might say that just as they treat carelessly and vandalize the project because, not liking or identifying with it they see no reason to cherish it, so they would see no reason to own it because they do not like it or feel it properly theirs. Surely no ordinary private investor would wish to buy such a set of buildings except as some sort of tax loss. A program for selling projects to the tenants would, one might think, only find eager customers in exactly those projects which now are relatively without problems—where the roses still bloom by the entries, and vacancies rarely occur. If the program were to put these into tenant ownership, the position of the tenants in those projects might, indeed, be a reasonably desirable one. But what of the tenants in the bad projects? Their financial burdens of wear-and-tear, vacancies, rent arrears, and basically unworkable financing formulae, are now uncompensated by any more viable parts of the system.

Organizing the tenants is a strategy of process which may move in the direction of several possible goals. It may be seen as a way of getting the tenant to be an active partner with housing authority staff in solving the problems of maintenance and management in the projects—creating a modernization program responsive to tenant needs and helping push the program through, or dealing with disruptive or delinquent behavior on the part of certain tenants. Or tenant organization may be seen as moving toward developing an adversary relationship with the housing authority, perhaps ultimately a somewhat formalized position as bargaining agent for the tenants vis-à-vis the public authority. Or tenant organization can be seen as moving toward the development of a political force, an organized low-income housing constituency, at the state or federal level.

All these strategies have their logic, and any given tenant organization is likely to emphasize more than one. But they very often tend to compete in practice. For example, the tenant committee, which works on the public authority's modernization program, has an attractive opportunity to develop control of a program with real money to spend and the possibility of producing some real and visible results. But the collaborative role competes with the adversary role. It is hard to maintain a collaborative framework and not be perceived by at least some embittered tenants as having sold out and become tools of management. Especially do these roles come into conflict when the question is one having to do with the system of rules and controls within the project system. Tenants who sit on a pre-eviction hearing board may find their own (often quite strong) desire to improve the environment by exercising sanctions on problem families coming into conflict with their position as representatives of the tenants, and their need for solidarity with the tenants' side of things. Thus while the idea of tenants controlling their own project has a certain inherent plausibility and a great deal of attractiveness even to a management which finds the job beyond them, this turns out to have serious practical difficulties in a setting in which tenants can exercise control only within a structure of administrative delegation. Similarly, a tenant organization which wants to build to a kind of broad-based strategy for influencing national legislation on public housing, must build a constituency at the local level by working on issues of immediate concern, and may, in the process, lose sight of its long-range strategy in deploying energy to win some immediate objectives.

The various strategies which move in the direction of decentralization and of community control of the public housing system raise, in addition, some very basic and very difficult issues. Even beyond the difficulties which arise when the tenants try to control a physical and institutional terrain bounded by a structure of external control, there are the problems which arise when, to some degree, they do make it their turf. The turf is limited. Along with the going control of a turf goes control over entry. In discussions with tenant groups it would appear that one of the first demands to arise under conditions of tenant control would be the expulsion of undesirable families. The people who are near the bottom see no reason why they should share what they have built with people lower still in the social system. Just as community groups in urban renewal areas have supported programs to get rid of the poorest families in their neighborhoods, so in public housing one may hear tenant leaders suggest that there should be limits, for instance, on the number of welfare clients admitted. Public housing cannot function effectively as both the housing of last resort for the people ejected by the rest of the city, and as a system controlled by, and in the interest of, the majority of its residents. The issues are hard to resolve. Poor people want to choose their neighbors as do middle-class people. But the urban environment must accommodate everyone.

Last, there is the strategy of improving public housing by getting more money for the system. The highly restrictive financial structure of public housing rests on the implicit assumption that once a federal subsidy has contributed to the building of the plant, the housing of low-income tenants can pay for itself out of rents. It cannot. There are ways of making money from the poor, but providing decent housing at prices the poor can afford is not one of them. If people, who receive incomes which will not pay for decent housing at the current prices of such housing, are to live in decent housing, the rest of us will have to help pay for it. As things now stand, the welfare system is helping to subsidize public housing by subsidizing its tenants, but it is not enough. Something more will have to come from the federal government (that is now clear), but it would appear that it will come only with difficulty and in insufficient quantities, responding to the structure of political forces around public housing which character, as noted earlier, the present program has had its part in creating.

A SELF–DESTROYING SYSTEM

This very basic problem of public housing—that it has become a special program for a special category of people not much valued in our society—was nicely stated several years ago by Catherine Bauer Wurster (1966):

> The most questionable assumption was the notion that slum rehousing should be established primarily as an independent program, with its own separate legislation and administrative machinery at both Federal and local levels, quite apart from other housing policies and the overall housing picture. This insured the segregation of the low-income slum dweller and fortified his isolation as a special charity case by permitting only public initiative and public landlordship with narrow rules of eligibility for any form of subsidized housing that might be needed. This also contributed to the segregation of upper income families in FHS schemes and to that lily-white suburbia that now presents such a critical problem.... Vested interests grew up and were institutionalized around each separate fragmentary program.

It is not simply true that there are things wrong with public housing. Public housing is a system which, by its very character, functions to create ghetto slums under public management.

Some years ago an artist named Jean Tinguely created in the sculpture garden of the Museum of Modern Art in New York a "self-destroying machine," made out of gathered scrap and junk parts including, I believe, an old upright piano. Set up and put into operation, it slowly knocked itself apart, burned itself up, and tore itself in pieces under the eyes of the invited and very elegant guests of the museum. At the end, it was unable to quite finish itself off, and the city firemen, who had been standing by to protect the museum, had to complete the job with their axes.

A system which is constructed in such a way as to be self-destroying should be dismantled, and the usable parts salvaged to build a system which works. Such a housing system has yet to be designed. We can already see, however, what some of its characteristics would be. It would not separate public housing projects from other housing for the poor. It would not separate programs for building public housing from programs for developing housing through other sorts of incentive mechanisms. It would not separate the problems of housing from the problems of income, and the problems of disparities of power, and of opportunity. It would not separate the problems of getting the buildings up from the problems of creating a structure of incentives and an on-going system of resources which will maintain buildings. It would not separate the problems of creating a dwelling environment for the poor, or for some group of the poor, from the problems of creating a dwelling environment for people who are not poor. It would be a system of much greater scale than the present public housing system. It would be part of a workable system for urban life.

NOTES

1. The Director of the Boston Housing Authority had asked me to conduct a short-term, informal investigation of the problems and needs of maintenance men in the system. Mind-boggled by the situation described in the following pages, I stayed on to work for six months for the agency, in a general effort to understand how such things might come to be.

2. Figures comparing family income project by project will soon be available for Boston, at least, but are not obtainable at time of this writing.

3. Aside from the ethical aspects of such policies, it should be noted that there are political problems attached as well. Sometimes a public housing system will attempt to have some tenant, seen as a nuisance, evicted from the project, only to find that the courts regard the action as prejudicial and refuse to issue the order. Or such policies may intensify the adversary position of tenants vis-à-vis management, thus increasing the managers' problems.

4. Unfortunately, the recent difficulties of Boston's infill housing program in gaining acceptance within local neighborhoods suggests that making the units smaller goes only part way in reducing public opposition, and that it is not only whites who are likely to object to having low-income families housed on their blocks.

REFERENCES

DEAN, J. P. (1966) "The myths of housing reform," in W. L. C. Wheaton, G. Milgram, and M. E. Meyerson (eds.) Urban Housing. New York: Free Press.

HARTMAN, C. W. and G. CARR (1969) "Housing authorities reconsidered." J. of the American Institute of Planners 25, 1 (January).

LIPSCHUTZ, E. (1967) "Some Effects of Layout and Administration on Tenant Relationships in Two Low-Rent Public Housing Projects." Radcliffe College. (Unpublished).

MOORE, W. (1969) The Vertical Ghetto. New York: Random House.

RAINWATER, L. (1970) Behind Ghetto Walls: Black Family Life in a Federal Slum. Chicago: Aldine.

SALISBURY, H. (1958) The Shook-Up Generation. New York: Harper & Row.

SCHORR, A. L. (1966) "How the poor are housed," in W. L. C. Wheaton, G. Milgram, and M. E. Meyerson (eds.) Urban Housing. New York: Free Press.

U.S. Housing Act of 1937, as amended.

U.S. News and World Report (1969) "Crisis in Public Housing." (November 24).

WURSTER, C. B. (1966) "The dreary deadlock of public housing," in W. L. C. Wheaton, G. Milgram, and M. E. Meyerson (eds.) Urban Housing. New York: Free Press.

10

Institutionalized Anglocentrism: Labeling Mental Retardates in the Public Schools

JANE R. MERCER

□ ALLEN IS A 15-year-old Negro boy with an IQ of 78 who has been in special education classes since he was 8 years old. His mother describes him in the following manner:

> He's a rebellious teenager. . . . I feel he pushes me to my limit. He's sick of a woman telling him what to do. Allen is very frustrated because he can't read. This has much to do with his problem. If you can't read, you can't do very much of anything. . . . I try to help him here at home. Around the neighborhood he's really a handyman. He fixes electrical appliances for the neighbors, moves things, mows lawns. He'll repair a lamp or something that's broken, if he wants to.
>
> I feel that if a child is put in a special education class in elementary school, by the time he gets to junior high he should be removed because there is a stigma that goes with a special class. Let's face it, children can be real cruel. I feel for the most part the youngsters

---

AUTHOR'S NOTE: *A more comprehensive and detailed analysis and report of the referral process, the clinical testing process, the assumptions and inferences of the clinician, and the pluralistic evaluation process discussed herein appear in my forthcoming volume entitled* The Eligibles and the Labeled.

> that are in those classes and retained suffer a great emotional handicap. It's as if they have a sign around their neck for everyone to read. Many of the teachers talk down to these youngsters. . . . When I signed for Allen to go into the slow class, I knew there were some retarded children in the group. This did not bother me, as these children have a right to learn just as anyone else. However, I did not realize that it was mixed to the degree that it is. I feel that it definitely inhibits Allen's learning. But, if he's taken out of the special class now, he will be so far behind I just wonder if he'll ever catch up.

John is a 16-year-old Negro boy who has been in special education classes for the past 8 years. He has an IQ of 83. When John's mother was asked to describe what he does on Saturdays and around the house and in the neighborhood, she gave the following reply:

> John works on weekends down at the service station as an attendant. . . . I would say he's a good mechanic. He likes to work on cars, changes the oil, helps with overhauling a car, and works on motorcycles. Sometimes he irons, washes dishes, mops the floor, cuts the lawn, sweeps off the driveway, goes to the store, runs errands, vacuums the rugs, makes his own bed, and things like that. He's never still too much. He likes to be outdoors and likes to ride motorcycles. He plays basketball and football on Saturdays, works on cars, and then goes to bed.

Bill, a Negro boy with an IQ of 66 who has been in special education for 6 years, has a mother who has the following feelings about special education classes:

> They told me Bill needed help in reading and if he was put in special class he would be able to get better help because the class is smaller. . . . Bill is being retarded in special education. He doesn't like being labeled as retarded. It's affecting him. He begs us to have him removed from that class. I feel his teacher is too old to teach, plus she told Bill she would see that he'll never get out of that class.
>
> We have to make Bill go to school because that class does not offer a challenge to him. What they do is repetitious–the same thing over and over. . . . He does not like school. We have to make him go. The

> only reason he consents to go is because we have been promised that he'll be taken out of that EMR class. The teachers have asked us to let them put another one of our kids in EMR. We said an emphatic "no!" because we knew what it was all about.
>
> I wish educators would stop lying to parents about this program. It's a conspiracy to keep the minorities down. They put as many as possible in these classes because it means more money for the school. Many times it's because of racial prejudice or a behavior problem. I do know it's most unfair.

Pete is a 14-year-old Mexican-American boy with an IQ of 79 who has been in special education classes since he was 10 years old. The mother described Pete's placement in special education class as follows:

> The school wanted to put Pete in special class because he was not learning. They thought he was mentally retarded. I took him to a doctor and he examined him and gave him tests with blocks or something. The doctor said there was definitely nothing wrong with him. I knew that too, but the school had to know it. All that is wrong with him is that he is hard-headed.
>
> He is like any other normal boy. . . . He likes school and would like to learn more, but he gets discouraged. Before this year (he was returned to regular class) nobody taught him anything and he wanted to know a lot. At school, all he learned was to play, paint, ceramics, paste, and, once in a while, they would spend time with some school subjects. I don't understand why teachers don't teach.
>
> Pete is a very bright child, he's always thinking and doing something; he built a two-room tree house that is just beautiful. He is good at anything that needs putting together. He makes cars with a motor, he makes them from old boxes, tires, wood, anything he can find. He makes cages for the animals. . . . He is very helping. Sometimes, when I am trying to do some plumbing but cannot do it, he knows how to fix things around here. He is good at plumbing and at figuring things out. He is not good for studies. He has a hard head.

Maria is a 13-year-old Mexican-American girl who has been in special education classes since she was 8 years old. She has a Full Scale IQ of 62. Her mother told the interviewer that Maria does not want to go to school and goes up to her room and hides until the school bus goes away. She knows that her mother cannot take her because her mother does not drive. Maria's mother said that she was very sorry that she had signed the papers because now Maria had to remain in the special education class while her sister went to another school. She said that she is not going to sign the papers for the other boys that the school wants to put in special education. She did it once but is not going to do it again.

Maria's mother said that she did not know at first that there were mentally retarded children in the special class that Maria attended. The school told her that Maria was behind in reading, they told her Maria was slow but not retarded, with a young mind about four years behind her age. She let Maria go into special class because she was going in a smaller class and would get extra help. Whenever Maria cannot do something, the family calls her an MR, especially since they found out that there are mentally retarded kids in Maria's class. Her brothers and sisters kid her about it. According to the mother this makes Maria feel bad, but no one means it–it's "done only in fun":

> Maria is no trouble, she is a good girl. . . . She is shy at school, but not here. I wish her teacher would see her at home. She is curious and loves to read. Many times I have to take books away from her; she reads too much and then gets headaches. . . . She makes dresses for her dolls, helps around the house, loves to go outside to play tether ball, and plays many different kinds of games. She is good at sewing. She has a Barbie doll and makes all her dresses. She has a box full of doll clothes for every occasion and uses the sewing machine. She babysits with my sister's children to earn spending money and helps around the kitchen. She cleans around the house and does a little bit of ironing and helps with the washing. She is very good at that.

These students all hold the status of mental retardate in public school. They have been placed in special education classes for the mentally retarded. Because they have been assigned the status of mental retardate they are presumed to suffer from a condition called mental retardation which has been defined by the Group for the Advancement of Psychiatry as

> a chronic condition present from birth or early childhood and characterized by impaired intellectual functioning as measured by standardized tests. It manifests itself in impaired adaptation to the daily demands of the individuals own social environment. Conversely, these patients show a slow rate of maturation, physical and/or psychological, together with impaired learning capacity [Group for the Advancement of Psychiatry, 1959].

Doll's classic definition of mental retardation includes six criteria: (a) social incompetence, (b) due to mental subnormality, (c) which has been developmentally arrested, (d) which obtains at maturity, (e) is of constitutional origin, and (f) is essentially incurable (Doll, 1941).

What have these students done to be labeled as persons with impaired learning capacity which is assumed to be of constitutional origin and essentially incurable? Who are the labeled retardates?

THE PROBLEM

Disproportionately large numbers of children of Negro and Mexican-American heritage are labeled mentally retarded by the public schools and placed in special education classes. This phenomenon appears to be true throughout the United States in communities with a sizable Mexican-American or Negro population.

MEXICAN-AMERICAN STUDENTS

For example, in the state of California in 1966, there were 86,781 children enrolled in special classes and 23,103 had Spanish surnames, 26.6% of the total (California State Department of Education, 1967). The comparable percentage has varied only slightly in subsequent years. Since approximately 13% of the children attending public schools in California have Spanish surnames, there were twice as many children with Spanish surnames in special education classes as would be expected from their proportion in the total population of the public schools. The ratio varied from county to county. In three of the counties with the highest overrepresentation in Spanish-surname children in special classes, the ratio was 2.48, 2.3, and 2.23. Although other counties have lower ratios, Spanish-surname children were consistently overrepresented.

When these statistics are presented as rates per thousand, the disproportions are more readily interpreted in terms of children. In one county, 51.8 Spanish-surname children in every thousand were in special education classes compared to 15.1 English-speaking Caucasian children (henceforth called Anglos). Comparable rates for some other counties with 12,000 or more Spanish-surname children enrolled were 21.2 to 7.1; 45.3 to 13.0; 44.5 to 13.2; 49.1 to 16.0; 25.5 to 8.8; and 32.4 to 9.9. The rates of placement for Spanish-surname children are consistently two to three times higher than Anglo rates.

NEGRO STUDENTS

In the state of California in 1966, there were 18,336 Negro children enrolled in special classes, 21.1% of the total. The percentage remained at 21% for 1967, and varied only slightly in subsequent years. Since approximately 8.0% of the children attending public schools in California are Negro, there were two-and-a-half times more Negro children in special classes than would be expected from their proportion in the total school population. Again, the ratio varied from county to county. In one county, the ratio ran as high as 4.6 times more Negro children in special education classes than would be expected from their proportion in that county's school population. Three other counties had over three times more Negro children than would be expected and most counties with more than a 5% Negro enrollment in school had twice as many Negro children as would be expected.

Again, when these statistics are presented as rates per thousand, they are even more meaningful. In one county, 108 Negro children per 1,000 were in special education classes compared to 16 Anglo children per 1,000. Other high rate counties had 51.3 Negro children to 13.0 Anglo children; 72.2 to 13.2; 62.2 to 15.1; 74.1 to 15.7. In every case, Negro rates of labeled mental retardates per 1,000 in the public schools ran more than three times higher than Anglo rates.

INSTITUTIONALIZED ETHNOCENTRISM

What accounts for these very large disproportions? Why are Mexican-American and Negro children labeled mental retardates at a much higher rate than their proportion in the population? Why were the children described in the case studies in classes for the retarded? This paper explores the general hypothesis that institutionalized procedures based on an ethnocentric definition of "normal" behavior can account for most of the difference in rates between ethnic groups.

Persons in every social system develop a set of shared expectations about each other's behavior. These expectations are an important part of the norms of the group and form a significant part of the common culture. Central to the culture of any society is its language and the world view expressed through that language and its literature.

Persons who conform to the expectations of a group, who are skillful in the use of the common language and its abstract concepts, and who are knowledgeable about the history and literature of the group, are usually accorded valued positions in the group. They are viewed as "intelligent." There is a corollary tendency to view anyone who does not conform to the behavioral expectations of the group, who does not comprehend the group's language, or know its history and literature, as unenlightened and, perhaps, unintelligent. In its most extreme form, this tendency becomes ethnocentrism—the belief that the customs and values and language of one's own group are superior to those of other groups and are the "right" and "good" standards by which the behavior of all persons ought to be measured.

The primary social institution in contemporary society for perpetuating the language and literature, value, and history of a society, is the educational system. In most societies, educational institutions have the task of establishing the role expectations and norms of the dominant group in that society as the legitimate canon of acceptable behavior. In the United States, the cultural patterns of that segment of the population consisting of white, Anglo-Saxon, Protestants has been the primary influence on the educational system. This cultural domination dates back to colonial times and has never been seriously threatened (Gordon, 1964). The study of the English language, English literature, Anglo-American political and social systems, and Anglo-American history, comprise the major portion of the public school curriculum. Training in a language other than English or acquaintance with the history or social systems of other cultural groups comes late in the educational sequence and tends to be peripheral. The public schools are highly Anglocentric and this tendency has been institutionalized in the curriculum and in the evaluation procedures used to determine which students shall have the valued statuses in that system.

The purpose of this paper is to explore two subhypotheses concerning the possible relationship between high rates of labeled mental retardation in the public schools for Mexican-American and Negro children and the institutionalized Anglocentrism of the educational system: (1) differentials in labeling may be related to differential referrals from teachers and principals; (2) ethnocentric clinical procedures which discriminate against persons who are not from a modal, Anglo background may account for the differentials.

THE REFERRAL PROCESS IN THE PUBLIC SCHOOLS

Disproportionately high rates of placement in classes for the mentally retarded of Spanish-surname and Negro children may result from a referral process in the public schools which discriminates against non-Anglo children. Proportionately more non-Anglo children may be placed in special education classes because teachers and principals may refer them at a proportionately higher rate than they refer Anglo children. To explore this hypothesis, we examined the referral processes in one medium-sized California school system whose procedures are similar to those used in other school systems of its size in the state. Like other districts in the state of California, it, too, had disproportionately large numbers of non-Anglo children enrolled in classes for the mentally retarded. In the year of the referral study, 45.3% of the children placed in special classes for the mentally retarded were of Mexican-American heritage, although only 11.0% of the population of the school district consisted of children from Mexican-American homes; 22.6% were Negro, although only 7.9% of the children in the district were Negro.

At the time of the study, the school district had approximately 25,300 students enrolled in 22 elementary schools, 6 junior high schools, and 3 senior high schools. In addition, there was a special school for physically handicapped children. Although the schools varied greatly in the size of their student bodies and the size of their administrative staffs, all schools depended upon the staff of the Pupil Personnel Department located in the central administrative offices of the district for psychological evaluations of children. The two primary responsibilities of the department were to organize and evaluate findings from the extensive group testing program of the district and to administer the individual intelligence tests which are required by state law before a child may be placed in a program for the mentally retarded. This department, directed by a psychologist with the rank of associate superintendent, controlled the movement of children into the many special programs offered by the school district. Any child who was removed from the status of normal student in the regular classroom to mentally retarded student in the special education classroom had to have his placement certified and legitimized by one of the three certified psychologists working for the Pupil Personnel Department. They, and only they, could certify a child as mentally retarded and eligible for special education.

In order to study the referral processes in operation, we investigated the characteristics of all 1,234 children who were referred to the Pupil Personnel Department for any reason during a single school year. We followed the steps by which a child was placed in a class for the mentally retarded to determine at what juncture in the referral process the disproportions so evident in final placement statistics appeared. Potentially, any child who is referred can be tested by a psychologist and found eligible for placement in a class for the mentally retarded. In actual practice, however, the 66.2% of the children referred by a teacher or principal were more likely to be tested than children referred from other sources. In the year studied, 31.0% of the children were referred as candidates for classes for the gifted, 29.8% were referred exclusively for behavior and disciplinary problems, 19.5% were referred for unspecified academic difficulties with no mention of possible retardation, and 8.3% were referred for evaluation specifically because the teacher-principal team believed they should be placed in a class for the mentally retarded. There was a consistent decline in the rate of referral as grade level increased.

We found that non-Anglo children were not overrepresented among the total referrals. Only 6.9% of the children sent to the Pupil Personnel Department for evaluation were Mexican-American and 10.3% were Negro.

Because school psychologists did not have enough time to administer individual intelligence tests to every child referred for evaluation, they had to decide which children to test. Any child who is not tested cannot be placed in a class for the retarded. When we examined the characteristics of the children psychologists elected to test, once again we found that non-Anglo children were not overrepresented. Only 7.6% of the children administered an individual intelligence test during the year of the referral study were of Mexican-American heritage and 9.5% were Negro.

Among those children tested, there were 134 who had an IQ of 79 or below and, hence, were eligible for placement in a class for the mentally retarded. Of these children, 32.7% were Mexican-American and 19.8% were Negro. Among the 81 children who were recommended for placement from this group, 40.9% were Mexican-American and 21.2% were Negro. Among the 71 who were actually placed, 45.3% were Mexican-American and 22.6% were Negro. Thus, there were four times more Mexican-American and twice as many Negro children among those who were eligible for classes for the mentally retarded because of their low IQs than we would expect from their percentage in the population tested. In addition, Mexican-American and Negro children with low IQs were somewhat more likely to actually be placed in special education classes than Anglo children with low IQs.

The discontinuities in the referral process are very distinct. Although teacher-principal teams referred Mexican-American and Negro children at about the same rate as their percentage in the population and proportionately fewer were given intelligence tests by school psychologists, about three times as many Mexican-American and Negro children appeared among those failing the intelligence test as we would expect from their proportion in the population of the school district. Subsequently, this disproportion increased so that four times more Mexican-American and three times more Negro children were placed in special education classes than would be expected from their percentage in the district because proportionately more Mexican-American and Negro children with low IQs were recommended for placement and were ultimately placed. It was at the point in the referral process when the intelligence test was administered that the sharp ethnic disparities first appeared. Disproportions in referrals appeared after the clinical process of intelligence-testing.

Evidence from another source supports this conclusion. There were 509 Mexican-American and 289 Negro children in the regular classes of the same school district who were administered a Wechsler Intelligence Scale for Children as part of a study conducted for the Public Health Service.[1] We found that 15.3% of the Mexican-American children and 12.4% of the Negro children scored 79 or below and thus would have been eligible for a class for the mentally retarded, if the teacher had referred them for evaluation and they had been tested. Only 1.2% of the 500 Anglo children similarly tested were found to have IQs of 79 or below. This indicates that many Mexican-American and Negro children who are potentially eligible for classes for the mentally retarded on the basis of IQ are not referred by teachers, but very few Anglo children who are eligible remain in the regular class. Thus, the preponderance of Mexican-American and Negro children in classes for the educable mentally retarded does not appear to be the result of overreferral from teachers and principals but rather results from the clinical testing process itself.

THE CLINICAL PROCESS IN THE PUBLIC SCHOOLS

Our evidence points to the clinical process rather than the referral process as the primary factor in disproportionate placement of Mexican-American and Negro children in classes for the mentally retarded. The primary instrument used to diagnose mental retardation is the standardized individually administered intelligence test, usually the Stanford-Binet LM or the Wechsler Intelligence Scale for Children (WISC). Although the clinical definition for mental retardation adopted by the American Association for Mental Deficiency defines a mental retardate as one who is subnormal in both intelligence and adaptive behavior, there are no recognized scales for measuring adaptive behavior and this dimension is not systematically evaluated as part of the clinical diagnostic process in the public school. In actual practice, a child's score on an intelligence test is the primary factor in his diagnosis as a mental retardate. If we are to understand the disproportionate placement of Mexican-American and Negro children in classes for the mentally retarded, we must focus specifically on the nature of intelligence tests.

THE STANDARDIZED INTELLIGENCE TEST

The IQ is ordinarily treated as a measure of an individual's intellectual capacity, his mental ability. Obviously, intellectual capacity cannot be measured directly because that would require assessment of the genetic component of performance, the genotype. An individual's genotype can only be expressed through behavior learned in a social and cultural setting, his phenotype. The intelligence test, of necessity, measures what a person has learned, his phenotype. On the basis of this performance, psychologists make inferences about the nature of the person's genotype. The clinician assumes that it is possible to make valid inferences about the genotype from a properly normed and administered intelligence test. Psychologists who make diagnoses of children in the public schools are constantly making inferences about the characteristics of genotypes on the basis of the performance of the phenotype. The logic behind these inferences is relatively simple but the assumptions are rarely if ever met in actual practice.

These inferences are based on the following sets of assumptions. If two persons have had an equal opportunity to learn certain types of cognitive, linguistic, and mathematical skills and to acquire certain types of information; if they were equally motivated to learn these skills and to acquire this information; if they are equally motivated to exert themselves in a test situation and equally familiar with the demands of the test situation; if they are equally free of emotional disturbances and anxieties which might interfere with their test performance; and if they are equally free of biological dysfunctions and organic difficulties which might interfere with their performance; then any difference between their performance on the test probably measures differences in their genetic intellectual endowment. Simply stated, if learning opportunities and all other factors are equal, those persons who learn the most and perform the best probably have greater innate mental capacity than those who learn and perform most poorly.

Of course, the major difficulty in applying logic based on these assumptions in actual test situations, especially situations which involve children from very different sociocultural backgrounds, is that "all other factors are never equal." Children from Mexican-American and Negro homes do not have the same opportunities to learn the types of cognitive, linguistic, and mathematical skills and to acquire the kinds of information covered in the Stanford-Binet LM or the WISC as Anglo children from middle- and upper-middle-class families. Therefore, the assumptions on which the standardized intelligence tests are based are not met for most Mexican-American and Negro children and valid inferences about comparative innate mental ability cannot be made on the basis of performance on these tests. When we consider that Mexican-American children were not even included in the samples on which the norms of the Stanford-Binet and WISC were established, and there were no Negro children in the normative sample for the Stanford-Binet, the inappropriateness of a direct comparison of scores of Anglo, Mexican-American, and Negro children is even more evident.

APPROACHES TO CROSS-CULTURAL EVALUATION

In general, there have been three different approaches to resolving the issues raised in cross-cultural intelligence testing: attempts to develop "culture-free" tests; modifying existing tests for cross-cultural applications; and developing culture-specific tests.

CULTURE-FREE TESTS

Serious attempts have been made to develop tests which are culture-free, such as the David-Eells Games, Raven Progressive

Matrices, and the Cattell Culture-Free Test. Among the major tests used in international studies have been the Goodenough Draw-A-Man Test (Harris, 1963) and the Raven Progressive Matrices while the Peabody Picture Vocabulary Test has been used domestically to test children from various subcultures (Torrance, 1968). Workers have attempted to develop culture-fair apparatus tests, mainly for low-level screening (Schwarz, 1963). Paper-and-pencil tests present special difficulties in cross-cultural testing. although work done in Africa by Schwarz (1961) indicates that basic adaptations in the number and kinds of test operations and the test format may make such tests feasible.

Many writers have discussed the inherent errors and difficulties in developing a culture-free test (Biesheuvel, 1952, Verhaegen, 1956). There are unknown complexities in comparing scores, cross-culturally, on any of these tests. For example, Dennis reports that scores on the Goodenough are positively correlated with modernization in 40 different cultures and subcultures which he studied (Dennis, 1966). Irvine (1966), working in African cultures, concluded that the Raven is not an unbiased test simply because it is based on a figural and low verbal content but is biased in different and undefined ways. Cross-cultural efforts have, at best, met with only limited success because an intelligence test can only measure phenotypic behavior, the product of learning in a specific social and cultural setting. Beyond simple reflex acts and basic organic processes, there are few if any human behaviors which are culture-free.

MODIFICATION OF EXISTING TESTS

Translation. One of the most common procedures in cross-cultural testing has been to attempt to modify either the language or administrative procedures when tests standardized in one society are used in making cross-cultural comparisons. Torrance (1968), in his program of international testing for creative development, modified existing tests by translating them into the native language of the subjects. Darcy (1963) reviewed numerous studies of the effects of bilingualism on the measurement of intelligence in which Spanish versions of standard intelligence tests were used. The Cattell

Culture-Free Test has both English and Spanish versions, as does the Stanford-Binet. Keston and Jiminez (1954) tested 50 Spanish-American children using both the English and Spanish versions of the Stanford-Binet and found the mean IQ on the English version 14 points higher than that on the Spanish version. They concluded that the difference resulted from the fact that the children's formal education was in English and that the development of their Spanish had essentially stopped when they entered the public schools. They concluded that the Spanish version of the Stanford-Binet was not appropriate because the quality of Spanish spoken in the area studied varied significantly from that used in the test.

The WISC has been translated and adapted to Puerto Rican Spanish. Norms were established on the basis of a representative sample of Puerto Rican school children. Morán (1962) found that his sample of Puerto Rican children still had a mean IQ of 88 with a standard deviation of 22 when he used the Spanish version. His study illustrates the difficulty of securing scores which can be interpreted clinically even when translated forms of standardized tests are used. In other cases alternate wording has been necessary to adjust for culturally different meanings. For example, Coyle (1965) found that the wording of one of the items in the Information Subtest of the WISC had multiple interpretations relating to sexual matters. He proposed a different phrasing which would prevent such misinterpretations in cross-cultural testing.

Thus, the translation approach has limited value. The content of the examination is culture-bound. The content of information questions and the concepts used in recognizing conceptual similarities are related to a particular society and language. The materials used to test comprehension and perceptional skills all reflect the content of the culture in which the test was developed. Simply translating the content of a test designed for persons socialized in one culture into the language of another culture does not eradicate the cultural differences. Persons from backgrounds other than the culture in which the test was developed will always be penalized. It is difficult to interpret the meaning of IQs when this is the case.

Emphasis on nonverbal tests. Another common adaptation of existing tests is to weight the nonverbal portion of the test more heavily than the verbal portion in cross-cultural testing. The rationale for this procedure is that performance tests are believed to be more culture-free than verbal tests which have a high language component. Most investigators have found, for example, that the Performance IQ of Mexican-American children averages 5 to 7 points higher than their Verbal IQ on the WISC. In our research in Riverside, we found that the Performance IQ ranges from 7.6 points higher for Mexican-American children whose family backgrounds most closely approximate the cultural configuration of the Anglo community to 11.6 points higher for children from homes which are not highly acculturated to the Anglo society. There were no differences for Negro children.

In a study of 235 bilingual Puerto Rican children in New York public schools, Darcy found an 8.3 point difference in favor of the nonlanguage test using the Pintner General Ability Test. She concluded that the language and nonlanguage tests were not measuring the same function to a large enough extent to warrant substituting one for the other. She recommends that both verbal and nonverbal scores should be used in making any diagnosis (Darcy, 1952). Johnson (1953) used the Goodenough Draw-A-Man Test and the Otis Self-Administering Test of Mental Ability in evaluating 30 bilingual Spanish boys. Although he found the mean Goodenough scores were over 12 points higher than the Otis scores, he concluded that there are many complex problems in making such comparisons and results are open to question. Studies of Welsh bilinguals have found a similar pattern of higher nonverbal scores among bilingual children. Jones (1952) has found that as "Welshness" of background decreased, mean IQ scores increased, a finding similar to that for Mexican-American and Negro children which will be reported later in this paper.

Most investigators report relatively high correlations between verbal and nonverbal scores but it is difficult to know the extent to which one may be substituted for the other. Factor analytic studies indicate that they are factorially separate dimensions.

Power tests. A third adaptation of the standardized testing situation which has been used in cross-cultural testing is to vary the speed and power components of the test. The rationale for this procedure is that persons from cultures which do not place a great value on the speed of performance do poorly on tests which are timed. Given adequate time, as in a power test situation, they would do significantly better. Some studies seem to confirm this hypothesis. For example, Knapp (1960) studied 100 Mexican-American, male immigrants using the Cattell Culture-Free Test, Form 2A and 2B. He found that the group averaged higher under power conditions and improved more than a comparable group of Anglos when the speed limits were removed.

However, Schwarz (1963) contends that there are unpredictable complexities when the speed-power factor in test administration is varied. In African samples, persons with little prior knowledge of tests varied greatly in the strategies which they used under the two conditions. This was especially true if the test required both speed and accuracy, a situation especially confusing to the uninitiated. He found that there is "no easy way of encouraging more thoughtful responses by the examinees who use only a fraction of the available time, or of speeding up those for whom exceptionally generous limits are not enough." He concludes that adequate power tests are, in general, the most difficult to devise.

We must conclude that any modifications in standard procedures in order to accommodate for cross-cultural differences is fraught with unknown biases, whether the modification be translation into the language of the subject, emphasis on nonverbal rather than verbal tests, or altering the procedures in test administration. Such modifications make it difficult to interpret scores produced under these varying conditions and leave the evaluator without a normative framework within which he can evaluate the meaning of a particular performance.

CULTURE-SPECIFIC TESTS

A third approach which has been proposed but has not been pursued to any great extent is to develop intelligence tests for each of the major cultural groups in American society. If this were done, each child could be tested using a test designed for persons from his linguistic and cultural background. While the approach is certainly worth further exploration, the creation and standardization of multiple intelligence tests would be expensive. It would be difficult to determine the cultural boundaries for each of the cultural groups for which such tests should be developed. The interpretation of the scores on such culture-speeific tests would also pose problems for the educator. While a person's score on a culture-specific test would indicate his relative position in the distribution of scores for other persons of his sociocultural background, it would not indicate his position relative to the larger society or his probability of success in the public schools. If nothing else, present intelligence tests are relatively good predictors of academic performance in the public schools.

Schwarz (1963) provides a detailed discussion of the complexities of developing specific tests adapted to varied cultural settings. Each test must be backed by adequate environmental supports. The test writer must be cognizant of the extent and nature of the environmental supports for each test before it is introduced. Test instructions, whether written, oral, or pictorial, must be adapted to the local vernacular and may need to be pantomimed, a difficult procedure for more advanced tasks. In summary, he concluded that the entire testing process requires a large number of cultural adaptations when culture-specific tests are devised. Even when such tests are developed, it is difficult to know how performance on them can be generalized beyond the culture for which they were designed. In this case, we have the same problem, in reverse, as we now have with the present tests standardized for Anglo-American society.

SOCIOCULTURAL PLURALISM

Our research has identified a fourth approach which tends to reduce the ethnocentrism of present testing procedures—pluralistic evaluation. Pluralistic evaluation modifies the normative framework by which the meaning of a particular score is interpreted rather than making changes in the form and content of the test itself. In a pluralistic evaluation procedure, a person is evaluated as intellectually subnormal only when he scores in the lowest 3.0% of his own sociocultural group on a standardized test.

PLURALISTIC EVALUATION OF INTELLIGENCE

The logic of the pluralistic evaluation of intelligence is based on a three-step process:

(1) The first step is to identify those sociocultural characteristics which account for the greatest amount of the variance in the IQs of Mexican-American and Negro children.

(2) Next is to develop a sociocultural index using those characteristics most highly correlated with IQ for placing children into categories according to the sociocultural grouping most closely approximating their background characteristics.

(3) Finally, the evaluator interprets the meaning of the IQ against two normative standards, the standardized norms of the test as published in the test manuals and the pluralistic norms based on the distribution of scores for persons from comparable sociocultural backgrounds.

Each aspect of this procedure will be discussed in turn.

Identifying sociocultural characteristics correlated with IQ for Mexican-American and Negro children. During the school year 1966-1967, 598 Mexican-American, 339 Negro, and 576 Anglo elementary school children in the Riverside Unified School District [2] were tested using the Wechsler Intelligence Scale for Children. The team of psychometrists who tested these children were given supplementary training, in addition to the training they had received in regular university courses in testing, to assure that they followed standardized administration and scoring procedures. The Mexican-American and Negro children in the sample included the total population of the three segregated minority elementary schools which then existed in the district. The Anglo children were a random selection of elementary school children in eleven predominantly Anglo schools. The parents of all of these children were interviewed, providing extensive information on the family background characteristics of all the children who were studied.

Seventeen of the family characteristics which were found to differentiate most highly between Mexican-American and Negro families and the modal sociocultural configuration of the community, as established in an earlier survey of representative households, were used as independent variables in a stepwise multiple regression analysis which used Full Scale WISC IQ as the dependent variable. In other words, we attempted to determine the amount of the variance in WISC Full Scale IQ which could be accounted for by the sociocultural characteristics of the child's family. Ten of the variables studied for Mexican-American children focused on characteristics of the head of household and were dichotomized as follows: born in Mexico vs. born in the United States; reared in Mexico vs. reared in the United States; reared in rural vs. reared in an urban area; citizen of the United States vs. noncitizen; Catholic vs. non-Catholic; white-collar occupation vs. blue-collar occupation; Duncan Socioeconomic Index Score for occupation 0-29 vs. 30 or higher; formal education eight years or less vs. nine or more; married and living with spouse vs. divorced, separated, or deceased; male head of household vs. female head of household. Six family variables focused on characteristics of the household: whether the household had moved two or fewer times in the past ten years vs. moved three or more times; owning or buying a house vs. renting; speaking English all the time or most of the time in the family vs. seldom or never; five or fewer family members vs. six or more; family structure consisting of parents and their children vs. all other types of family structures; 1.4 persons per room or less vs. 1.5 persons per room or more. The seventeenth variable used in the regression was the number of years of schooling the mother expected the child to complete. Responses were dichotomized, a high school education or less vs. more than a high school education. The same variables were used for Negro children except that language spoken in the home and citizen vs. noncitizen were omitted; items dealing with place where head of household was born and reared were dichotomized Southeast vs. all other states; and religion was dichotomized Baptist or Methodist vs. all others.

All seventeen variables were correlated .39 with Full Scale IQ, .42 with Verbal IQ, and .30 with Performance IQ for Mexican-American children. Together they accounted for approximately 15% of the variance in Full Scale IQ, 18% of the variance in Verbal IQ, and 9% of the variance in Performance IQ. Sociocultural variables were correlated .44 with Full Scale IQ for Negro children, .46 with Verbal IQ, and .35 with Performance IQ. Together, they accounted for approximately 19% of the variance in Full Scale IQ, 21% of the variance in Verbal IQ, and 12% of the variance in Performance IQ.

We noted earlier that some clinicians use Performance IQ as the primary score in assessing the measured intelligence of non-Anglo children because they believe that this score is less influenced by sociocultural factors than the Verbal IQ. Our findings support this belief, but they also show that this expedient does not eliminate cultural factors entirely. The correlation between sociocultural factors and Performance IQ is statistically significant ($p < .01$) and even 9% and 12% of the variance cannot be ignored.

For Mexican-American children, overcrowding is the primary indicator, having a linear correlation with Full Scale IQ of .24. The next variable in order of its contribution to the multiple correlation coefficient is the number of years of schooling the mother expects the child to complete. It is correlated .23 with the Full Scale IQ and adds significantly to the multiple correlation coefficient, raising it from .24 to .32. Education of the head of household, use of English in the home, whether the family owns or is buying its home, and whether the head of household was reared in an urban or rural area appear in that order.

In brief, Mexican-American elementary school children with higher IQs tend to come from less crowded homes and have mothers who expect them to have some education beyond high school. They have fathers who were reared in an urban environment (over 10,000 population) and have a ninth grade education or more. They live in a family which speaks English all or most of the time and is buying its home.

The primary variables for Negro children are similar to those found for Mexican-American children. Instead of overcrowding, the size of family emerges as the most important single variable for Negroes ($r = .21$). School expectations for the child appears as the second most significant variable after the common variance with size of family is taken into account. Marital status of the head of household, socioeconomic index score of the occupation of the head of household, and whether the family is buying or renting its home appear in that order.

In brief, the Negro child with the highest IQ comes from a family that has characteristics similar to those of the modal configuration of the community. He comes from a family with less than six members; has a mother who expects him to get some college education; has parents who are married and living together in a home which they either own or are buying; and he has a father who has an occupation rated 30 or higher on the Duncan Socioeconomic Index (Reiss, 1961).

Thus, the more the family is like the modal sociocultural configuration of the community, the higher the IQs of the Negro and Mexican-American children on the WISC. Clearly, sociocultural factors cannot be ignored in interpreting the meaning of a standardized intelligence test when evaluating a child from a non-Anglo home.

Developing a sociocultural index for classifying children by family background. The findings from the multiple regression were used to group each Negro and Mexican-American elementary school child who was given the WISC into one of five groups according to the extent to which his family background conformed to the modal configuration for the total community of Riverside. Each child was given one point for each of his family background characteristics which were like the dominant society on the five primary sociocultural variables predicting Full Scale IQ for his group.

Mexican-American children. A Mexican-American child received one point for each of the following: if he came from a family with 1.4 or fewer persons per room; if his mother expected him to go to school beyond high school; if the head of household had a ninth grade or higher education; if his family speaks English all or most of the time at home; and if his family owns or is buying its home. If his family was similar to the modal configuration on all five characteristics, he received a score of 5. If his background was similar to the dominant configuration on four characteristics, he received a score of 4, and so forth.

There were 25 Mexican-American children who came from backgrounds that included all five characteristics, only 4.2% of the total group. When Anglo children were scored, using the same variables and scoring procedures, 68.7% scored 5 points. The large difference between the percentage of Mexican-American and Anglo children who came from socioculturally modal family backgrounds emphasizes how few of the Mexican-American children in the public schools are likely to come from sociocultural backgrounds which correlate with high performance on the WISC, a test developed and standardized on the general population of the United States. However, the mean IQ for the group of 25 children whose families were most like the dominant cultural configuration, were within a standard error of the norms for the test—mean Full Scale IQ was 104.4; mean Verbal IQ was 100.6; and mean Performance IQ was 108.2. Those Mexican-American children with cultural backgrounds most similar to the Anglo mode did as well on the WISC as the Anglo children on whom the tests were normed.

The 174 children with scores of 4 had a mean Full Scale IQ of 95.5, a mean Verbal IQ of 91.8, and a mean Performance IQ of 100.5. All scores were approximatley nine points lower than those for the group most like the dominant society, a half standard deviation difference.

The 126 Mexican-American children with three sociocultural characteristics similar to the modal configuration for the community had a mean Full Scale IQ of 89.0; the 146 with two such characteristics had a mean Full Scale IQ of 88.1; the 127 with no or one characteristic had a mean Full Scale IQ of 84.5. The group least like the modal sociocultural configuration for the community is one standard deviation below the Full Scale norm for the general population. The drop of 20 points in Verbal IQ is even more precipitous, 1.3 standard deviations. In other words, about half the 127 children coming from backgrounds least like the dominant Anglo society would qualify as retardates if only an intelligence test were used for diagnosis. In fact, several of these children had been

TABLE 1

SOCIOCULTURAL GROUPINGS AND IQ FOR MEXICAN-AMERICAN AND NEGRO CHILDREN

	Resemblance to Modal Sociocultural Configuration of the Community					
	Most 5	4	3	2	Least 1-0	All Groups
Full scale WISC IQ	(n=25) (4.2%)	(n=174) (29.1%)	(n=126) (21.1%)	(n=146) (24.4%)	(n=127) (21.2%)	(n=598)
Mexican-American mean	104.4	95.5	89.0	88.1	84.5	90.4
Standard deviation	10.4	12.1	11.8	11.6	11.3	12.7
	(n=17) (5.0%)	(n=68) (20.0%)	(n=106) (31.3%)	(n=101) (29.8%)	(n=47) (13.9%)	(n=339)
Negro mean	99.5	95.5	92.8	87.1	82.7	90.5
Standard deviation	12.1	11.3	11.0	10.5	11.4	—

labeled mentally retarded and were in special education classes at the time of our study.

Negro children. The situation is just as dramatic for Negro children. There were only 17 in the group which had all socioculturally modal characteristics, 5% of the total. When Anglo children were scored using the same variables and procedures used for Negro children, 39.6% had all five of the favorable background characteristics and 35.2% had four characteristics. Only 5.4% of the Anglo children had less than three of the characteristics favorable to high IQ. As with Mexican-American children, there are many more Anglo than Negro children in Riverside who come from homes that are like those of the modal culture in these five characteristics.

The 17 Negro children who did come from homes that resembled the modal cultural configuration in all five characteristics, however, had an average IQ of 99.5 exactly at the national norms for the test. Thus, Negro children who come from backgrounds even roughly comparable to those of persons in the Anglo core culture do just as well on the Anglo-oriented intelligence tests as the Anglo children on whom the norms were based. Negro children in each sociocultural group did equally well on Full Scale IQ, Verbal IQ, and Performance IQ.

The mean Full Scale IQ of the 168 children with four socioculturally modal characteristics dropped 5 points below the national norms to 95.5. The 106 children with three such characteristics had a mean Full Scale IQ of 92.8 while those with only two such characteristics had a mean of 87.1. The 47 children who had none or only one of the five socioculturally modal characteristic had a mean IQ of 82.7. The average child in this group would qualify as a borderline retardate. Here, again, we find over one standard deviation difference in IQ between the children most like the sociocultural mode and those least like the sociocultural mode in family background. Sociocultural characteristics can account for approximately one standard deviation difference within both the Negro and Mexican-American group.

Interpreting the meaning of the IQ from a sociocultural perspective. A pluralistic, sociocultural perspective evaluates each child in terms of the sociocultural group to which he belongs. For example, the Mexican-American child who has only one sociocultural characteristic that is like the dominant configuration for the community would fall into the group we have called 0-1 in Table 1. For him an IQ of plus or minus one standard deviation, 11.3 points, from 84.5 is within the normal range for his sociocultural group using standard statistical definitions. An IQ of 73.2 to 95.3 is "normal" for a child from his sociocultural background. An IQ between 61.9 and 73.2 would be "low normal" but not intellectually subnormal, if we regard only the lowest 3% of the population as intellectually subnormal. At the other end of the scale, an IQ between 95.3 and 106.6 would indicate "high normal" ability. An IQ above 106.6 would indicate very superior performance for a child from such a nonmodal cultural background. Only an IQ below 61.9, the lowest 3% for his group, would be diagnosed as subnormal. The same procedure would hold for Negro children.

A culturally aware pluralistic interpretation would thus evaluate the intelligence of each person only in relation to others who have come from similar sociocultural backgrounds and who have had approximately the same opportunity to acquire the knowledge and skills needed to answer questions on an intelligence test designed for an Anglo-American society. If the person scores more than one standard deviation above the mean for his group, then he probably has high normal ability, even if his actual IQ is 100–average by the standard norms of the test. Conversely, a Mexican-American child who manages to achieve a score of 75 on an IQ test when he comes from an overcrowded, Spanish-speaking home in which the father has less than an eighth grade education and was reared in a rural area, and his mother does not expect him to go beyond high school, is well within the normal range for persons, like himself, who have had little exposure to cultural materials needed to pass an Anglo intelligence test. His educational program can be planned on the assumption that he is a person with normal learning ability who needs special help with English as a second language and emphasis on socialization to the ways of the dominant society. His educational opportunities are greatly reduced if he is labeled mentally retarded, a developmentally arrested person with a chronic, constitutional mental subnormality who is not capable of acquiring any facility in intellectual matters.

A PLURALISTIC, SOCIOCULTURAL PERSPECTIVE

A pluralistic evaluation defines a person as intellectually subnormal only if he scores in the lowest 3% of his own sociocultural group. Evidence from our study indicates that there are very significant sociocultural differences within the Mexican-American and Negro groups and that neither ethnic group can be treated as a homogeneous category. Persons need to be located by sociocultural group within ethnic group.

Specifically, in a pluralistic IQ evaluation, the WISC is administered and scored in the usual fashion. The Full Scale, Verbal, and Performance IQs are calculated using procedures published in the manuals. Questions are asked in an interview with the parents to determine the extent to which the sociocultural background of the child conforms to the modal sociocultural configuration of the community. Each child is then assigned to the sociocultural group within ethnic group to which he belongs on the basis of family characteristics. The IQ of the child is interpreted according to the norms for his sociocultural group as shown in Table 1.

Discrepancies between the child's placement according to the standard norms and according to the sociocultural norms are interpreted as indicative of the amount of supplementary assistance he will need to achieve in the public school. If he scores in the lowest 3% on intelligence for his sociocultural group, then he would be evaluated as intellectually subnormal. He might be placed in a class for the mentally retarded if he is also subnormal in adaptive behavior for persons from his sociocultural group. However, if he scores within the normal range for his sociocultural group, he would not be evaluated as mentally subnormal even though his score might be below the norm for the larger community. Instead, the child would be treated as a person from a nonmodal sociocultural background who has not had the opportunity to acquire the knowledge and skills needed to achieve in the dominant society. He would be placed in programs specifically designed to provide for his educational needs and would be treated as a child of normal intellectual potential who needs special programs to prepare him for participation in the mainstream of American life. More pluralistic, sociocultural interpretations of the meaning of scores on all types of standardized tests would provide a sounder basis for educational decision-making and reduce the institutionalized Anglocentrism which now ignores sociocultural differences and evaluates children from all types of backgrounds against a single Anglo norm.

NOTES

1. Data in this report have been collected under the auspices of the following grants: Public Health Service Research Grant No. MH-08667, from the National Institute of Mental Health, Department of Health, Education, and Welfare and Public Health Service General Research Support Grant No. 1-S01-FR-05632-02, from the Department of Health, Education, and Welfare, Socio-Behavioral Study Center in Mental Retardation, Pacific State Hospital, Pomona, California; Public Health Service Grant No. PH43-67-756; McAteer Grant No. M8-14A and M9-14 from the California State Department of Education, Office of Compensatory Education.

2. We wish to thank key persons from the Riverside Unified School District for their support and advice in our research efforts: Past Superintendent of Schools, Bruce Miller; Superintendent of Schools, Raymond E. Berry; Albert Marley, Director of Pupil Personnel Services; Former Director of Pupil Personnel, Richard Robbins; Mabel C. Purl, Director of Research and Testing. From the Alvord School District we acknowledge with gratitude the help of Robert W. Hocker. Director of Pupil Personnel; and J. Martin Kaeppel, Psychometrist.

REFERENCES

BIESHEUVEL, S. (1952) "The study of African ability." African Studies: 105-117.

California State Department of Education (1967) "Racial and ethnic survey of California public schools. Part I: Distribution of pupils, Fall 1966." Sacramento.

COYLE, F. A. (1965) "Another alternate wording on the WISC." Psych. Reports 16, 3 (Part 2): 1276.

DARCY, N. T. (1963) "Bilingualism and the measurement of intelligence: review of a decade of research." J. of Genetic Psychology 103: 259-282.

––– (1952) "Performance of bilingual Puerto Rican children on verbal and on non-language tests of intelligence." J. of Educational Research 45: 499-506.

DENNIS, W. (1966) "Goodenough scores, art experience, and modernization." J. of Social Psychology 68 (April): 211-228.

DOLL, E. A. (1941) "The essentials of an inclusive concept of mental deficiency." Amer. J. of Mental Deficiency 46: 215.

GORDON, M. M. (1964) Assimilation in American Life. Oxford Univ. Press. pp. 72-76.

Group for the Advancement of Psychiatry (1959) "Basic considerations in mental retardation: a preliminary report." Report No. 43: 7.

HARRIS, D. B. (1963) Children's Drawings as Measures of Intellectual Maturity. New York: Harcourt, Brace: 367.

IRVINE, S. H. (1966) "Toward a rationale for testing attainments and abilities in Africa." British J. of Educational Psychology 36 (February): 24-32.

JOHNSON, G. B., Jr. (1953) "Bilingualism as measured by a reaction-time technique and the relationship between a language and a non-language intelligence quotient." J. of Genetic Psychology 82: 3-9.

JONES, W. R. (1952) "The language handicap of Welsh-speaking children: a study of their performance in an English verbal intelligence test in relation to their nonverbal ability and their reading ability in English." British J. of Educational Psychology 22: 114-123.

KESTON, J. J. and C. K. JIMINEZ (1954) "A study of the performance on English and Spanish editions of the Stanford-Binet Intelligence Test by Spanish-American children." J. of Genetic Psychology 85: 263-269.

KNAPP, R. R. (1960) "Effects of time limits on the intelligence test performance of Mexican and American subjects." J. of Educational Psychology 51: 14-20.

MORAN, R. E. (1962) "Observations and recommendations on the Puerto Rican version of the Wechsler Intelligence Scale for children." Pedagogia, Rio Piedras 10: 89-98.

REISS, A. J., Jr. (1961) Occupations and Social Status. Glencoe, N.Y.: Free Press.

SCHWARZ, P. A. (1963) "Adapting tests to the cultural setting." Educational and Psych. Measurement 23, 4: 673-686.

--- (1961) "Aptitude tests for use in the developing nations." Pittsburgh: American Institute for Research.

TORRANCE, R. P. (1968) "Testing the educational and psychological development of students from other cultures and subcultures." Rev. of Educational Research 38 (February): 71-76.

VERHAEGEN, P. (1956) "Utilite actuelle des tests pour l'etude psychologique des Autochtones Congolais." Rev. de Psychologie Appliqueem 6: 139-151.

11

Institutional Persistence and the Paradox of Reform: The Case of Migrant Farm Labor

GARY S. GOODPASTER

□ IN ONE OF the introductory scenes of Fellini's *Satyricon*, there appears a vast, gray-brown building, a ziggurat with apartment rooms terraced pyramidically upward around a gloomy and cavernous atrium. Fellini knows his Dante, and one can see in this image, reversed and now projecting skyward, the descending levels of the *Inferno*. As two of the movie's central figures move upward rectagonally along the slanting walkway toward the telescoping recesses of this tower, they pass, in the apartments, all forms of human bestiality, wickedness, and agony, portrayed with all the interest and picaresqueness that only Fellini is capable of. As they ascend, the visual images accumulate and the cacophonous babel grows to such a crescendo that, for a brief instant, the viewer is admitted to omniscience: he, as it were, perceives human life in all its manifestations simultaneously, with the knowledge of God. Attending this divine insight is an incredible sense of overwhelming hopelessness and forgiveness, a genuine *tout comprendre est tout pardonner*.

Would that there were a Fellini to similarly integrate and present the harsh and the prosaic agonies of migrant farm worker existence, to accumulate and distill them with the aesthetic power needed to evoke an implacable sense of divine outrage and a comparable will to act. But we do know about the injustice, human waste, degradation, and agony associated with the system of migrant farm labor. Periodic exposés disclose familiar evils and abuses, stimulate public shock and outrage, and call for substantial reforms. But to our jaded sensibilities, news is theater, and ritual shock is a satisfactory way to give moral attitudes exercise, somewhat like taking them for a walk as one would a dog. Public anger is not sustained; eventually things get back to politics as usual, and the guiding perceptions of the fundamental reasons for the migrants' problems are obscured by fractional reforms which deal primarily with symptoms.

Although the problems of the migrant agricultural laborer are becoming well known and although serious attention is finally being given them, no currently proposed solutions to the problems seriously address the social and economic forces which create the need for and perpetuate migrancy as a way of life. Current and proposed ameliorative efforts are inadequate, underfinanced, and operate at a level and in a set of circumstances which make them inefficient and ineffective. Ultimately, these efforts have little impact because the system of migrant farm worker peonage is a national one, feeding off extreme poverty in the rural South, economically pitting farmers against farm workers in a market in which farmers have far superior advantages of political power and vested interest. It survives because of the need for cheap farm labor and the more intense need of the poor for employment, and because it is impossible to solve a multifaceted, interstate, national problem through disparate, uncoordinated local efforts.

The same rural poverty and unemployment which create the large surplus farm labor pool underlying the migrant farm labor system also contribute to the "urban crisis," for many of the unemployed farm laborers move to the cities to find work and improve their lives. It is clear, for example, that the rural Texas counties responsible for most of the Texas migrant stream also generate a substantial outmigration of rural poor Mexican-Americans to large urban areas and to farming areas of substantial agricultural employment (Migrant and Seasonal Farm Worker Powerlessness, 1969: 1035).[1] Notwithstanding, population growth is such that the population in these same areas of outmigration is actually increasing. To the degree that the need for agricultural labor is increasingly declining overall, this rural-urban exodus is encouraged.

The flow of rural poor to the cities exacerbates urban problems. What was rural poverty becomes urban poverty, involves difficulties of adjustment from rural to urban existence, and enhances the maintenance of ethnic and group solidarity. The poverty, related low social status, and race of these urban immigrants join to reinforce racial and class stereotypes, maintain division in the cities, swell welfare rolls, and contribute to the white middle-class exodus to the suburbs. Viewed from this perspective, the tie between rural and urban problems is clear, and the national policy of pouring funds into the cities without a comparable devotion of resources to the alleviation of rural poverty is simply irrational, although perhaps politically sound. Indeed, one of the convictions arising from the study of migrant labor, rural poverty, and the conditions of the cities is the close interrelation of the problems, the piecemeal nature of current policies and solutions, and the national failure to establish comprehensive policies and programs. Indeed, many proposed reforms appear to assume the indefinite survival of the migrant farm worker system and offer ameliorations which, if successful, will have a tendency to preserve it rather than end it. But past experience and knowledge of the way in which the system operates reveal that there is no way to save the migrant system of labor and to bring equality, dignity, and well-being to the migrants and their children.

MIGRANTS AND THE CONDITIONS OF MIGRANT LABOR

Agricultural farm laborers range from permanent farm employees through locally based seasonal farm workers, who may or may not have other employment, to the interstate migrant farm worker, who moves long distances to work short periods during critical labor need times in agriculture, such as harvesting. In 1965, over 3 million persons did farm work in the United States. Of these, 1.9 million were classified as noncasual, doing sufficient farm work for it to be regarded as a major source of income for them (Migratory Labor

Legislation, 1967: 332).[2] In 1967, approximately 446,000, or fifteen percent of the farm work force, were persons who migrated. Of these, twenty-five percent were Mexican-American (Migrant Powerlessness, 1969: 983)[3], of whom 276,000 left their home county to do farm work. When dependents of migrant farm workers are included in their numbers, as they must be since migrants often travel with their families, the number of persons migrating yearly to engage in farm work totals over one million.

The present farm labor institution employing these people in the United States has been created and maintained by specific historical, cultural, and economic circumstances and conditions which must be grasped in order to understand how the system operates and why it is inherently difficult to ameliorate or end it.

The essential motive force of the migrant stream is rural poverty in south Texas, Florida, and Southern California. The major migratory stream originates in Texas and flows to the West and Midwest. Smaller streams originate in Florida, flowing to the North and to the adjoining southern states, and in Southern California, from which it flows into the northwestern states. When the agricultural season ends, migrant workers will have worked in 900 counties in 46 states. (Migratory Farm Labor Problem in the United States, 1969: 1)[4], for an average overall work period of 85 days and an average annual wage of $922 (1969 Report: VII).

Unemployed or underemployed, many of these rural poor are forced to undertake poorly paying farm jobs at great distances from their homes. In their coming and going, traveling and working, migrants are, for most of the American public, legions of invisible men, isolated and unknown. Many of these hidden people develop a certain expertise in farm labor, partly out of economic necessity since more money is made by the worker skilled in particular crop operations, and farm work often becomes a family tradition, a damning inheritance.

The labor market in which they seek work is not coherent or rationally administered. They become aware of job opportunities in a variety of ways. Some travel yearly to the same areas and work for the same farmer. Others are recruited by processors or crew leaders. Yet others travel on rumor. In their travels, they are readily subject to control by labor entrepreneurs who provide job opportunities, transportation, and some other elemental services, usually at a high cost. The work they seek is seasonal, but intermittent, often as short as a week or ten days in any one place—the length of a particular harvest season. But even when on the job, work is dependent on the weather and often on the availability and condition of mechanical equipment.

In wages and in working and living conditions, the migrant farm worker suffers an exploitation similar to that undergone by factory workers during the industrial revolution. Now, as then, the worker subsidizes the social system responsible for his misery and the conditions of his labor. Many of the farmers for whom he works have little bargaining power in the agricultural market; to increase their income, they must lower costs. They do so by providing poor housing, disobeying laws protecting migrant labor, and by cutting labor costs, occasionally even by cheating. Work done is often paid for on a piece-work basis, thus avoiding payment for nonwork time on the job or payment for inefficiency. The farmer is indifferent to the efficiency or capability of his workers: anyone can work and does: the old, infirm, the young, the sick, men and women. The work is backbreaking and only the strong and able prosper modestly on piece-rates. For many workers, piece-work translated hourly is a poor wage, often averaging little more than $ .50 an hour. Although migrant families travel together partly because of family solidarity, it is also an economic necessity: wages are so low the entire family must work. Children travel with their parents and are put to work in the fields at an early age. They get a poor education; indeed, they learn well only uprootedness, transience and farm work: they learn to be farm workers.

Because migrants are poor, lower class, uneducated, and often of a different race than their employers, and because they compete with the farmer for the same dollar, common human sympathies are often withheld from them and they are readily victimized. Because they are different, mobile, and transient, no cognitive dissonance arises in the minds of local residents; highly suspicious and hostile attitudes are reinforced by class and racial differences. Local responsibility for migrant problems is rationalized away on the ground that the problem is not a local one. If any threat of migrant militancy or organization appears, vulnerable and fearful farmers become the incarnation of violent territorial defensiveness. As might be expected, these migrants, tools to work the earth, victims, peons, suffer many injustices and have a great many legal, and social problems in addition to their abysmally low income.

Housing for the migrant farm laborer, whether publicly provided or provided by individual farmers or associations of farmers, is grossly inadequate, overcrowded, and unhealthy. Often invited to a state through a state employment service which is federally required to certify the adequacy of housing and health conditions, migrants find only bad and unhealthy conditions. When privately housed, the migrant is subjected to coercive threats of retaliatory eviction if he attempts to enforce wage rights and agreements or if he reports housing conditions to a state health department. If he is successful in reporting unhealthy, unsafe housing, the housing is condemned and he has no place to live. Without a house, of course, he cannot work. Even were retaliation not a problem, authorities charged with ensuring satisfactory housing are often lax in their duties or lack the resources or cooperation necessary to carry them out.

In other areas of concern, various authorities responsible for providing social welfare services act only either when absolutely compelled to do so or without comprehensive and coherent programs. In none of these problem areas has consideration been given to thorough and frequent consultation with migrants themselves, to learn about their problems, to discover their needs, to hear their desires, or to understand them generally. When states do take some responsibility for the lives of migrant workers, there is little coordination or mutual planning by state agencies responsible for administering programs relating to migrants. These deficiencies and the lack of positive and ungrudging state programs of aid, coordination, and enforcement of law, contribute substantially to the misery of migrant life and ultimately intensify conflicts between farmers and farm workers.

Since there are no overall mandatory clearinghouse services for agricultural labor opportunities, too many migrants enter or are invited into areas where the labor market may already be filled. In addition, the use of both legal and illegal alien labor permits farmers to escape some stringencies of existing law, takes already scarce farm jobs from domestic agricultural laborers, and helps to keep wage rates down. These kinds of job competition, together with the activities of inadequately controlled crew leaders who often use a number of devices to extract money from migrant workers, reduce migrant wages substantially, making an already less than marginal income even more inadequate. The meagerness of migrant income requires whole families to work, but wage and hour limitations for women and children, when regulated, are not enforced.

If the agricultural worker is injured or killed, no workmen's compensation covers him. If he falls ill or is disabled, if for some reason his family is left without him, welfare residency rules preclude welfare assistance or medicaid. Emergency welfare funds are often denied on the ground that his state of origin does not reciprocate. The same low income and the same compelling circumstances force the migrant to apply for food stamps or commodities, but the Dickensian administration of these programs is parsimonious and raveled in red tape, requiring work time to be spent in trip and application after fruitless trip and application.

Because of family solidarity, low wages, and relatively large debts requiring the entire family to work, the migrant travels with his entire family, and his children must move from school to school. They are subjected to discrimination and, invariably, because they are not given the special treatment and education their circumstances require, they fall behind. But they are expected to fall behind and remain there. Schools grudgingly receive and educate them, feeling they can do nothing in a few weeks' time, resenting the irregularities created, and the special adjustments called for. Even when school administrations draw on specially earmarked federal funds for migrant education, they often attempt to divert them to other, unrelated purposes.

Because of the life style he is forced to adopt, because of his low income and inadequate diet, and because his housing is bad, the migrant and his family are frequently ill. Yet hospitals require large precare deposits, which cause familial financial crises, often resulting in large unpaid debts, court suits, default judgments, and ultimately, perhaps, in family break-up.

Migrant suffering subsidizes our national agricultural economy and the local merchant economy. Local businessmen and shopkeepers in rural communities profit from the food, goods, and services sold the migrant. When additional costs of fraud, high interest rates, sharp practices, and court fines are added, the migrant worker often leaves behind perhaps as much as eighty percent of what he earns. Many families, after an entire season of agricultural work, leave for home with no money at all.

THE MIGRANT SUBCULTURE

General poverty and the nature of circumstances of farm work, the persistent and dominating fact of transience, of not belonging anywhere, create coping life styles characteristic of the migrant subculture.

As Harvard research psychiatrist, Dr. Robert Coles, stated (1969 Report: 14):

> In order to adapt to such unusual facts of environment, migrants turn their isolated, mobile life inward, becoming guarded and suspicious toward outsiders but, in compensation for a rootless life, exceptionally close-knit with their young children. They tend to be unusually warm and stimulating with their infants, and rather lax about disciplining them. They so treat them that there appears to be significantly less hostility among the children; in contrast, hostility and suspicion are channelled toward other families as well as the world in general, which is seen as unfriendly and punitive. Families thus become separated from families, even within the migrant culture, so that the price for cohesion within the family is isolation and alienation from others.

The social consequences of migrant life are severe and their natural and understandable adaptive responses are facts to be reckoned with:

> Again and again one hears that migrants are unreliable, unkempt, lacking in thrift and tidiness. Worse, they tend to be willfully destructive of property, heavy drinkers, quarrelsome with one another, and generally a discouraging and depressing lot, unresponsive to aid, sullen before advice, ill suited even for more money or better working and living conditions. . . . Many migrant families don't want their children to go to school, don't care whether they receive their "shots," and don't, certainly, pay attention to the amenities of property care or civil behaviour with one another, to the value of moderation to alcoholic intake, sexual indulgence, and child bearing, to the conventions of marriage and divorce, or to the refinements of budgeted expenditures for sensible and hopefully "permanent" possessions [1969 Report: 16].[5]

The situation of migrant children is particularly poignant and critical, for children raised in migrancy learn to be migrants, and as adolescents and young adults fill the labor pool the system continues to draw on. Compared with middle-class American children, their lives are strained and their imagination and vitality are repressed. Dr. Coles describes the impact of such a life on the migrant child:

> He learns that each day brings toil for his parents, back-breaking toil: bending and stooping and reaching and carrying. He learns that each day means a trip: to the fields and back from the fields, to a new county or on to another state, another region of the country. He learns that each day means not aimlessness and not purposeless motion, but compelled, directed (some would even say utterly *forced*) travel. He learns, quite literally, that the wages of work is more work rather than what some of us call "the accumulation of capital." He learns that wherever he goes he is both wanted and unwanted, and that in any case, soon there will be another place and another. I must to some extent repeat and repeat the essence of such migrancy (the wandering, the disapproval and ostracism, the extreme and unyielding poverty) because children learn that way, learn by repetition, learn by going through something ten times and a hundred times and a thousand times, until finally it is there, up in their minds in the form of what me and my kind call an "image," a "self-image," a *notion,* that is, of life's hurts and life's drawbacks, of life's calamities–which in this case are inescapable and relentless and unremitting [Migrant Powerlessness, 1969: 451-452].

We tolerate and even maintain this self-perpetuating system of farm labor which is based and feeds on extreme poverty, the relentless denial of opportunity or means of escape, and the channeling of a deprived childhood. While it would be easy to go on and show how federal and state laws operate to support this system of peonage, essentially through farm labor recruitment subsidies and discriminatory farm worker exclusions from labor protection laws and the toleration of rural poverty, it is more important to point out that the injustices reflect economic and power realities. The farm worker is poor and powerless for a hundred reasons, but a significant portion of our agricultural economy relies on that powerlessness. Our laden tables, our high standard of living, the livelihood of many farmers are subsidized by migrant life. Every injustice dealt the migrant protects a coin in our purse. What he pays in body and soul ultimately to provide a commodity, we retain in our material well-being.

But one can understand from an examination of the migrant subculture that in some sense the farm work employment market has created or met a labor force fitting its demands. But this sort of structural-functional conclusion, looking at the self-perpetuating interaction cycle of migrant culture and the conditions of farm work does not explain how the institution of farm labor developed to this state nor does it in any helpful way explain why it is currently maintained as it is.

THE FARM LABOR ECONOMY

As the issues are often cast in congressional hearings on the plight of the migrant farm worker, every extra pound of farm worker misery is an extra ounce of farm worker profit. The system does act that way, pitting farm entrepreneur against the mere seller of labor. The purchase of manual labor receives the benefit of the dire economic necessity, reinforced by ignorance and culture, which drives the migrant on his vagrant diaspora. This appropriation of the migrant labor product, however, is one of the very basic elements which has gone into the creation and maintenance of the present structure of the American farm economy.

While American farming has always relied heavily on hired labor, the depression enhanced the creation of an agricultural structure dependent on very cheap human labor, and the current migratory labor problem should be viewed with this in mind. Countless small farmers, tenants, sharecroppers, illiterates, marginally surviving on their small holdings in the best of times, were forced into the vast depression labor pool. Speculators, investors and well-heeled farmers consolidated holdings and used displaced Okies, Kentuckians, and other members of the depressed rural population to produce agricultural products cheaply, but at substantial profit. When depression radicalism appeared—generated by intense poverty and righteously incensed by wretched labor and living conditions imposed by wealth—agricultural strikes and labor violence grew. Agricultural entrepreneurs and growers began to seek more tractable labor, and Mexican-American farm laborers began to appear in Michigan and Ohio in the thirties.

The coming of World War II, followed by the Cold War, Korea, and Vietnam, solved many of the depression poverty problems and wholly diverted attention from the economic problems of the thirties. But, insofar as it related to the farm worker, the economic structure of agriculture did not change; indeed, the continued use of cheap agricultural labor was necessitated, and even indirectly approved, by Congress as an essential part of the war and postwar effort. War manpower needs drained off the more adaptive members of the farm labor population into the service and war industries. The population left behind for farm work, the rural unemployed, the rejects, the unfit, the infirm, the severely uneducated and poor, was amplified with a replacement farm labor work force of Mexican-Americans, Indians, and blacks.

Throughout all of this process, the immense productivity of American agriculture, together with the individualistic and competitive character of the agricultural market over which farmers have no collective control, meant that agricultural profit and farm capital was extracted by reducing costs, especially labor costs, rather than by rising prices. The result, for the impoverished farm laborer, was always intense job competition and incredibly poor wages and living conditions.

With major changes in cast and some developments in plot, this economic play continues to run. Poverty, profit, and the structure of the agricultural market continue to be the dominant themes. While it is relatively easy and safe to generalize in this fashion, certain specific apparently operative forces can be pointed to, particularly as reasons for the migration of Mexican-American farm workers and as a demonstration of how a labor surplus beneficial to the American farmers is created.

The use of foreign labor in the border areas of Texas, New Mexico, Arizona, and California, either in the form of green carders (that is, Mexican aliens holding alien registration cards who work in the United States but reside in Mexico), and illegal aliens generates economic necessity which feeds the migrant stream. Conservatively estimated, there are 39,000 green carders doing agricultural labor, and in 1967-1968, 41,425 illegal Mexican aliens working in agriculture were apprehended and deported (1969 Report: 63-64). The total number of legal green carders is probably over 100,000 (Migrant Powerlessness, 1969, pt. 5-B: 2531). This foreign labor is used extensively in border areas of the United States, the same areas which generate a significant portion of the migrant stream. For example, it has been estimated that 88,700 domestic farm workers residing in south Texas must migrate yearly to other parts of the country because of the lack of local employment (Migratory Labor Legislation, 1967: 61).

It would be a serious mistake, however, to judge the effect of foreign labor employment upon the native farm worker employment and poverty solely by the number of foreign laborers employed in agriculture. It is also necessary to take into account those foreign laborers working in agriculture-related industries, such as packing and processing, and those working in the unskilled and semiskilled labor market, since there is direct competition with native farm workers in these labor markets. In addition to the unemployment, the increase of the labor supply from the use of foreign labor also reduces wage rates. It is surely a startling fact that the unemployment in Texas border cities is almost ninety-five percent greater than in Texas interior cities and that wages for seasonal farm work in Texas border areas are more than thirty percent less than those paid in the rest of the state (1969 Report: 65). Consequently, even when a native farm worker can get employment, he may have to accept lower wages than the native labor market would dictate. Furthermore, foreign labor is often used to break agricultural strikes, and its use has interfered substantially with agricultural organizing, for administrative procedures used to keep foreign labor out during labor disputes are wholly ineffective. Finally, the use of this low-cost labor may in fact overstimulate the production of fruits and vegetables and ultimately have a depressing effect on farm prices (Migrant Powerlessness, 1969, pt. 5-B: 2774).[6]

The situation in the southeastern United States is not exactly comparable, although some laborers have been imported into Florida from Puerto Rico and the British West Indies. The conditions of rural poverty, unemployment, deprivation, and hunger for this area, however, are well known, and it is obvious that these are the conditions which create the large farm labor pool there.

In perceiving the migrant labor problem as an economic problem, it would be a serious mistake to neglect racial and class discrimination as a fundamental cause of the circumstances of the migrant farm worker. While racial discrimination is often mentioned in the literature as a problem of the migrant farm worker or even as the source of other specific problems he has, it has not often been considered as lying at the very heart of the national system of agricultural farm labor. But poverty and racial discrimination are the incessant furies hounding migrant workers which together produce economic gains for farmers and processors, but also for whites generally. A large percentage of migrant farm workers come from minority groups, principally Mexican-Americans, blacks, Indians. The institutional and individual racism that they experience at home through white monopolies in the distribution of job opportunities, educational inputs, and other opportunity resources contribute to their poverty and inflate their need to seek work far beyond the home community. Rural racial enclaves become labor export markets for low-cost labor. In that sense, the economics of the use of migrant farm labor is dependent on racial discrimination. In this connection, it should also be noted that unemployment in farm labor, which tends to be quite high because of contingencies, falls essentially on minority group members. Racial discrimination tends to ensure that a disproportionate number of minority group members will be employed in industries where the risk of unemployment is high or where higher wages would have to be paid where whites are generally employed.

Certain features of, and half-truths and myths about, our agricultural economy also have great bearing on the current condition of farm labor and on proposed solutions to some of the economic and social problems of farm workers.

In the past few decades, there has been a substantial growth of large and profitable commercial farming interests and a corresponding decline in small family farms. While this development has proceeded apace, the comparatively small farmer still exists in fairly large numbers. For the farm worker, these two separate kinds of farm economy have important employment consequences. The labor problems of the farm worker working for an agribusiness concern as contrasted with the farm worker working for a small farmer or family farm, while having surface similarity in terms of conditions, are in reality fundamentally different. The agribusiness employer has incomparably greater capital and market resources than the small farmer, whose economic position may often be precarious. The agribusiness employer is also able to take advantage of economies of scale and agricultural technologies to a far greater extent than the small farmer. The agribusiness employer is, therefore, in a far better position to improve the wage and labor conditions of the farm worker and in general to rationalize the contingencies of farm labor economy. One of the clearest impressions one gains when attempting to deal with farm labor problems on behalf of the farm laborer is the differential capacity of agribusiness interests and the small farmer to respond to demands for improvement.

Lacking any kind of price control in the agricultural market, this class of small farmers survives only through maximization of profit through all means available. One of the chief means is reliance upon the bent backs and misery of the migrant farm laborer. It is to the advantage of the farmer to capitalize on the conditions of the agricultural labor market created by extreme poverty. In doing so, the farmer is not necessarily malign, for he has many ready justifications and rationalizations to explain the circumstances of these poor migrant people, none of which lays the problem at his door. Why should the problem be laid there, particularly when one farmer's benefits given to farm workers are canceled just down the road or in another state? Why should the individual farmer accept the responsibility to redress all the evils laid upon a stranger's back?

To further show how the unregulated competitive character of the agricultural market operates to injure the farm worker, take the example of state housing and health regulation for migrant farm workers. Aside from the cost of shelter and health improvements placed upon the farmer, which he will resist intensely, state enforcement staffs generally lack sufficient resources either to enforce at all or to enforce fairly against all farmers. Thus, within one state, some farmers will view themselves unfairly treated vis-à-vis other farmers, particularly because they compete with them in the same agricultural market. If their costs of production are made higher, and this same increase is not imposed on other farmers similarly situated, there will be resentment and attempted widespread evasion. Beyond this, however, is the existence of the same problem interstate. That is, the disparity between various state housing and health regulations and their differing patterns of enforcement among the states. One state may enforce stringently, another not at all. In these circumstances, a farmer living in a state which strictly enforces such laws must perceive himself at an economic disadvantage as against a farmer similarly situated in a state which does not enforce them at all.

In this context, a most instructive irony arises upon a reading of congressional hearings on migratory labor problems. In opposing ameliorative legislation for our farm workers, large agribusiness interests rely very heavily upon arguments based on the vulnerability of small and marginal farmers to farm worker economic demands backed by work stoppages. As long as the analysis of farm labor problems remains on this level, as long as proponents of legislative remedies for these problems allow opponents to identify all farm interests as one, there is little likelihood of change.

Approaching these problems from the farm worker side, it can also be seen that there are distinct classes of farm workers, the locally based farm worker and the migratory farm worker, each having problems with surface similarity, but whose fundamental economic position is very different. Again, a differential perception and analysis is required, and it is necessary to see that legislation and other remedial efforts designed to assist one class of farm workers will not necessarily assist the other.

THE IMPACT OF AGRICULTURAL TECHNOLOGIES

In addition to the processes discussed, the development of agricultural technologies has been an important factor in the state of the agricultural economy, the development of agribusiness, and the status of the farm laborer. The mechanization of farm labor tasks and the development of pesticides and herbicides has favored farmers with capital and has greatly and increasingly reduced the need and demand for farm labor.

The threat and promise of mechanization has also been an important restraint on efforts to improve the lot of farm workers. Invariably, when proposals to assist farm workers are made in state legislatures and Congress, farming interests always assert that it would be unfair and irrational to impose costs of improving farm labor conditions on the farmer because the farm crops will be mechanized within five to ten years. In other words, just over the horizon there is an end to the farm labor problem. Conscience postponed is conscience assuaged, and in this fashion, for example, many legislatures are convinced that it would be economically irrational to force farmers to provide adequate farm labor housing when housing costs are so expensive and when they would have to be amortized over a period of five years. While these claims are very difficult to rebut when made, they have in fact been false. The same claims have been made time and again since the thirties; each time some costly piece of remedial farm labor legislation was proposed. Nevertheless, the claims continue to be made.

The displacement of farm labor by mechanization and the use of chemicals have, of course, occurred. One of the undeniable major historical displacements occurred in wheat harvesting following the development of the combine. But the development of agricultural technologies for a particular crop does not mean that they can be developed for all crops, nor that they can be made cheaply enough for many farmers to use. In addition, the technologies may not be satisfactory, either because they do not do the jobs properly, have undesirable consequences for the product, or because they are expensive to maintain and operate. For example, sugar beet-thinning machines are very expensive and will not do a proper job unless accurately set, which requires a fair amount of attention. Some growers have even found it necessary to use farm workers with such machines. Tomato-picking machines require a standardized product, not necessarily desirable from a consumer's point of view. Pesticides and herbicides are now found to have serious biological and ecological consequences.

Beyond this, which goes to show that agricultural technology is not the complaisant genie waiting to end farm labor problems, it is more important to see how the promise and threat of technology give farmers monopoly control in the farm labor market.

Failure to impose higher labor costs on farmers through remedial legislation—say, for example, by requiring farmers to pay a minimum wage—means that farmers are more free to adopt or not adopt an agricultural technology than they would be otherwise. As rational economic calculators, they would resort to a technology only when increased productivity justifies its cost. When the technologies are expensive, as they now are for many labor-intensive crops, farmers can rely on cheap farm labor to maintain profits until the time comes when the technology is less expensive, more reliable, and more satisfactory. As long as they can rely on cheap labor, farmers are not impelled to demand the needed technology, are not forced to make critical economic decisions.

There is another important economic consequence of this decisional luxury in agriculture. While smaller farmers may not be able to utilize a new technology, large farming interests can afford and make efficient use of expensive machines and processes because of economies of scale. The large farmers compete with small farmers in the same market. Since the large farmer utilizes a technology, the smaller farmer is placed at a competitive disadvantage. To make profit, or to create capital to obtain the technology he needs so as to be competitive, the small farmer is forced to cut his costs as much as he can. As far as he is concerned, this means cutting his labor costs. For the farm worker, this means low wages and miserable living and working conditions.

But a vicious circle can develop. If prevailing farm laboring conditions are poor, it may be advantageous for large farming interests to make extensive use of cheap farm labor, to capitalize on the conditions in which the farm labor market finds itself. Such interests may also defer the acquisition of a new technology or combine the use of a technology with the use of farm labor in a mix that creates the most profit.

In any case, the following conclusions appear warranted. The rate of development of agricultural technologies is in part within the control of farming interests. Farmers have this control because of the existence of a large pool of cheap farm labor. Agricultural technology and the competitive market structure of agriculture promote the exploitation of farm labor, indeed, may exacerbate it. Finally, agribusiness can profit from the labor exploitation of small farmers resulting in part from competitive conditions it creates.

For the farm worker, however, the important feature of this process is the lingering maintenance of the farm labor system. The continued persistence and slow decline of the use of farm labor depends on the features described above. In human terms, the misery of farm laborers continues to be drawn out. Under these circumstances, it is impossible to say how long the institution will last. It is possible, however, to see that there are no incentives in the system to change it or to improve conditions for farm workers.

On the other hand, it should also be clear that the effect of enforcement of old and new law intended to improve the lot of farm laborers will be to drive out the small farmer. This will lead to consolidation of farming units into larger units more able to utilize the new technologies efficiently. But the processes will be slow and painful and no provision has been made for the farm workers who ultimately will be sacrificed.

THE PROVISION OF LEGAL ASSISTANCE TO MIGRANT WORKERS

The essential focus in this section is upon the use of legal services and techniques to assist migrant workers; and in the next section, upon proposed legislative reforms. Since the same social and economic conditions obtain for all remedial efforts, however, much of the discussion is applicable across the board. The themes remain constant: poverty, racial discrimination, and labor exploitation as sources of farm worker problems; belief in amelioration, protection of vested interests, self-deception, and, most important, the failure to address rural poverty through national job creation and income-transfer policies as causes for the persistence and continuation of these problems. My major concern, however, is to point out the limitations on various kinds of efforts and to highlight the elements of what appears to be, from the point of view of an activist lawyer, the only available strategy to provide migrant farm workers fundamental assistance.

Significant, but sorely inadequate and generally uncoordinated, efforts are currently under way to assist migrant farm workers as a class to achieve justice, some measure of well-being, and a chance for improvement. Most of these, such as legal services, are OEO-funded, but a number of them derive their support from churches in the form of social action or migrant ministries or from unions working to organize farm workers. Although these organizations exist, there are great limits to what they can do.

It is inherently more difficult to provide effective legal and social assistance to migrant farm workers than it is to provide them for many other impoverished groups in our society. For many migrants there are serious language, cultural, and educational barriers, together with a general timidity and reticence. Fear of retaliation for legal action taken is endemic: fear of the malice of a powerful crew leader; physical fear of beatings; fear of the police; of the Anglo; fear that jobs will be lost; that they will be thrown out of their houses. With the fear is an admixture of worldly cynicism: they have tried white middle-class institutions, such as the hospitals and the police, and found them wanting; they've been in court; they know welfare and OEO: what good would a lawyer do them? For these and other reasons, the migrants are often fearful and unreliable clients, sometimes accepting a course of action to solve the immediate problem of dealing with the attorney rather than fully intending to pursue it.

Migrants are hard to reach. They are often inaccessible, in effect imprisoned on a farm or in a private camp and not allowed visitors, particularly social service types or organizers, who are driven off with guns or arrested and charged with trespassing. They lack leaders and are unorganized. Perhaps because of the conditions of their life, their travel, transience, atomization, their extreme need for work and housing, and their extreme vulnerability to retaliation, they are as yet virtually unorganizable.

When a case is brought on behalf of a migrant client, there are often more than the usual difficulties in attempting to prove it. Migrants need aid immediately, are here today, but must be off tomorrow; the law dawdles. If there are witnesses, they may also be gone at time of trial; indeed, the client may be gone then. Promises to return will not make them easier to find: earning a living is far more important than bringing and winning a case.

Sometimes cases cannot be brought because of certain retaliation: loss of job or housing or both. Can the migrant be blamed if he is unwilling to fight a possible retaliatory action in court? To protect the meager income and poor dwelling of the migrant, it may be necessary to wait until his work is done and he is paid. But then he may likely move on to other farm work in another state where there may be no resources to assist him.

It is also possible that legal action taken to benefit the migrant farm worker may actually hurt him seriously. For example, federal and some state laws and regulations relating to health and safety standards provide the basis for strong legal attacks on substandard farm labor housing. The likely immediate result of legal action will simply be the closure of substandard housing. If such attacks are brought across a broad front, much farm labor housing may be closed. The ultimate result, therefore, may be the loss of a substantial amount of farm labor housing for the migrant worker. While there is a gain in some sense in the elimination of bad housing, there is a real loss for the migrant farm worker who cannot work without housing for himself and his family and who subsidizes his meager income through cheap housing. Of course, dangerous housing should be eliminated, but some housing regulations may go beyond health and safety to include amenities. To lose housing for the sake of an amenity not going to health or safety, simply because there was a basis for a good legal attack, is a luxury migrant workers cannot afford. Thus, even when there is solid law backing the farm worker, it may not be wise or helpful to use it just because it exists.

There are, however, ways in which to attempt to escape the paralysis of circumstances. Aside from legislative reform efforts, some of which, as they relate to migrant laborers will be discussed in the next section, the most important devices utilized by legal services attorneys to give impact assistance to the poor are test case litigation and the use of legal pressures to assist in the organization of the poor.

In test litigation, a law thought to create or perpetuate an injustice is attacked in the hope that striking it down will significantly assist the client groups. This may or may not actually turn out to be the case. Test cases are often difficult to bring, particularly for migrant workers because of their extreme vulnerability to retaliation and because of their short residency in an area. In addition, while there may be a proposition of law which can be challenged, it may be impossible to create the conditions for bringing a test case. The other side may acquiesce and settle rather than allow a general unjust practice to be challenged. In such circumstances, duty to the client may require settlement. But this means that to end the general injustice, it may be necessary to undertake a systematic pursuit of many similar individual cases. This is a very difficult effort when there are so many varied demands for legal assistance and when resources are scarce.

Even when a test case is successfully brought and won, there is the substantial problem of enforcement in individual cases. For example, it may be very easy to establish that farmers must provide sugar-beet workers with a statement of their wages, yet virtually impossible to see that farmers do so. Most important, however, as long as the basic economic and political parameters of the system do not change, law reform litigation cannot reach the basic constants, and even major courtroom victories may not have great practical value. For example, farm workers do have some basic wage and working conditions protection under the Wagner-Peyser Act.[7] Under this law, the United States Employment Service acts in conjunction with state employment agencies to provide farm placement services for farmers and farm workers. In this system, after first determining compliance with federal labor and housing regulations, the state employment agency forwards farmer employment requests interstate to a state having a labor surplus. The regulations which must be complied with provide, inter alia, that it must be shown there is a need for farm workers, that the wages offered are not less than prevailing wages in the area for domestic farm workers, and that other terms and conditions of employment are not less favorable than those for domestic agricultural workers. In addition, relatively stringent federal agricultural housing, safety, and sanitary standards must be met by employers who use the farm placement service.

In the past, ensuring compliance with the applicable regulations was honored in the breach. In *Gomez v. Florida Employment Service*[8] the practice of noncompliance was challenged. In that case, the state farm placement service forwarded a clearance order interstate without having ensured compliance with the federal regulations. The court held that the Wagner-Peyser Act and its regulations implied civil remedies for their violation both against state officials and against the farmer requesting the workers. But the victory was hollow, for use of the Wagner-Peyser system of interstate labor recruitment is voluntary. Faced with the threat now inherent in use of the system, farmers will, as they have, turn to channels of farm labor recruitment other than state farm placement services.

This unsatisfactory ultimate result of a significant piece of test litigation for migrant farm workers is not merely a case of additional loopholes which made the favorable decision practically ineffective. More important, it demonstrates the politics and economics behind the migrant farm labor system which test case litigation has difficulty in reaching. As better said elsewhere:

> The problem of unjust laws is almost invariably a problem of distribution of political and economic power; the rules merely reflect a series of choices made in response to these distributions. If a major goal of the unorganized poor is to redistribute power, it is debatable whether judicial process is a very effective means toward that end. This is particularly true of problems arising out of disparities of wealth and income. There is generally not much doctrinal judicial basis for adequately dealing with such problems, and lawyers find themselves developing cases whose outcomes are peripheral to the basic issues that these problems raise [Yale Law Rev., 1970: 1069, 1077].[9]

For similar reasons, many social, legal, and legislative reforms for the benefit of migrant farm workers, at least when established on a piecemeal basis, are simply border skirmishes. This is so because the system of migrant farm labor has sufficient flexibility to circumvent reforms, because the problems of migrant farm workers have no attackable locus, and because reforms often attack symptoms and do not reach the heart of the problem, which lies in the economic system.

Litigation and legal pressure, whether of a test nature or not, are also a way in which to promote the creation and strengthening of viable groups or organizations of the poor, thus giving them some pressure or power base. The basic idea is that "prevailing arrangements, or power patterns, can only be altered by power" (Sherrard and Murray, 1968: 200). Legal pressure is used to support goals of a nascent group or to frame and bring issues around which group political consciousness may form.

Arguably, even this use of legal services will not be of great assistance to the interstate migrant farm workers since their vulnerability to reprisal, their transience, and their lack of even elements of community mean that the social conditions requisite to create an organizational base do not exist for them. This is only partly true, for interstate farm workers on occasion are gathered together in farm labor camps for significant lengths of time. Experience has indicated that when this is so, an organization can quickly coalesce and have significant power in specific situations. For example, a major rent strike, which was successful, was conducted in a farm labor camp in eastern Colorado; and a short but also successful strike by Navajo farm workers at the height of the sugar beet work season was conducted from a labor camp in western Colorado. Aside from the demonstration that it is possible, in certain circumstances, for interstate migrant workers to organize themselves and invoke legal pressures to achieve set goals, the most important result of such action, however, is not the settlement achieved. It is rather the development of group consciousness and the incremental attitudinal changes in individuals created by successful group action and the aggressive use of legal pressure. And here, successful group action is not necessarily defined by winning an issue but often merely by confronting the man.

Yet quickly formed migrant organizations quickly dissolve, as migrants go their various ways to different states and other employments. Lacking an organization to plug into where they next arrive, and lacking any overall, nationwide organization to which migrant workers might belong and which would be represented in every locality in which they work, it is difficult to assess long-term benefits of spontaneous but guided organization. One might suggest that there is new perception, greater militancy, more willingness to engage in similar activity in the future; that the seeds had been planted. What is needed then is continual stimulus and support for further activity along the same lines, with the long-run hope that genuine organizations with real strength and resilience will ultimately develop.

However, the major problem in giving organizational assistance to the interstate migrant farm worker is, in addition to his transience, the lack of significant and coordinated organizational activity in the areas which originate the interstate migrant farm worker stream. Some efforts have been made to create migrant unions, and there is a substantial amount of organizational effort among migrants, say, in south Texas, but the organizations are very weak and disconnected and frequently undergo internal power struggles that are debilitating.

To the degree that such organizations orient themselves only to organizing and supporting migrants while they are in the stream, as often appears to be the case, a grave mistake may be made. Such organizational effort is bound to become diffuse and create insuperable resource and coordination difficulties as migrants go their separate ways. If migrant problems in foreign states, on the fields and farms, are but symptoms of underlying economic and power realities, then originating state organizational effort should be directed to the local home base of migrants, attacking those arrangements which make them poor and which force them to migrate North to seek work during the agricultural season. This is particularly important because the system of interstate agricultural farm labor is dying anyway. But it is dying slowly, and in its slow dance of death, many migrants will be injured irrevocably. To the degree that legal and social services are unable to assist home-based migrant organization, either because there is no program in the area or for some other reason, the ultimate impact of this assistance for migrants will not be great enough to change matters for them appreciably.

It should be pointed out that test litigation and the use of legal pressure to catalyze the development of groups also have an important spinoff. There is, in many areas, particularly in large cities, a great reservoir of liberal or reformist sentiment which can be stimulated to action in support of farm workers. The clear crystallization of a significant injustice in a law case or a legal pressure action, if properly framed, can generate pressure for change on the part of groups that already have power and can lead to coalition creation among such groups. This joining in common cause can also have a significant impact on the stimulation and maintenance of organizations by migrant workers. This benefit, however, is also a danger, for liberal groups have a clear managerial tendency and tend to evanesce when the issues get really difficult or too close to home. In addition, reliance on such groups may tend to weaken rather than enhance farm worker organizations, which should use, but not depend on, outside support, even in the case when the outside organization is a group of legal service attorneys.

But the organizational support activity, so essential to the effective use of law to aid the migrant farm worker, tends to undercut the OEO legal services programs providing such assistance:

> When we started we assumed that withdrawal of funds wouldn't matter if we could first build a political base among the poor. This takes a good deal more time, more trust, more community work than we were willing to recognize. To have carried out this approach would have involved focusing on a small number of offices with seven or eight attorneys in each and an enormous amount of supervision, both in political perspectives and in lawyering. It would have involved four or five years as we developed contact, sophistication, and skills. This can't be done by a poverty program with year-by-year funding because definable, explicit results are constantly required to justify the program's existence. The attacks on such a program far exceed in intensity and effectiveness the capacity of the program to build a deep political base to counteract it. Coalitions can be created that enable the program to survive, but a focus and emphasis on political community organizing probably can't be done with government funds at all [Yale Law Rev., 1970: 1111].

The recent firing of top OEO legal services officials, apparently for political reasons—the fired officials described their discharge as evidence of the capture of OEO legal services by conservatives and southern politicians—clearly reveals the limits upon aggressive reform through government instruments (New York Times, 1970). Nevertheless, notwithstanding all the infights necessary to ensure survival, one must work with what there is, even if it turns out to be little. That is always the difficulty with using institutional means of engaging in what is fundamentally a political fight, as I think an examination of proposed legislative reforms for migrant farm workers will confirm.

LEGISLATIVE REFORM FOR FARM WORKERS

There have been many proposals, intensely resisted by farming interests, for legislative reforms to assist farm workers by increasing their income and improving their working conditions. It has been recommended, inter alia, that farm workers be included under the protections of the National Labor Relations Act; that the Fair Labor

Standards Act be amended to include more farm workers under the federal minimum wage law; that the Social Security Act be amended to make violations relating to farm workers more amenable to punishment and to recognize the migrant farm worker as a "federal" person for welfare purposes; that farm workers be included under workmen's compensation and unemployment compensation laws; that the provisions of the Wagner-Peyser Act, which create certain protections for migrants recruited interstate under the auspices of the U.S. Department of Labor, be made mandatory; that the Pure Food and Drug Act be amended to prevent transportation in interstate commerce of food produced in violation of health and safety standards; and that there be increased appropriations for migrant health, education, child care, and legal services (see 1969 Report).

The essence of most of these ameliorative proposals is the provision of more income to the farm worker; the enhancement of his ability to further his labor goals through organization; the reduction of hazards in farm work; a substantial increase in social welfare services; and the provision of the legal authority and resources necessary to enforce laws and ensure equal justice. Enactment of such proposals is necessary, and adoption undoubtedly would ameliorate the work and living conditions of many farm workers; nevertheless, I seriously question the sufficiency of the proposals to really assist the migrant farm worker. The problems of agricultural labor should not be viewed as being all of one piece. As stated previously, there are distinctly different kinds of agricultural labor, and the remedies for the problems of one will not necessarily benefit the others. A number of the major proposals for legislative change appears to be designed principally to benefit locally based farm workers working for large farming interests. The distinct interests of the migrant farm worker are obscured in these proposals. While there may be genuine benefits for him, his problems are not necessarily being solved.

On the microeconomic side, the effect of many of these suggested reforms will be simply to impose greater costs on farmers. This will force them to seek other means to cut labor costs, such as hiring alien labor or resorting to mechanization, if that is an alternative, or escaping the cost by resorting to nonlabor-intensive crops. In addition, some presently marginal farmers can be expected to be forced out of business, while others yet will become marginal farmers if they cannot make some cost adjustment to their benefit. In any case, the result will be to decrease the number of available jobs for migrant farm workers. If there is such a decrease in the job market for agricultural farm labor, the pool of surplus farm labor will be increased, and those farmers in some way able to escape the application of the reforms to themselves will reap the benefits. The reforms, while perhaps aiding some migrant farm workers, will operate to the serious detriment of many others.

Aside from such criticism, it must be noted that much of the remedial legislation hangs together: there are so many loopholes in the law that many different laws must be changed to bring substantial benefits to farm workers. This makes passage of the legislation difficult; it also means that some vital items may be sacrificed. This would mean that enacted changes may have less impact than they would have otherwise. Many of the proposals also create the same contingencies of nonenforcement that now exist with present law because of the heavy reliance on third-party interventions to protect the farm worker. Inasmuch as there are very important and powerful interests involved, and because farm workers do not have a natural administrative and law enforcement constituency, one could expect rather systematic nonenforcement or the adoption of techniques to elude enforcement. Evasion of legal rules and requirements, such as that which has occurred in desegregation, would probably take place, even if on a greatly reduced scale. This would then require legislative, judicial, and administrative tinkering, which would delay the realization of full benefits. Besides this, the influence of politics on administration is well known, and administrative delay, division, and inertia are all likely to make them paper reforms.

The most important proposal, as well as the one which receives the greatest resistance, particularly from agribusiness interests, is the inclusion of farm workers under the National Labor Relations Act.[10] The inclusion would undoubtedly give great impetus to the organization of farm workers and provide important governmental protections for union activity as well as remedies for employer interference and recalcitrance. But some of the possible consequences of this legislative change should be pointed out. Offhand, it would seem that the NLRA inclusion would principally help farm workers working on large farms, particularly for agribusiness interests, and seasonal farm workers, that is those having other work during the remainder of the year and those living in the area of the farm work. The principal reasons are that, relatively speaking, these are the easiest groups to organize, have potentially greater resources to support union activity and resist the importation and use of state labor, and that the agribusiness concerns have sufficient financial depth and market control to respond favorably, even if unwillingly, to organized demands.

The more serious difficulties in organizing migratory farm workers have been discussed above, and it is clear that organizing them would be a very long and painful process. In addition, the ability of the small farmer to meet organized demands is unclear, and many are so wary of labor difficulties that they would rather switch to nonlabor-intensive crops than deal with organized farm workers. Beyond this, the existence of organized farm labor would prompt mechanization and the consolidation of small farms into larger units. Therefore, while it is difficult to assess the effect of the NLRA inclusion on true migrant farm workers, it would appear that the benefits to them are not as great as to other kinds of farm workers while the possible adverse effect is greater.

A second legislative proposal of importance to farm workers is to provide them with federal minimum wage protections. This particular proposal, while essential, is also illustrative of the general inadequacy of current ameliorative approaches to solve migrant farm worker problems. For this reason, the proposal is worth discussing at some length.

The 1969 Report of the Senate Subcommittee on Migratory Labor, a group of senators on the whole quite favorably inclined to assist migrant workers, recommended:

> (a) providing a *gradual* increase in the agricultural minimum wages *over a period of years* until the industrial minimum is reached; (b) expanding coverage of the Fair Labor Standards Act's provisions by *gradually* including more employees working on farms by lowering the number of mandays required during the calender quarter . . . [1969 Report: 51].

In 1967, migratory farm workers who ranked lowest in income in our economy, averaged 85 days of work for an income of $922 or $10.85 a day. Given the fact that there are 240,000 farm workers making less than $1.00 an hour, this emphasis on a *gradual* increase simply cannot be justified. It is apparent that the subcommittee made this recommendation to ease the blow to farmers. Giving the subcommittee the best of intentions, it may be a political device to divide the farm pressure bloc, but even were there an immediate raise of migrant farm worker wages to the minimum wage level, farm worker income would still fall far short of the national poverty guidelines (which, one should remember, are somewhat arbitrary and low). Taking the 1967 averages, for example, the addition of $ .60 per hour for farm workers making only $1.00 an hour would have increased their income by an average of $520 a year ($ .60 per hour for ten hours a day for 85 days). This would amount to a sixty percent increase in income, but the total income would yet be only $1,442. Given our seemingly omnipresent inflation, a gradual increase to this point would mean a far smaller gain in real income by the time the minimum wage was reached.

The subcommittee also believes that an increase in the minimum wage would not result in greater unemployment for farm workers, basing its conclusion on minimum wage farm worker coverage now required by law on farms utilizing more than 500 man-days of farm labor a quarter. But the subcommittee also states that "the long-term trend of declining employment, attributable to employers seeking to cut down on the use of labor by accelerating their adoption of the newer technologies, continued and thus had an indirect effect on employment" (1969 Report: 51). Why an increase of farm worker wages to the minimum wage level would not have an effect on farming decisions to adopt the new technologies or consolidate small farms, the subcommittee does not say. One may surmise that, like other items of proposed remedial legislation such as the NLRA inclusion, the major benefits would result to those farm workers who are not migrants. This is perfectly understandable, since most farm workers are not migrants. But it should also be understood that legislation intended to help farm workers generally will not necessarily help migrant farm workers.

None of this criticism is intended to suggest that provision of the minimum wage to farm workers would not be a significant step having substantial benefits for them. It does point out, however, that another result of this line of amelioration is to postpone the demise of the use of migratory labor in such a way as to ease the economic difficulties of the farmer and allow him time to make transitions to the new farming technologies or allow further consolidation of small farms into large enterprises. The real economic burden of the increase to the minimum wage will, therefore, fall on migrant farm workers as a class—although individual farm workers will benefit—since there will be a likely decrease in migratory labor employment.

If that is true, one wonders whether there is a real net gain either in the elimination of poverty or in its incidence among migrants.

Many of the other legislative proposals reveal a similar inconsistency of attitudes and approaches toward farm labor problems. The guiding principle in the recommendations appears to be a combination of a desire not to affect existing interests adversely, and a desire to better the life circumstances and working conditions of migrants through social service subsidies. It is based on the assumption that the system of migrant farm labor must continue for some time. The contradiction in this approach lies in the hope to mitigate the evils of the system while allowing it to continue. Thus, for example, the subcommittee expresses great dismay that federal food stamp and commodity programs and federal farm labor housing and welfare programs have not been graciously administered or used to benefit migrant farm workers. Even though the subcommittee appears to perceive that the reasons for this lie in the structure and administration of the programs, it adopts as recommendations what amount to enhanced continuations of the programs. Consequently, even though past experience virtually dictates that even a reformed system will be inhuman, the subcommittee is willing to tolerate it, albeit hoping for the best.

What should be clear is that much of the proposed legislation, while humanitarian, directs itself to the relief of symptoms. It would not be unfair to conclude that the tenor of many legislative proposals intended to benefit migrant workers either assumes the continuation of the system of migrant labor, or it is a political compromise based on notions of what Congress might be convinced to accept. Even in the latter case, there are indications that the recommendations would not fare at all well in Congress. If the net result of all such reforms, however, is ultimately to phase out the system of migrant farm labor over a relatively long period of time because of the effect of social service subsidies and the creation of a larger impoverished labor pool than currently exists, the only real gain will be the slow elimination of the farm labor system. This would be a desirable result, to be sure, but that delay will have enhanced and maintained the evil and the pain of the system while only very slowly eliminating it.

To give the subcommittee its due, however—for the 1969 Report is an excellent one—it does recognize rural poverty, migrant mobility, and the surplus of farm labor as being the essential problems of the migrant worker. At the same time, it suggests the development of a comprehensive rural manpower development program, the creation of jobs in rural areas, and income maintenance programs as keys to the real resolution of the plight of the migrant farm worker. But the subcommittee looks to this as something to be developed in the indefinite future. Additionally, in proposing a transition to an agricultural economy not dependent upon migrant workers, the subcommittee revealingly chooses to give primary protection to farming interests:

> The need to develop a local, nonmobile seasonal farmwork force must be given top priority to achieve a revitalization of the rural economy that will benefit the migrant farmworker and the agriculture industry also. The agriculture industry must be guaranteed an adequate work force to harvest crops [1969 Report: 107].

This willingness to reconcile interests should not obscure the fact that in the contest among the farm workers, farmers, and the American consumer, the interests of the latter two have been deemed paramount. This is perhaps in part due to an unwillingness to resort to any schemes which are not voluntary, "free enterprise"-incentive designs for instituting change such as, for example, direct government creation of jobs. In any case, the cost of transition to an agricultural economy which does not rely on migratory labor is essentially to be borne by migratory labor. I wish to emphasize the character of the choice of recommendations made, for had the concern of the subcommittee been primarily for the plight of the migrant farm worker, the top priority programs would have been the creation of jobs and income maintenance programs in those areas where migrants originate, thus decreasing the need for migration.

This statement of the problem has the merit of exposing what is at stake economically, what the political and power realities are. Functionally viewed, some part of the reason for ameliorative change rather than a termination of the system of migrant farm labor is to assuage conscience, and to provide some social welfare benefits, while continuing to supply cheap, necessary labor to American farmers and farming interests. The expectation is that in the long run, perhaps twenty or thirty years, because of changes in the structure and technology of farming, there will be no further need for migrant farm labor. The burden and cost of the transition, however, falls on the back of migrant labor. There is thus no current national policy to end the need for migrancy as a way of life because to do so would be deemed to injure American farming interests and the current structure of the economy, and because to do so would seriously conflict with the rules of the game to how changes are induced in our society.

If it is understood that the problems of migrant workers are rural poverty and racial discrimination, then many of the proposed ameliorative laws and the postponement of rural job creation amount to a recognition that nothing can be done. Yet it would be far better analytically, if not politically, to emphasize the problem that way. So framed, it can be seen that the problems of migrant workers can be solved only as other national poverty problems can be solved, by the end of governmental and institutional racial and class discrimination and by governmental creation of jobs and transfer of incomes.

CONCLUSION

How can middle-range social and legal efforts and attacks be upgraded to the point where either the overarching institution begins to change or fundamental legislative change is deemed necessary? Realizing that it is not yet possible to obtain through Congress the fundamental economic changes which are essential, the best that can be done is to adopt a strategy which maximizes current gains and improvements, while highlighting the basic economic character of the problem, for there is a vast public education program which must be carried out. Such a strategy is not unique to migrant farm worker problems, but rather is basic to efforts to eradicate poverty generally, and is similar to that adopted by environmentalists.

Legal and social services and organizational efforts should continue to be used to provide assistance for specific problems of migrants and to alleviate some of the harsh and inhuman consequences of migrant status. Constant legal and group pressure on the migrant system of labor is also a way to weaken the system by making it inefficient, expensive, and generally unpleasant to continue. In this sense, the positive and aggressive handling of migrant problems can accelerate the end of the migrant system of labor in the same fashion as has the use of legal and group pressures in the welfare arena. The active and concerted promotion of interests overwhelms financial and administrative capacities to deal with them and demonstrates the unworkability of the system when powerlessness can no longer be relied upon. At the same time, it is just this kind of action which, by continually exposing the conditions of migrant life, leads to coalition-creation among reformist groups and to greater enforcement of existing law.

Even a marginal change at success in this strategy requires a national program and the allocation of more legal and organizational resources to migrants and their problems. At a minimum, it is essential that there be a national migrant farm worker clearinghouse which would coordinate information and activities of local organizations and assistance programs. There is a need also for vigorous, organizationally oriented legal and social service programs in all states originating and using substantial migrant labor. In this strategy it is essential constantly to keep in mind and to work publicly for general national programs of income transfers and governmental creation of jobs. Any such changes which help the poor as a class will help the migrant. The organization of the rural poor, with particular focus on local job development and limiting competing labor supplies, is a necessity. In particular, continuing efforts must be made to stop the use of green carders and illegal aliens.

However, for reasons articulated above, the prospects of substantial and relatively quick change for migrants are poor, even if this structure were realized today. Success in this strategy will undoubtedly reduce the need for migrant farm labor, thereby increasing the surplus labor pool of poor persons. For this reason, it is essential to see the migrant struggle as just one expression of the struggle of the poor generally. If a quicker death for the migrant system of labor contributes to that ultimate goal, it will be a happy passage.

NOTES

1. Hereafter referred to as Migrant Powerlessness.

2. Casual workers, such as students and housewives, averaged only 9 days' work and earned $66.00. The 1.9 million noncasual workers averaged 137 days' work and earned $1,045. Whites averaged 142 days and earned $1,170, while nonwhites averaged 125 days' labor and earned $747 (Migratory Labor Legislation, 1967: 332).

3. United States Civil Rights Committee Staff Report on Farm Workers, reprinted in Migrant Powerlessness (1969).

4. Hereafter referred to as 1969 Report.

5. A 1966 U.S. Department of Labor survey reveals the use of commuter labor in many occupations and "revealed a very common pattern of firms employing alien commuters paying lower wages than . . . firms employing U.S. workers" (Migrant Powerlessness, 1969, pt. 5-B: 2635, 2687-2690).

In Texas, average 1966 hourly wages for seasonal farm workers in reporting border areas was $ .76 (Migrant Powerlessness, 1969, pt. 5-B: 2631).

6. Saladini: The Short-Term Socio-Economic Effects of the Termination of Public Law 78 on the California Farm Labor Market of 1965-1967, reprinted in Migrant Powerlessness, (1969, pt. 5-B: 2774).

7. 29 U.S.C. § 49 (1964).

8. 417 F. 2d 569 (5 Cir. 1969).

9. This statement was made by Gary Bellow, who is the former California Rural Legal Assistance Deputy Director and now professor of law at the University of Southern California.

10. 29 U.S.C. § 151 (1964).

REFERENCES

BRITTON, R. L. (1963) "Open sky sweat shops." Houston Law Rev. 1: 131.

CHASE, J. B. (1967) "The migrant farm worker in Colorado: the life and the law." Colorado Law Rev. 40: 45.

Davis Law Review (1970) "Legal Problems of Agricultural Labor." University of California, Davis 2.

GIVENS, R. A. (1965) "Legal disadvantages of migratory workers." Labor Law J. 16: 584.

Hastings (1968) "The constitutionality of the NLRA farm labor exemption." 19: 384.

--- (1968) "Privileged entry onto farm property for union organizers." 19: 413.

--- (1968) "Unlawful agricultural working conditions as nuisance or unfair competition." 19: 398.

HOFFMAN, J. and R. SELTZER (1968) "Migrant farm labor in upstate New York." Columbia J. of Law and Social Problems 4: 1.

Maine Law Review (1970) "The farm worker: his need for legislation." 22: 213.

McDANIEL, I. G. and L. L. GORDON (1963) "Agricultural labor relations: the other farm problem: a rebuttal." Stanford Law Rev. 15: 616.

McELROY, R. C. (1967) The Hired Farm Working Force of 1967. U.S. Department of Agriculture Economic Research Service, Report No. 148.

McWILLIAMS, C. (1942) Ill Fares the Land. New York: Barnes & Noble.

METZLER, W. H. (1955) Migratory Workers in the Atlantic Coast Stream. U.S. Department of Agriculture cir. 966, January.

--- and F. O. SARGENT (1960) Migratory Workers in the Midcontinent Streams. U.S. Department of Agriculture Agricultural Research Service, Production Research Report No. 41.

Migrant and Seasonal Worker Powerlessness (1969) Hearings before the Subcommittee on Migratory Labor of the Senate Committee on Labor and Public Welfare, Ninety-First Congress, First and Second Sessions.

Migratory Farm Labor Problem in the United States (1969) Report of the Senate Committee on Labor and Public Welfare, Ninetieth Congress, Second Session.

Migratory Labor Legislation (1967) Hearings before the Subcommittee on Migratory Labor of the Senate Committee on Labor and Public Welfare, Ninetieth Congress, First Session, Part 1.

MORRIS, A. P. (1966) "Agricultural labor and national labor legislation." California Law Rev. 54: 1939.

MYERS, R. (1959) The Position of Farm Workers in Federal and State Legislation. National Advisory Committee on Farm Labor Pamphlet. New York.

New York Times, 1970 (November 24): 20.

PARKINSON, H. S. and K. A. HARPER (1961) "The law and migrant agricultural workers." Dicta 38: 341.

RAPTON, A. (1967) Domestic Migratory Farm Workers, Personal and Economic Characteristics, U.S. Department of Agriculture Economic Research Service, Report No. 121.

RUMMEL, C. A. (1968) "Current developments in farm labor law." Hastings 19: 371.

SPRIGGS, K. (1969) "Access of visitors to labor camps on privately owned property." University of Florida Law Rev. 21: 295.

Stanford Law Review (1961) "Agricultural labor relations: the other farm problem." 14: 120.

THUROW, L. C. (1969) Poverty and Discrimination. Washington, D.C.: The Brookings Institution.

SHERRARD, T. D. and R. C. MURRAY (1968) "The church and neighborhood community organization," in B. J. Frieden (ed.) Urban Planning and Social Policy. New York: Basic Books.

Yale Law Review (1970) "The new public interest lawyers." 79.

12

Local Variations in Urban Law Enforcement

HARLAN HAHN

□ ONE OF THE MOST FUNDAMENTAL ISSUES in the field of urban affairs, and in many other disciplines, is the subject of decentralization. In fact, several of the distinguishing characteristics of cities—including the density of population settlements as well as their location in a definite geographic area—would seem to make diversity an especially central concern in the study of urban communities. Municipal problems and policies not only display many comparable features, but they also exhibit significant local variations. Political leaders, therefore, are constantly confronted by what might be termed "the dilemma of the one and the many," or the task of developing unified social policy from the separate and disparate interests of numerous localities.

Perhaps in no area is the problem of community similarities and differences more clearly exemplified than in the field of law enforcement. As a basic means of regulating public conduct, both the law and its administration presumably must satisfy rigid standards of total uniformity and equality. The provisions of the criminal code may comprise the ultimate foundations of orderly social life

(Holden, 1969). As a result, citizens have a right to expect that legal requirements will not vary from one community to the next and that they will be impartially applied to all members of society.

Paradoxically, however, law enforcement operations also are profoundly influenced by the principles of American federalism. Whereas the enactment of most criminal statutes is a prerogative of the states, the enforcement of the laws is primarily a responsibility of local jurisdictions. Even though there has been extensive collaboration among national, state, and city law enforcement agencies, policing usually is regarded as almost a strictly local function (Grodzins, 1966: 89-124). Ironically, therefore, law enforcement practices may be more vulnerable to the effects of diverse community characteristics than many other types of public policy. Although the fundamental tenets of organized society would appear to require absolute uniformity and impartiality in the administration of the law, the delegation of this task to municipal governments may introduce extensive variability into police activities.

While numerous studies have sought to explore community differences on a broad range of municipal policies, few attempts have been made to examine local variations in police services. Law enforcement practices, as well as other local issues, however, are the products both of divergent local circumstances and of major similarities that transcend community boundaries. Therefore, the discovery of the extent to which police operations represent either comparable or relatively distinctive city characteristics, might become an appropriate goal of urban researchers. A comparative approach to the study of law enforcement functions could provide an important method of examining several major questions in the fields of local politics or urban affairs.

Since police practices form a principal means by which public standards are imposed upon social conduct, law enforcement operations may reflect commonly held moral values. As an extension of public authority that stretches from the highest levels of political dominion through national, state, and local institutions to the individual citizen, police activities exemplify the basic restraints upon public behavior that comprise the foundations of the social order. At the same time, however, they also are highly susceptible to the effects of numerous mediating influences at each level of governmental jurisdiction. The regulations that policemen enforce, therefore, may consist either of fundamental norms to which nearly all citizens subscribe or of the pervasive mores of widely dispersed sectors of society. Research on police practices offers a means of investigating the association among municipal characteristics, governmental structure, community culture, and urban policies.

DECENTRALIZATION AND POLICE PROBLEMS

The comparative study of local law enforcement activities, not only is closely related to several basic theoretical issues in the field of state and urban politics, but it also contains some important implications for the understanding of pressing social controversies. Perhaps the two most critical problems confronting police departments are the mounting threat of crime and the intensification of violent social conflict in urban areas. Ultimately, the development and the resolution of both problems may be strongly influenced by the decentralized nature of police practices.

Initially, the apparent growth of urban crime rates may provoke increased public pressure—often approaching the level of hysteria—for the adoption of municipal policies to eliminate the threat of lawlessness. In some localities, politicians might be tempted to react to civic alarm by advocating expedients, such as unwarranted police "crackdowns" on suspicious groups and the suppression of civil liberties, that could seriously undermine widely respected legal principles. In other communities, public officials may be prepared to adopt a relatively sober, coordinated approach to the problem of crime. The predominant response of urban areas to this popular movement may depend upon the nature of local political leadership, the composition of city populations, and other similar factors. In any event, the accelerating public fear of crime raises the potential danger that conduct which is legally permissible in one community might be severely punished in another, and that persons living in different sectors of society may be subjected to radically discrepant standards of behavior. Since law enforcement is normally regarded as a municipal responsibility, perhaps the principal effects of expanding public anxiety about crime will be evident at the local level.

Similarly, local diversity in police behavior could have a critical impact upon the course of urban conflict or violence. Initially, the outbreak of major community upheavals may be the product of accumulating tensions between policemen and the residents of the areas in which such incidents have erupted. Presumably, the level of public confidence in police officers tends to vary in different cities, and the occurrence of violence may reflect the character of prevailing relationships between local residents and law enforcement agencies. In addition, however, the civil violence that cities experienced during the 1960s, as well as the sporadic combat that urban areas continue to endure, may have been significantly shaped by the nature of American federalism.

In a fundamental sense, the development of violent conflict in predominantly black ghetto neighborhoods has represented a form of "localized rebellion" against the police and other representatives of public authority. While not all city residents or black citizens were prepared to participate in a sweeping revolutionary force to defy existing political institutions, the presence of a relatively large proportion of persons, living in highly segregated pockets of society, who were willing to employ violence as a means of achieving desired social objectives, may have been sufficient to ignite the upheavals. Ironically, the joint effects of residential segregation and American federalism may have provoked the explosion of ghetto riots while preventing them from emerging as a national revolutionary movement. The latter consequence may have been precisely what the founding fathers intended when they created the federal system. By superimposing territorial boundaries upon the landscape of America, James Madison hoped to quarantine potentially disruptive political groups and to preclude them from capturing the principal center of governmental authority. As Madison wrote in his famous tenth essay in *The Federalist Papers,* "The influence of factious leaders may kindle a flame within their particular States, but will be unable to spread a general conflagration through the other States." The principles of federalism, therefore, provide a basis for interpreting both the occurrence and the relatively scattered or diffuse nature of ghetto revolts.

In addition, of course, the issues posed by the growing fear of crime and by the eruption of urban violence are highly interrelated. In this period of rapid social change, increasing public demands for "law and order" and for an end to "crime in the streets" actually may reflect a fear of groups that appear to threaten, however vaguely, major social disruptions as well as a desire to return to an earlier epoch of alleged harmony and stability. Many Americans are turning to the police as a bulwark of defense against both new or disturbing social movements and the dark shadow of crime. The growth of urban anxiety about crime and violence, however, also appears to represent mounting public insecurity concerning the adequacy of police protection and the capacity of police officers to repress violent disorders.

Perhaps the most striking aspects of the popular fear of crime are reflected in the erosion of confidence in police forces among white segments of the population. In an important analysis of national survey data, Feagin (1970) found that slightly more than half of white Americans were not prepared to rely upon the police to defend their homes from criminal attack and that this lack of trust in police protection was significantly related to gun ownership, to the opinion that policemen should adopt tougher measures to control civil violence by, for example, shooting a few rioters, and to the belief that urban rioting represented an effort by blacks "to take over the cities." The findings prompt some vexing questions about the extent to which America actually may fulfill Weber's criteria of the state as a community that maintains a monopoly on the legitimate use of physical force. In addition, the evidence suggests that local law enforcement agencies may be faced with a rising tide of white resentment and with the danger of armed attempts to usurp police prerogatives.

A precipitous decline in public respect for police officers has developed among black Americans, as well. Numerous surveys of ghetto residents throughout the country have demonstrated extensive black criticism of police behavior (Kraft, 1966; McCord and Howard, 1968; Campbell and Schuman, 1968). In many areas, this hostility has extended to other facets of the legal process. A study in the Twelfth Street ghetto of Detroit, where the riots of 1967 originated, revealed that an overwhelming proportion of the residents in this neighborhood were convinced that injustice and inequality pervaded both the law and its enforcement or administration. Persons who expressed a conventional faith in the impartiality of the legal system formed a small and nearly imperceptible segment of this ghetto community (Hahn, 1969a, 1969b).

Perhaps this mounting loss of trust in law enforcement agencies among both white and black citizens reflects a major growth of public dissatisfaction with existing political institutions. In many sectors of society, people not only are demonstrating strong disapproval of incumbent political leaders, but they also are voicing serious reservations about the capacity of established political processes to resolve urgent social problems. Since police officers comprise the most immediate and tangible representatives of relatively remote governmental bodies, they may be compelled to assume both the advantages and the burdens of general popular attitudes toward the political system. Citizens who hold a favorable image of the state might exhibit a great deal of deference to policemen serving as the local agents of political authority, but persons who regard the sources of the law as illegitimate may be reluctant to obey the orders of law enforcement officers. The emergence of escalating public concerns about the adequacy of police protection from crime in white segments of the community and about the impartiality of law enforcement practices in black ghettos can seriously undermine prevailing perceptions of the legitimacy of police activities. As a result, law enforcement agencies increasingly may be forced to confront a fundamental crisis of public confidence in police authority.

The examination of police conduct as an extension of public authority based upon a belief in the political legitimacy of the state seems to provide a more comprehensive perspective for assessing police practices than other commonly accepted concepts of law enforcement responsibilities (Hahn, 1971a). Police duties are not confined solely to the imposition of basic restrictions upon social conduct. Nor are they limited to the enforcement of the law and the preservation of public order. Policemen engage in a large number of actions, including many that are not strictly related to law enforcement. Their functions may be defined to include the granting of protection, or the effort to control crime by imposing restraints upon public behavior; the maintenance of order, or the enforcement of fundamental social norms and values; the exertion of authority, or the supervision and management of social interaction; and the provision of services, or the willingness to respond to personal or family needs and emergencies. Although nearly all local law enforcement agencies may share in each of those obligations, the conduct of the duties also might vary greatly in different communities.

The purpose of this essay, therefore, will be to explore the impact of community differences upon the performance of the four basic police functions encompassed by the concepts of protection, order, authority, and service. Since major public attention has concentrated on the role of police officers in controlling crime and civil violence, however, special emphasis will be devoted to those specific law enforcement duties. In addition, this chapter will focus on *intracity* as well as *intercity* differences in police operations. Perhaps the major domestic crisis emerging in America is the polarization of divided segments, especially black and white areas, of the urban population. As a result, an effort will be made to examine police practices not only in separate localities but also in different sectors of the community.

This study also will seek to assess the policy implications of differential police practices both within and between cities. Perhaps two of the most salient controversies affecting law enforcement agencies are reflected in the movements for police "professionalism" and for establishing "community control" of police forces. While the former plan might tend to promote uniform law enforcement activities, the latter proposal may contribute to a close working relationship between police officers and the public. The comparative study of police conduct, therefore, could provide a useful approach to the evaluation of both programs.

DECENTRALIZATION AND POLICE ORGANIZATIONS

Although a major source of local diversity in police practices may be found in the principles of American federalism and in the classification of law enforcement as a local responsibility, the problems of decentralization in police work also are affected by several major characteristics of law enforcement organizations. Police departments possess many features that tend to differentiate them from other local government agencies. A discussion of the structural attributes of police forces, therefore, may constitute a necessary prerequisite for the investigation of local variations in law enforcement operations.

One characteristic that may exert an important initial impact upon the performance of police duties is the recruitment of law enforcement officers. Several studies have indicated that most policemen emerge from working- or lower-middle-class origins and that they often represent a family tradition in police work (Niederhoffer, 1967: 36-37; McNamara, 1967: 192-193). In addition, members of some ethnic groups, such as the Irish, may occupy a disproportionate share of the positions in police departments (Wilson, 1964). By contrast, other ethnic and racial minorities, including blacks and Mexican-Americans, often are seriously underrepresented in the ranks of major urban police organizations. In part, the relatively low-status backgrounds of most law enforcement officers may be a product of the modest economic and personal incentives offered by police work. Yet, this attribute of police personnel could tend to both enhance and impede local discrepancies in law enforcement activities. The relative absence of heterogeneity among police officers in most cities might inject a type of uniformity into law enforcement functions by insulating the police from disparate community influences and pressures. In addition, similarities in the occupational expectations of policemen may tend to promote the growth of a narrow or restrictive police viewpoint on public conduct. As a result, there is a potential danger that the rules enforced by police officers may represent the values and perspectives of a single stratum or segment of the population rather than the sentiments of the entire community or society.

Perhaps even more important than the recruitment of police officers, however, is their socialization. After policemen have joined the force, they normally are subjected to extensive pressures from experienced officers who may succeed in totally redefining police tasks and objectives at this critical stage of their careers (McNamara, 1967; Niederhoffer, 1967). In the process of acquiring the predominant skills and values of the force, patrolmen also develop an intense loyalty to the organization and a strong sense of estrangement from the external environment in which they must operate. The net effect of the powerful socializing influences that permeate police ranks is to produce an exceptional orthodoxy in attitudes and behavior. In fact, the solidarity features of police organizations may be one of the most striking traits of law enforcement agencies.

In addition, police contacts with members of the community in circumstances that do not entail law enforcement duties often are confined primarily to other policemen. As a result, police officers may tend to develop distinctive perspectives on public behavior. A study related to this issue was conducted in Illinois communities by Clark (1965) on the attitudes of policemen, the public, and community leaders toward six legal offenses involving Sunday blue laws, racial prejudice, gambling, drunkenness, prostitution, and obscene literature. He found that police officers were somewhat less likely than the public as well as most other groups to regard each of those acts as "morally wrong," although the greatest differences between policemen and the public emerged on the blue laws and racial discrimination issues. On the other hand, more police officers than members of the public believed that the police should take action against each of those activities, except for racial prejudice and blue laws violations. In addition, with those two exceptions, policemen were more likely to invoke the power of arrest in each of the situations than the people wished or than the police felt the public desired. Although relatively little comparative research has examined the extent to which socialization may differ in local departments, this process might tend to intensify police resistance to major changes in policy and to increase the gap between policemen and the community. The conduct of law enforcement officers, therefore, could be shaped both by a collective "police subculture," that encompasses diverse communities, and by distinctive local traditions that may arise in separate police forces.

Although extensive socialization may instill a common body of information or experience in police procedures and routines, at least one factor that might mitigate this influence is the vast amount of discretionary authority bestowed upon law enforcement officers. Unlike most private or governmental organizations in which top executives enjoy greater freedom of action than their subordinates, police departments, as Wilson (1968a: 7) notes, tend to allocate increasing discretion "as one moves *down* the hierarchy." The rookie patrolman on the street may have a wider scope of alternatives available for handling his job than the highest police administrator. Perhaps the most critical aspects of police work arise from their encounters with the public. In managing the situations that they confront in the community, police officers may adopt a broad range of actions that include neglecting the event as well as invoking the most stringent penalties of the state. Police chiefs and supervisors, by contrast, can be subjected to relatively restrictive legal and political responsibilities. The discretionary power vested in the average cop on the beat, therefore, may tend to exacerbate the diversity of police activities. Major discrepancies in law enforcement practices might develop both between and within police departments.

The problem of police discretion also is compounded by the relatively low visibility of most police actions (Goldstein, 1960). In contrast to the conduct of many other government employees, the behavior of policemen, especially in their contacts with civilians, frequently is not amenable to careful surveillance either by their organizational superiors or by responsible political leaders. The treatment accorded citizens by police officers may be one of the most crucial aspects of the relationship between the public and political authority, but those interactions often occur in the absence of disinterested witnesses or in other circumstances that may preclude effective supervision. As a result, charges of police brutality or of the failure to provide adequate protection from crime frequently are difficult to evaluate or adjudicate. In a limited number of events that result in an arrest or trial, the actions of policemen may be the focus of intensive scrutiny by the courts. In the preponderance of incidents that do not become the basis for subsequent legal action, however, a detailed investigation of the case might be impossible. The difficulty of gathering adequate information about the activities of policemen on patrol in the community, therefore, also may tend to promote the growth of significant disparities in law enforcement practices.

Variations in police operations, however, not only are encouraged by the intrinsic—and perhaps irreversible—characteristics of law enforcement agencies, but they also are enhanced by the implicit or explicit policies of public and police officials. Criminal statutes in most states, for example, require the police to seek the full enforcement of the laws. Yet, a literal adherence to that policy probably would provoke a massive popular outcry. Neither the public nor the lawmakers expect police officers to impose the penalty of arrest upon each infraction of the legal code. In fact, the pattern in most areas may conform to the underenforcement rather than either the overenforcement or the full enforcement of the laws. Consequently, police practices in different communities or even in separate neighborhoods of the same city, may range from serious underenforcement to excessively zealous overenforcement, or to some intermediary position, without any statutory basis for their actions. Perhaps the most serious implication of this problem is the reluctance of political leaders and police departments to make the dictates of underenforcement or of selective enforcement a definite matter of public policy. Although police organizations may not be able to fulfill the mandate of a full enforcement policy, decisions concerning the type of law enforcement to be imposed upon the community at least might be submitted to normal political processes for resolution in the same manner as other public issues are determined. The failure of police departments to satisfy the criteria of full enforcement, or, perhaps more important, to establish clear law enforcement policies, probably are major sources of local variations in police conduct.

The relative absence of detailed guidelines for the use of police officers in enforcing the law as well as the increasing complexity of legal regulations also may encourage policemen to substitute their own moral values for the requirements contained in the statutes. As Banton (1964: 146) notes, the actions of a law enforcement officer usually are "governed much more by popular morality than by the letter of the law; most often morality and the law coincide, but when they do not, it is usually morality that wins." Moreover, the moral values that policemen acquire and that they impose upon public behavior frequently are the predominant norms of the community in which they reside. As a result, police officers may develop highly divergent definitions of their law enforcement responsibilities.

Paradoxically, law enforcement agencies possess many characteristics that may tend to both promote and hinder the development of significant differences in local police operations. Initially, discretion in police forces is widely dispersed among personnel at the lowest level of the organization, and some of the most critical decisions may be made by those officers in relatively unsupervised situations on the basis of personal moral predilections rather than carefully formulated departmental policies. On the other hand, police departments also are quasi-military institutions in which authority is supposed to be hierarchically structured in clearly defined chains of command. Those conflicting attributes may create serious internal strains that could significantly affect the adoption of major changes in police routines. In addition, however, the quasi-military nature of police forces can provide a partially self-adjusting mechanism for overcoming the institutional lethargy that might otherwise afflict law enforcement agencies. A strong police administrator may have at least some of the necessary means at his disposal for implementing relatively firm policy directives within the organization.

Another factor that may have a major impact on community differences in law enforcement activities is the relationship between police executives and civilian political leadership. In fact, in one essay on police organizations, Bordua and Reiss (1966) identified two basic dimensions of this problem—the tenure of the chief and his relationship to the mayor—as significant factors that might be employed in differentiating police departments. They hypothesized that security of tenure for the chief would produce organizational subordination based upon the principle of command, while insecure tenure would yield a "personalized" form of institutional control. In addition, they argued that a relationship of strict accountability between the police chief and the mayor might be associated with a bureaucratic form of organization, but the absence of this relationship could be related to a network of "feudal loyalties" between the chief and the mayor or local influentials. As a result, police administrators frequently may be compelled to rely upon personal charisma both to gain control of the organization and to fulfill their responsibilities to public officials. The administrative style adopted by police chiefs, as well as their relationships with community leaders, therefore, may produce additional diversity in local law enforcement practices.

Perhaps the most widely accepted proposal for creating general uniformity in local police practices has been represented by the mounting drive toward police professionalism. Since the professional characteristics of police departments often have been regarded as a primary source of city differences in law enforcement practices, professionalization also has been viewed as an important means of improving the quality of police work. On the one hand, this movement has reflected a growing desire for the elimination of corruption and favoritism, for raising standards of police performance, and for an increased emphasis upon human relations in the conduct of law enforcement duties. Hence, spokesmen for police professionalism have sought to gain higher salaries and increased education for law enforcement officers and to end the excessive violence and racial prejudice that have often been characteristic of police organizations. On the other hand, the trend toward professionalization also has been occasioned by an increased sense of expertise and self-esteem among police officers and by a corresponding deprecation of external influences. The intense opposition of police forces to proposals for civilian police review boards during the 1960s, therefore, appeared to be consistent with some of the major tenets of professionalism. Basic to the impetus for professionalization was a prevalent belief among police officers that they should be freed from outside interference and that they should be allowed to pursue their central mission—the prevention and control of crime—using their own methods and judged by their own criteria. This focus, of course, may tend to deemphasize the many other crucial duties that policemen are required to perform. In addition, police professionalism might tend to encourage the development of uniform law enforcement procedures in widely dispersed communities, but it also may foster the growth of a "police subculture" that could divorce policemen from universally shared and respected values in civilian society.

INTERCITY DIFFERENCES IN LAW ENFORCEMENT

Although the organizational attributes of law enforcement agencies seem to provide numerous opportunities for the growth of local variations in police practices, attempts to describe those differences have often seemed to be as difficult as efforts to develop typologies of cities. In one study that sought to characterize the role of local government as providing life's amenities, promoting economic growth, maintaining traditional services, or arbitrating among conflicting interests, Williams and Adrian (1963: 206-208) found that cities displaying the former two characteristics rated higher on police expenditures than communities reflecting the latter two attributes. Despite this evidence, however, relatively few additional studies have examined the association between community characteristics and a broad range of police practices.

Perhaps the most common method of classifying police departments has been based upon the concept of professionalism. In an early essay, Wilson (1963) distinguished between "professional" police organizations based upon universalistic standards of conduct and so-called "fraternal" or "system" codes that tend to enforce particularistic values or norms. He argued that "professionalism" might be associated with "cosmopolitan" police officers, a faith in the moral validity of the law, bureaucratic forms of authority, a reluctance to employ force except in self-defense, and a desire to project an image of honesty and impartiality. By contrast, he asserted that police codes founded upon the system may be related to a "local" orientation among police recruits, a belief that policemen rather than legal rules comprised the principal source of social controls, the personal exercise of authority, a propensity to use violence to instill respect for the police, and an effort to demonstrate that law enforcement officers are simply human rather than highly skillful or efficient. In addition, Wilson (1968b) revealed in a comparative study of the handling of juvenile offenders in two communities that professionalism was highly related to the equal treatment of white and black young people and to the imposition of penalties according to the rules rather than through the exercise of personal discretion.

Subsequently, however, Wilson (1968a: 140-226) identified three distinct forms of police conduct that he termed "the watchman style," which emphasized the preservation of public order rather than law enforcement duties; "the legalistic style," which focused upon the vigorous and impartial enforcement of the laws; and "the service style," which stressed both order maintenance and law enforcement functions without necessarily exacting the full penalties of the law. In many respects, this formulation was remarkably parallel to the earlier effort by Klonoski and Mendelsohn (1966) to relate the characteristics of community politics to local processes of criminal justice that yielded three types of systems including "the exchange model," in which legal treatment is dispensed on a quid pro quo basis; "the elite-bar model," which subjects all persons to relatively rigid and arbitrary standards of conduct; and "the communal system," in which the allocation of justice is closely related to the distribution of political influence in the community. Although each taxonomy is primarily heuristic, they are predicated on the proposition that police practices reflect critical variations among communities.

Perhaps the most persuasive evidence of community differences in law enforcement operations, however, was reported by Wilson (1968a: 271-277) in an analysis of arrest rates compiled from FBI files and the structure of political institutions in 146 American cities. The data revealed that municipalities with highly professionalized council-manager governments recorded the highest arrest rates for larceny and intoxication but that partisan mayor-council cities disclosed the highest arrest rates for disorderly conduct. Most significant, however, the strong association between arrest rates and the form of urban governments suggest that manifestations of police work might be closely related to other community variables such as population characteristics, prevailing moral values, public attitudes, and similar local features.

In addition to the correspondence between community attributes and arrest rates, local variations might be found in other facets of the protective function performed by law enforcement officers. Undoubtedly the largest share of police time is devoted to patrolling city streets in an effort to protect the lives and property of residents. Although both the techniques of patrol and the ratio of policemen to the population vary widely from community to community, relatively little information is available concerning either the effects of different patrolling procedures or their relationship to urban characteristics. Major issues affecting patrol work, such as the protection granted commercial or residential areas, and the amount of resources devoted to detecting or solving various types of crimes, may be less the product of overt political decisions than of local networks of influence and decision-making. As police departments encounter increasing demands for security from the threat of crime, additional comparative research on the performance of protective duties could contribute significantly to an understanding of law enforcement problems.

A major source of local diversity in law enforcement practices results from the responsibility of maintaining public order. In addition to drunkenness and disorderly conduct, which are the two principal grounds for arrests, policemen must assume the task of regulating other "crimes without victims" including gambling, vice, and narcotics violations. Experience with such laws clearly indicates that what may be permissible in one community could be considered totally unacceptable in another locality. Many cities, for example, acquire extensive reputations as "wide open" towns, while other urban areas are noted for their civic virtue. In large measure, distinctions depend upon prevailing public—and especially police—concepts of morally proper or intolerable conduct. The authorization to prohibit behavior that does not cause harm or injury to another party, however, not only creates opportunities for the perpetuation of divergent standards of law enforcement, it also may provide the impetus for corruption within police departments. Since policemen supposedly are obligated to impose restrictions upon activities that may be desired by powerful or extensive local interests, elaborate systems of graft might be required to enable law enforcement officers to serve those incompatible ends. As a result, police departments may differ in the honesty as well as in the efficiency or effectiveness of their operations.

Another primary basis for differential law enforcement activities is the duty of exerting police authority. Although this function may overlap with other jobs, it could encompass activities ranging from the management of social interaction to the control of highway traffic. In another impressive demonstration of community differences in police behavior based upon a survey of 508 American cities, Gardiner (1968) found that the volume of traffic citations issued per 1,000 registered automobiles in a single year extended from less than 100 in twenty-four percent of the communities to more than 300 in eighteen percent of the towns. The number of traffic tickets distributed in various cities is not only related to population mobility and local requirements governing court appearances by police officers, but also may be strongly influenced by the policies and expectations of police administrators (Wilson, 1968a: 95-99).

Appreciable community variations in traffic law enforcement suggest that major local distinctions also might become evident in other aspects of the administration of police authority. As Black and Reiss (1967: 27) have noted, perhaps the major responsibility of a police officer in any contact with the public is to gain control of the encounter and to order the social situation for himself. The techniques that policemen employ to assert their authority obviously could differ both within and among law enforcement agencies. Just as police forces can gain notoriety for their tolerance of a "wide open" town, they also may become known for their "tough" or aggressive behavior. Although physical harassment or brutality may be more difficult to measure than the issuance of traffic violations, the former incidents also might be closely associated with local differences in police conduct.

A final major function of law enforcement agencies consists of providing community services through intervention to alleviate personal needs or emergencies such as resolving family troubles, transporting sick or injured persons to a hospital, and similar acts. Although many policemen tend to deprecate such tasks as unwelcome interference with their principal mission of controlling crime, those activities may consume the largest amount of active police duty in most communities. Early studies in Syracuse (Cumming et al., 1965; Wilson, 1968a: 130-131) disclosed that the greatest number of calls made by residents for police assistance request services that are unrelated to criminal matters. A subsequent analysis of the calls received by police departments in Detroit, New York, and St. Louis indicated that this trend also may prevail in other urban areas (Bercal, 1970). Since the ability to provide vital services during periods of crisis may be a major source of active and enduring popular support for law enforcement agencies, the response to citizens' requests for those services could be a critical factor in shaping local sentiments regarding the police. Additional comparative research, therefore, is needed on police reactions to widespread public demands for community services.

Increasingly, major local variations in police practices have assumed significance not only because they may represent marked departures from the standard of absolute uniformity in the enforcement of the laws, but also because they might be related to the emergence of local hostility and even violent attacks upon the police.

As the riots of the 1960s seemed to demonstrate, intensive animosity toward policemen, concentrated in relatively limited segments of society, apparently was sufficient to spawn a crisis with widespread ramifications for law enforcement generally. Since the outbreak of riots—at least at dramatic levels of intensity—did not appear to be uniformly distributed among all cities, those incidents prompted several efforts to investigate the association between community structures and the occurrence of urban violence.

Some evidence has suggested that community differences might be related to the outbreak of violence. In fact, a comparative study by Lieberson and Silverman (1965: 897) of matched pairs of American cities concluded, "Riot cities not only employ fewer Negro policemen, but they are also communities whose electoral systems tend to be less sensitive to the demands of the electorate." Since police officers have played a major role in the precipitating incidents that triggered most major urban riots, the growth of local tensions between community residents and law enforcement officers may have been an important contributing factor in the outbreak of violence. As the Kerner Commission (1968: 206) observed, "Almost invariably the incident that ignites disorder arises from police action." A study reporting a cumulative scale of the intensity of seventy-five urban riots in 1967 disclosed that the level of violence was directly associated with police preparedness for disorders (Wanderer, 1969). In addition, the actions taken by communities to alleviate local conditions in the aftermath of riots seemed to be related to the degree of centralization in municipal governments (Hahn, 1970a).

INTRACITY DIFFERENCES IN LAW ENFORCEMENT

In addition to the diversity that has marked police practices in separate communities, the activities of law enforcement officers within cities also have reflected major discrepancies. Perhaps the most commonly recognized traditional source of differential police action was the distinction between relatively affluent and impoverished urban neighborhoods. During the late nineteenth and early twentieth centuries, police officers in American cities may have primarily been concerned with protecting the economic interests of dominant businessmen and property owners. Within lower-class ethnic enclaves, policemen maintained a paternalistic surveillance of the neighborhood, but the outbreak of crime in wealthy or middle-class areas of the city may have provoked a full mobilization of police resources to apprehend the culprit. Historically, prosperous and influential residents probably received more police protection than working-class citizens in urban areas, and the distribution of police services was divided primarily along social and economic lines.

As increasing numbers of black Americans migrated to Northern cities, however, public perceptions of law enforcement, as well as many other urban issues, began to change dramatically. The principal opposing forces in the expanding controversy over the role of the police in urban areas have been black and white residents of the community. Numerous surveys have revealed that, while most whites have retained favorable opinions of police actions, black citizens as well as other minorities usually have been highly critical of police behavior (President's Commission on Law Enforcement and Administration of Justice, 1967: 144-149; Bayley and Mendelsohn, 1969). An early survey (Marx, 1967: 36-37) revealed that black citizens in New York, Chicago, and Atlanta were less likely to object to police conduct than those living in Birmingham. Yet, black disapproval as well as white commendations of the police have seemed to spread to nearly all sections of the country. As a result, racial cleavages apparently have replaced social and economic distinctions as a principal basis of conflict concerning law enforcement operations in American cities. In fact, the growing public debate about police activities has threatened to separate black and white Americans into warring camps.

Perhaps one of the major sources of black and white disagreement has been centered upon the protective function of police officers. While white Americans generally have expressed a consistent demand for increased protection from crime and for harsh or stringent law enforcement measures, black attitudes often have appeared to represent a troubling enigma. Black citizens not only have displayed intense grievances about police brutality or harassment, but they also have voiced strong complaints regarding the inadequacy of police protection in crime-ridden ghetto neighborhoods. Whereas the former sentiments would seem to imply a desire for a relaxation of police surveillance in urban ghettos, the latter view has appeared to reflect support for increased police patrols. In part, perhaps this apparent paradox has been promoted by the increasingly prevalent and supposedly professional policy of "preventive patrolling," or the practice of stopping and questioning large numbers of allegedly suspicious persons in areas with high crime rates (Fogelson, 1968). While this procedure has tended to produce many hostile and abrasive encounters between black citizens and policemen regarding serious violations of the law, it also has probably diverted police manpower from the investigation of the many so-called minor crimes that plague ghetto neighborhoods (Hahn and Feagin, 1970).

In addition, black perceptions of police protection also have been shaped by differential law enforcement operations in urban areas. The survey in the Twelfth Street area of Detroit revealed that most black residents of this neighborhood believed that economic distinctions tended to spawn appreciable inequities in police protection, but large proportions of people felt that unequal protection was a product of racial rather than socioeconomic considerations. Perhaps most important, however, perceived inequalities in police protection also were significantly related to the unwillingness of persons to cooperate with law enforcement officers (Hahn, 1971b). The results of the research seemed to suggest that the removal of major inequities in police practices may constitute an important means of gaining public trust and confidence in police officers in predominantly black urban ghettos.

Although increasing black as well as white demands for police protection eventually may exert their strongest impact upon local rather than state or national politics, relatively few comparative attempts have been made to examine city differences in the intensity of those sentiments. The study by Feagin (1970: 800-802), however, revealed that black citizens living in large cities were less prone to rely upon the police for protection from crime than those who resided in smaller cities, but whites in major metropolises expressed more confidence in police protection than their counterparts in small towns. Since the probability of crime usually has displayed a linear increase in relation to the size of cities, the findings suggest that, while the willingness of black citizens to adopt self-defense as a form of protection was directly associated with the risk of crime, it was inversely related to the threat of victimization among whites. The mounting fears that grip white Americans who suffer relatively little exposure to crime have seemed to emerge as a serious danger to the prospects of achieving both racial harmony and public respect for police authority.

Ironically, even though the white middle class forms a major source of anxiety about crime, this segment of society also may be the principal beneficiary of police protection. A similar statement probably could be made about the function of law enforcement officers in preserving public order. Black complaints about unjust police practices are not confined to incidents of physical brutality and violence, but they also encompass cases of discriminatory treatment for minor infractions such as drunk or disorderly conduct as well as other forms of petty harassment. While many white suburbanites who disrupt the public order may be subjected to a reprimand and transportation in a squad car to their places of residence, members of minority groups and other lower-class persons find themselves subject to fines or imprisonment for similar offenses. In addition, many illegal enterprises such as vice, gambling, and other "crimes without victims" that would not be tolerated in other sectors of the city often are allowed to flourish in ghetto neighborhoods. In fact, those activities may be permitted to operate with explicit police recognition and supervision in a system sustained by bribery and other forms of corruption. Differential law enforcement in predominantly black and white areas of the city, therefore, may enhance the difficulty of promoting confidence in the integrity of police officers among black urban residents.

Another critical area of controversy between black and white citizens has entailed the imposition of police authority upon public conduct. Although the exercise of this responsibility may vary widely in different communities, law enforcement officers apparently have not been highly sensitive to local variations in public appraisals of police authority. In a survey of the policemen who patrol ghetto neighborhoods in 13 American cities, Groves and Rossi (1970) found that police perceptions of ghetto hostility were less closely related to city differences than to the individual attributes of law enforcement officers, although this generalization was more applicable to the attitudes of white than of black patrolmen. While black policemen may have acquired a greater understanding of the actual character of public assessments of police authority than their white colleagues, the judgments of law enforcement officers apparently are influenced more by personal characteristics than by potentially significant local disparities in ghetto perspectives.

The exercise of police authority within cities also has been closely related to conflict over the issue of police brutality. Although some persons even have seemed to refuse to acknowledge the existence of police mistreatment, accounts of such incidents have been widely circulated—and believed—in predominantly black urban ghettos. A survey that asked residents of the Watts area about six forms of police brutality including the display of a lack of respect, insulting language, rousting and frisking, unnecessary force in arrests, and beating up suspects in custody also discovered that nearly two-thirds believed that those practices occurred in the area, at least one-fifth claimed that they had seen them happen, and a small percentage said that they had experienced such actions (Raine, 1967). Apparently, perceptions of police brutality have been shaped more by personal traits than by local variations in law enforcement practices. A survey of white and black attitudes in 15 American cities disclosed that reports of experienced police abuse apparently were influenced more by individual attributes than by city differences, but this study also indicated that white and black statements about police misconduct were highly interrelated by city, implying that the respondents may have encountered similar local experiences (Schuman and Gruenberg, 1970). While divergent local law enforcement procedures probably have had an impact upon the treatment accorded both white and black urban residents, perceptions of police abuse generally have been more prevalent among black than among white segments of the community. As a result, the issue of police brutality also has been a major source of controversy between white and black citizens.

A final police function that has reflected important differences between white and black sectors of the city is the provision of community services. Unlike many other police duties, however, the performance of this work usually offers greater advantages to lower-class and minority groups than to higher-status white persons. While relatively affluent members of the community may possess the resources to seek private assistance from clergymen, physicians, attorneys, and psychiatrists, the police often constitute the only source of help readily available to many less prosperous citizens to alleviate personal needs and emergencies. Despite the tendency of policemen to disparage their service responsibilities, those duties could enable law enforcement to render invaluable aid and comfort to those groups from which they have encountered the strongest resistance and animosity. Ironically, the function that has become highly unpopular among police officers also may be the one task that might permit the police to gain badly needed public respect and support.

The allocation of police services among different racial groups and classes within cities also may exert a major impact upon the growth of public animosity toward law enforcement officers. Perhaps the most visible and the most serious manifestations of this enmity are reflected in the outbreak of urban violence. Although black antipathy concerning police behavior apparently encompasses ghetto areas throughout the country, some evidence indicates that those sentiments were especially prevalent in neighborhoods that exploded in major riots (Feagin and Sheatsley, 1968; Governor's Select Commission on Civil Disorder, 1968: 32-36; Sears, 1969: 524). In addition, other data suggest that unfavorable opinions of the police are related to the propensity to support riots, to participate in the upheavals, and to endorse violence as a means of securing progress (Murphy and Watson, 1967: 46-48; Bayley and Mendelsohn, 1969: 177-181; Hahn, 1969c). Extensive ghetto hostility toward law enforcement officers not only raises the threat of massive urban disorders, but it also may seriously impede subsequent police efforts to control riots and to reimpose their authority in the aftermath of violence (Hahn, 1970b). As a result, the task of gaining public respect for police duties in predominantly black ghettos may be one of the most important objectives confronting urban law enforcement agencies.

Despite the importance of this goal, the effort to earn the confidence of ghetto residents may be an exceptionally difficult task. A major reorganization of police responsibilities, rather than limited programs to improve police-community relations, might be required to overcome the intense resentment of law enforcement that has developed in most ghetto neighborhoods. In addition, the growth of public demands for enhanced police protection among whites who are relatively insulated from the risk of crime could subject police departments to increasingly severe political pressures and strains.

Law enforcement officers might become embroiled in a divisive conflict between black and white urban residents that could seriously undermine public trust in police authority between both sectors of the community.

IMPLICATIONS FOR FURTHER STUDY

The mounting problems that confront law enforcement officers seem to create a need for enhanced public support of police duties. As policemen become increasingly involved in grave public disputes, they may require widespread approval from all segments of society to sustain their activities. Major changes in law enforcement policies and practices, therefore, could be necessary both to secure this support and to prevent the police from acting as the servants of only one party to a controversy.

Perhaps most important, serious consideration might be given proposals to remove the inequalities that seem to permeate police work both between and within communities. Abstract standards of law enforcement would seem to provide little justification for the markedly divergent procedures in separate localities that affect the protection of lives and property, the maintenance of public order, the exertion of police authority, and provision of community services. In addition, significant discrepancies in police practices within cities may be a principal source of the serious public animosity that has been focused upon law enforcement officers, especially in urban ghettos. The elimination of major inequities in police operations, therefore, could form an important means of developing increased public respect for law enforcement agencies.

In seeking to achieve impartial procedures, police administrators may encounter formidable obstacles arising from the diffuse and fragmented character of their organizations. Yet, the quasi-military nature of law enforcement agencies may provide police supervisors with a valuable means of accomplishing significant policy changes. If the directives of police chiefs can exert a strong impact on the issuance of traffic citations, they also may have a powerful effect upon the performance of other police functions. With the assistance and support of local public officials, police chiefs appear to possess the necessary resources for implementing major innovations in law enforcement policies.

The adoption of significant changes in law enforcement activities, either to equalize police practices or to attain other objectives, however, may be contingent upon the resolution of a major paradox affecting police departments. Increasingly, law enforcement agencies are confronted by opposing movements to "professionalize" and to "decentralize" police operations. While the former plan may tend to promote relatively uniform police conduct both between and within communities, it also could heighten the isolation and estrangement that has developed between police officers and the public. The latter proposal—to establish neighborhood control of police organizations—probably would enhance the growth of local diversity in law enforcement practices, but it might strengthen the relationship between police officers and residents of the community. The conflict between the policies of "professionalization" and "decentralization," therefore, may be one of the most critical dilemmas confronting police agencies.

There appear to be implicit dangers in both programs. Initially, the development of police professionalism could contribute to relatively parochial standards of law enforcement attitudes and behavior that might ultimately become a law unto themselves. Since police organizations already possess many characteristics of a self-contained community, the granting of additional powers of self-regulation to law enforcement officers might encourage the police to neglect their responsibility to the public and to political authorities. By contrast, the creation of police forces under community control may exacerbate many of the problems that have previously resulted from the decentralization of law enforcement duties. Not only would the administration of the law be subjected to widely divergent moral values, but the increasing fear of crime also might become the basis for efforts by some segments of the community to resist intrusions by other sectors of the population. Although Waskow (1969) has argued that decentralization may offer an important means of regaining democratic control of law enforcement agencies, the adoption of community control in white neighborhoods, for example, could permit those groups to employ police powers against what they might term alien or threatening black residents of the city.

As a result, the decentralization of law enforcement units might inject additional inequality and conflict into police operations.

There is a third innovation in law enforcement policy that might be studied. Since the largest proportion of public demands for police assistance and the preponderant share of active police duty are devoted to the performance of community service activities, those responsibilities could be elevated to the status of a major function of law enforcement agencies. Police departments might be regarded as governmental bodies whose mission encompasses responding to personal needs and emergencies as well as imposing legal controls upon public conduct. The strictly law enforcement obligations of the police thus might become one component of a broad range of duties that would be assumed by extensively reorganized law enforcement bureaus. Although the acceptance of this proposal may require a major revision in prevailing concepts of the scope of both police and government action in community life, it also would permit police departments to implement relatively uniform standards of law enforcement and to adapt their community service role to differing local needs and circumstances. In addition, an increased emphasis on the community services performed by law enforcement agencies might enable the police to gain a valuable source of public support, especially among segments of society that have been highly antagonistic toward police authority.

The increasing crisis of public confidence in the police seems to demonstrate a need for major changes in law enforcement policies. Further research may be required to investigate the possible effects of such changes. As an initial step, attention might be focused upon the responses of police organizations to public requests for community services both in different sectors of the city and in separate localities. In addition, studies might profitably be conducted on the relationship between the activities of law enforcement officers and environmental conditions or the demographic attributes of cities; community attitudes, including the sentiments of dominant segments of the population; factors affecting the role of political leadership; and the characteristics of police organizations. By employing this approach, efforts could be launched to examine the process by which different local values eventually are transferred to police behavior. Most important, the study of police conduct might be related to broader efforts to identify critical distinctions both within and between urban areas. Law enforcement activities play a crucial role in the economic, social, and political life of cities, and they can no longer be justifiably neglected or ignored. As the growing involvement of police officers in major social controversies has indicated, a careful appraisal of police practices may be a necessary condition for achieving a thorough understanding of emerging urban problems.

REFERENCES

ABERBACH, J. D. and J. L. WALKER (1970) "The attitudes of blacks and whites toward city services: implications for public policy," pp. 519-538 in John P. Crecine (ed.) Financing the Metropolis: Public Policy in Urban Economics. Beverly Hills, California: Sage.

BANTON, M. (1964) The Policeman in the Community. New York: Basic Books.

BAYLEY, D. H. and H. MENDELSOHN (1969) Minorities and the Police. New York: Free Press.

BERCAL, T. E. (1970) "Calls for police assistance: consumer demands for governmental service." American Behavioral Scientist 13 (May-June-July-August): 681-691.

BLACK, D. J. and A. J. REISS, Jr. (1967) "Patterns of behavior in police and citizen transactions," pp. 1-139 in Studies in Crime and Law Enforcement in Major Metropolitan Areas, Vol. 2. Washington, D.C.: Government Printing Office.

BORDUA, D. J. and A. J. REISS, Jr. (1966) "Command, control, and charisma: reflections on police bureaucracy." Amer. J. of Sociology 72 (July): 68-76.

CAMPBELL, A. and H. SCHUMAN (1968) "Racial attitudes in fifteen American cities," pp. 1-69 in Supplemental Studies for the National Advisory Commission on Civil Disorders. Washington, D.C.: Government Printing Office.

CLARK, J. P. (1965) "Isolation of the police: a comparison of the British and American situations." J. of Criminal Law, Criminology and Police Science 56 (Fall): 307-319.

CUMMING, E., I. CUMMING, and L. EDELL (1965) "Policeman as philosopher, guide and friend." Social Problems 12 (Winter): 276-286.

FEAGIN, J. R. (1970) "Home-defense and the police: black and white perspectives." American Behavioral Scientist 13 (May-June-July-August): 797-814.

--- and P. B. SHEATSLEY (1968) "Ghetto resident appraisals of a riot." Public Opinion Q. 32 (Fall): 352-362.

FOGELSON, R. M. (1968) "From resentment to confrontation: the police, the Negroes, and the outbreak of the nineteen-sixties riots." Political Science Q. 83 (June): 217-247.

GARDINER, J. A. (1968) "Police enforcement of traffic laws: a comparative analysis," pp. 151-172 in James Q. Wilson (ed.) City Politics and Public Policy. New York: John Wiley.

GOLDSTEIN, J. (1960) "Police discretion not to invoke the criminal process: low-visibility decisions in the administration of justice." Yale Law J. 69 (March): 543-594.

Governor's Select Commission on Civil Disorder (1968) Report for Action. Trenton, New Jersey.

GRODZINS, M. (1966) The American System. Chicago: Rand McNally.

GROVES, W. E. and P. H. ROSSI (1970) "Police perceptions of a hostile ghetto: realism or projection." American Behavioral Scientist 13 (May-June-July-August): 727-743.

HAHN, H. (1971a) "The public and the police: a theoretical perspective," in Harlan Hahn (ed.) The Police in Urban Society. Beverly Hills, California: Sage.

--- (1971b) "Ghetto assessments of police protection and authority." Law and Society Rev.

--- (1970a) "Civic responses to riots: a reappraisal of Kerner Commission data." Public Opinion Q. 34 (Spring): 101-107.

--- (1970b) "Cops and rioters: ghetto perceptions of social conflict and control." American Behavioral Scientist 13 (May-June-July-August): 761-779.

--- (1969a) "Philosophy of law and urban violence." Soundings 52 (Spring): 110-117.

--- (1969b) "Ghetto sentiments on violence." Science and Society 33 (Spring): 197-208.

--- (1969c) "The political objectives of ghetto riots." Presented at the annual meeting of the American Political Science Association, New York.

--- and J. R. FEAGIN (1970) "Riot-precipitating police practices: attitudes in urban ghettos." Phylon 31 (Summer): 183-193.

HOLDEN, M., Jr. (1969) "The quality of urban order," pp. 431-454 in Henry J. Schmandt and Warner Bloomberg, Jr. (eds.) The Quality of Urban Life. Beverly Hills, California: Sage.

Kerner Commission (1968) Report of the National Advisory Commission on Civil Disorders. New York: Bantam Books.

KLONOSKI, J. R. and R. I. MENDELSOHN (1966) "The allocation of justice: a political approach." J. of Public Law 14 (Winter): 323-342.

KRAFT, J. F. (1966) "The attitudes of Negroes in various cities," pp. 1383-1423 in Vol. 6 of Federal Role in Urban Affairs, Hearings before the Subcommittee on Executive Reorganization, Committee on Government Operations, U.S. Senate, Washington, D.C.: Government Printing Office.

LIEBERSON, S. and A. R. SILVERMAN (1965) "The precipitants and underlying conditions of race riots." Amer. Soc. Rev. 31 (December): 887-898.

McCORD, W. and J. HOWARD (1968) "Negro opinions in three riot cities." American Behavioral Scientist 11 (March-April): 31-34.

McNAMARA, J. H. (1967) "Uncertainties in police work: the relevance of police recruits' backgrounds and training," pp. 163-252 in David J. Bordua (ed.) The Police: Six Sociological Essays. New York: John Wiley.

MARX, G. T. (1967) Protest and Prejudice. New York: Harper & Row.

MURPHY, R. J. and J. M. WATSON (1967) The Structure of Discontent. Los Angeles: UCLA Institute of Government and Public Affairs.

NIEDERHOFFER, A. (1967) Behind the Shield. New York: Doubleday.

President's Commission on Law Enforcement and Administration of Justice (1967) Task Force Report: The Police, Washington, D.C.: Government Printing Office.

RAINE, W. J. (1967) The Perception of Police Brutality in South Central Los Angeles. Los Angeles: UCLA Institute of Government and Public Affairs.

SCHUMAN, H. and B. GRUENBERG (1970) "City as an explanatory variable in the study of racial attitudes." Amer. J. of Sociology 76.

SEARS, D. O. (1969) "Black attitudes toward the political system in the aftermath of the Watts insurrection." Midwest J. of Pol. Sci. 13 (November): 515-544.

WANDERER, J. J. (1969) "An index of riot severity and some correlates." Amer. J. of Sociology 74 (March): 500-505.

WASKOW, A. I. (1969) "Community control of the police." Trans-action 7 (December): 4-7.

WILLIAMS, O. P. and C. R. ADRIAN (1963) Four Cities. Philadelphia: Univ. of Pennsylvania Press.

WILSON, J. Q. (1968a) Varieties of Police Behavior. Cambridge: Harvard Univ. Press.

--- (1968b) "The police and the delinquent in two cities," pp. 173-194 in James Q. Wilson (ed.) City Politics and Public Policy. New York: John Wiley.

--- (1964) "Generational and ethnic differences among police." Amer. J. of Sociology 69 (March): 522-528.

--- (1963) "The police and their problems: a theory." Public Policy 12: 189-216.

Part IV

COMMUNITY POLITICS: CONTROL AND PARTICIPATION

The chapters in this section suggest varied approaches to what has become one of the central issues of our age; community and control in large-scale urban bureaucratic structures, structures largely unresponsive to those they presumably exist to serve. Few issues are more salient to the facts of urban form and process than are those of community control, and the public's participation in the decisions affecting their lives. Given the peculiar character of the subject, it is not surprising that few other topics construed to be in the proper realm of the social sciences are so readily personalized or politicized.

The section opens with Marvin Surkin's detailed discussion of the power issues hidden behind the rhetorical shields of professionalism and liberal bureaucracy in New York's recent controversy over decentralization of the Ocean Hill-Brownsville school district. His analysis points up the inevitable obfuscation of issues and the inevitable conflicts which attend any attempt to effect community control of local affairs in an urban setting, especially when matters of race, mobility, and the maintenance of advantaged position are involved. Untangling the threads of the Ocean Hill-Brownsville controversy, Surkin, while extremely critical of a self-serving teachers' union, the mayor's office, and the Ford Foundation (the last two seeking, he argues, only to diffuse an urban time bomb) observes that "our cities are only pseudo-communities, and for alienated, disenfranchised blacks and Puerto Ricans, the concept of community is unrealistic." This fact, he maintains, stands massively in the way of the political efficacy of the poor in their conflicts with powerful and semantically agile opponents like those encountered in the New York case.

In Chapter 14, Seeman, Bishop, and Grigsby set out to test other aspects of community and control, summarizing the results of several studies of black and white residents of the Los Angeles metropolitan area. Their interest explicitly is not in political features of this problem, but in understanding its conceptually distinguishable elements and its measurable role in the urban context. The focus of their study is on the urbanite's experience of mastery, personal efficacy, and community, rather than on the collective responses of urban dwellers to the deprivation of resources, power, or status. Comparing a variety of psychological measures of powerlessness and the like, the authors raise "some questions about a 'package' conception of urban alienations–the kind of bundle of alienations that writers often assume to be typical of the urban reality." They doubt that there is such a package.

In a comprehensive analysis, Dale Marshall evaluates the types and consequences of participation by brown, black, and poor people in the resulting politics of federal legislation enacted to combat poverty. She focuses specifically on the Office of Economic Opportunity and its progeny such as the Community Action Programs. Her review is of critical importance because it is a detailed look at policy action based on assumptions considered in the arguments of Surkin, and Seeman, Bishop, and Grigsby, albeit from very different perspectives. Juxtaposed, these three pieces summarize almost all the conflicts and convergences among the political, social scientific community, and policy features of the 1960s which have made it such a singularly problematic period for everyone.

Thelton Henderson has, as an attorney, experienced all the frustrations of attempting meaningful legal action on behalf of the poor through the medium of meagerly funded federal programs such as that of the OEO's Legal Services Program. As a result of the failure he and other lawyers have experienced, he is moved to question the legal profession's traditional role in relation to aggravated social change. In the concluding chapter in this section he analyzes the failure of the legal profession to effectively respond to the legal needs of a previously unserved people who have exploded upon the contemporary legal scene with new demands for civil rights, new grievances (or old ones newly expressed), and unprecedented challenges to official discretion. His prescription for the legal profession stretches the role of lawyer and the Bar's conception of appropriateness far beyond traditional form. He concludes that a radical change in these roles and the locations in which they are enacted is as critical to the survival of the legal profession as it is to the improved prospects of the poor.

These political, practical, and social-psychological interpretations of community and control, while in some respects incompatible in their focus and implicit or explicit recommendations, give a fairly rounded picture of both the nature of the problem and the recent approaches to it.

13

The Myth of Community Control: Rhetorical and Political Aspects of the Ocean Hill-Brownsville Controversy

MARVIN SURKIN

[The attitudes of Negroes toward whites is] a reaction that was produced by the society; and I think that it is the society that produced this that should be attacked, not the reaction that developed among the people who are the victims of that negative society.

Malcolm X,
Autobiography

More than ever before men now live in the shadow of the state.

Ralph Miliband,
The State in Capitalist Society

□ IN AMERICA there is a serious predilection to play games with words, their meanings, and the social relations they describe. We all have become—social scientists as well as the man in the street—in some sense semantic acrobats.

Consider our easy characterization of passersby as "egghead," "hippie," "square." Words seem to lose meaning, or perhaps meanings are in need of ways to be described, embellished, adorned. Whatever the case may be, it is becoming increasingly evident that the situation in language usage reflects the current crisis in social relations. This is nowhere more true, yet less discussed, than in regard to the community control movement and its panoply of meanings, interpretations, and structures.

In the struggle over control of the schools in Ocean Hill-Brownsville in Brooklyn, "community control" became the battle cry for all the interested groups related to the area. Sometimes referred to as "community involvement," "local control," or "decentralization," the community control movement encompassed the widest array of advocates who, in one way or another, understood that what was needed to revive the decrepit schools in Ocean Hill-Brownsville was institutional reform with some degree of local involvement. For residents of the Ocean Hill-Brownsville ghetto who became involved in the struggle over the schools, and organizations like CORE and the Black Panther party that supported these efforts, the fight for community control was a political struggle. The ultimate goal was radical institutional change and a redistribution of power and wealth. For the United Federation of Teachers (UFT), the Board of Education of New York City, or the Ford Foundation, community control was meant to be (though there were wide differences among these interests) more of an administrative than political change, a reform of the schools, not a revolution within them.

On the surface, these differences of meaning and interpretation need not be irreconcilable. Traditionally, what one group or class had called revolution has been its opposite for another; and conflict and compromise are the mainstays of a democratic society. Yet, after more than three years of continuous efforts to bring community control to Ocean Hill-Brownsville, the general goal remains allusive. Even as late as June 1970, two years after the Ocean Hill-Brownsville situation precipitated a major crisis in the city school system and three strikes by the UFT, the Board of Education's renewed promises of community control were met with derision by the local district representatives. The Board of Education, as was reported in the New York Times (1970) "promised to 'strongly recommend' that the Ocean Hill-Brownsville demonstration district in Brooklyn and the Intermediate School 201 complex in East Harlem be maintained as special units" when they become part of the new community school districts, the latest city plan for decentralization (i.e., community control). In addition, the new community school districts are to be given the extra funds necessary to maintain the present level of operation of the two experimental districts. The response to this announcement by Reverend C. Herbert Oliver, chairman of the Ocean Hill-Brownsville governing board, reveals the sense of division between promises and achievements, between rhetoric and reality, between establishment views and those of the insurgents. Oliver (1970) said, "To us this is just another piece of paper from central headquarters. In the past, we have gotten letters that say one thing and mean another."

The purpose of this analysis is to show that the struggle over community control, both its meaning and its practice, reflects the difficulty faced by established powers to make the institutional life and organizational structures of American society function effectively and efficiently, as well as the difficulty those calling for radical change or new institutions have in achieving their ends. The battle for the schools in Ocean Hill-Brownsville will be used as an example of the general crisis of the established organizational structures in urban America to relate to the new demands and needs of the people. What we shall examine is the process in which each interest group involved in Ocean Hill-Brownsville talked the same language to describe varying views of how to change the schools, all of which served to conceal and perpetuate their established, vested interests. We will indicate, in addition, that in the face of these hegemonies, the Ocean Hill-Brownsville insurgents were ill-prepared—in theory and practice—to spell out the meaning of community control for themselves and to devise ways of achieving their goals.

While community control embodied an embryonic vision of a socially reorganized, politically transformed America, the movement lacked organization and strategy to contest the established powers. The community control movement in Ocean Hill-Brownsville was recognized by the black insurgents as a social reorganization necessary for any political or economic transformation of American society. Existing institutional life and organization, it was believed, could not produce meaningful change but only reproduce bureaucracy and reinforce existing power relations. But the theoretical elaboration of what community control would mean as the base for political, social, economic, and educational change, or how an institutional reorganization would be the necessary first step toward community control remained vague during the struggle over the schools. Moreover, that the Ford Foundation and Mayor Lindsay understood the connection between institutional organization and power and wealth was of no avail to the residents of Ocean Hill-Brownsville, since, as the events described below demonstrate, their interests were fundamentally in conflict. Hence, this analysis attempts to show that the community control movement, as it emerged in Ocean Hill-Brownsville, was a promising step in the right direction. It correctly reflected the real interests and needs of the insurgents, but it lacked the means, theoretical as well as practical, to prolong the controversy to the point of realizing its objectives.

THE INSTITUTIONS AND THE SCHOOL CRISIS

In the Ocean Hill-Brownsville controversy over control and reorganization of the eight schools in the district, an impressive array of institutional structures was put to the test. Not only was the destiny of the Ocean Hill-Brownsville schools and their children involved, but the major institutions of New York City appeared to many to be embroiled in a life and death struggle. Could the Board of Education bureaucracy bend to the new demands of insurgent groups or neighborhoods wanting quick and effective changes? Could the conflict between the Mayor's office and a largely autonomous Board of Education be offset? Could the teachers and their powerful union, the UFT, overcome parochial interests in light of the rising demands of poor, politically unrepresented, and undereducated blacks and Puerto Ricans? Could a *private* agency, the Ford Foundation, with its significant financial assets, orchestrate change by skirting the power struggle within the various institutional bureaucracies and bypassing or mediating the serious differences between *public* and *representative* agencies and the people of Ocean Hill?

In response to these questions, the history of the Ocean Hill-Brownsville dispute is a negative experience for those most affected. Whether because of bureaucratic red tape, the struggles of the different power interests, or misunderstanding, the real interests of Ocean Hill-Brownsville were lost in the struggle, their basic needs and demands—political and economic as well as educational—were not achieved. The history of this institutional failure is itself a portrayal of the current malaise of the cities. According to the record, each institution involved in the school controversy favored community control or decentralization as the best way to improve public school education and resolve the conflict in the ghetto. Yet the rhetoric in each case was far removed from the reality.

The central Board of Education claimed that "all members of our board are committed to the principle of decentralization of operations" (New York City Board of Education, 1969: 17). As early as October 1966 and again in February 1967, the board announced its plans for establishing experimental, decentralized districts. "Through one or more of these demonstration projects, we hope that new techniques and new approaches will be developed for teacher training and instruction and for increasing parental and community involvement and [will] thus strengthen our educational program" (New York City Board of Education, 1969: 18). The board spoke of involvement, not decision-making power or influence; decentralization, not community control; but they went on record as favoring movement in the direction of what the black community was demanding.

In spite of these proposals, the board was plagued by a history in which their semi-autonomous status had built vested interests in maintaining the status quo in education at all costs. This had become increasingly the case as the board developed into a large, cumbersome bureaucracy—leaderless, inefficient, conservative, and unmoving. David Rogers (1970: 129-130) after five years of extensive study of the New York City schools, came to the depressing conclusion that

> it is a "sick" bureaucracy, incapable of reforming itself from within. Its structure and day-to-day operations regularly subvert the realization of its stated goals, prevent any flexible accommodation to changing client demands, and makes orderly administration little more than a pleasant fantasy. . . .

> Indeed, this is a system that is strangled in red tape; mired in inertia, incompetence, and corruption; insufficient; insulated from its clients and from most outside institutions; and fragmented into power blocs . . . which veto new ideas and, by failing to coordinate them, prevent the efficient use of resources. It is, in addition, leaderless. . . . it is highly inbred, allowing protectionist power blocs inside to develop and solidify their baronries and vested interests against client demands for change; it is then accountable to nobody but itself; and it victimizes almost everybody who comes in contact with it, superintendents and lay boards who . . . are regularly emasculated by the professional staff, and whose efforts are absorbed by the cumbersome workings of the system.

Thus, when the Board of Education entered into the school controversy in Ocean Hill-Brownsville, the results were not unexpected. One of the conclusions of the New York Civil Liberties Union (NYCLU, 1969) Report on the school controversy, is that the central board stood in the way of community control by "consistently refusing to define the authority of the local governing board" (NYCLU, 1969: 104). When the experimental districts were established at IS 201 in Harlem, Two Bridges in the lower east side of Manhattan, and Ocean Hill-Brownsville in Brooklyn in the wake of the failure of integration, busing, and other efforts to overcome the inequality and poor performance of the schools, the proviso of the Board of Education called for "somewhat autonomous" experimental schools. However, throughout the crisis, over a two-year period, "somewhat autonomous" was never clearly delineated. According to the NYCLU (1969: 106-107) Report, "For whatever reason, the board simply never defined the powers of the local governing boards of the experimental districts."

This view of the central board is given credence by the board's own advisory and evaluation committee report, the Niemeyer Report, which explained that the provisions for evaluation of the experimental program were to be made by the local board but the evaluation itself, and therefore the decision-making power, was to be maintained by the central board. Formal powers were granted to the local board, according to Niemeyer (1969: 75), but substantive powers went unchanged.

> The [local] board shall make provisions for periodic evaluation of the total program. Such evaluations will include the project administrator, principals, teachers, community workers, etc. This is not to be construed as meaning the [local] board will do the evaluating. Existing Board of Education procedures for evaluating teachers will remain intact.

In addition, the board was reluctant to grant Rhody McCoy the full right to make appointments and transfer personnel, which he was entitled to as district administrator. McCoy requested the transfer of nineteen of the "most uncooperative" teachers in the Spring of 1968. His request was denied by Superintendent Donovan, and he proceeded to send notices of transfer to the nineteen teachers in May. As the NYCLU (1969: 115) Report concludes, "Apparently it was clearly within McCoy's authority to transfer personnel *within* his district until he actually tried to exercise it." A rash of strikes by the UFT began the following Fall.

This unmoving attitude of the central Board of Education converged with the militant, antidecentralization stance of the UFT against the forces of liberal reform represented by Mayor Lindsay and the Ford Foundation and against the Ocean Hill-Brownsville district itself. This alignment proved sufficient to scuttle the experiment in community control, though both the board and the UFT claimed the outcome a victory for decentralization.

In the case of the UFT, they had come out for decentralization as early as 1966 (Harrington, 1969: 231), but they carefully masked their opposition to change. While the board relied on bureaucratic red tape to fight change, the UFT employed legal maneuvering and generated a dangerous scare campaign about antisemitism in Ocean Hill-Brownsville among its largely Jewish constituency. The NYCLU (1969: 117) Report describes the "due process' issue as a myth: "There is no question that under present standards, the United Federation of Teachers created the due process issue out of thin air." The union employed this issue to its fullest, even taking their case to the legislature in Albany in their effort to defeat community control. The NYCLU (1969: 117) reported that "almost precisely at the time that the UFT decided to create the due process myth, UFT representatives were in Albany lobbying against community control. It certainly seems abundantly clear that the due process issue as used by the UFT was nothing but a smoke screen behind which the effort to discredit and destroy community control could go on."

At the base of the union's position was an important, if questionable, consideration of the "right to teach" and the whole character of upwardly mobile professionals.

> Tenure is a precious teacher right. Tenure gives teachers the security they need to teach honestly, free from community pressures. Under the tenure concept, a teacher can be dismissed only for cause after a hearing on the basis of charges brought against him *by other professionals who are competent to evaluate professional performance. Under the Bundy Report, charges could be brought against a tenured faculty member by a community board of laymen with no professional expertise.* This proposal is anti-professional. It would encourage local vigilantes to constantly harass teachers. No teacher with professional integrity could teach in such a district [United Federation of Teachers, 1967].

This mode of defensive professionalism is a mainstay of the teachers union. The point is, however, that economic considerations and social class structure underly most arguments that support this mode of professionalism. As one teacher succinctly put it, "The New York teacher is a member of an aspiring economic class first, and a teacher second," who, like most lower-middle-class whites in the city, has become "paranoid" about unionism, about job security, and the fear of black power in almost any form (Harnett, 1969: 210-211). These are natural ingredients not only for racial conflict but also for motivating the calling of three strikes and paralyzing the entire New York City school system. For instance, John O'Neill, a vice-president of the UFT who was fired by Albert Shanker over his opposition to the union's position, asserted that the strike method could produce no positive results and was based on a misunderstanding of, and insensitivity to, what was going on in the inner-city ghetto. In O'Neill's view (Ferretti, 1969: 310-311):

> Basically, the problem is that the strike weapon is irrelevant to the dispute in Ocean Hill-Brownsville. The conflict there is political and social . . . the thrust of the black and Puerto Rican people against

> the "ins." The struggle of the poor to get into the middle class; all the while president Shanker tries to sell the incredibly simplistic notion that this is a simple labor-management dispute in eight shops . . . our rash action in Brownsville has turned a brush fire into a conflagration sweeping across the city.

In short, for both the central board and the UFT, power and bureaucracy stultified any efforts at compromise and any hopes of substantial change.

However, the entrance of the Ford Foundation into the school crisis with its private endowment and substantial resources brought a new diminsion to the controversy. Ford did a pilot study for Mayor Lindsay, *Reconnection for Learning,* referred to above as the Bundy Report, for its chairman, McGeorge Bundy. The Mayor was interested in cutting into the power of the largely autonomous central board with the hope of bringing change to the school system by establishing a more direct liaison between the school districts and the city executive, and by then being in a position to appeal directly for more state funds for education. The Ford Foundation, for its part, was for the liberal solution to the decaying schools and chaotic, inflammable inner city.

Shanker of the UFT attacked the upper-middle-class motivations of big corporations such as Ford whose executives live in the suburbs, send their kids to private schools, and think up ways to integrate and cajole the minorities in the cities.

> What you have is people on the upper economic levels who are willing to make any change that does not affect their own position. . . . What if you said to them "Give 20% of Time, Inc., or U.S. Steel to the blacks"? [Leman et al., 1969: 20]

What Shanker saw clearly, even though he was opposed to community control, was that the Lindsay-Bundy solution would not serve the interests of lower-middle-class teachers and, in retrospect, only questionably serve the interests of the people of Ocean Hill-Brownsville. One teacher's view correctly emphasizes the interconnection between the racial issue and the class interests of those involved. Patrict Harnett (1969: 212) criticizes the Lindsay-Bundy scheme because

> This approach is being carried out by an upper-middle-class establishment but is revealing a built-in contempt for teachers for having normal interests, goals, and fears of any similar economic group. As such, what is being served is racial polarization, not community control for black and Puerto Rican people.

Shanker even went further in his understanding of the Lindsay-Bundy proposals. He always emphasized black extremism and teacher protection, though he correctly perceived that what Lindsay and Ford wanted was not only to break the UFT and the central board, but to pacify the black and Puerto Rican ghettos (Leman et al., 1969). As Shanker saw it, "Bundy brings his Vietnam experience to the city. He was one of the superhawks, and by the time he came to New York he saw the blacks as the Vietcong who would destroy the city unless we built up loyal cadres as we failed to do in Vietnam" (Leman et al., 1969). Shanker, of course, had his own interests to defend. But what is certainly true is that Lindsay and Ford wanted to defuse the powderkeg in the inner city and bring peace and order. Jason Epstein (1968) writes that

> In the ghettos of New York as, a decade ago, in the Mekong Delta, an angry and insurgent population feels that it has exhausted its last political options and is now ready for violence, even if violence means suicide. . . . The city is thus faced with a classic revolutionary situation. The problem for Mayor Lindsay and Mr. Bundy is to keep the peace but the present strategy is the opposite of what it had been in Vietnam. There we strengthened the mandarins. The plan now is to weaken them [the UFT and the Board of Education] and to offer a form of self-government to the indigenous population.

The goal, therefore, is not community control; it is merely the means to reestablish order and make manageable the most difficult and complex social problems. In short, because the stakes are high, reform is necessary, a price must be paid. The pattern should by now be well-known: a little bit of reform or a little bit of repression is the popular liberal way of dealing with insurgency. Thus, Lindsay and Bundy called for "a community school system," a federation of "largely autonomous school districts," and a central educational agency.

The Bundy Report suggested thirty to sixty local districts with broad local power, and the Ford Foundation offered initial funding of the Ocean Hill-Brownsville experimental district to the tune of $44,000. Mayor Lindsay strongly supported this plan. This beginning of community control in Ocean Hill-Brownsville gave confidence to the local board and life to the experiment. However, when the going got rough due to the intransigence of the UFT and the central board, the militance of the Ocean Hill-Brownsville people, and the crippling strikes and mounting public controversy, both the Mayor and the Ford Foundation backed down. Ford denied the local board additional funds while the Mayor gave in to the UFT in order to end the strikes. Meanwhile, the Mayor resorted to either closing the schools of the district or calling in the police to keep them open, according to the different stages in the agreements reached with the UFT and the central board.

The upshot of this survey of the role of established organizational structures in the urban crisis is first to underscore the difficulty any of these established centers of power have in achieving even the most modest reform. Bureaucracy, racism, and class interests defined the decision-making process, circumscribed the struggle for community control, and ultimately determined the outcome. American institutions, in the last analysis, respond to power and wealth, not social demands; and the whole process of institutional life is embroidered by cumbersome bureaucracies such as the New York City Board of Education and the petty interests of influential baronries such as the UFT. As long as this remains the case, and as long as blacks and Puerto Ricans remain marginal to that institutional process, largely outside that institutional structure, then the hope that these institutions can be made truly responsive to the demands of blacks and Puerto Ricans (e.g., by creating community control of schools and permitting the degree of power and financing necessary to achieve it) will remain a myth.

THE MYTH OF COMMUNITY CONTROL

The institutions we have analyzed above were never in favor of community control if that were to mean giving in to the potentially revolutionary demands of the ghetto insurgents. The Mayor and the Ford Foundation wanted some reform but not a revolution; the UFT and the Board of Education were fundamentally against any change in the direction of community control. The interests of a considerable segment of the New York City black and Puerto Rican ghettos not only wanted community control but were ready to fight for it. The struggle for community control symbolized the demand for more power, more wealth, and better education. This struggle turned a social problem into a heated political issue and reflected the growing national crisis in urban America. What is clear is that these poor, unrepresented and undereducated blacks and Puerto Ricans wanted a redistribution of power and wealth—a revolution, if necessary. But what needs to be more carefully considered is what community control meant for them and how they expected to achieve it.

Rhody McCoy called it the "Year of the Dragon," (1969) a kind of historic awakening of an oppressed, invisible people in search of dignity, power, and social change. The idea of community control of schools was simple enough:

> they [white, middle-class teachers and administrators] have run the schools long enough because they have neither educated the children of the ghettos nor responded to the social, cultural and political demands of their parents; we (blacks and Puerto Ricans) want to run our schools, hire our own teachers, teach our history and involve our own people in the planning, development and administration of the schools, in short, to begin taking control of our own lives and the things that affect us. Community control is a step in the direction of taking control of *all* the institutions that affect our lives and putting an end to our own subordination and oppression. Community control is the road to our own liberation.

McCoy eloquently states the case for community control in this way:

> In Ocean Hill-Brownsville, there are people groping in the dark; they are people who for a long time have felt themselves outside the mainstream of public concern. The city takes no notice of them. These people are obscure, unnoticed—as though they do not exist. They are not censured or reproached; they simply are not seen. They are the invisible residents of a demoralized, poverty-ridden inner city. To be ignored or overlooked is a denial of one's rights to

> dignity, respect, and membership in the human race. These residents have been frustrated at every turn in their attempt to reverse the process. . . . The parents of Ocean Hill-Brownsville are determined to have a permanent voice in matters pertaining to their schools and to have it now. They are dedicated to the goal of joining forces to bring about better educational results for their children. The community, generally acknowledged by the various social and political forces as, in every respect, the most disadvantaged in the city, has finally risen to demand a change—to make history. And it is demanding not only change but also a share in bringing about that change [McCoy, 1969: 52-53].

McCoy states elsewhere in more muffled tones that

> The stage is set for a revolution. People—black people want control over their schools for self-determination, for building of a strong self-image, for individual and community development, for restoration of confidence in education, for economic stability, for recognition and for survival. Community control means community growth and development, and the school is the hub of this growth [McCoy, 1970: 171].

Not only do the outs want in, meaning more power, more wealth, and better educational facilities, but they want to decide specifically what they want, how they want it, and when they want it. This, in capsule form, is community control as seen from the inside.

Another view of community control often left out of consideration is detailed in a recent study by Harold W. Pfautz (1968: 13-39). Professor Pfautz's major argument is that the new demand for community control of the schools has emerged precisely at the time when the status of the black community is most questionable. While he agrees that the schools are largely dysfunctional for the masses of the black underclasses, he relates this to the fact that "literally hundreds of thousands of black residents of most of our major cities are functionally outside the dominant polity and social system" (Pfautz, 1968: 14). To begin with, our cities are only pseudo-communities, and for alienated, disenfranchised blacks and Puerto Ricans, the concept of community is especially unrealistic. Moreover, schools, whether controlled by local boards or centralized bureaucracies, tend to reflect the dominant ethos and ideology, not the community or would-be community, especially if that community falls outside of the dominant polity and social system. When related to the elitist structure of American pluralism (Bachrach, 1967), the pseudo-nature of the community looms as a major stumbling block to the demands of the underclasses in regard to schools as well as elsewhere. Thus, tinkering with the educational system, as Pfautz rightly argues, will not resolve the racial fault in the social structure (Pfautz, p. 15). Nor, for that matter, will such tinkering seriously alter the white middle-class value system and elitist control of the schools. For Pfautz, therefore, "the community school concept today is less an idea than an ideology; it has become part of the rhetoric of a social movement," whose style, content and concern is more political than education (Pfautz, p. 14).

Though Pfautz's conclusions need to be seriously challenged because of the implicit racism and the limitations of his explicit liberal reformism, his analysis and critique of the black "community" and the community school idea are worth careful attention. He puts his finger on precisely those factors which are lacking in the attempts of the black community to achieve reorganization of the schools *plus* power *plus* wealth.

According to Pfautz, "every society is something of a 'community' [although] every community is not necessarily an adequately functioning society" (Pfautz, p.17). For instance, he writes that an "administrative" political community characterized by "school districts" is a symbolic or spatial community with a legal, political or administrative structure. It does not necessarily function as a psychic community in the natural, ecological sense (Pfautz, p. 17). Part of the problem in Ocean Hill-Brownsville from the outset was that the symbolic or administrative change in social or community structures–i.e., decentralization–was intended to be for those involved in "ecological" change; i.e., political, social and educational change. The black advocates of community control wanted administrative change to be the foundation for a more far-reaching social transformation. However, even if they could have assumed greater control of the finances for the schools with the obvious political implications this might have, the nature of the community, hence the roots of the new society, would still have been in doubt.

As Harold Cruse, a renowned black scholar, has argued, what Americans, both white and black, suffer from most is an identity crisis. Racial warfare of the cities, class opposition, absentee ownership, political gerrymandering, broken neighborhood patterns, separation of work place and home, make it increasingly difficult to think of the American city as a community, or a community of communities.

The solution for the crises of the cities will require overcoming this identity crisis as well as devising new social forms as the basis for community development. For the blacks, and the underclasses generally, this process will be more difficult and complex. As Pfautz notes (1968: 22), the high rate of physical mobility, the fundamental lack of economic resources with its debilitating political consequences (Pfautz, p. 24) and the mixed attitude of lower-class blacks to middle-class blacks as well as toward the ghetto itself (Pfautz, p. 24) signify "the essentially underdeveloped status of the local black community as a functioning social group" (Pfautz p. 22). In other words, the preconditions for the development of community and community power are generally lacking in the black ghetto. The way Harold Cruse put it is perhaps very much to the point when he remarked that what black Americans suffer from is not too much capitalism but too little capitalism. This point suggests that the bourgeois revolution is still necessary before the socialist revolution becomes possible. Or, put differently, social change itself requires organization, power, and wealth, in addition to a vision—in this case, of a new community.

Obviously, this situation becomes more acute when blacks begin to make demands on society without the means to achieve them; or when large corporations or City Hall make new promises in elaborate administrative packaging with little substance. This was basically the situation in the Ocean Hill-Brownsville struggle. In fact what may be the case is that decentralization, rather than fostering community development and community control, will facilitate social regulation. "Decentralized districts are more manageable and easier to manipulate; dissident groups can be isolated, uprisings quelled and local police forces brought to occupy a community (as they did in Ocean Hill-Brownsville) without affecting adjoining areas" (Leman et al., 1969: 21).

In short, community control as defined by the people of Ocean Hill-Brownsville is no panacea for the black ghetto. As Pfautz concludes, "there will be no significant institutional changes in our educational systems without previous and accompanying economic and political changes in society and in our local communities" (Pfautz, 1968: 14). In addition, it will be necessary for blacks to develop further analyses of their own communities and of what will be required to achieve community control. To move in the direction of community and community control will require a program for social change as well as the revolutionary fervor which to a high degree already exists.

Pfautz and other liberal social scientists criticize the black militants for their lack of programmatic means to achieve community control or black power, and they suggest more integration or more centralized power to bring about change. For Pfautz, power does not reside in the neighborhood (Pfautz, 1968: 49), and there is no status in the ghetto (Pfautz, p. 51). Therefore, the hope that remains is that the liberal version of community control (e.g., the Bundy Report) might provide a base for long-term reform such that "it would give blacks a tangible stake in the American political system" (Altshuler, 1970: 199). In other words, the rhetoric of change is supposed to pass for the reality. "By giving them systems they considered their own, it would—hopefully—enhance the legitimacy of the whole system in their eyes" (Altshuler, p. 199).

Other forms of this liberal version dominate not only social science scenarios for reform, but official views of welfare state politics. In the 1960s liberal social scientists such as Daniel P. Moynihan (1968: 28) shared with the government the view that "commitment . . . and an honorable desire to be helpful" were required to combat racism and social inequality. This commitment led Moynihan to suggest that the "tangle of pathology" *in* the Negro family structure should be (and could only be) combatted with the assistance of the white world (Rainwater and Yancey, 1967: 42). In the 1970s pragmatic action is now to be replaced by pragmatic inaction (Sennett, 1969). Moynihan, now a Nixon special advisor on urban affairs, has recommended "benign neglect" for the inner city, only a short Nixonian step from the views of Edward C. Banfield, political scientist and Nixon advisor, who believes that there is something in the nature of all poor people as a class—both black and white—which makes them poor and keeps them that way. His conclusions, which have apparently been well received by the Nixon administration, are merely the latest version of the liberal/conservative strategy for the pacification of the poor and the black by "reform" (Banfield, 1970).

For many blacks in the ghettos this ameliorative solution is apparently too late and too little. As they see it, there is (potential) power in the neighborhood as well as status in the ghetto. While the resources available for these people are severely limited, and effective programs for realizing radical social change are allusive, it must nonetheless be seen that the community control struggle in Ocean Hill-Brownsville is part of a revolutionary process of social change in which a people begin to use whatever resources they have to assume a margin of control or influence over their own lives. To counsel them to conform, to integrate, to quit the ghetto, or to adjust to the values and social structures of white, middle-class America is to offer a form of subordination which fewer blacks are willing to accept quietly or tolerate. On the contrary, Philip Green (1970: 258) suggests that

> In the absence of a will to real social integration, then, decentralization (with its corollary, community control) presents itself as the only alternative to repression. It is such an alternative, first, because it and it alone is the program upon which the disaffected community has fixed its sights; and second, because it is a cure logically related to the disease that has turned out to be inherent in the liberal conception of public education.

But perhaps it is Huey Newton who best characterizes both the hope and the futility of bringing about these changes in American society without which survival for black Americans remains at stake. In a recent interview, Newton contrasted "revolutionary suicide" and "reactionary suicide." "When a corrupt regime strips a man of all dignity and crushes his spirit, and he kills himself . . . or if he does nothing, and is killed by that reactionary set of conditions through a spiritual kind of death—this is reactionary suicide." But "when we take action, when we take our destiny in our own hands, and try to change that reactionary set of conditions by going in a head-on conflict with those forces, then this is edging upon suicide, but I call it revolutionary suicide, because there's a small chance we'll actually succeed, I mean individually succeed" (Newton, 1970: 3).

Huey Newton reminds us of the dimensions of the task before us for which the Ocean Hill-Brownsville community control movement has played a significant role. If such efforts are to make inroads in restructuring American institutional life, and redistributing power and wealth, it will be necessary to recognize the "inside" of the process of social change, as expressed by Huey Newton, Rhody McCoy and the parents and children of Ocean Hill-Brownsville, as well as to devise models of community control and strategies to achieve it. This is all the more imperative today, since the history of the Ocean Hill-Brownsville controversy over the schools is another reminder of the incapacity of America's dominant ideology and its institutional structures—along with the liberal social sciences—to point the way toward adequate, or even viable, social change for America's underclasses.

REFERENCES

ALTSHULER, A. A. (1970) Community Control: The Black Demand for Participation in Large American Cities. New York: Pegasus.

BACHRACH, P. (1967) The Theory of Democratic Elitism. Boston: Little, Brown.

BANFIELD, E. C. (1970) The Unheavenly City. Boston: Little, Brown.

BERUBE, M. R. and M. GITTELL (1969) Confrontation at Ocean Hill-Brownsville: The New York School Strikes of 1968. New York: Praeger.

EPSTEIN, J. (1968) "The politics of school decentralization." New York Rev. of Books, June 6: 26.

FERRETTI, F. (1969) "Who's to blame in the school strike," in M. R. Berube and M. Gittell (eds.) Confrontation at Ocean Hill-Brownsville: The New York School Strikes of 1968. New York: Praeger.

GREEN, P. (1970) "Decentralization, community control, and revolution," in P. Green and S. Levinson (eds.) Power and Community. New York: Vintage.

HARNETT, P. (1969) "Why teachers strike: a lesson for liberals," in M. R. Berube and M. Gittell (eds.) Confrontation at Ocean Hill-Brownsville: The New York School Strikes of 1968. New York: Praeger.

HARRINGTON, M. (1969) "An open letter to men of good will (with an aside to Dwight MacDonald)" in M. R. Berube and M. Gittell (eds.) Confrontation at Ocean Hill-Brownsville: The New York School Strikes of 1968. New York: Praeger.

LEMAN, B., M. JOSEFOWICZ, and T. GOLD (1969) "Decentralization: strategy to reorganize the cities," Leviathan 1, 3 (June).

McCOY, R. A. (1970) "The formation of a community controlled school district," in H. M. Levin (ed.) Community Control of Schools. Washington, D. C.: The Brookings Institution.

--- (1969) "The year of the dragon," in M. R. Berube and M. Gittell (eds.) Confrontation at Ocean Hill-Brownsville: The New York School Strikes of 1968. New York: Praeger.

MOYNIHAN, D. (1968) "The professors and the poor." Commentary 46, 2 (August).

New York City Board of Education (1969) "Policy statement on decentralization," in M. R. Berube and M. Gittell (eds.) Confrontation at Ocean Hill-Brownsville: The New York School Strikes of 1968. New York: Praeger.

New York Civil Liberties Union (1969) "The burden of blame." pp. 104-119 in M. R. Berube and M. Gittell (eds.) Confrontation at Ocean Hill-Brownsville: The New York School Strikes of 1968. New York: Praeger.

New York Times (1970) June 9: 1.

NEWTON, H. with J. BANCROFT and D. FENTON (1970) Hard Times 87 (September 7-14).

NIEMEYER, J. (1969) "Final report of the advisory committee on decentralization," in M. R. Berube and M. Gittell (eds.) Confrontation at Ocean Hill-Brownsville: The New York School Strikes of 1968. New York: Praeger.

PFAUTZ, H. W. (1968) "The black community, the community school, and the socialization process: some caveats." Prepared for the Conference on the Community School. Washington, D. C.: The Brookings Institution,

RAINWATER, L. and W. YANCEY [eds.] (1967) The Moynihan Report and the Politics of Controversy. Cambridge: MIT Press.

ROGERS, D. (1970) "The New York City school system: a classic case of bureaucratic pathology," in A. T. Rubinstein (ed.) Schools Against Children: The Case for Community Control. New York: Monthly Rev. Press.

SENNETT, R. [ed.] (1969) Classic Essays on the Culture of Cities. New York: Appleton-Century-Crofts.

United Federation of Teachers (1967) The United Federation of Teachers Looks at School Decentralization.

14

Community and Control in a Metropolitan Setting

MELVIN SEEMAN
JAMES M. BISHOP
J. EUGENE GRIGSBY, III

□ THE POPULAR IMAGE of Los Angeles is that it is Americana writ large. Whether or not that is so, this particular metropolis is an embodiment of the major drift in contemporary social change—namely, increase in scale (meaning essentially, the number and scope of people in interdependent relation). As Greer (1962: 207) remarks concerning urban development, "the entire process may be circumscribed by one phrase—the increase in societal scale." We need not elaborate upon the troubles, large and small, that scale supposedly generates. For our purposes, the most significant trouble is captured in a slogan that has recently emerged from the urban turmoil: the popular phrase is "community control."

We are *not* concerned with that idea as a political program or an ideological doctrine. We intend to make no political recommendations about community control as such. Our interest in that phrase lies chiefly in the way it can point one's attention. The idea of community control, when pulled apart, bares two quite independent problems—and both of them are at the heart of our theories about what has happened in the course of urbanization. Whatever else the increase in scale has accomplished (e.g., opening up leisure opportunities, and freeing us from residence near the work place), it has complicated two crucial features of social life: the *sense of community,* and the *sense of control.* The thesis we often hear is that the metropolitan life style involves both a lost (or at least, severely attenuated) sense of community, and an erratic (or at least, badly fragmented) sense of control over the important decisions. Furthermore, though bad enough in themselves, attenuated community and fragmented control are held responsible for additional evils: they are said to lead to ethnic hostility, civil disorder, negative self-images, generalized despair and distrust, and the like.

These assertions set up a massive agenda for comprehending the attitudes and behaviors that constitute the metropolitan life style. That agenda includes not only some difficult definitions (e.g., what do we mean by life style or by community), but also some concerted research that can speak with authority about the supposed "eclipse of community" and "loss of control." Such evidence concerning Los Angeles is not available, and we will have to settle for approximations, merely hoping at the close to be clearer on four major points: (1) the meaning of community and control (especially in relation to life style); (2) the social topography of community and control in Los Angeles (especially their distribution across race and class lines); (3) the relation of community and control to each other; and (4) their potential consequences in the life of the metropolis.

ON THE MEANING OF COMMUNITY AND CONTROL

One way to think about life style is in terms of three major domains: work, family, and friends. These are, in some degree, ad hoc—the "rounds of life" that an observer like Liebow, in his report on Negro men (1967), might focus upon. But these domains tie into more systematic descriptions, especially when we note that, in American life, friends means largely an ethnically segregated (and often locality-based) set of relations. In that degree, the work-family-friends trio is more or less coordinate with Shevky and Williams' (1949) identification of the "social areas" of Los Angeles on the basis of three major axes: (1) social rank (mainly occupation-education); (2) urbanization (mainly "familism" as reflected in apartments versus family-type housing, employment of women, and childlessness); and (3) segregation (i.e., ethnic or racial concentration in a locality).[1]

In each of these domains of life style (call them work, family, and friends-ethnicity), the separate problems of community and control are visible. Thus, we are interested in the sense of community at work—what is sometimes called "occupational community" (Lipset, et al., 1956), referring to the feeling of camaraderie, social integration, and colleagueship with coworkers; and we are also concerned about the sense of control at work—indeed, the lack of control over the work process has been a central theme in studies of alienated labor which derives from the Marxian viewpoint (Blauner, 1964). The same duality of interest applies to the family, which can be seen both as a basis for the sense of community (i.e., belonging, attachment, solidarity), and as an arena for exercising control over decision-making (Baltzell, 1968). And the neighborhood-locality-ethnic group likewise becomes the arena for community (friendship, neighboring, and the like) and control (as in the recent struggle in Ocean Hill-Brownsville over control of the schools, compare with Rogers, 1968). In principle, then (the limitations of available data aside), our task becomes one of examining community and control in each of these three major domains.

But what do we mean by community and control? Both of these terms seem immediately understandable, yet elusive; with respect to each we think that a three-way distinction is helpful. Concerning control, we need simply to distinguish, as we do in the analysis of social classes, between objective control (i.e., actual influence over decisions in some specified sphere), reputed control (nicely illustrated by the many studies of reputation for community leadership), and subjective control (that is, the individual's sensed capacity to influence events in various realms). It is the last which is our focus here: we will have little to say about political structures of control or reputed influentials, and more to say about feelings of mastery or powerlessness.

Regarding community, it should be apparent already that we do not restrict that notion to its territorial meaning. There are, we suggest, three basic types of common bond to which the word regularly refers, these three being interlaced but by no means the same thing. Community means either:

(1) a common *fate,* determined by objective interdependence and interaction—e.g., a community defined by economic exchange (the European Economic Community), political unity (e.g., the communities of Long Beach or Hollywood), professional interdependence (the professional community of physicians, lawyers, sociologists, and so forth, and their accompanying organizations), or differential treatment (as with the ethnic community, whether scattered or enclaved in the barrio or ghetto);

(2) a common *identity,* determined by personal sentiments of belonging and attachment—what we commonly mean when we speak of the sense of community, but by no means tied to locality;

(3) common *aims,* determined by agreement in values—i.e., a community more or less in the sense of shared norms and goals.

Obviously, these three—the community of fate, feeling, and faith—are interdependent in complex ways, and we shall have something to say about each of them; but our primary interest is in the community of feeling—the signs of solidarity whether with a work group, a locality, a family, a friendship clique, or a professional body. Thus, for both community and control, our major interest lies in the social psychological experience within the metropolis: community means largely sentiments of attachment; and control means largely sensed capacity to control.

THE RESOURCE STUDIES

What follows, then, is an effort to organize some of the data from recent social surveys in Los Angeles in terms of the perspective outlined above. We have used the data from four different studies, all of them post-Watts, involving both black and white respondents. These studies are not directly comparable, and they were not carried out with the express intention of examining community control; hence our indices sometimes are bare approximations of what they should be. Nevertheless, these studies frequently asked similar questions, were all conducted by interview, and sampled reasonably well in a wide range of areas in the metropolis. A brief word to identify each study:

(1) The study by Morris and Jeffries (1969) was carried out in six areas of Los Angeles (e.g., Pacific Palisades, Long Beach, Reseda), selected so as to represent three levels of socioeconomic status and different degrees of racial exclusiveness. The sample was entirely white, both male and female, with a total of some 500 interviews.

(2) Ransford's (1968) study was executed by Negro interviewers for an entirely black sample of employed males (n=295), chiefly in the Watts, Crenshaw, and Baldwin Hills areas of Los Angeles (again covering a wide range of socioeconomic status).

(3) Seeman's study (1967) involved an exclusively white sample (n = 400) of employed males residing in eight areas of Los Angeles (e.g., Woodland Hills, Inglewood, Bell Gardens), selected so as to represent the range of socioeconomic status, and to exclude heavily ethnic areas (i.e., Negro, Oriental, Spanish, or Jewish concentrations).

(4) TenHouten's (1970) study involved intensive interviews with 286 black and white families selected by a sampling procedure calculated to control for age, socioeconomic status, and racial composition of neighborhood. In all cases, interviewers were matched for race with those interviewed—the husband, the wife, and the two oldest children being the ones questioned.[2]

THE DISTRIBUTION OF COMMUNITY AND CONTROL

Treating community in terms of sentiments of attachment, it is reasonable to ask first about what might be called the distribution of community in Los Angeles—i.e., the social topography of the sense of community when this refers to a variety of experiences and commitments reflective of solidarity not necessarily based on locality. The evidence, we think, supports two general conclusions about the sense of community: (1) the dominant impression is one of uniformity across socioeconomic or racial lines; and (2) though it is not an easy judgment to make, in many ways it looks as though some of the widespread worry about the problem of community in a metropolis such as Los Angeles is misplaced. For example:

(1) If one considers sense of community in general (that is, social attachment in general) as an item such as "I often feel lonely," only

ten percent of those asked agree with such a sentiment, and this is true for both manual and white-collar workers. If such a general question is asked in a somewhat less stringent way (*"Sometimes* I feel very much alone in the world"—as Morris asked it), approximately thirty percent will agree (again, with no significant manual versus nonmanual difference).

(2) If a question about social support or involvement is asked in still another way, of both blacks and whites (as TenHouten put it: "I am very popular with my friends"), neither income nor race makes a difference. The overwhelming choice is positive: less than twenty percent are concerned about unpopularity (for example, in the black low-income group, nineteen percent express such concern; for the black high income, seventeen percent).

(3) If one makes the question specific to an occupational community, the figures again reflect fairly high engagement in a social circle. When asked, "Do any of your close friends do the same kind of work that you do," fifty percent of both manual and nonmanual workers (white) have close friends in the same type of work. On the related question, "Do any of your friends work at the same place you do," fifty-three percent of both manuals and white-collar workers say yes—and there is evidence, too, that these friendships involve a good deal of off-the-job socializing.

(4) Another way to look at the extent of community is in terms of what values people will sacrifice for it. Table 1 shows what happens when we ask both blacks and whites how strong, in effect, their familism and friendship ties are (as compared with the pull of occupational advancement). The most striking thing in Table 1 is the uniformity across race and occupation: most people (roughly seventy percent) are *not* willing to give up family association for advancement; and people (meaning about eighty percent of them) are not so worried by or bound by friendship ties that they would allow such ties to interfere with an improvement in the job situation.

Two things ought to be stressed regarding these data. First, the uniformity displayed across race and class regarding loneliness, popularity, occupational community, and family or friendship values is by no means something to be taken for granted—especially in view of the prevalent argument which posits substantial difference in family values among blacks (as in the Moynihan thesis concerning the instability of the lower-class black family). Second, these data in themselves do not tell us very much about the meaning or

TABLE 1

Choosing the Values of Family and Friendship over Occupational Advancement for Black and White Samples in Los Angeles (in percentages)

Item	Choice	Blacks: Manual Workers (n=124)	Blacks: Nonmanual Workers (n=105)	Whites: Manual Workers (n-154)	Whites: Nonmanual Workers (n=243)
1. Family vs. advancement I would probably turn down a promotion if it meant I would have to be away from the family a lot.	agree (family)	64	65	73	77
2. Friends vs. advancement I wouldn't let my friendship ties in a community stand in the way of moving on to a better job.	agree (advance)	72	84	73	80

NOTE: Black data (Ransford, 1968); white data (Seeman, 1967).

significance of community; they merely give a rough and very partial portrait of it. But they begin to develop a major theme that will take shape as we proceed; namely, the view that it is not the sense of community that is the central problem, but the sense of control.

Regarding the sense of control, then, what is the topography like? Tables 2a and 2b present evidence on generalized feelings of powerlessness, based on varied ways of asking the question and on three different samples (both black and white). The dominant impression here is one of difference rather than uniformity, and the difference has two interesting features:

(1) it is more by race than by class—manual and nonmanual whites (or high income versus low income) do not differ in powerlessness, while the low-status blacks are generally high in powerlessness;

(2) it seems to be, at least to some extent, a *discriminating* difference between blacks and whites—i.e., the difference occurs on those powerlessness items that realistically reflect the life experience of blacks (thus, there is no difference between blacks and whites on a question that concerns the inevitability of war, but there is a substantial difference where the questions concern likelihood of success on the job market or influence on social policy).[3]

TABLE 2a

Choosing the Powerless Alternative on Three Forced-Choice Items for Black and White Samples in Los Angeles (in percentages)

Item	Blacks		Whites	
	Manual (n=124)	Nonmanual (n=105)	Manual (n=154)	Nonmanual (n=245)
1. (a) Becoming a success is a matter of hard work; luck has little or nothing to do with it.	–	–	–	–
(b) Getting a good job depends on being in the right place at the right time.	42	29	22	22
2. (a) It is only wishful thinking to believe that one can really influence what happens in society at large.	48	35	22	26
(b) People like me can change the course of world events if we make ourselves heard.	–	–	–	–
3. (a) Wars between countries seem inevitable despite the efforts of men to prevent them.	62	67	65	65
(b) Wars between countries can be avoided.	–	–	–	–

NOTE: Black data (Ransford, 1968); white data (Seeman, 1967).

TABLE 2b

Agreeing with Two Powerlessness Items, for Black and White Samples in Los Angeles, by Income (males only) (in percentages)

Item	Blacks		Whites	
	Low Income (n=76)	High Income (n=72)	Low Income (n=44)	High Income (n=94)
1. Planning only makes a person unhappy, since your plans hardly ever work out anyway. (agree or strongly agree)	43	21	27	19
2. When a man is born, the success he is going to have is already in the cards, so he might as well accept it and not fight against it. (agree or strongly agree)	38	25	17	10

NOTE: Black and white data (TenHouten, 1970).

These are, however, fairly generalized questions about control. As with community (where we asked not only about generalized loneliness, but also specifically, about the occupational community), we can ask specifically about the individual's sense of control at work. In doing so, we are focusing on what is commonly called alienation in work: for Marx, lack of control over the work process was a crucial element in alienated labor, and lack of control is the hallmark of routinized work–i.e., work in which the individual cannot try out his own ideas, or make independent decisions about the work process.

The nature of control in the work sphere is suggested in Table 3, where the data are presented from three different studies using the same questions. The important differences occur where we would expect to find them:

(1) the manual workers, whether white or black, are considerably more alienated than the nonmanuals (in that degree, these items appear to be describing a reality about work control that we can easily recognize);

TABLE 3

Choosing the Alienated (Low-Control) Alternative in Describing Their Work Circumstances, for Black and White Male Workers in Los Angeles (Three Independent Studies) (in percentages)

	Black Sample		White Samples			
	Ransford Data[a]		Seeman Data[b]		Morris Data[c]	
Item	Manual (n=129)	Non-manual (n=105)	Manual (n=154)	Non-manual (n=243)	Manual (n=65)	Non-manual (n=120)
1. Does your job really give you a chance to try out your ideas on your own? (no)	48	16	26	12	46	20
2. On an ordinary work-day, do you have the opportunity to make independent decisions in carrying out your tasks, or is it rather routine work? (routine)	54	11	29	10	32	15

a. 1968
b. 1967
c. 1969

(2) race does not make a consistent or thoroughgoing difference (the black nonmanuals are no different than their white counterparts—roughly ten to twenty percent of the nonmanuals throughout admit to low control); yet

(3) it is true that the black manual workers express uniquely low control (fifty-four percent describe their work as being thoroughly routine).

As usual, such simple distributions using rather simple measures raise some not-so-simple questions. Two are of particular interest. First, do these figures reflect a high level of work alienation or not? And second, are we simply talking about job satisfaction under the deception of control? These questions are of interest primarily because of the unstated assumptions that commonly lie behind them—first, the assumption that work experience is perhaps *the* central fact in the individual's life style (hence the level of work alienation is equally central); and second, the assumption that life style readily becomes a matter of *generalized* satisfaction or dissatisfaction—a kind of thematic positive or negative judgment that pervades the varied domains of the person's experience.

Obviously, the level of work alienation discovered is dependent on the form of the question. Using a more stringent measure than ours, Wilensky (1964) reported that "the incidence of alienation is low" (roughly eighty-five percent in his sample showed *no* signs of work alienation, as indexed by the possibility of realizing one's "prized self-image" in the work role). But where, as here, as many as fifty percent of the manual workers describe their job as routine and confining, we are inclined to regard the potential reservoir of discontent about work life as being rather full. That potential discontent, however, is by no means a certainty. For one thing, the worker may treat much more matter-of-factly what our assumptions tell us is problematic for him (e.g., he has learned to adjust to the fact that a job is a job). For another thing, the importance of work as a phenomenon in its own right may be less powerful a determinant of discontent in a consumption-oriented metropolitan milieu (depending, of course, on the access that various groups of workers—e.g., black and white, manual and nonmanual—have to the consumption resources).

This brings us to the second question posed above: Is this work alienation simply job satisfaction? We are inclined to think not, both on theoretical and factual grounds. Theoretically, there is no reason to assume that intrinsic engagement in work—i.e., finding it rewarding in itself through the creativity, control of process, or sheer interest involved—is the same as expressing satisfaction with a job, since that satisfaction may revolve around such extrinsic matters as pay, security, work surroundings, or sociability. The evidence, furthermore, supports such a distinction between job satisfaction and work alienation. For the black sample described in Table 3, a straightforward job satisfaction question was asked ("Taking into consideration all the things about your work, how satisfied or dissatisfied are you with it?"), along with seven questions on work alienation much like the two shown in that table. The total score on work alienation correlated only -.02 with work satisfaction among the manual workers (the r was -.35 for the white-collar blacks). One might conclude that manual blacks cannot afford the luxury of applying standards about intrinsic reward to their judgment of the job! In any case, alienation in work (meaning basically, low control involving low engagement) and job satisfaction are not the same thing.

THE RELATION BETWEEN COMMUNITY AND CONTROL

What is the relation between these two elements that so easily get put together in the phrase community control? It would be tedious to explore all the possible interconnections of the indices we have presented thus far. That would involve intercorrelating the four community measures with each other—familism, friendship commitment, occupational community, and loneliness; and then showing the association of each with the two kinds of control—generalized powerlessness and low work control. For our purposes here, Tables 4a and 4b get at the heart of the matter. They show the relation between the four community measures on the one hand, and two control indices on the other hand (an item on work control, and an item on generalized powerlessness).

The tables provide some basis for arguing three things:

(1) If we think of community in terms of family and friends (which is what Table 4a is about), we would do well to think of them rather

TABLE 4a

Choosing the Community Alternative (Family or Friends) for Respondents Who Feel Personal Control at Work, and Personal Control in Social Affairs Generally (Manual Workers Only) (in percentages)

Community Measure (respondents who choose:)	Blacks				Whites			
	Can Try Out Own Ideas at Work		Can Influence Social Affairs		Can Try Out Own Ideas at Work		Can Influence Social Affairs	
	Yes	No	Yes	No	Yes	No	Yes	No
Family over advancement	78 (n=67)	49 (n=61)	71 (n=65)	55 (n=62)	78 (n=113)	63 (n=40)	77 (n=118)	62 (n=30)
Friends over advancement	29 (n=73)	28 (n=61)	21 (n=66)	36 (n=77)	23 (n=114)	32 (n=41)	24 (n=120)	30 (n=34)

NOTE: The table is read as follows: Of the 67 blacks who said that they could try out their own ideas at work, 78% said they would not take a promotion that involved being away from the family a good deal; of the 61 blacks who said they could *not* try out their own ideas at work, only 49% chose family over advancement. The data for blacks are from Ransford's study (1968) and the white data from Seeman (1967). The exact wording of the items is given in Tables 1, 2a and 3.

TABLE 4b

Those Who Do and Do Not Feel Personal Control at Work and in Social Affairs Who Report That They are Not Engaged in an Occupational Community or are High in Loneliness (Manual White Workers Only) (in percentages)

Community Measure	Can Try Out Own Ideas at Work		Can Influence Social Affairs	
	Yes (n=115)	No (n=41)	Yes (n=121)	No (n-34)
Occupational community (low)	47	49	53	47
Loneliness (high)	11	7	8	18

NOTE: The table is read as follows: Of 115 manual workers who indicated that they could try out their own ideas at work, 47% reported that they were not involved with friends at work or had friends in an occupation similar to their own; of those (n=41) who claimed they could not try out their own ideas at work, 49% reported uninvolvement in an occupational community. The data are from the study by Seeman (1967).

more separately than we are wont to, since they seem to work differently; indeed, to some extent, they work in opposite ways (as is specified next).

(2) The most consistent result in Table 4a is that low familism goes with the sense of powerlessness. People who say that their present work provides them little opportunity for control, or people who say that they have little influence in social affairs generally, are more ready to give up family solidarity for work advancement. That result is quite consistently demonstrated in Table 4a (for both blacks and whites, though the difference is greater for blacks); and it seems there is a kind of realism in it: it is reasonable to say that a low-control job and an unsure generalized control make one willing to pay more, as it were, for the improved job chances that do develop. But note that the readiness to give up friendship ties is quite another matter. On the whole, powerlessness makes little difference there. If anything, the more powerless are *more* committed to their friendship community.

(3) Except as noted, the connection between community and control in these data is weak. In Table 4b, control in either sense (at work or in social affairs) has little to do with whether the individual is engaged in an occupational community or not, or whether he experiences a sense of loneliness to any significant degree. Though it is plain that our measures do not fully explore the sense of community, due to limitations of the data (e.g., there are no indices here of neighborhood attitudes or visiting patterns), it seems clear that the tendency to think of a tight bundle of urban alienations—which includes both the loss of community and the loss of control as part of a general malaise—is not supported here.

For that matter, even *within* the measures of control, the evidence argues against a package conception. The items concerning control over the work process (which we take to be indices of work alienation) do not correlate highly or consistently with those which concern control over social affairs (the indices of generalized powerlessness). For both black and white samples where total scores on work alienation and powerlessness are available, the r's are quite low between these two brands of control (e.g., .03 among manual whites; .01 among manual blacks).

THE CONSEQUENCES OF LOW COMMUNITY AND CONTROL

Why be concerned about the levels of sensed community and control? One can argue for their importance in their own right—that the opportunity for control at work is a facet of the creative life, and that a sense of community is also in itself an element of the good life. But the further argument is that they make a difference; they have consequences (when they are missing) that are bad for the society and bad for the individual.

Table 5 presents some of the evidence on such supposed consequences of low control and lost community. It deals with black manual and nonmanual workers, and shows how their scores on social powerlessness and on work alienation correlate with various presumed outcomes of these two brands of alienation.[4] The outcome measures need a word of clarification: (1) the index called "identification with Negroes" is based on four items, such as: "I look at everything from a Negro point of view," or "The idea of complete integration bothers me because Negroes would have to give up some of the sense of belonging together"; (2) "social distance" refers to a four-item scale ranging from marrying to eating with whites; (3) "white hostility" refers to three items on the order of, "Sometimes I hate white people"; (4) "occupational striving" is an 8-item index of willingness to give up a series of values for the goal of upward mobility (e.g., the "family" and "friends" items referred to earlier, plus six others on risking one's health, peace of mind, political freedom of speech, and so on); and (5) "self-hate" refers to two items which insist upon the in-group's own blame (e.g., "Negroes blame whites for their position but it's really their own fault").

There are two things to be noted about the data in Table 5. First, and most impressive, it is generalized social powerlessness rather than work alienation that consistently correlates with such outcomes. For both manual and nonmanual workers, those who express high powerlessness are more hostile to whites, less willing to interact with

TABLE 5

Correlations Between Two Indices of Control (Two Forms of Alienation) and Various Outcome Variables, for Manual and Nonmanual Black Workers in Los Angeles (Male Workers Only)

	Control Indices	
	Powerlessness	Work Alienation
Identification with Negroes		
Manual	.47	-.04
Nonmanual	.37	.17
Social distance (from whites)		
Manual	.35	-.03
Nonmanual	.30	.01
Hostility toward whites		
Manual	.32	.04
Nonmanual	.34	.18
Occupational striving		
Manual	-.26	.53
Nonmanual	-.07	.21
Self-hate		
Manual	-.09	-.03
Nonmanual	.12	.14

NOTE: The n for the manual workers was 172, and for the nonmanuals, 122. The data are from the study by Ransford (1968).

them socially, and (perhaps most interesting) identify more with an expressly Negro point of view.[5]

The second fact of interest in Table 5 is a cautionary one—namely, that the generally low correlations for work alienation are apparently not a function of difficulties with the measure itself. Work alienation *does* correlate highly with occupational striving (Table 5): especially among the black manual workers, those who are engaged in highly routinized, low-control work are more ready to give up a wide range of important values in the interest of improving their work situation. Interestingly enough, and in keeping with our earlier strictures about the package conception of alienation, powerlessness works in the opposite way from work alienation: high powerlessness apparently inhibits striving ($r = -.26$); but high work alienation encourages it ($r = .53$).

One further point about Table 5 should be noted; it concerns what is there called self-hate. What is noteworthy is that though powerlessness is rather consistently predictive it does not correlate with anything and everything: it does not, for example, correlate with self-hate. And neither does work alienation. It seems important to note these low r's for self-hate because it is so readily assumed that the individual's problems with community and control generalize in a wide swath: it is asserted that they make problems for his self-image, for his family, and for his relations with others in his social world. What these data on self-hate seem to say is that this kind of generalization may have some very severe limits.

If, as Swados (1962: 111) argues, routinized workers are sometimes "furiously angry with themselves for what they are doing, or desperately hunting for work," our data suggest that the alienated black workers are, indeed, hunting for better work. How desperately they are hunting for work we don't know; but desperately enough to be ready to give up a good many commonly held values for better work. That, at least, is what the .53 correlation between alienated labor and occupational striving suggests. But it seems that these workers are certainly not angry enough at themselves to blame the Negro group for their troubles.

One might well argue that such a degree of generalization is expecting too much—that being angry or dissatisfied with oneself means unhappiness with *self,* not with the group. It should be pointed out that self-hate (Table 5) means very much what Lewin (1948) meant by Jewish self-hate; namely, indicting one's own group for its troubles. But even on the dissatisfaction-with-self argument, the data are not supportive. Morris asked a direct question on satisfaction with self ("Are you apt to be a person who is dissatisfied with himself" or not?); and we find that the two questions on control over the work process are not related to self-satisfaction as indicated by response to this item. Nor is generalized social powerlessness related to this index of self-esteem.

We are led to the conclusion that so far as likely outcomes are concerned, control at work is considerably less significant than powerlessness in social affairs.

It would be good to know how the various measures of community (e.g., loneliness, or low-occupational community) relate to these same outcomes for the black sample on which Table 5 is based, but these data are not available. However, we can distinguish (as in Table 1) those who are oriented toward familism or toward friendship solidarity, and determine whether these community indices are related to the outcomes under discussion here. For the most part, they are not; but what emerges (and is shown in Table 6) is further evidence that family and friends is a phrase that hides more than it reveals. These two kinds of commitment relate quite differently to the measures of social distance (toward whites) and resistance to thorough integration (with whites). Holding occupation constant (i.e., manuals only), the familism choice is associated with relatively *low* social distance—only forty percent of those who are family-oriented express such distance, versus sixty-eight percent of those who are committed to advancement. But the choice of friends over advancement is associated with relatively *high* social distance. The conclusion that suggests itself is not a new one, but one that is

TABLE 6

High on Social Distance Toward Whites, and on Opposition to Integration, for Those Choosing Family and Those Choosing Friends over Advancement (manual workers only)

Community Choice	n	High On: Social Distance %	High On: Opposition to Integration %
Family advancement	82	40	15
	47	68	31
Friends advancement	38	69	32
	97	43	16

NOTE: The black data are from Ransford (1968). The table is read as follows: among those manual workers (n=82) who chose family over advancement (see Table 1 for the complete item) 40% scored high in social distance toward whites; for those (n=47) who chose the advancement alternative, 68% scored high.

frequently forgotten by community enthusiasts; namely, that some kinds of community can come closer to a form of ethnocentrism (Wood, 1959).

Thus far, we have been concerned largely with consequences of community and control for a black sample. The consequences for a white sample are exhibited in Table 7. The *prejudice* score is a combined measure of anti-Negro and anti-Semitic attitudes; the *anomia* index is the well-known Srole scale, which is perhaps best interpreted as an index of despair and distrust; the *striving* measure is a 14-item scale of values that would be given up in the interest of advancement; and *job satisfaction* is a single-item overall job evaluation.

Of particular interest in Table 7 is how unpredictive is a measure like occupational community. The correlations are modest throughout, but it is nonetheless the case that membership in an occupational community tells nothing about prejudice, anomia, or even job satisfaction, while powerlessness is at least minimally predictive.

Since the anomia score is perhaps the most general index of disaffection that we have, this measure was cross-tabulated with the four indices of community; i.e., with familism, friendship, solidarity, occupational community, and loneliness. In no instance did the anomia differences approach significance (e.g., among the nonmanual workers who say they are lonely, seventeen percent score high in anomia; for those who say they are not lonely, eleven percent are high in anomia).

The relative unimportance of loneliness (and the relative importance of powerlessness) show up again in Morris' data (Morris and Jeffries, 1969). In that case, a prejudice score, much like the one involved in Table 7, shows no connection with his index of loneliness (a result which is sustained when occupation and education are controlled as well as for the total sample). But powerlessness does make a difference; endorsement of an item which reads "What

TABLE 7
Correlation of Two Indices of Control (Powerlessness and Work Alienation) and an Index of Community (Occupational Community) with Various Outcome Variables, for Manual and Nonmanual White Workers in Los Angeles

	Prejudice		Anomia (Srole Scale)		Occupational Striving		Job Satisfaction	
Control and Community Measures	Manual	Non-manual	Manual	Non-manual	Manual	Non-manual	Manual	Non-manual
Powerlessness	.21	.18	.26	.42	.06	.16	-.18	-.16
Work alienation	.22	.18	.07	.17	.15	.11	-.39	-.28
Occupational community	-.02	.00	-.02	-.13	-.01	-.02	-.09	-.00

NOTE: The n for the manual workers was 157; for the nonmanuals, 243. The data are from the Seeman sample (1967).

happens to me is my own doing" (i.e., endorsement of the mastery point of view) is associated with *low* antagonism. Of those who choose the powerless alternative seventy-six percent are high in ethnic antagonism, whereas fifty-three percent of those who endorse the mastery view are high in ethnic antagonisms.

A gross summary of this discussion of consequences would suggest that there is a rank order of significance for the control and community variables we have explored. The control aspect is more telling than is community; and within the control domain, powerlessness is more consequential than work alienation. Within the community domain, the loneliness variable seems more discriminating in general than occupational community. Perhaps this simply says that the weaker variables of work alienation and occupational community are more nearly *descriptions* of life style related to work, while powerlessness and loneliness are *attitudes* about the respondent's circumstances in general. Even so, given the importance that is often assigned to work life (meaning either control or creativity in it, or social engagement in it), the evidence on the relative insignificance of work alienation or occupational community raises some interesting questions about some standard views of work as an important element in contemporary life styles. It is difficult to fit the notions of familism and friendship solidarity into this rank order; but at the very least one can say that these two seem to work rather differently as dimensions of community.

THE COMMUNITY OF FATE AND VALUE

The fact that these data deemphasize the importance of community, in the sense of "belonging," and highlight the importance of control, hardly dismisses the problem of community altogether. That is so for two reasons. First, it must be remembered that the two additional conceptions of community outlined earlier–the community as common fate and as common value–have not yet been examined. Second, one may easily construe the findings we have reported as saying that control is not as independent of community as we have implied. The clearest indication of this is found in what we have reported about intergroup relations. In-group selectivity and out-group rejection (that is, social distance, prejudice, in-group identification and the like) are, after all, expressions of belonging; and we have shown that such measures of ethnic community are among the most consistent correlates of powerlessness. For blacks, the greater the sense of powerlessness, the sharper the identification with the black community. And among whites, the greater the sense of powerlessness, the sharper the exclusionary attitude toward minorities.

But, then, we are no longer talking about solidarity in the sense of family, friends, coworkers, or loneliness. We are approaching a less sentimental definition of boundaries—a boundary where the community of common fate (i.e., being black) and the community of common identification (i.e., identification with a black community) is joined; and where, apparently, control over that fate (the feeling of powerlessness or mastery) is an ingredient in how such identifications work. There are numerous ways in which the community of common fate can be defined—e.g., by race; class; territorial political subdivision; neighborhood; industrial firm of employment—and we shall have very little more to say about such communities of common fate. But we hazard the guess that none of them becomes of any serious interest to the people concerned until the problem of control comes up. This is what Greer (1962: 100, 108) refers to as the "contingent commitment" of Americans to the local politically defined area, and his suggestion is that "even work is a commitment of limited liability." Our data on the relative insignificance of work alienation would seem to be congruent with his view.

What about community defined as value consensus? That is an idea which is sometimes intertwined with the community of common fate—for example, we come to speak, as Greer (1969: 108) does, of "the neighborhoods of the like-minded." We suspect that on this aspect of community, too (as with the overemphasis on belonging), the portrait of metropolitan trouble is overdrawn—which is to say that we overdraw, as Keller (1968: 55) puts it, "the danger that the more heterogeneous, individuated population that must now fend for itself can no longer achieve the consensus needed for collective life." That would be a very difficult thesis to test; certainly we cannot more easily test our doubts about it, especially with the available data. Suffice it to say that we are more impressed with the many uniformities in value and attitude that we find than with the heterogeneities. We are not so much concerned with whether the overall level of agreement is high, but with the fact that the agreement is high across the major lines of cleavage that are so fundamental in our thinking about urban life: race, sex, income, occupation, education, and area of residence.

TABLE 8a

Rating Various Accomplishments in Life (for the Child) as Important or Very Important, for Black and White Fathers and Mothers (in percentages)

Accomplishment	Value	Black Fathers (n=148)	Black Mothers (n=141)	White Fathers (n=138)	White Mothers (n-138)
1. Reaching higher social standing	Status	72	58	53	41
2. Doing what gives most personal satisfaction	Satisfaction	76	84	86	91
3. Being outstanding in a chosen field	Recognition	87	89	82	73
4. Being a respected citizen in the community	Citizenship	93	96	89	89

TABLE 8b

Choosing Various Values as Preferred Achievements for Oldest Child in High School, for Black and White Fathers, by Income Level (in percentages)

Respondent Would Like Him (Her) To Be Remembered As:	Black Low Income	Black High Income	White Low Income	White High Income
1. Brilliant student	33	32	39	25
2. Most popular	0	1	9	4
3. Athletic star (cheerleader)	5	4	5	3
4. Leader in activities	8	11	5	16
5. Most respected by teachers	43	49	39	48

TABLE 8c

Husbands High on Conjugal Power for Six Family Decisions, by Race and Socioeconomic Status (Based on Responses by Both Husbands and Wives) (in percentages)

Family Decision:	Black Low SS	Black High SS	White Low SS	White High SS
1. Job husband could take	90	92	93	91
2. What car to buy	46	56	51	44
3. Buying life insurance	35	41	40	41
4. Where to vacation	87	89	91	86
5. Money to spend on food	54	46	47	46
6. Apartment or house to take	70	72	73	70

NOTE: Percentages refer to those who responded that the "husband always" or "husband more than wife" made the final decision on the family issues named. The data of Table 8c are reproduced from TenHouten's paper (1970).

Agreement about what? We have used the data from all four of our resource studies to address this question in Tables 8, 9 and 10. These tables refer to value agreement between fathers and mothers, blacks and whites, low- and high-income respondents, manual and nonmanual workers, residents of localities as different as Reseda, Long Beach, and Baldwin Hills. Without being formal or exhaustive, we have tried to incorporate in these tables, concerning value agreement, the conception presented at the outset. We concentrate on the domains of work (e.g., occupational striving, desired careers), family (attitudes about children and family loyalties), and friends-ethnicity (loyalties to friends and ethnic attitudes); and obviously within those domains, the questions bear heavily on the two dimensions of control and community.

We are by no means limited to the data in these three tables in thinking about the question of value consensus—e.g., as noted earlier, the data of Table 1 show considerable congruence across race and occupation; and TenHouten (1970) reports many uniformities in family structure and attitudes beyond those reproduced in Table 8c. Our thoughts based on these data include the following:

(1) The cross-race, cross-class, cross-community similarities are all the more impressive for the range of convictions involved and their overall consistency (note the considerable differences in educational standards for the communities in Table 9, and the cross-community similarity in values despite that fact). It makes one begin to have some second thoughts about the meaning of that popular phrase "internalizing middle-class values." One would have thought that striving was one of those values, but the suburban residents of Reseda are no different from their counterparts in urban-central Leimert Park in that regard (see items 3, 5 and 9 in Table 9); nor is materialism (at any rate, the confession of it in verbal form!) much different across these neighborhoods (see item 4), or trust, or loneliness, or social concern (e.g., items 6, 7 and 12 in Table 9). The data are consistent with some of the doubts that have been raised by Bennett Berger (1966), Leonard Reissman and Ktsanes (1959), and others about our typical images of suburbia—images which stress the striving and conformism of suburbanites or their propensity for taking on new values that go with urban mobility.

(2) Though it is probably true that a portion of the similarity in these tables stems from the fact that the value statements are sometimes bland and stereotyped, it is nevertheless true that many of these items concern quite specific agreements (about specific occupations, or specific family decisions, for example), and the responses are not so

TABLE 9
Agreeing with Various Value Items in Six Los Angeles Localities, Men and Women (Morris Data)[a] (in percentages)

Item	Baldwin Hills (n=99)	Bell (n=100)	Leimert Park (n=87)	Long Beach (n=104)	Pacific Palisades (n=100)	Reseda (n=93)
1. Do you think people can generally be trusted (yes)	90	82	92	81	92	89
2. The majority of Negroes in Los Angeles have respect for law and order (agree)	82	73	80	70	84	81
3. I am a person who tries to get ahead (vs. content to stay where he is) (choosing get ahead)	78	74	75	83	71	80
4. I am a person who likes the material comforts of life (vs. doesn't pay much attention to comforts)	77	68	69	79	75	76
5. The most important thing to look for in a job is whether it will help you get ahead (vs. enjoying your work is more important than getting ahead)	37	39	31	39	17	30
6. Most of the time a person should look out for himself, even when it may put others at a disadvantage (vs. looking out for others first)	54	56	33	52	43	54
7. People should be left alone and made to help themselves more (vs. every effort should be made to give aid to people who are in need or not as well off as others)	41	30	41	30	36	43
8. I sometimes feel very much alone in the world (agree)	34	37	31	30	39	34
9. If a better job should come along, a person should take it even if he has to break off with his old friends with whom he has worked for a long time (vs. do not break ties)	84	83	80	70	76	85
10. I am a person who is satisfied with himself (vs. dissatisfied with self)	70	69	69	62	63	70
11. The way things are today in the world, people cannot count on really getting to know each other very well	46	39	51	53	35	46
12. A person's first concern should be for others (vs. self)	61	47	53	53	45	57
Completing high school only (no advanced education in college)	27	81	48	74	20	52

a. 1969

stereotpyed as to screen out some very intelligible disagreements. It is striking that Table 8a, for instance, shows agreement on everything except the value of social status. The black parents (and especially the fathers) rate the achievement of higher social status as being more important than do the white parents, which is hardly surprising given the social standing from which they start. This does not seem like a vague and bland value commitment about the child, since it matches what blacks and whites say they would give up in the interest of their own occupational improvement (see the data already presented in Table 4a, where the blacks in poor work circumstances show marked concern about job advancement). In general, the data suggest that if one were to look for domains where value differences *would* appear, two ready candidates suggest themselves: for the black samples the issues are social status and power (see, for example, the differences between black and white manual workers on item 7 in Table 10, concerning who should make "the basic decisions on political and social questions").

(3) With respect to the value consensus community as it appears in these data on Los Angeles, the major segmentations of race, class, and locality do not establish discriminable value communities; there is, rather, a fundamental similarity. There are, to be sure, individual variances: on some issues, the population divides almost 50-50 in its opinion (but on numerous others the agreement hovers around eighty percent). Beyond such unexplored individual variations, there *are* major differences on some of the themes that have been central in this paper. But these are not value differences. They are differences in the person's expectancies or in his description of his situation. We refer to the fact that, for example, the communities in Table 9 do show differences in powerlessness and in work alienation—e.g., agreement with the statement that "becoming a success is a matter of hard work" (rather than luck) ranges from fifty-nine percent in Pacific Palisades to eighty-three percent in Long Beach. Moreover, on the work control item which asks whether the individual has the opportunity to make independent decisions, the range of those who claim they cannot make such decisions at work is from only nine percent in upper-middle-class Palisades to forty-four percent in Leimert Park.

Thus, we are drawn back in this discussion of value consensus to our theme of the centrality of control. Ellison (1969: 341) seems to be drawn in much the same way when he remarks: "My problem is not whether I will accept or reject American values. It is, rather, how I can get into a position where I can have the maximum influence upon those values." The point is that what often looks like dissensus is more nearly a struggle for control not too far removed from

TABLE 10

Agreeing with Various Value Items, for Black and White Samples, by Manual vs. Nonmanual Occupation (in percentages)

	Black		White	
Item	Manual (n=124)	Nonmanual (n=105)	Manual (n=154)	Nonmanual (n=245)
1. I really prefer to put my roots in solid in a community rather than move as the chances for advancement come along.	36	32	54	40
2. A person must be willing to put off having children for a while, if he wants to be ready to take advantage of the opportunities for advancement.	*	*	38	33
3. I wouldn't let being a stranger for a while keep me from moving every so often to a higher position in a new community.	*	*	72	83
4. High mobility attitude (summary score on the choice of advancement over other values).	32	36	19	22
5. Negroes should live around their own people.	7	11	*	*
6. My outlook toward life is more white than Negro.	14	16	*	*
7. Actually, the basic decisions on political and social questions in this country ought to be made by the specialists and experts.	24	35	41	48

*Starred items were not asked comparably in the Ransford (1968) and Seeman (1967) studies, from which these data are derived. See Table 1 for two further items of this type which were included in both studies. The total scores on mobility attitude (item 4 above) are not directly comparable since they were based on a different number of items in the two studies (8 items in Ransford's case; 14 items in the Seeman study).

Ellison's struggle—getting into a position to achieve what are the commonly recognized values or to influence the subtleties of their shading in American life. It is not easy to tell how much of the racial confrontation, women's liberation, and student militance is to be read in this way, but the implication is that such a reading would not be too far from the mark. It is a reading which argues that the value community is not the first-order problem. Rather, the emphasis is on the powerlessness-mastery dimension and on the community of common fate—on the common fate that race, sex, and student status severally signify, and the problem of control over that fate which is created by the special circumstances in which these groups find themselves in American life.

We are hesitant, however, to press the value similarity too hard. For one thing, there are numerous studies that report relevant differences—for example, the many studies of social class differences in values (everything from the lower-class inability to defer gratification to its emphasis on security rather than creativity in work). Second, though one can say that the power struggle is over common values, the struggle itself generates differences in modes of operation (e.g., the use of violence) that divide the value community. Third, though we have sampled a reasonable range of value commitments, our questions barely begin to cover the diverse domains in which serious value differences among major social segments might be found. We have no questions that tap orientations to fundamental structural change in the American social system, and we have little information about what might be called specific interests (as distinguished from values, norms, or expectancies—interests such as transportation, noise, pollution control, housing priorities, and the like) that might well differ by locality, race, and class. Finally, despite the value consensus, there remain quite importance differences that are truly a matter of style—aspects of life style that have fallen outside the net that can possibly be managed by this paper and these data, including language, leisure, culture, and expressive styles so characteristic of the pluralistic populations of Los Angeles and other metropolitan areas.

SUMMARY

Leaving qualifications and elaborations aside, what we have said is essentially this:

(1) When we think of life style, we are concerned with three domains of action: work, family, and a complex of friends-ethnicity-neighborhood.

(2) In each of these domains, there are two recurring themes which are central: the theme of community and the theme of control.

(3) It is important for our conception of community to divest it of the connotation of a bounded territory; we think of it in three related ways, none of them requiring such territoriality: the community of common *fate,* the community of sentimental attachment or *solidarity,* and the community of common *value.*

(4) When we try to get an estimate of these sentiments of attachment (e.g., loneliness, commitment to family, the sense of being bound to an occupational friendship group), we find (a) little difference in such

sentiments across major social lines, and (b) little evidence of a metropolitan malaise on the matter in Los Angeles.

(5) As for the sense of control, it is distributed rather differently. And it is quite important to keep in mind the distinction between feelings of control on the job (what is here called work alienation), and feelings of control over one's destiny in society (called powerlessness). In general, the black manual worker is distinctively high in both of these alienations: he feels generally more powerless than others; and he reports his work as being more routinized than others. As one might expect, the white manual worker tends to say his work is relatively routinized, as well. The proportion of workers who report lack of control at work seems high, though that is hard to evaluate (both in itself and in terms of its likely consequences).

(6) We tried to determine whether our indices of community (e.g., loneliness, or occupational community) were closely tied to feelings of efficacy—whether, in short, the alienation of lack of community was bound up with the alienation of lack of control. On the whole, the associations are weak. We raised some questions about a package conception of urban alienations—the kind of bundle of alienations that writers often assume to be typical of the urban reality. We do not think it is.

(7) Even within the dimension of control (leaving community aside), there are serious doubts about such a package: the sense of control at work is not closely tied to the sense of generalized social control. Within the domain of community, our data suggested that family and friends, as two aspects of community, work in different ways. We found that sometimes commitment to friends goes with ethnocentric attitudes about race (which is not the case with commitments about family ties). The evidence raised doubts about the assumption that community is uniformly a positive good.

(8) These alienations (both of community and of control) are presumed to have serious consequences in the life of the person and the metropolis. We sought to examine, as best we could, what some of these consequences might be—consequences like intergroup hostility, low self-esteem, and so on. We find that generalized powerlessness is more important than the community variables or work alienation. Put otherwise, knowing whether a man is in a job that provides little control, or whether he is tied in with coworkers in an occupational community, tells very little about what his prejudice, striving, or ethnic identification will be like. If anything illuminates these, it is generalized powerlessness.

(9) When we look at community as indicated by value consensus, we found a rather high level of agreement across major social lines on a

> wide variety of beliefs. We found agreement, that is, across territorial lines, across race, occupation, sex, and income. We argued that this similarity too often is underplayed in the prevalent concern about the lack of community that derives from value dissensus.

In conclusion, we ought to make it clear that the data and the analysis we have presented are of broader import than may initially appear. We have confined our attention to Los Angeles, but obviously the area is taken as a prototype of urbanization, and comparative studies (including the collation of available evidence from other urban centers, and new studies on the issues reviewed here) are very much in order. We have confined ourselves, also, to two basic varieties of alienation; the loss of community and of control, but there are other forms of alienation (compare with Seeman, 1959), and each of them engages values that are important in American life—for example, the values of mastery, creativity, social acceptance, consensus, and mutual trust. The implicit argument is that a more comprehensive and systematic examination of the various forms of alienation commonly attributed to urban life can tell us a good deal about "the metropolitan enigma" (Wilson, 1968).

NOTES

1. Schnore (1967: 124) comments in another way on the importance of these three domains when he remarks: "There are three main principles of residential segregation operating in the modern community in the United States. These are (1) type of household, (2) socio-economic status, and (3) racial-ethnic status"—i.e., family, work rank, and ethnicity.

2. Unfortunately, the results of two major studies—one among Mexican-Americans in the Los Angeles region, and another among Japanese in the area—were not accessible to us for the kind of analysis required. We would like to thank Professors Morris, Jeffries, TenHouten, and Ransford for making their data available and consulting with us on their use.

3. For those who have doubts about the forced-choice pairings in Table 2a, it is instructive to note that the Morris sample responded to item 1a in that table ("Becoming a success . . .") in an agree-disagree format, and the overall endorsement (seventy-six percent) is very similar to the endorsement level shown in the table. Similarly coordinate results were obtained in TenHouten's samples.

4. The powerlessness score derives from twelve forced-choice items of the kind already presented in Table 2a. The work alienation score is based on a 7-item scale which includes the two items on control at work presented in Table 3, plus five others which seek to index routinization versus engagement in the work process (compare with Seeman, 1967). Obviously, we are dealing here with correlates rather than with outcomes or consequences in any literal sense.

5. Clearly, there are many other potential consequences of powerlessness that might be of interest. For example, in TenHouten's data (based on a different black sample), high powerlessness was associated with low militance on civil rights (determined by simple

self-ratings on degree of militance). That finding is interesting in light of the data shown in Table 5, where high powerlessness goes with hostility and social distance toward whites. Thus, powerlessness seems to be associated with a somewhat curious blend of (a) low militance, but (b) greater potential hostility and explosiveness. Perhaps the fact that *low* powerlessness goes with militance implies the distinction between planned, if relatively radical, pressure for change with anticipated efficacy in one's action (a stance from which the powerless retreat), versus readiness to be violent when the circumstances seem to demand or allow it (which the powerless find more congenial).

REFERENCES

BALTZELL, E. D. [ed.] (1968) The Search for Community in Modern America. New York: Harper & Row.

BERGER, B. M. (1966) "The myth of suburbia," pp. 167-178 in R. L. Warren (ed.) Perspectives on the American Community. Chicago: Rand McNally.

BLAUNER, R. (1964) Alienation and Freedom: The Factory Worker and His Industry. Chicago: Univ. of Chicago Press.

ELLISON, R. (1969) Interviewed by R. P. Warren, in "Leadership from the periphery," pp. 339-356 in D. W. Minar and S. Greer (eds.) The Concept of Community. Chicago: Aldine.

GREER, S. (1969) "Urbanization and social character: notes on the American as citizen," pp. 95-127 in H. J. Schwandt and W. Bloomberg (eds.) The Quality of Urban Life. Beverly Hills, Calif.: Sage.

--- (1962) The Emerging City. New York: Free Press.

KELLER, S. (1968) The Urban Neighborhood. New York: Random House.

LEWIN, K. (1948) Resolving Social Conflicts. New York: Harpers.

LIEBOW, E. (1967) Tally's Corner: A Study of Negro Streetcorner Men. Boston: Little, Brown.

LIPSET, S. M., M. TROW, and J. S. COLEMAN (1956) Union Democracy. Glencoe, Ill.: Free Press.

MORRIS, R. T. and V. JEFFRIES (1969) "The white reaction study," pp. 480-601 in N. E. Cohen (ed.) The Los Angeles Riots: A Socio-Psychological Study. New York: Frederick A Praeger.

RANSFORD, H. E. (1968) "Isolation, powerlessness and violence: a study of attitudes and participation in the Watts riot." Amer. J. of Sociology 73 (March): 581-591.

REISSMAN, L. and T. KTSANES (1959) "New homes for old values." Social Problems 7 (Winter): 187-195.

ROGERS, D. (1968) 110 Livingston Street. New York: Random House.

SCHNORE, L. (1967) "Community," pp. 82-150 in N. Smelser (ed.) Sociology. New York: John Wiley.

SEEMAN, M. (1967) "On the personal consequences of alienation in work." Amer. Soc. Rev. 32 (April): 273-285.

--- (1959) "On the meaning of alienation." Amer. Soc. Rev. 24 (December): 273-285.

SHEVKY, E. and M. WILLIAMS (1949) Social Areas of Los Angeles: Analysis and Typology. Berkeley and Los Angeles: Univ. of California Press.

SWADOS, H. (1962) "The myth of the happy worker," pp. 106-113 in E. Josephson and M. Josephson (eds.) Man Alone: Alienation in Modern Society. New York: Dell.

TenHOUTEN, W. (1970) "The black family: myth and reality." Psychiatry. Forthcoming

WILENSKY, H. (1964) "Varieties of work experience," pp. 125-154 in H. Borow (ed.) Man in a World of Work. Boston: Houghton-Mifflin.

WILSON, J. Q. (1968) The Metropolitan Enigma: Inquiries into the Nature and Dimensions of America's Urban Crisis (rev. ed.) Cambridge, Mass.: Harvard Univ. Press.

WOOD, R. C. (1959) Suburbia: Its People and Their Politics. New York: Houghton-Mifflin.

15

Public Participation and the Politics of Poverty

DALE ROGERS MARSHALL

□ THE IDEOLOGY OF public participation in the political process which emerged in the United States in the 1960s is an important influence on the racial aspects of urban life. The ideology has been linked with many government programs that use participation as a strategy for dealing with urban problems, including poverty, model cities, housing, health, and education programs (for reviews of the many federal social programs that have incorporated provisions for participation, see Mogulof, 1970a; Mathews, 1968). Such projects, intentionally or unintentionally, have resulted in the formation of new neighborhood organizations by blacks, browns, and the poor. The formation of these groups may be one of the "great structural changes in the political system of American urban society" (Bell and Held, 1969: 144). The change may institutionalize the involvement of blacks, browns, and the poor in political bargaining on urban issues just as the Wagner Act institutionalized the involvement of labor unions in economic bargaining (Bell and Held, 1969: 144).

In spite or perhaps because of the importance of participation in our era, discussions of the issue have been marked by rhetoric rather than reason, by advocacy rather than analysis. The proponents of participation (e.g., Bell and Held, 1969; Walker, 1966: 285-295; Altshuler, 1970) are convinced of the desirability of increased participation, and the opponents (e.g., Lowi, 1967; Moynihan, 1969a; Rustin, 1970; see also Backrach, 1967) stress the threat posed to the political system by certain types of participation. Underlying the debates are conflicting assumptions about human nature, the political system, and social change.

The debates may indulge our need for drama in the political arena, but they are becoming monotonous and reaching the point of diminishing returns. (Why is it that social scientists, so quick to deplore the polarization perpetuated by politicians, seem to resort to more subtle but similar name-calling tactics in their own journals?)[1] The importance of the topic requires more sober and systematic analysis than it has received. Social scientists must give the issue of participation the reasoned attention it deserves.

This chapter considers the role of public participation in one social program—the war on poverty. The poverty program is of concern here because it has gone the farthest in institutionalizing participation and has been studied more than the other program (Mogulof, 1970a: 84-85). Patterns found in poverty programs seem likely to foreshadow future developments in other participation programs. The ideology of public participation will first be placed in historical perspective, then the role of public participation in the war on poverty will be discussed in terms of its origins, implementation, and results. The effort is to synthesize and analyze the evidence available in the literature, and to indicate the implications of the evidence for the dispute about the value of participation and for urban reforms.

THE PUBLIC PARTICIPATION IDEOLOGY IN HISTORICAL PERSPECTIVE

Advocates of participation frequently express shock that a democratic country such as the United States has had so little participation in the government by the people. But evaluation of the previous level of participation depends largely on the perspective taken. Certainly democracy is based on the idea that individuals should have a voice in the government, but which individuals and what kind of voice?

The answers to those questions have changed throughout history; the meaning of government by the people has been modified in every age. Once it was an argument used by nobles to wrest power from monarchs. Later it encompassed conflicting interpretations of who should be eligible to vote. The long American struggle to widen the franchise to include all adult white males, regardless of property holdings, then adult Negro males, then all adult females, has been too quickly forgotten. Current considerations about lowering the voting age to eighteen and eliminating English literacy requirements indicate that the issue of who should vote will continue to be debated. Historically in the United States the focus shifted from "who votes" to "what items do people get to vote for?" The first half of this century was marked by lively concern with increasing citizens' ability to participate directly in legislation as marked by new provisions for initiative, referendum, and recall, and by the lengthening of ballots as citizens were given the chance to vote on an increasing number of positions and issues (for review of historical roots of current demands for participation, see Altshuler, 1970: ch. 2).

Thus, taking a historical perspective on participation, one finds that current democratic governments have more participation by the people than their originators envisioned. In one sense, what is shocking is not how little participation we have, but how far the democratic idea of participation has come. In another sense, however, participation has not developed fast enough to keep pace with the rapidly changing governmental system, especially the dramatic increase in the size and importance of the executive branches of government.

Herbert Kaufman (1969: 3-14) discusses the causes of dissatisfaction with the present arrangements for participation. He explains that certain segments of the population do not feel they can receive fair shares of the system's benefits through the existing institutions; they feel "only the powerful get attention, and that the already powerful are helped by the system to deny influence to all who now lack it" (Kaufman, 1969: 4-5). The demands of the dissatisfied for increased participation or representation focus mainly on the executive branches of government. They want to have a louder voice in how the policies of government are implemented. "Power to the people" is not primarily a demand for letting more people vote on more issues, but for administrative changes such as community control of agencies, and participation by the poor, blacks, and browns, and other disaffected groups in the operation of the agencies.

Just as proponents of participation are shocked by the absence of participation, opponents of the idea are aghast at the demand for more participation. They incredulously respond that elected officials *are* the mechanisms of public participation and representation. In their view, traditional forms of electoral participation are adequate and they urge disaffected groups to work through the existing party system and to take an active part in elections (Mogulof, 1970a: 33). To them the new demands for participation in administration threaten to eliminate the advantages achieved by previous administrative reforms which stressed neutral competence and strong executive leadership as antidotes for corruption (Kaufman, 1969: 3-4). The demands are seen as challenges to professional experts and governmental officials.

This fear of the intrusion of the public into new areas of government is supported by certain social science generalizations which stress the value of apathy and the dangers of high levels of political participation (Litt, 1970: 145; Milbrath, 1965). Evidence includes studies which link the masses with authoritarianism and indicate that when they are mobilized, conflict and extremism—detrimental to the political system—increase (Lipset, 1960: 86-126).

But the shock of the opponents of participation is not any better founded than the shock of the proponents. The latter tend to ignore the past and the former tend to ignore the reality of change as a part of that past. Involvement in politics has been a recurring strategy by which disadvantaged groups make their claims for improved position within society (Litt, 1970: 144), and the strategies have always taken new forms as the setting has changed. The current demands by blacks, browns, and the poor raise the issue of administrative distribution in a society where this is central. As Litt says (1970: 149-150):

> White immigrant politics emphasized the representative phase of political life while contemporary Negro politics most fundamentally

raises the issue of distribution in a society where the federal government is the key to the distribution of values.

The values of neutral competence and strong executive leadership are challenged by increased participation, but these values emerge as challenges to previous values and it is naive to expect them to be immune to change. Corruption led to attempts to achieve competence and leadership in the form of civil service and strong executives. These reforms now reveal that corruption is not the only evil—lack of responsiveness to major segments of society may also be harmful. When professionals are asked to yield to the judgment of amateurs, their egos are wounded because they are "convinced that they themselves are the best judges not merely of their own competence but also of what is best for the people for whom they perform services" (Hughes, 1958: 3, 54). The demand for increased participation can best be seen as a new stage in the reform cycle, a stage which stresses the value of representativeness. Kaufman would call it a return to an earlier stage of the cycle but this is misleading unless one is aware that representativeness now focuses on administrations rather than legislatures.

The ideology of participation, then, is part of the continuing evolution of the democratic idea that people should have a voice in government. The ideology calls for participation in administration by those dissatisfied with traditional forms of participation. It suggests new approaches to participation and representation because many currently do not feel their voices are being heard. Burkeans are undoubtedly correct in pointing out the dangers of aroused people, but failure to develop new institutions to meet new demands also endangers the system (Litt, 1970: 153). The ideology of participation forcefully raises the issue of whether or not the theory of democratic participation should be broadened to find positive value for types of participation which once seemed to be disruptive (Litt, 1970: 149; see also Coser, 1956; Dahrendorf, 1959).[2]

ORIGINS OF PARTICIPATION IN THE WAR ON POVERTY

Although historically the ideology of participation can be seen as part of the continuing development of the democratic idea, it is instructive to look at the proximate causes of its embodiment in federal programs in the 1960s, particularly in the war on poverty. No definitive history has yet been written but a variety of influences are usually cited.

There is agreement on the primacy of the executive rather than the legislature in formulating war on poverty legislation, including the requirements for participation (see Bibby and Davidson, 1967; Donovan, 1967: ch. 1; Levitan, 1967: 15-16; Marshall, forthcoming: ch. 1). The Shriver Task Force developed a poverty proposal which was a compromise between the desires of the various governmental departments and constituencies with interests in welfare programs. The community action component of the proposal (Title II), which included the participation provisions, was favored by the Council of Economic Advisors and the Bureau of the Budget. It was the only truly innovative part of the poverty package.

Disagreement centers on exactly where the participation clause originated and what the proponents meant by participation. Title II, as originally passed by Congress in 1964, simply stated that community action programs were to be "developed, conducted, and administered with the maximum feasible participation of residents of the areas and members of the groups served."[3] One view suggests that Richard Boone and the staff of the President's Committee on Juvenile Delinquency wrote the language and got the Shriver Task Force to insert it into the legislation, hoping it would form a basis upon which the poor could organize and promote social change. Others contend the language was simply intended to ensure participation by southern Negroes in community action programs (for treatments of the origins of the participation clause, see Rubin, 1967: 5-18; Donovan, 1967: 41-43; Bibby and Davidson, 1967: 236; Marris and Rein, 1967: 215; Kravitz, 1968; Moynihan, 1969a: 236; Sundquist, 1968).

Whatever the intentions of the drafters were, many social developments of the 1950s and 1960s could have influenced the choice of the language and its implementation. The most direct models were the Grey Areas projects of the Ford Foundation and the activities of the President's Committee on Juvenile Delinquency. These community projects sought to promote change through a variety of strategies, one of which was the involvement in the planning of people from the target areas (for detailed description of the two programs, see Marris and Rein, 1967: 164). These programs grew out of disillusionment with previous reform attempts such as urban renewal which had increased the distress of the poor by displacing them from their homes (Marris and Rein, 1967: 15).

The Grey Areas and President's Committee projects had also been influenced by other trends in our society. The most obvious of these was the growing civil rights movement and the supporting white student groups which challenged the adequacy of traditional institutions and developed the strategy of mass protest to dramatize their demands. Grass roots participation was used to confront the perceived complacency of the established leadership (Lipsky, 1969; Litt, 1970: ch. 9). A second trend of importance for the participation idea was Saul Alinsky's approach to community organization which emphasized organizing the poor to gain power by conflict with the existing power structure (Moynihan, 1966). A related trend was the social work tradition of trying to involve low-income people in neighborhood self-help projects both in the United States and in community development foreign assistance programs (Kramer, 1969; see also O'Donnell and Chilman, 1969: 1-10 for reference to earlier government programs).

All commentators agree that participation provisions in the war on poverty legislation were simply not the result of direct pressure from the poor or other potential clients of the program. They were conceived by professional reformers and inserted somehow into the legislation with almost no public or private discussion of the possible implications (Kramer, 1969: 9; Bibby and Davidson, 1967: 236; Moynihan, 1969a: 25-36). But once the legislation was passed the provisions for maximum feasible participation became the focal point of struggles for power in Washington and in local communities throughout the country. These struggles revealed conflicting interpretations of the meaning of participation in poverty programs and of the causes of poverty (Moynihan, 1966). One interpretation sees participation as a way of helping the poor adjust to the system and another sees it as a way of changing the system by organizing the poor (Mogulof, 1967: 12-20). The former views poverty as a result of the deficiencies of the poor; the latter links poverty with structural failings.

It is not necessary to repeat here the details of the struggles for power and the resulting changes in poverty legislation (Haddad,

1965; Rubin, 1967; Donovan, 1967). Suffice it to say that amendments to the Economic Opportunity Act after 1964 became increasingly specific about the meaning of maximum feasible participation, who should participate, and how. The amendments also toned down the emphasis on organizing the poor to push for social change by sharply cutting the amount of discretionary funds available to community action programs.

We will turn now to consider evidence on the implementation of the poverty legislation in local communities. Who actually participates and how?

IMPLEMENTATION OF PARTICIPATION IN POVERTY PROGRAMS

The most important point to remember about the implementation of participation in poverty programs is the great variety that exists. Many observers fall into the trap of generalizing about the war on poverty from the experience of just a handful of cities (Moynihan, 1969a). Yet the poverty legislation gave rise to over 1,000 community action programs (CAP). Some are in the largest metropolises, others in the smallest hamlets; some in rural, others in urban areas. And they are operated by diverse types of community action agencies (CAA). Some of the agencies are governmental units, others are private organizations; some cover only one political jurisdiction, others have jurisdictions that include the territories of many governmental units.

Comparative studies of some of the programs have been done but none of them pretends to deal with a representative sample of all CAP.[4] One generalization supported by all the studies is that differences in the implementation of participation are associated with differences among the communities in which CAA are located (Brandeis, 1969: 15). Size of the city, size of minority population, and political factors seem to make a difference. The Brandeis study (1969: 17) distinguishes cities by population over and under 150,000 and by minority population over and under ten percent. The political factor considered by the Brandeis study is whether the city government has a mayor and a council or a council-manager. The political factor Greenstone and Peterson (1968) find important in large cities with mayor-council forms is the strength of the reform movement as measured by the power of political parties. Kramer (1969: 180) deemphasizes the importance of the form of city government and the size of the minority population and stresses the variable of political capabilities, by which he means the ability of minorities to mobilize around an issue. Mogulof (1970b: 151) considers a factor not investigated by the others; he finds the roles played by federal and local staffs an important variable—the degree to which they serve as advocates for the poor.

The studies available do not permit any firm conclusions about the relative importance of these variables in explaining differences in implementation, but they do suggest some of the variables which should be tested in further studies. And they do highlight the diverse ways in which participation in the poverty program was implemented.

WHO PARTICIPATES

One of the basic questions about participation which was not clearly answered in the original poverty legislation was: Who participates? Is it the public, or specifically the poor? Some local administrators feel participation in poverty means a partnership of all elements of the community: public agencies, private welfare organizations, and minorities (see Los Angeles Times, 1968; also Miller and Rein, 1969). As evidence, they point to the passages in the legislation calling for mobilization of *all* segments of a community and *broadly* representative boards of directors. They stress the word “feasible” in the phrase maximum feasible participation of the poor and say it is not feasible to run a poverty program unless broad public participation is maintained.

However this interpretation of the meaning of maximum feasible participation is not ascendent. Public participation in poverty has become a euphemism for participation by the poor, blacks, and browns or by their chosen representatives. Participation in the poverty program has come to mean participation of potential clients; participation of the disadvantaged (Mogulof, 1970a: 2). The rationale behind this position—which may seem too narrow to some—is that the advantaged citizens can be heard in the traditional democratic institutions, while the disadvantaged citizens must be especially provided for in innovative programs or else these programs, too, will be dominated by the white middle and upper classes (Mogulof, 1970a: 40).

Even though participation refers primarily to the involvement of the poor, blacks, and browns, nothing in the federal guidelines prohibits these clients from selecting as their representatives in poverty programs people who are not poor, black, or brown. OEO guidelines say "although representatives of the poor need not themselves be poor, they must nonetheless be selected in a manner which insures they truly represent the poor" (CAP, 1968). And, regarding the selection process, the guidelines permit the participation of residents of poverty *areas* who are not necessarily poor themselves. The guidelines say all residents of poverty areas "may participate in the selection process, but special emphasis and attention must be given to insuring that those residents who are poor participate fully in the selection process" (CAP, 1968). Yet some CAAs chose to restrict participation in poverty elections to those who were actually poor themselves and to insist that all candidates meet the poverty criteria (Marshall, forthcoming: ch. 2). This move may be seen as a genuine attempt to obtain the full participation of the disadvantaged or as an attempt to ensure that the electorate would be inexperienced and would not be able to choose experienced leaders, thereby rendering participation meaningless (see Marris and Rein, 1967: 167).

Studies of the types of poor, blacks, and browns who are actually recruited into poverty programs show that in general they are not the so-called hard-core unemployed. The most alienated members of the lower classes have not been involved. But some lower-class individuals, who had not participated in previous social service programs, did participate in poverty programs. It is impossible to prove that these people would never have become involved in traditional programs, but poverty programs undoubtedly facilitated their participation.[5]

HOW DO THEY PARTICIPATE?

The poverty legislation did not specify who participates or exactly how people should participate in the program. Title II[6] called for participation in "the development and implementation of all programs and projects" in order to "best stimulate and take full advantage of the capabilities for self-advancement and assure that those programs and projects are otherwise meaningful to and widely used by their intended beneficiaries." Some administrators concluded that this participation requirement could be met by merely having poor people on advisory groups. However, OEO guidelines became increasingly specific about the diverse forms of participation which must be provided for by the CAA (see CAP, 1968).

The forms include the following: employment in CAP programs, membership on advisory committees for poverty projects, on neighborhood-based organization, on area councils, and on the principal representative board of the CAA. These forms of participation involve different assumptions about the goals of participation and the causes of poverty. Based on these forms of participation, Mogulof (1970a: 7) and Kramer (1969: 4) have developed typologies going from the least to the most intense kinds of participation. A synthesis of their typologies shows four main kinds of participation: employment, advice-giving, shared power in policy-making, and social action.

Participation of the poor by means of employment in poverty programs is in some ways the least controversial mode of participation (Pearl and Riessman, 1965). The poor are hired as paraprofessionals in poverty programs, assisting teachers, social workers, medical personnel, and other professionals especially in their relations with the community. The two main goals of hiring the poor are to change the individual and the agency employing him. It is hoped that the individuals will gain skills and economic benefits, and that the agency will become more responsive to the needs of the poor. In addition, the goal is to by-pass the rigid barriers of professionalism which limit entry to those with formal degrees (Kramer, 1969: 18-19). Large numbers of poor people have been employed as aides by CAAs and their delegate agencies. Many of the positions are low paid, without stable career possibilities, and in fields such as education, so they have been filled very often by women. The results of this form of participation will be discussed later in this chapter (Senate Subcommittee, 1967).

The second type of participation involves the poor as advisors to poverty programs. This advisory participation takes place in many different organizational settings. For example, agencies having contracts with CAAs to run poverty programs (these agencies are referred to as delegate agencies) often form policy advisory committees of the poor to assist with the programs. In addition, OEO encourages the utilization of a wide variety of neighborhood-based organizations composed primarily of neighborhood residents to assist CAAs "in the planning, conduct, and evaluation of components of the community action program."[7] Consultation with both of these types of advisory client groups is seen as a way of improving social services and building support for poverty programs (Kramer, 1969: 11-12).

In many urban centers a further form of advisory participation evolved as CAAs delegated limited responsibilities to area councils which were subsidiary arms of the principal CAA board (Senate Subcommittee, 1967; Kramer, 1969: 12; CAP, 1968: e, 1; Peterson, 1970).[8] Once created, these area councils typically try to increase the intensity of their participation; they want to gain more decision-making responsibilities, switching from advice-giving to shared power in policy-making (Kramer, 1969: 18-19). By 1968, OEO guidelines provided for the creation of area councils with specifically delegated powers (CAP, 1968: e, 1, a). But such councils were not required, so some CAPs did not create them. In small cities this absence has not concerned OEO, but pressure has been put on the larger cities to organize area councils where they have been absent (see Marshall, forthcoming: ch. 2).[9]

Some area councils have a high proportion of low-income members, but others are composed mainly of upwardly mobile, highly motivated minority people who no longer meet OEO criteria for poverty. Kramer finds that no more than a third of the members of the area councils in the five CAAs he studied meet OEO poverty criteria although more than that consider themselves poor. But communitywide generalizations should not obscure the wide variety in the proportion of low-income members in each area council—some councils are composed totally of low-income people and others have less than one-fifth low-income members (CAP, 1968: e, 1, b; Kramer, 1969: 191).[10] Kramer finds that most area council members were in their thirties or forties, and were either black or brown. He mentions that women predominate in a number of councils (for support of Kramer's findings, see Wayman et al., 1967-1969).

The third kind of participation considered here is shared power. The poor sit as equal members on the policy-making board or boards of the CAA. By 1967, OEO guidelines required that at least one-third of the members of boards of directors be representatives of the poor, one-third should be government officials, and the rest representatives of other important elements of the community, such as private welfare organizations. One idea behind this form of participation is that a new coalition will be created, a new coordinating body, which will force the three main segments of the community to work together for the first time. But the poor on the boards have demanded an increasingly large share of the power over policy-making. They want to move from membership in a coalition to control of a coalition (Kramer, 1969: 6-10).

Studies of the poverty representatives on CAA boards support and expand upon the impressions of area council participants. Poverty representatives tend to be younger than other board members and include a higher proportion of women and minority-group members. Many are divorced or separated (Brandeis, 1968, 1969: 56; Marshall, forthcoming: ch. 3; Bowen and Masotti, 1968: 89-100). Their level of previous organizational and political participation varies with the type of selection process used to recruit them; it is high in election systems which encourage organizational ties and low in those which do not (Marshall, forthcoming: ch. 3; Bowen and Masotti, 1968; White, n.d.; Peterson, 1970).

The fourth and most intense form of participation—social action—has also been the most controversial. Early guidelines of OEO said, helping the poor develop "autonomous and self-managed organizations which are competent to exert political influence on behalf of their own interest" is promising strategy (OEO, 1965). This interpretation of participation was the one most influenced by the ideas of Saul Alinsky (Kramer, 1969: 13-18) and least linked with CAA structures; i.e., it did not require the creation of any specific councils or any certain proportion of poor members. Critics of CAAs often talk as if social action is the most prevalent result of poverty programs (Moynihan, 1969a), yet other critics blame CAA for the absence of social action. For example, Herbert Krosney (1966: 16-17) says that when social action proved too radical, other less significant forms of participation, such as membership on boards, were substituted for it. Any of the neighborhood-based organizations connected with CAAs could have formed the nucleus for social action programs, but these organizations usually concentrated on social service rather than on social action (Kramer, 1969: ch. 9). More than half and perhaps as much as two-thirds of the CAA funds have been channeled to established agencies rather than newer neighborhood-based organizations (Senate Subcommittee, 1967, VI: 911).

In both the advisory and shared-power modes of participation, the question arises as to how the poor participants are to be selected. Originally many of the representatives were selected by the public or private agencies most active in creating the CAAs (Greenstone and Peterson, 1968, cite Chicago). But as OEO objections to this procedure became firmer, elections became the primary selection method. Various forms of elections developed for the area councils and the central board.[11] The most popular form involves elections in the communities for the members of the area councils (either direct election by residents of poverty areas whether or not they are poor [e.g., Philadelphia], or indirect election by organizations within the poverty neighborhood [e.g., New York]). Then each area council elects its own representative to the central board of directors. Los Angeles is the only major metropolis which holds a direct general election to select poverty representatives for the central board in which all poverty people are eligible to vote regardless of whether they live in poverty areas (and only poverty people are eligible to vote; Senate Subcommittee, 1967; AAA, n.d.; Marshall, forthcoming: ch. 2).

In community after community, the forms of elections have been subjects for debate between governmental officials and existing neighborhood groups. The arrangements finally agreed upon are an important indicator of the relative power of the contending forces. Peterson finds that in New York, where the neighborhood groups had the most power, only organizational representatives could vote in the neighborhood conventions to select the area council. In Philadelphia, the neighborhood groups were less powerful, so a direct election of the area councils was arranged as a compromise between the groups and the mayor. Direct elections avoided charges of mayoral dominance while still preventing organizations opposing the mayor from getting easy control of the selection process (Peterson, 1970).

The low voter turnouts in direct elections received much national publicity but should not have surprised anyone familiar with the literature on voter participation (Milbrath, 1965; Lipset, 1960; Lane, 1959). Low-income groups typically participate less actively in politics than higher-income groups. Even the privileged groups show very low turnout in local elections lacking in issues, partisan labels, or established cleavages (Lee, 1960: 120). Poverty elections not only lacked these features but also were an entirely new phenomenon which often required people who wanted to vote to attest to their poverty. The quality of the participation is more disturbing than the quantity. There is little evidence that direct elections for either area councils or CAA boards resulted in effective representation (Lee, 1960: 159, presents criteria for effective representation). Often the elected representatives were not chosen by nor could they speak for any organized segment of the community. Since their selection was primarily the result of a politics of acquaintance, they lacked a power base (Lee, 1960: 120; Kramer, 1969: 195-197; Marshall, forthcoming: ch. 2).

The method of selecting poverty representatives for area councils and CAA boards appears to be an important factor in the impact they have. Peterson (1970) finds that selection procedures controlled by neighborhood groups were linked with high power for these groups and a demand for major change in governmental services for poor people. Direct elections were associated with less influence and with demands for particularistic benefits rather than universalistic changes.

RESULTS OF PARTICIPATION– KINDS OF CONFLICT

Given the implementation of participation described above, what were the results? What kinds of conflict occurred, and what impact did the participation have on society?

Three levels of conflict or interaction can be distinguished in poverty programs, from the most latent to the most overt. They can be called advisory, coalitional, and adversary (this terminology is based on Mogulof, 1970a: 17). The first level–advisory conflict–is typically associated with CAAs in which area councils are either set up belatedly after federal prodding or are delegated very limited responsibilities. The Brandeis study (1969: 28-29, 21) found this type of interaction[12] in small cities or large cities with a small black population. The coalitional pattern of conflict is most likely found on CAA boards where poverty representatives sit as equal members in the policy-making but do not have strong power bases. This form of interaction can coexist in the same community where advisory conflict is taking place.

The most intense form of conflict is an adversary relationship in which area councils dominated by poverty representatives contend actively to achieve some measure of separate authority or control over poverty programs and to increase the number of poverty representatives on the central board (Kramer, 1969: 182-187). The Brandeis study (1969: 21) associates this type of conflict with large cities having large black populations. Kramer's (1969: 181-182) description of communities in which conflict takes the form of a fight seems to fit this description closely (Kramer, 1969: 181-182; compare with findings of Greenstone and Peterson, 1968, and Peterson, 1970, that conflict increases where low-income groups are organized). He says it is most likely to occur where there is a concentration of political power and leadership among the minority population experienced in organizing their own community. Kramer uses San Francisco as an example where such conditions exist and Oakland as an example where they do not.

Mogulof (1970a: 84-87) contends that the trend as of 1968 was from coalitional to adversary forms of participation. He acknowledges that CAPs were being downgraded by the decrease in discretionary funds, but argues that poverty legislation and OEO guidelines were increasingly requiring forms of participation which lead to adversary relationships and control rather than coalition. The examples he cites include: the 1967 requirement that representatives of the poor make up one-third of the policy board (as opposed to merely sitting on advisory groups); that area councils have at least 50% representation of the poor; and that CAAs contract with neighborhood groups for program operation.

However, comparative studies of CAPs present evidence which challenges the association of CAPs with adversary conflict. The Senate

Subcommittee studies (1967) show that only a few CAPs were characterized by confrontations; most stressed the provision of service. Advisory rather than adversary forms of participation were most typical. Kramer (1969: 236-237), too, finds that little militant confrontation occurs in CAP, and the Brandeis study (1969: 45) shows that only seven of the twenty cities studied had adversary patterns of conflict. The seven were the large cities with large black populations. A concluding chapter of the study takes up the charge made by Moynihan and others that CAP triggered violent conflict among minorities. It shows that even in the seven cities where adversary conflict prevailed, the causal role of community action varied; often community action, rather than initiating conflict, merely became enmeshed in preexisting conflicting forces. The following pattern is suggested:

> Where the mobilization of black protest had just begun the CAA became an important part of the development of an organizational base for protest. Where black protest had not emerged and where it had already become a fullfledged political movement, the CAA . . . had little or no significance for protest organization in the black community [Brandeis, 1969: 61].

The study stressed the importance of local forces shaping the types of conflict which resulted from participation in poverty.

The available evidence is not sufficient for an assessment of the prevalence of each type of conflict or the importance of the explanatory variables. It does, however, indicate the existence of complex variations and the danger of simplistic explanations. The evidence underscores the need for continued efforts to understand the nature of the conflicts and to integrate the data with existing knowledge of community conflict (see Coleman, 1957). The present studies of conflict in poverty programs are hampered by limited samples, short-time perspectives, and diverse uses of terms. The disagreement between Moynihan and the Brandeis study may be partly an artifact of differing frames of reference. The problem of cumulating the results of fragmented studies on types of conflict is obvious. This problem is magnified when one turns to a consideration of the impact of participation.

RESULTS OF PARTICIPATION–IMPACT ON PROGRAMS, AGENCIES, AND PARTICIPANTS

Evaluating the results of social intervention is always hazardous. Agreement is usually lacking on what actually happens. The conflicting testimony often leads the public to assume that deception is rampant; after all, why cannot the experts agree on the number of jobs for poor people which result from a given program? But even without the intent to deceive, observers do not report consistent findings, often because they use different indicators. A job may mean an initial matching between a client and an opening, or the appearance of the client for the first day of work, or his continued work for a specified length of time. And a poor person in one agency's eyes may not be a poor person in another's eyes.

If it is difficult to determine what happens, it is much more perplexing to discern why it happens. In social action, all the forces operating in a social system may be relevant to the outcome. It is impossible to clearly differentiate the effects of participation in poverty programs from the effects of other aspects of community action programs, and from other community forces. Kramer (1969: 239) puts it well:

> The ambiguity of the goals of resident participation; the difficulty of separating out the influence of other aspects of the CAP; the effects of other complex and often unknown variables; the relatively short duration of the program; as well as the elusiveness of reliable substantive evidence inevitably make any conclusions highly tentative.

Yet the existing studies do present some conclusions about the impact of participation in poverty which can serve as a stimulus for further work.

Anyone who expected that the innovation of involving the poor in poverty programs would result in a miraculous cure for social ills will surely be upset by the evidence. Virtually all analysts agree that the problems of poverty have not been significantly lessened in the years since the legislation was passed in 1964 (Kramer, 1969: 256; Senate

Subcommittee, 1967). Hard-core poverty has not been substantially reduced. Programs have been implemented, clients have been processed, slots have been filled, reports have been filed, but both participants and observers are at times overwhelmed by the realization that no big dents have been made: the problems remain.

Some immediately conclude that the poverty program and participation are failures, that the whole effort should be given up. They forget to ask whether this alternative might not lead to even worse results. They forget the relationship of poverty programs to both the national economic outlook and the levels of expectation among the poor and the affluent. The fact that the number of unreached poor remains high may be due to the extension or redefinition of poverty through population increase, economic decline, or rising expectations, rather than the failure of poverty programs to reach the poor. To stop trying to reach the poor who can be reached just because many have not been touched ensures that the problems will increase even more rapidly than before.

Thus, the need arises to determine what has been accomplished even though the accomplishments have been meager. What effect have the various types of participation had on the content of the poverty program, on the agencies involved in the program, and on the individuals and groups involved in the programs?

There is widespread agreement that participation by the poor has not had a substantial impact on the content of the poverty program or on the distribution of power-over-poverty policy. This generalization appears to hold true for each type of participation; indigenous employment, advice-giving, shared power, and social action. None of these has overcome the powerlessness of the poor (Kramer, 1969: 257, 261; Brandeis, 1969: 22; Peterson, 1970; Marshall, forthcoming: ch. 4). The Brandeis study (1969: 32), for example, finds that CAAs continue to "reflect the interests of established organizational and governmental leadership" and do not emphasize increasing the power of minorities. The data show that low-income residents had little impact on major CAA program decisions such as the structure and membership of CAAs, the choice of delegate agencies, the funding of programs, and the creation of neighborhood organizations (Brandeis, 1969: 50).

However, evidence of at least limited influence by the poor participants on poverty programs does exist. First, the poor are employed in agencies, and are sitting in groups as peers where they were not previously located, and are organizing to bring about changes in their communities (Brandeis, 1969: 54). These are new opportunities which create the possibility of influence on programs even though these influences have not yet been large enough to be easily measured. Second, the presence of the poor in these settings does bring new parties into the policy process; parties which formerly were merely the subject of social negotiations now are involved in the transactions (Mogulof, 1970b: 122). Access for the poor is increased even if power is not. A potential for more influence in the future may have been created by the changes currently seen in structure and process.

Realization of this potential has occurred in some special instances. Kramer (1969: chs. 7, 8) gives examples of the success of area councils in gaining broader powers than originally delegated to them and of the success of groups demanding greater representation on CAA boards and councils. The Brandeis study (1969: 51-55) finds that participation of the poor did have regular and continuing impact on CAA policy in programs characterized by adversary conflict. The poor on these boards had in impact on the selection of personnel, program, and budget. Similarly Mogulof (1970b: 80) describes one western CAA where the poverty representatives are united, determine the board priorities, and control matters internal to the agency which do not need outside resources or sanction. For example, they were able to hire a black director unpopular with the public agencies. Peterson (1970), too, finds an example of impact on programs. New York area councils, unlike those in Philadelphia or Chicago, had some influence over poverty resources such as staff, funds, and program content and were able to distribute funds to independent neighborhood groups rather than the traditional welfare agencies. This finding suggests the Brandeis conclusion associating impact with large cities of large populations does not adequately account for variations among large cities. More work is needed to determine whether the political variables considered by Kramer, and by Greenstone and Peterson, well help account for these variations.

In Los Angeles, the examples where poverty representatives succeed in having their position accepted are all cases where the public and private agencies are willing to let that position prevail. If they wanted to, these agencies could block the poverty representatives (Marshall, forthcoming: ch. 4). This veto capability must be looked for in all cases where it appears that poverty representatives are able to influence CAA policy. One must ask whether the poverty representatives are able to exercise influence over the opposition of the established agencies or whether the agencies are willing to allow the poor influence in certain narrowly circumscribed spheres which are not of major significance.

Many who argue that the poor are exerting at least some influence over policy admit to the latter case. For example, Mogulof (1970b: 81) comments on the tendency of established agencies to deemphasize the importance of CAAs. They send junior men as their representatives and dissociate themselves from the wrangling of the CAAs. The poverty representatives are permitted more and more control over less and less. Local government officials are often happy to be able to pass the buck to CAAs; one commented on how nice it is that when there are crises in poverty programs, the pickets go to the CAA offices rather than to city hall (see Marshall, forthcoming: ch. 4). The utility of CAAs as a scapegoat may help explain why, even after the 1967 amendments permitted local governments to designate the agency which would administer the CAP, almost all of them chose to designate the existing CAA (which had been recognized unilaterally by OEO) rather than to run the programs themselves (see Yankelovich, 1969).

The evidence, then, on the impact of participation on the content of poverty programs suggests only marginal influence. What impact has participation had on the agencies involved in the program? Evidence on the increasing control of poverty representatives over less, suggests that the picture of participation's impact on other agencies is not bright. This impression is supported by most available analyses.

Neither the presence of indigenous employees, nor the monitoring of delegate agencies by advisory and area councils and CAA boards, nor the pressuring of agencies by social action groups have resulted in major changes in the basic orientations of the agencies (Kramer, 1969: 240). Most CAA have emphasized providing services rather than changing established institutions (Senate Subcommittee, 1967, III: 798-799; IV: 913-914). These results suggest the assumption—that participation and pressure from constituents can change bureaucracies—needs reconsideration; the model may be effective in legislatures but it is not effective in administrative organizations (Kramer, 1969: 239).

Most of the changes which did take place in agencies were related to federal requirements rather than pressure from the poor. Thus, poverty money has stimulated traditional agencies to expand and decentralize services to the poor and hire nonprofessional aides (Senate Subcommittee, 1967, IV: 913-914; Kramer, 1969: 240-247; Marshall, forthcoming: ch. 5). Pressure from indigenous aides was less influential than hoped; it may have changed the attitudes of some agency personnel and modified the handling of specific cases, but little evidence of this influence is available (Kramer, 1969: 244). Explanations for the meager impact include the cooptation of the aides, their small numbers and lack of training, and high turnover. Similarly, the requirement that agency programs receiving poverty funds be approved by boards including representatives of the poor has not resulted in major changes in the practices of the delegate agencies. But delegate agencies are now forced to submit their programs to review and public criticism which may have significance in the future (Kramer, 1969: 246). The Brandeis study (1969: 52) did find that CAAs with adversary conflict had area councils which succeeded in changing the programs of agencies such as school boards, housing authorities, and city and county governments.

What accounts for the relatively small impact of participation on the content of policy and on the agencies involved? A variety of factors have been operating. First, the national ear-marking of funds serves to limit the ability of boards to make decisions. Second, the narrow definition of the CAA's role merely as overseer of poverty programs rather than as advocate for the poor on all matters of public policy has decreased the area in which CAAs were operating (Brandeis, 1969: 56). Third, the initial plans for the structure and membership of CAAs were made by the more organized and powerful local interests, such as the private social welfare agencies and the public agencies; at first, representatives of the poor, civil rights groups, and welfare rights groups were not involved (Brandeis, 1969: 27). The CAAs were structured to protect the interests of the established agencies,[13] and it is not surprising that the poor participants were not able to counteract the structural bias. If the prevailing interests set up the kind of organization they want, and the rules, and the selection procedure for participants, they do not have to directly manipulate the decisions made by the participants. The results are largely predetermined.

It is against this background that all comments on the inadequacies of the poverty participants must be viewed. The situation in which they were placed was one which minimized their influence. Thus it is not surprising that they had little impact. In an impossible situation, significant influence may not be possible. Yet the impact of participation was also hindered by the behavior of the participants. They lacked experience in lobbying effectively within agencies, area councils, and boards. Their impact was further hampered by internal strife among poverty representatives and by their interest in meeting their particularistic needs rather than in pushing for broad social changes (Kramer, 1969: 236-237; Marshall, forthcoming: chs. 5,6; Peterson, 1970).

When one turns from the impact of participation on policy and agencies to the impact on participants, the evidence is not nearly as negative. There is growing awareness that poverty programs may have their most significant effect on the individuals and groups involved.

Mogulof (1970a: 19) refers to the Pygmalion stories about community participants in poverty programs. Kramer (1969: 247-256, 267) talks about the sociotherapeutic effects of participation. Poor people get jobs or positions in groups which lead to jobs. They also exhibit increases in political education, information about how the system works, and in organizational skills and feelings of political efficacy (Marshall, forthcoming: chs. 5, 6; Mogulof, 1970a: 19; Kramer, 1969: 249; Zurcher, 1968). Thus, participation serves as a training ground for indigenous leaders (Freedman, 1969: 6; Wayman, 1967-1969: 23). Interviews with poverty representatives in Los Angeles document the changes in participation and self-image which occur; people who before joining poverty organizations were not active in their community, subsequently increase their participation in community groups and start viewing themselves as leaders in their communities. They have been activated, not deactivated (Marshall, forthcoming: chs. 5, 6).

Unfortunately, most of the evidence on individual socialization is impressionistic. More systematic attention must be given to participation as a means of socialization. One example of the need for more work concerns the issue of the relationship of participation-to-affect for the system. Some observers feel that participation reduces hostility and alienation (Mogulof, 1970a: 19; Zurcher, 1968). Others have found an increase in hostility when poverty programs and influence of the poor fail to meet rising expectations (O'Donnell and Chilman, 1969: 4; Kramer, 1969: 268; Marshall, forthcoming: ch. 7). The latter finding calls into question the view that participation necessarily serves to "cool off" trouble, to neutralize the threat of militance. Both those who want to minimize conflict and those who want to increase it need more information on the conditions under which participation is linked with positive and negative evaluations of the system.

In any case, participation has had an impact on organizations within poverty communities. The war on poverty gave minority organizations a new focus; the CAAs were new power centers with missions which forced them to relate more directly to minority organizations than other power centers did. Kramer (1969: 248-250) describes how minority organizations fought to win favorable positions in this new resource allocation system. The struggle stimulated the development of community groups and increased their experience in political action. And even though the poor did not gain control over poverty policy, blacks, browns, and their organizations did have a larger say in poverty programs than in previous programs. Some CAAs became the special preserves of blacks and browns who claimed to be able to speak for the poor better than other interests could, even though they might not be poor themselves (Kramer, 1969: 248). Thus blacks, browns, and their organizations achieved increased influence over the patronage and prestige available through CAAs. Some feel this activation of minority groups does not compensate for the previously mentioned downgrading of CAPs by established interests; others say the insignificance of the beachhead does not dim the importance of having a beachhead (Kramer, 1969: 250).

Evidence of the activation of new leaders and groups (see also Mogulof, 1970b: 165-171) is often met with the charge that the activation is phoney because it has merely bought off potential opponents of the system. Examples are given of the continued powerlessness of the poor in agencies and on poverty organizations and of the moderate style of the poor, blacks, and browns involved. The assumption is that all the moderates are militants who have been cooled down. The limited evidence we have suggests many of the individuals and organizations involved had never been militant. Instead they have reached a higher level of activity than they ever exhibited in the past. Of course, no evidence can tell us what they might have become if the poverty war had not attracted their energy. But in view of the age and style of these participants, it is hard to imagine them in violently militant groups as long as other alternatives are still open to them. (If society becomes so polarized that the only alternatives are supporting the white-dominated status quo or extremist groups, the situation would change.)

Another charge against the significance of participation comes from the other side. Participation is called phoney because the poor, blacks, and browns do not participate effectively. They are described as ignorant of ways of solving their problems, not attending meetings, talking endlessly and rhetorically, bickering, self-interested, and manipulated by special interests. There is no answer to these charges; after all, they have been used as excuses throughout history for denigrating government by the people. Yet the same criticisms apply to the quality of middle-class participation in the political process. Who cannot attest to the dreariness of city council or welfare agency meetings? Is rhetoric, wordiness, bickering, and self-interest absent? Is attendance never a problem? There certainly is no reason to expect participation by the poor to be "exercised better or worse than the norm established by the traditional holders of power" (Freedman, 1969: 6). Some proponents of participation romanticize the process, but this is no reason to judge the results by their unrealistic standards.

In summary, evidence on the results of participation confirms neither the hopes of its proponents nor the fears of its opponents. Participation has not significantly reduced poverty or changed established institutions; nor has it resulted in widespread confrontations between the poor and the established forces in communities. Yet participation by new groups and individuals has initiated some changes. The long-run effects of these small-scale changes may be very important. It is too early to tell, but both the proponents and opponents should watch for them carefully because they may still vindicate some of the predictions.

IMPLICATIONS OF THE EVIDENCE FOR THE DISPUTE

The opponents of participation are not quieted by the evidence. Their dislike of participation involves preferences which lead them to evaluate the data negatively in light of other considerations. However, their opposition to participation cannot be taken as proof that they do not care about the plight of the poor. They feel their proposals will be more effective in helping the poor than participation per se.

Moynihan, for example, says participation is not as important a strategy for social change as such economic measures as fixed full employment or income maintenance. He considers participation to be a result of economic change, not a cause. This fits in with his view that because opportunities for desirable social change are limited, actions must be moderate and should avoid potential conflict which could destroy society's gradual progression. There are many difficulties raised by the preference for economic measures. Can an enlightened elite be relied upon to institute economic changes which will benefit the poor, or is pressure needed from the poor? Are economic measures such as fixed full employment feasible without radical changes in the economic system? The difficulty of achieving full employment in our current system is underscored at times of increasing unemployment and economic crisis—what good is job training when the number of jobs is declining? Even if economic measures were possible, would they be effective since higher income does not necessarily decrease hostility if relative economic position is not improved and higher aspirations are not met? (Moynihan, 1969a: 193; Nie et al., 1969)[14]

Other opponents see participation as a threat to cherished goals. Lowi sees it as a perpetuation of interest group liberalism which strengthens conservative forces working against the realization of the common good. Merely creating new interest groups will not reach the goal of denying special interests and achieving an enlightened social order. His solution is not to increase participation in administration but to decrease it, to isolate administration from group pressures and to limit participatory democracy to legislatures. Lowi (1967: 20-24) argues for strengthening the profession of public administration and the independence of the executive. Like Moynihan, he places great faith in the ability of elites to perceive and foster the common good. Evidence that the caliber of participation by the poor is no worse than participation by the middle class does not alter his negative evaluation of both forms of participation.

Other opponents see participation in CAA as a threat to local government because it increases fragmentation (Moynihan, 1969b:

10-12). Evidence that local governments have not been hurt by CAA does not ease their suspicion this possibility still exists. The Advisory Commission on Intergovernmental Relations (1966), in contrast, argues that CAA offers a valuable opportunity for local governments to become more responsive. By-passing local government may increase fragmentation in the short run but in the long run, if it strengthens citizen activity, local governments will be improved. The Commission rejects the assumption that every new center of power represents a diminution of local government power, and, instead, assumes that ineffectively wielded power in the long run will undermine the power of local government. Therefore the highest priority is to improve performance.

Other opponents of participation in poverty equate it with radical decentralization or community control[15] which they see as a threat to the creation of metropolitan governments and central standards. Rustin (1970), for example, opposes community control as just another form of states' rights which allows groups to avoid compliance with standards he values more than the competing value of self-determination (see also Riker's [1964] argument against federalism on similar grounds). Others feel community control is a step backward when the real need is to create larger governmental units to end the duplication and gaps caused by multiple governmental units in metropolitan areas (for review of these arguments, see Altshuler, 1970: 45-53; also Kristol, 1968). Evidence on the results of participation in poverty does indicate a tendency for poverty groups to demand increasing control over poverty programs, so opponents of community control have grounds on which to fear participation in poverty.

While the evidence cannot change the preferences of opponents of participation, it should be given serious consideration by the proponents. Those who assumed participation was a quick way to achieve specific social change goals need to reassess that assumption. The changes will not be quick and may not be in the direction of the desired goals. For example, the poor may be less interested in changing the system than in acquiring upward mobility into the system. And pressure for change from the poor may not be effective in the short run since traditional institutions with firm power bases will not abdicate at the first cries from the newest claimants for power. Those who saw participation as a way of changing individuals can find more comfort in the evidence, but easy assumptions—that participation decreases alienation—are challenged. Activated individuals may in the long run increase the level of conflict in the system rather than decrease it.

My own reassessment of participation of the poor still gives it high priority. Elitist views of change do not seem satisfactory; previous reforms touted in the name of the common good have served to benefit middle-class interests (see citations in Lineberry and Fowler, 1967). For both pragmatic and ethical reasons, we must still favor the development of "compensatory power" for those disadvantaged in the political system (Freedman, 1969: 5). The continued development of a political system depends on its ability to adapt to changing demands. And demands are being made by the poor, blacks, and browns, and there is no way to make them disappear. They must either be satisfied or repressed.

There is still merit in the position that a certain amount of apathy is functional for a system, that not everyone needs to participate. But can anyone argue it is functional for large groups with strong claims to feel no channels are open to them for meaningful participation? To feel they are discriminated against on every issue and have no chance of being heard? Surely the democratic idea ought to mean people must feel they can participate if they want to. When this feeling is absent among many over a long period of time, the democratic system is certainly in trouble. Attempts must be made to open channels of power to nonparticipants so they will be able to change the system enough to feel they can be heard. Of course there will be dangers and waste. Conflict may increase, but open and flexible systems can accommodate conflict productively (see Coser, 1956: ch. 9). Conflict among people with a stake in the system can be therapeutic for the system (Altshuler, 1970: 195-197). Thus the advantages may outweigh the disadvantages. Participation may provide the poor, blacks, and browns with valuable training for future more effective political action as well as a stake in improving rather than destroying the political system (Kramer, 1969: 249).

PARTICIPATION'S CONSEQUENCES FOR REFORM

Participation has been viewed here as a new stage in the reform cycle, one which stresses representativeness. What implications does this stage have for other major urban reform proposals?

A belief in the importance and value of participation of the poor in poverty programs does not necessarily entail the view that participation alone will solve poverty or racial problems. The evidence discussed above suggests the success of participation will depend on the implementation of reforms in employment, housing, welfare, and health care. But the evidence also suggests that the success of other reforms is linked to increasing participation. Given the level of hostility which exists, no delivery of services may be possible unless there is participation (Altshuler, 1970: 221). Thus multifaceted approaches are called for to meet the diversity and complexity hidden behind the simple label of poverty (Kramer, 1969: 140).

Participation also has implications for the creation of metropolitan governments. The evidence suggests plans to centralize power must incorporate provisions for decentralization to ensure that the voices of minorities will be heard. It is possible to strengthen communities without isolating them from a larger metropolitan system (Altshuler, 1970: 50-53, for description of program for combining both; see also Mogulof, 1970a: 108-114; Freedman, 1969: 6). But this approach will require an appreciation of the importance of a process which has often been lacking when the efficient provision of services is stressed.

The study of participation also guards against the false belief that it is the ultimate reform. We can see some of the problems which will result, such as a growth in disparities, favoritism, protection of values repugnant to some, and paralysis of action due to bickering. These problems will give rise to demands for the creation of uniform standards and the removal of politics from urban programs so as to overcome the limitations of the participation stage in the reform cycle (Kaufman, 1969). But an awareness that the participation trend will lead to new problems which will require future reforms should not recommend inaction.

The issue of participation by the poor, blacks, and browns in the political system has been raised in a most compelling manner. All signs indicate this major strategy for reform is having an important influence on racial and urban problems in the United States, and should not be ignored. Whatever, one's personal assessment of participation, the need to gather more evidence on the nature of that influence is clear. A great deal more work must be done by social scientists to fill in the gaps in our knowledge which become painfully apparent in any review of the existing evidence. Only exploratory research has been done on the patterns of implementation and impact and on the variables which explain these patterns. The studies reviewed here suggest that blacks, browns, and the poor have been activated by participation in the proverty program. We need to test this finding and specify the actual results of activation. If this activation proves to institutionalize the involvement of blacks, browns, and the poor in the bargaining over the shape of our political system, then the importance of the strategy of participation will have been matched by the importance of the results achieved.

NOTES

1. See exchange between Walker and Dahl on the elitist theory of democracy in American Political Science Review, June 1966, and between Wilson and Banfield and Wolfinger and Field in American Political Science Review, December 1966.

2. Coser and Dahrendorf challenge the view that conflict is disruptive.

3. 78 Stat. 508, sec. 202, 3.

4. Greenstone and Peterson (1968) consider the nation's four largest cities: New York, Chicago, Los Angeles, and Philadelphia. The Brandeis study (1969) looks at twenty cities with populations between 50,000 and 1,000,000; the Senate Subcommittee (1967) includes thirty-five case studies of a variety of types of CAPs; and Kramer (1969) looks at five CAAs in the San Francisco area.

5. All the consultant reports of Senate Subcommittee (1967) consider patterns of resident participation in CAAs.

6. 81 Stat. 691, sec. 201, a, 4.

7. 81 Stat. 693, sec. 211(c) and CAP (1968) memo 81 part e, 3.

8. Senate Subcommittee shows that most CAAs had area councils by 1967; Kramer calls them target area organizations. Peterson calls them neighborhood councils.

9. Los Angeles is the largest city without area councils. Marshall describes the rationale for the absence of councils and of the pressure from OEO to create councils.

10. OEO guidelines in 1968 required for the first time that at least fifty percent must be representatives of the poor but not necessarily poor.

11. Selection for advisory councils and neighborhood-based organizations is more informal—usually by attendance at meetings or by invitations to join from friends who are already active.

12. Brandeis divides this category into two, calling one limited participation, the other advisory.

13. For description of how existing local forces capture urban renewal programs, see Greer (1966).

14. Nie shows that increase in organizational participation is more important than social status as a stimulus to political participation.

15. Lipsky (1969: 102) distinguishes between administrative decentralization and radical decentralization; the latter he defines as "placing authority to the maximum extent possible in the hands of those who are affected by institutions or programs." This latter definition is the one used here.

REFERENCES

Advisory Commission on Intergovernmental Relations (1966) Intergovernmental Relations in the Poverty Program: A Commission Report. Washington, D.C.: Government Printing Office.

ALTSHULER, A. (1970) Community Control: The Black Demand for Participation in Large American Cities. New York: Pegasus.

AAA, American Arbitration Association (no date) Study of CAP Elections. (mimeo)

BACKRACH, P. (1967) The Theory of Democratic Elitism: A Critique. Boston: Little, Brown.

BELL, D. and V. HELD (1969) "The community revolution." Public Interest (Summer).

BIBBY, J. and R. DAVIDSON (1967) On Capitol Hill. New York: Holt, Rinehart & Winston.

BOWEN, D. and L. MASOTTI (1968) "Spokesmen for the poor: an analysis of Cleveland's poverty board candidates." Urban Affairs Q. 4 (September).

Brandeis University, Florence Heller Graduate School for Advanced Studies in Social Welfare (1969) Community Representation in Community Action Programs. Report no. 5 (March). (OEO grant no. CG68-9499 A/2.) (mimeo).

--- (1968) Board Members of Community Action Agencies: An Analysis of Interview Data. Report no. 2 (August).

COLEMAN, J. S. (1957) Community Conflict. New York: Free Press.

CAP, Community Action Program (1968) Memo 81 (February 15).

COSER, L. (1956) The Functions of Social Conflict. New York: Free Press.

DAHRENDORF, R. (1959) Class and Class Conflict in Industrial Society. Stanford: Stanford Univ. Press.

DONOVAN, J. (1967) The Politics of Poverty. New York: Pegasus.

FREEDMAN, E. (1969) "Citizen participation: a training tool for leadership." Public Management 51, 7 (July).

GREENSTONE, D. and P. PETERSON (1968) "Reformers, machines, and the war on poverty," in J. Wilson (ed.) City Politics and Public Policy. New York: John Wiley.

GREER, S. (1966) Urban Renewal and American Cities: The Dilemma of Democratic Intervention. Indianapolis: Bobbs-Merrill.

HADDAD, W. (1965) "Mr. Shriver and the savage politics of poverty," Harper's Magazine 231 (December).

HUGHES, E. (1958) Men and Their Work. Glencoe, Ill.: Free Press.

KAUFMAN, H. (1969) "Administrative decentralization and political power." Public Administration Rev. (January/February).

KRAMER, R. (1969) Participation of the Poor: Comparative Community Case Studies in the War on Poverty. Englewood Cliffs, N.J.: Prentice-Hall.

KRAVITZ, S. (1968) "The community action program in perspective," in W. Bloomberg and H. Schmandt (eds.) Power Poverty and Urban Policy. Beverly Hills, Calif.: Sage.

KRISTOL, I. (1968) "Decentralization for what?" Public Interest (Spring).

KROSNEY, H. (1966) Beyond Welfare: Poverty in the Supercity. New York: Holt, Rinehart & Winston.

LANE, R. (1959) Political Life: Why and How People Get Involved in Politics. New York: Free Press.

LEE, E. (1960) The Politics of Nonpartisanship: A Study of California City Elections. Berkeley: Univ. of California Press.

LEVITAN, S. (1967) "Planning the anti-poverty strategy." Poverty and Human Resources Abstracts 2 (January-February).

LINEBERRY, R. and E. FOWLER (1967) "Reformism and public policies in American cities."Amer. Pol. Sci. Rev. (September).

LIPSET, S. (1960) Political Man. New York: Doubleday.

LIPSKY, M. (1969) "Radical decentralization: a response to American planning dilemmas." Institute for Research on Poverty, reprint 28.

LITT, E. (1970) Ethnic Politics in America. Glenview, Ill.: Scott, Foresman.

Los Angeles Times (1968) "Poverty war director assesses four and one-half year fight," by Jack Jones. (August 23)

LOWI, T. (1967) "The public philosophy: interest group liberalism." Amer. Pol. Sci. Rev. LXI (March).

MARRIS, P. and M. REIN (1967) Dilemmas of Social Reform: Poverty and Community Action in the United States. New York: Atherton Press.

MARSHALL, D. (forthcoming) The Politics of Participation in Poverty: A Case Study of the Board of the Economic and Youth Opportunities Agency of Greater Los Angeles. Los Angeles: Univ. of California Press.

MATHEWS, V. (1968) "Citizen participation: an analytical study of the literature." Washington, D.C.: U.S. Department of Justice, Community Relations Service (June).

MILBRATH, L. (1965) Political Participation: How and Why Do People Get Involved in Politics. Chicago: Rand McNally.

MILLER, S. M. and M. REIN (1969) "Participation, poverty and administration." Public Administration Rev. (January-February).

MOGULOF, M. B. (1970a) Citizen Participation: A Review and Commentary on Federal Policies and Practices. Washington, D.C.: The Urban Institute (January). (mimeo)

--- (1970b) Citizen Participation: The Local Perspective. Washington, D.C.: The Urban Institute (March). (mimeo)

--- (1967) "A developmental approach to the community action program idea." Social Work XII (April).

MOYNIHAN, D. P. (1969a) Maximum Feasible Misunderstanding. New York: Free Press.

--- (1969b) "Toward a national urban policy." Public Interest (Fall).

--- (1966) "What is community action?" Public Interest (Fall).

NIE, N. et al. (1969) "Social structure and political participation: developmental relationships." Amer. Pol. Sci. Rev. (September).

O'DONNELL, E. and C. CHILMAN (1969) "Poor people on public welfare boards and committees–participation in policy making." Welfare in Review VII, 3 (May-June).

OEO, Office of Economic Opportunity (1965) Community Action Workbook III, A.7. Washington, D.C.

PEARL, A. and F. RIESSMAN (1965) New Careers for the Poor: The Non-professional in Human Service. New York: Free Press.

PETERSON, P. (1970) "Forms of representation: participation of the poor in the community action program." Amer. Pol. Sci. Rev. (June).

RIKER, W. (1964) Federalism: Origin, Operation, and Significance. Boston: Little, Brown.

RUBIN, L. (1967) "Maximum feasible participation: the origins, implications, and present status." Poverty and Human Resources Abstract 2 (November-December).

RUSTIN, B. (1970) "The failure of black separatism." Harper's (January).

Senate Subcommittee (1967) on Employment, Manpower, and Poverty of the Committee on Labor and Public Welfare. Examination of the War on Poverty: Staff and Consultant Reports, 90th Congress, 1st session. Washington, D.C.: Government Printing Office.

SUNDQUIST, J. (1968) Politics and Policy. Washington, D.C.: Brookings Institute.

WALKER, J. (1966) "A critique of the elitist theory of democracy." Amer. Pol. Sci. Rev. XL (June).

WAYMAN, J. C. et al. (1967-1969) The War on Poverty in San Diego. La Jolla, Calif.: Western Behavioral Sciences Institute.

WHITE, E. (no date) Articulateness, Political Mobility, and Conservatism: An Analysis of the Philadelphia Anti-Poverty Election. (mimeo)

YANKELOVICH, D., Inc. (1969) Study of the Effects of Sections 210 and 211 of the 1967 Amendments to the Economic Opportunity Amendments. Vol. I (February). (contract with OEO)

ZURCHER, L. (1968) Social-Psychological Changes Among OEO Indigenous Leaders As a Result of Maximum Feasible Participation. Presented to the Society for the Study of Social Problems, American Sociological Association Meetings, Boston, Mass., August 25.

16

New Roles for the Legal Profession

THELTON E. HENDERSON

□ FOR A GENERATION, interlocking problems of race and poverty in America have steadily worsened, despite a score of laws and a hundred governmental programs and good intentions. Today, despite the current popularity of concern about ecology, the problems of race and poverty in our core cities have reached emergency proportions and constitute our number one domestic crisis. We seem paralyzed by its dimensions.

THE PROBLEM

As the great migration of rural Blacks trickled and then flooded into our Northern cities following World War II, what was once dismissed as a Southern problem became a national problem. Lack of access to decent housing, lack of decent jobs, and inferior educational facilities and opportunities, were the realities facing rural Blacks who had come North looking for a better way of life.

Northerners, when faced directly with the presence in large numbers of the Southern refugees, were no more tolerant or understanding than their Southern racist counterparts. After access to hot dog stands, toilets, and drinking fountains had been grudgingly conceded to the Blacks, nothing remained to be conceded except the important things: jobs, education, housing. As Blacks began to apply pressure for reform in these important areas, and as the "white power structure"[1] became more and more resistant to these pressures, the hurried flight to the suburbs began, leaving our urban centers to people who absorbed far more tax money than they paid. The reaction was predictable. Mounting frustration and bitterness–a feeling of hopelessness–exploded in Watts, Harlem, Newark, Detroit, and countless other places in the summer of 1967, bringing unprecedented racial disorders to American cities, and with them shock, fear, and bewilderment to the nation.

At about this time the so-called crisis in "law and order" began. The things this phrase should mean–law for all, order for all–like the full benefits of American life, remain remote from the millions of slum inhabitants.

Whatever may be meant by others who use the emotion-packed term "law and order," it shall here refer to the inability of our legal system effectively to handle the plethora of problems being hurled at it by the discontented elements of our society. This crisis is largely an expression of the loss of faith of these discontented elements in getting meaningful results by working within the system (see Report of the National Advisory Commission on Civil Disorders, 1968: 5).

Richard C. Wade (1969), commenting on urban violence, has explained this loss of faith as follows:

> The urban slum has never been a very pleasant place, and it was tolerable only if the residents, or most of them, thought there was a way out. To American immigrants generally the ghetto was a temporary stage in their incorporation into American society. . . . Soon they were dispersed around the metropolitan area, in the suburbs as well as the pleasant residential city wards. . . .
>
> It was expected that Negroes would undergo the same process when they came to the city. . . .But the ghetto proved to be not temporary. Instead of colored residents disbursing (sic) in the second generation, the ghettos simply expanded. Block by block it oozed

> out into the nearby white neighborhoods. Far from breaking up, the ghetto grew. . . .What had been temporary for other groups seemed permanent to Negroes. . . .
>
> As a result, they comprise now a growingly alienated and embittered group. The middle-class Blacks are now beginning to turn their attention to organizing among the poor in the worst parts of the ghetto. Their children make up the cadres of Black militants in the colleges. And when the riots come, they tolerate the activity even though they usually do not themselves participate. In short, the fact of the ghetto forces them to identify with race, and not class. When the riots break, they feel a bond with the rioters, not white society. This has not been true of the emerging middle class of an immigrant group before [Wade, 1969: 24-26].

Edgar and Jean Cahn (1964: 1317) have described the law and order crisis in terms of the conflict now faced by the "rule of law." The Cahns (1970: 1005) traced the origin of the conflict to at least three major sources which have caused the breakdown of our traditional legal institutions:

1. *The rights explosion.* The creation in recent years of a vast and still growing array of legally vested rights which are enshrined in statutes and court decisions which are not honored in practice. The developing case law regarding the rights of juveniles, tenants in public housing, persons accused of crime, welfare recipients, and minority group members, have challenged the capacity of the legal system to fulfill promises made by creating "rights."

2. *The grievance explosion.* Grievances which do not attain the status of legal rights, but nonetheless in the minds of those perceiving the grievances, urgently require redress. An obvious example would be the policeman or perhaps the schoolteacher who uses insulting language when addressing Blacks. A less obvious example is the demand of Blacks to be addressed by titles of courtesy in appropriate situations.[2]

3. *The mounting challenges to official discretion.* The increasing scrutiny and challenges to previously unrevealable, discretionary actions by officials who administer major government grant programs; e.g., in the field of poverty, welfare, housing, transportation, and urban renewal, all vitally important to the daily existence of the poor.

As if to underscore the Cahns' view that our traditional adjudicative machinery is about to be overwhelmed, Chief Justice Warren E. Burger has recently stated that unless changes in our judicial system are made "very, very soon," the United States Supreme Court's work "either will break down or so deteriorate in quality that the Court will no longer be fulfilling its historic role" (San Francisco Chronicle, 1970).

Clearly, then, dramatic, imaginative, and perhaps drastic steps must be taken by the legal profession.

Many who read this article will argue that the nature of the American system is such that it is not particularly suited to rapid or short-term solutions to problems such as racial conflict in America. The American political and legal processes, it is said, are understandably slow. The pluralistic nature of the society makes negotiating, bargaining, and compromising, the primary means of pursuing one's goals. When talking about race relations and the current crisis, however, such a protracted and tedious process is a luxury the society can no longer afford. We must rethink our views about the legitimate ways to raise issues and some appropriate ways to go about resolving them.

The Report of the National Advisory Commission on Civil Disorders (1968), the Kerner Commission, shocked many Americans, not so much by the recitation of facts and statistics on the imperfections in the society, as by daring to cite "white racism" as a major barrier to political and social change. The commission predicted that the United States was moving toward two societies, one white and one Black. It would have been more accurate to say that we are now divided into those two societies, and to point out that we must move from that sad position or face increasingly violent lives in which only some can find liberty, justice, and order; i.e. law and order.

Despite such dire predictions and warnings, complacent whites, an obdurate Congress, and a passive Bar find ways to avoid and evade the problem, leaving those most concerned with the problem to wonder why history's most affluent, technologically advanced society cannot act to solve the vices of our urban centers. The answers would seem to be a meld of ignorance and indifference, bigotry and callousness, escapism and sincere confusion. But the inescapable truth is that so far America has not wanted to move. On that point, there is, indeed, an American consensus—spelled out rather clearly in the way a democratic society allocates its resources. America spends $75 billion for defense but only $7 billion on welfare for the poor, $17.4 billion for tobacco and liquor but only $1.6 billion for the war on poverty, $3.2 billion for cosmetics and toiletries, but only $400 million a year for the training of unemployed adults (Newsweek, 1967).

TRADITIONAL ROLE OF THE LEGAL PROFESSION

It is not fortuitous that lawyers have played an important part both in constitutional history and in the actual workings of our democracy. Lawyers are peculiarly equipped, by training and experience, to take the lead in the vigorous and frank discussions of society's problems. They have long functioned as architects as well as artisans of social reform—redesigning, reforming, and creating not only legal institutions, but social, economic, and political institutions as well.

The lawyer's most significant asset, however, is the unique advocacy orientation of his profession—one to which our legal system and the Canons of Legal Ethics commit him. Other professionals such as social workers and educators are institutionally given the role of mediating between their employers and their clients. The lawyer need not be apologetic for being a partisan, for identifying. That is his function (Cahn and Cahn, 1964: 1317).

The urban crisis is significantly legal in nature, not in the vague sense that we are dealing with a batch of laws or with the need for Congress to pass new forms of remedial legislation. It is legal in the sense of de Tocqueville's observation (1969) that "in a modern democracy social problems become translated into legal problems—if the democracy coheres."

It is legal in nature because it is intimately connected with such legal concerns as municipal taxation, metropolitan finance, and interlocal city contracts. Ghetto education today is fairly engulfed in such controversies as the legality of de facto segregation, the constitutionality of unequal state aid vis-à-vis city and suburban school districts; and the federal enforceability of metropolitan multidistrict school systems. The whole question of providing decent low-income housing is run through with legal considerations ranging from the contract authority of local public housing units to act independently of federal parent agencies, through the intricate legalities of mortgage financing, Federal Home Administration (FHA) insurance, and "turnkey" construction, to the ancient legal doctrines of real property law and landlord-tenant relations.

THE FAILURE OF THE LEGAL PROFESSION

Despite the traditional role of the lawyer in legal reform, lawyers have been strangely negligent in handling the affairs of those least able to pay the price for such representation.[3] Now, in the face of the urban crisis, the legal profession has been found ill-prepared and disorganized, with no particular plan of attack.

In initially addressing itself to the urban crisis, the legal profession did exactly as we would expect practicing attorneys to do—focus on serving the needs of the individual client. Local Bar Associations vowed to do more for the poor, and their individual members determined to take cases involving poor people who normally could not afford their services. Primarily, however, the profession relied upon Legal Aid bureaus which had been set up locally.[4] Among the proudest of the Canons of Ethics is the one that obliges lawyers to accept clients on the merits of their cases, without regard to their ability to pay the lawyer's fee. Psychologically, the Legal Aid bureaus relieved lawyers of their guilt at failing to live up to this code. Practically, the bureaus gave lawyers a place where they could refer people whom they could not simply kick out the door. In reality, the bureaus were woefully inadequate (see Caplovitz, 1967).

In taking this approach, the profession has failed to devise new techniques badly needed for what is largely a new situation; and the profession has failed to summon forth old techniques which have proved useful in meeting other domestic crises in the past. Some of the necessary jobs are not very different from what lawyers have been doing all along for government, for business, for those who can pay them well. These jobs include lobbying, organizing constituents as was done with labor in earlier years, negotiating, mediating, and generally radicalizing their approach to the problem and then participating meaningfully in its solution.

To date the major direct response of the profession to the urban crisis has been through the Legal Services Program (LSP), administered by the Office of Economic Opportunity (OEO). The Legal Services Program was incorporated into a package submitted by President Lyndon B. Johnson as "1966 Amendments to the Economic Opportunity Act of 1964."[5] The bill also encompassed youth employment, the Job Corps, Headstart, health centers, adult education, rural loans, small business loans, and VISTA for a total administration request of $1.75 billion, legal services to get $25 million. The strategy of the administration appeared to be to tie all the War on Poverty programs together so that the controversial projects would benefit from those which were noncontroversial.

LSP was set up with the underlying assumption that enforcement of legal rights through litigation could significantly improve the situation of poor people (Bamberger, 1966). Thus began a series of Legal Aid offices across the country, staffed by young but dedicated attorneys. The feeling of being in the vanguard of impending institutional change accounted for the sense of excitement and crusade expressed by so many of the early participants in the program. Soon they were inundated by divorces, wage attachments, automobile repossessions, and the like.[6] After five years of coping with staggering and demoralizing caseloads,[7] LSP attorneys have come to suspect that a cruel joke has been played upon them—that someone in Washington, D.C., knew all along that a budget of $25 million, funding 551 offices and manned by 1,157 lawyers (Edson, 1966) could not possibly begin to solve the legal problems of the nation's poor in the nation's core cities. Indeed, Carol Silver, Director of the Berkeley, California, Neighborhood Legal Services Program, has written that "it is physically impossible for legal services, legal aid and other programs established to give civil, legal assistance to the poor to serve all, or even a majority, of those unable otherwise to afford legal representation" (Silver, 1968: 217). Miss Silver has further speculated that the need of the poor for civil, legal assistance as we now know it, surpasses the capacity of all the lawyers in the United States combined to render such assistance (Silver, 1968). Most legal services attorneys with whom I have talked readily agree with this opinion.

My own experience as a legal services attorney bears out this notion. Upon opening the East Bayshore Neighborhood Legal Center in East Palo Alto, California, October 1966, I imagined that my experience would be roughly the same as during my three years in private practice, except that my clients would be poor rather than middle class. Joined by one other attorney, recently graduated from the Stanford Law School, and ten law school students, I prepared to solve the legal problems of the poor in East Palo Alto. During the first week, my assistant and I saw seventy-four clients, and the waiting list had already grown to four days. It became immediately obvious that we could not possibly see everyone in the community who needed legal services and still find the time to research our cases, go to court, and otherwise do the things that attorneys are supposed to do for their clients. Perhaps the reader will better understand the dilemma when it is pointed out that half the practitioners in New York City handle no more than fifty cases a year, and only two percent take more than five hundred cases a year (Wald, 1965: 50).

NEW GOALS AND DIRECTIONS

The foregoing is fairly typical of the early experiences of the legal services attorneys. With a few notable exceptions, such as the California Rural Legal Assistance office,[8] virtually all of the legal services offices decided to adopt this "service function"—the provision of legal services to all indigent applicants on an individual basis, without regard for broader reform.

It was sometime later that LSP officials in Washington and LSP attorneys fully recognized that if the Legal Services Programs were to restrict themselves to the service function, their value as part of the War on Poverty would be slight and they would be making only a limited contribution toward the goal of equal justice; that to focus on the individual especially in view of the manpower and budgetary problems, is to largely ignore the greater problems now strangling our cities; that to file bankruptcy petitions for poor people does not solve the economic conditions that create the need for bankruptcy.[9]

The adoption of "non-service functions," primarily law reform, aimed at improving the legal position of all the poor, is a step in the right direction. In order to restore the overall integrity of the adversary system of law, OEO-funded Legal Services Programs are now required to list as one of their goals the reform of substantive law in the interest of the poor (Office of Economic Opportunity, 1964). The law reform device which is most often used, and which most easily arises in a Legal Aid office, is the test case. Despite its popularity, however, the test case has some disadvantages. It is a long process and always involves the risk of an indecisive or adverse result. Additionally, after the test case has been won, the fact that a principle of law has been established does not automatically assure that people's conduct will change. The neighborhood law office must continue to seek out cases in which the same issue is raised in order to keep the pressure on until the new rule becomes established in practice. Finally, poor clients neither know nor care if their cases are convenient vehicles for establishing points of law.

Clearly, then, in addition to greater emphasis upon such things as law reform, we need to develop new kinds of legal rights in situations that are not now perceived as involving legal issues. The legal profession must look beyond its court-oriented functions and imaginatively and aggressively pursue the rights of the disadvantaged in other arenas where they, the lawyers, can use their peculiar skills to advantage.

The following suggestions are not to be viewed as limiting the possibilities for legal action. It is far easier to express dissatisfaction than to accomplish solutions in a society encouraging freedom of expression and which allows accommodation of divergent views. There are many levels of commitment, many levels of frustration, and the individual attorney must be guided in most instances by his own sense of urgency, as well as by his circumstances—an attorney who is also a state legislator may be most effective in pushing for an effective Ombudsman bill. A young, radically oriented law school graduate may most easily become a part of a "social change" legal office.

SOCIAL CHANGE LAW OFFICES

While certainly not furnishing the complete answer to the problem of our urban crisis, an important new role for the legal profession is suggested by Ralph Nader's approach to dealing with legal problems—an independently funded office of professionals, including lawyers and persons from other disciplines. These practitioners, who are outside government and corporate law practice, have as their prime goal the promotion of significant social change. The activities of these lawyers, coupled with their sense of commitment and willingness to make personal sacrifices, have led publicists to call them the new "public interest lawyers" (New Public Interest Lawyers, 1970: 1069). The professional role filled by these lawyers and their offices differs from the typical private practitioner, whose primary motivation is profit. It also differs from the government lawyer who, while charged with protecting the public interest, often serves a particular favored segment of the public. This sometimes works contrary to the interest of the public at large.

Many observers believe that the new public interest lawyers will be the gadflies and catalysts for a major reallocation of legal resources, which will stimulate the expression of important interests and values in our society and further political, economic, and social equality (New Public Interest Lawyers, 1970: 1069).

Before rushing headlong into the area of public interest law, however, lawyers would be well advised to heed the warning of Edgar and Jean Cahn (1970: 1005-1006):

> In the current romance between public interest law devotees and pollution, there is danger of a major moral default by the legal profession. Mayor Hatcher correctly observed that the environment issue had done what Alabama's George Wallace had not been able to do—"distracted the attention of the nation from the pressing problems of the black and poor people of America." Given the current unresponsiveness of the political system to ethnic minorities, the allocation of public interest law resources to majoritarian, middle-class white concerns is contrary to the public interest. The political system can respond to these concerns without siphoning off the limited, special and constitutionally distinctive resources of the legal profession.
>
> The constitutionally insulated role of legal advocacy carries a concomitant obligation to recognize that, while many causes may be legitimate, the allocation of scarce legal resources to a "public" cause where other avenues of redress are clearly open is profes-

> sionally and institutionally immoral. As the Supreme Court stated in *NAACP v. Button:* "(U)nder the conditions of modern government, litigation may well be the sole practicable avenue open to a minority to petition for redress of grievances." There are fashions in all things, including fashions in righteousness, but for the legal profession to succumb to the most recent fashion would be a betrayal of its historical and constitutional role as the advocate of disenfranchised members of our society.

The Cahns go on to argue that "the allocation of scarce legal resources away from the poor, the Black, the Mexican-American, the Puerto Ricans, and the Indians deprives them of the only political weapon available to them" (Cahn and Cahn, 1970: 1005). If this is so, and I believe it to be, then concentration on majoritarian issues such as pollution may add to an already high-burning flame, leaving minorities with a far greater sense of frustration than they now have, and giving greater credibility to those whose motto is "Burn, Baby, Burn." Indeed, the Report of the National Advisory Commission on Civil Disorders (1968) has warned us that:

> Many Negroes have come to believe that they are being exploited politically and economically by the white "power structure." Negroes, like people in poverty everywhere, in fact lack the channels of communication, influence, and appeal that traditionally have been available to ethnic minorities within the city which enable them—unburdened by color—to scale the walls of the white ghettos in an earlier era. The frustrations of powerlessness have led some to the conviction that there is no effective alternative to violence as a means of expression and redress, as a way of "moving the system." More generally, the result is alienation and hostility toward the institutions of law and government and the white society which controls them. This is reflected in the reach toward racial consciousness and solidarity reflected in the slogan "Black Power."

This fear is only valid, of course, to the extent that the public interest lawyers and law firms do, in fact, divert attention away from the poor, the Black, and the other dispossessed minorities. There is no denying the fact that such lawyers are capable of providing much-needed legal expertise and direction in areas of importance to minority groups. They tend to think in broader social, economic, and political terms, and the types of lawsuits they file reflect this philosophy. They tend to be committed to causes rather than to individual clients, and, therefore, they bring to their case a greater degree of knowledge about the cause they represent than does the lawyer with the traditional practice who waits for the client to come into his office. In attacking broad problems and pursuing long-term goals, public interest lawyers have engaged in a wide range of strategies and activities, use of propaganda and the press, mobilizing community demonstrations, and organizing citizens' lobbies and community groups. Though none of these activities is novel in itself, the programmatic use of various strategies and various forums by a single lawyer or law firm on behalf of broad social issues has given new dimensions to the notion of "legal representation."

The move of LSP toward what is being called law reform was an important step toward more effectively dealing with urban problems. A large number of private law offices with social change as their main objective are springing up in various communities, carrying the law reform concept to its next logical step. For descriptive purposes these offices will be referred to as "social change law offices." As in the case of LSP, most of these offices are underbudgeted and undermanned, and they are likely to remain so. They are also manned primarily by young, idealistic attorneys more committed to effecting social change than to enhancing their financial status (Stanford Law Review, 1970).

To render maximum services on extremely slim budgets, any social change law office broadly conceived must concentrate heavily on efficiency. Some ways of increasing efficiency involve heavy costs in terms of other important goals. For example, as we have already seen with LSP, mass production of cases defeats the imperative of close client relations necessary to accurately follow their wishes. Other methods of increasing efficiency, however, seem fairly worthwhile. Increased use of paraprofessionals[10] to perform legal work is a necessity, both to increasing the amount of legal work that can be turned out, given the chronic shortage of attorneys in this area, and also as an approach toward decreasing the potency of the myth associated with attorneys and the law. Full use of volunteer help, including recruitment, both among lawyers and clients, serves to broaden the base of support for the office as well as providing real sources for use in serving client's needs. Specialization, both in terms of procedure and subject matter, can greatly increase the effectiveness of the social change law office, and should ideally be pursued not only within such offices but also among offices in a given area. Finally, strong efforts must be made to keep expenses down. This implies that all people working in the office receive relatively low salaries. This serves an additional goal of decreasing the distance between the standard of living of legal workers and the clients they serve.

There can, of course, be as many types of social change law offices as there are legal or quasi-legal tasks that must be performed. The multitude of possibilities, however, seem to break down into three basic types: (1) the community law office, (2) the research office, and (3) the referral office.

Community Law Offices

The community law office engages primarily in traditional legal work through the courts, before administrative agencies, and in giving general legal advice to clients.

The community law office can take an infinite variety of specific forms. Since OEO Legal Services Program has been discussed elsewhere, the primary emphasis here will be on establishing a self-sufficient community law office in an urban area. Urban areas break down into identifiable neighborhoods, districts, and community groups. One district, of manageable size—e.g., 50,000 or fewer persons in the constituent area—should be chosen as the general clientele. The structure of the office should be egalitarian and specialized, insofar as possible. Equality of salary, dignity, and amount of distasteful work[11] should be primary goals, lest the office mirror internally the evils in the constituent community it seeks to change.

Specialization must stress use of volunteers, recruitment of volunteers in all areas, use of clients to work on their own cases, and increase the extent to which nonlawyers perform work on cases and projects. Persons in the office should choose one or more areas in which to develop expertise. In addition, as more community law offices develop, it should be possible to arrange office specialities among the different offices in one particular area. One office may concentrate on tenant-landlord problems for the entire area. Another variety of interoffice specialization is in terms of representing one large community group exclusively.

The most difficult task for the community law office is becoming economically self-sufficient (New Public Interest Lawyers, 1970: 1069; Drake, 1970: 4). Such an office, as will be seen, promises little return through traditional fees. Most efforts to date have revolved around the attempt to get established law firms, government, foundations, and private contributions, to support community law firms and the public interest law firms discussed elsewhere herein. A central difficulty arising from the need to seek financing from sources other than clients is that the source of funds may place explicit or implicit conditions on their use which make the lawyer less effective in his advocacy. And the search for funds is itself a time-consuming process which may interfere with legal work.

The two approaches suggested here provide maximum freedom from the controls over what work the office does except those restraints that arise within the community itself. As stated, it seems to be a fact of life that whoever pays your bills can control, to a greater or lesser extent, your activities.[12] Since the major goal of a community law office is to serve the community, conflicts are minimized if the money needed to run the office can be drawn from the community. This posture has an additional advantage: it will stimulate the community to set its priorities carefully, rather than regarding the office as a general resource to be expended on a first-come, first-served basis without regard to community benefit.

On the other hand, the drawbacks of this posture are self-evident. The chosen clientele will be primarily poor, although some groups that are not so poor but have corresponding advantages of political and social activism will be represented. The community cannot afford to pay much, at least under current restrictions, ethical and otherwise. When and if group legal services arrangements become possible, in terms of legal ethics, then even a poor community can pool its meager economic resources to keep an office going if they see the need for one. In addition, even under present ethical restrictions on lawyers' activities, if a group is organized primarily for some purpose other than giving legal services to its membership, it can, as a group, retain an office to represent its members. Therefore, a coalition of community-group clients represents one possible source of funds.

A second general approach, not mutually exclusive of the first, is to do general lawyer work for general clients and charge the going legal rate[13] for the services. This income can be used to support the primary/social change activities of the office. This approach can be successful only where the office is run on a minimum budget, and even so, the problems of deciding what fee to set loom large. How much should office survival be taken into account in turning away individuals who cannot afford to pay? What about setting fees in between the going rate and the free work? Are the salaries of the office workers too high? It is perfectly proper for those who can afford the going rate for legal services to subsidize those who cannot, but the posture is difficult for the persons involved in setting fees.

The solution—and there appears to be one—lies in striking a combination of the approaches suggested here that is (a) acceptable to the community, (b) acceptable to the persons manning the community law office, and (c) successful enough to pay the rent and other necessary expenses. If the office budget is made public within the community, and the basis for charging fees is made clear, then, after some amount of discussion and constant struggle, the system can evolve into a successful form.

This model of a community law office may seem unrealistic and idealistic to those not attuned to the new type of lawyer graduating from our better schools. In the unlikely, middle-class community of Palo Alto, California, for example, the "law commune" of Henner, Wolpman, Constantinides and Cohen has, for a year, been operating much along the lines suggested here. Four attorneys and five recent graduates, primarily from the Stanford Law School, established a community law office with the primary goal of serving constituencies that could not otherwise obtain legal services. Salaries are roughly $300 per month for single persons, attorneys and staff alike, with necessary increases for those with families to support. The work of the office consists of general legal work on fee-bearing cases, the minimum necessary to support the office on an extremely slim budget, and a large proportion of free legal services including criminal defense work, occasional test cases, and legal advice and work for local community organizations. Their clients include the unaccredited Mid-Peninsula Free University, primarily attended by white liberal-radical young persons in the Stanford University community; Nairobi College, which is accredited and all Black, and the local tenants union. The firm is run cooperatively, with staff exercising a full voice in decision-making alongside the attorneys. It seems important to add that one of the five recent graduates in this firm was the President of the Stanford Law Review, a position usually guaranteeing employment with one of the top New York or Washington, D.C., law firms at a starting salary of at least $15,000.

I am aware of at least three other such community law offices in the San Francisco Bay Area, all operating generally along the lines described above, and ranging in size from two to eight attorneys. These offices have had many enquiries from other attorneys and students, I am informed, and it seems safe to assume that similar offices, if they are able to survive financially, will arise in other urban areas.

Research Offices

Research offices include everything from projects such as those associated with Ralph Nader, through more traditional lobbying offices that deal directly with legislation, to "watchdog" offices geared to oversee in detail the operations of federal and state government agencies that affect the poor.

With the startling success of Ralph Nader's research teams both in maintaining near unanimity of national support and in recruiting highly skilled students and graduate student volunteers during the summers, that mode of operation has taken hold. Numerous institutions now exist along similar lines, including Edgar Cahn's Citizens Advocate Center, the Washington Research Project, and the Center for Law and Social Policy. It is important to note, however, that all of these organizations are based in Washington, D.C., the seat of national government. Only one of these makes efforts to stay closely in touch with a chosen minority constituency. The Washington Research Project, run by Marian Wright Edelman, sees its task as bringing continuing pressure and skill to bear on Congress and agencies affecting the poor, especially poor Blacks in the rural South (New Public Interest Lawyers, 1970: 1080). It appears very difficult, however, to maintain close relationships with poverty community groups in the South from a base in Washington, D.C.

The other organizations in Washington appear to use their own judgment concerning the needs of the poor and of minorities.[14] While in many cases their judgment may be correct, in some it may be wrong, in others they may be concentrating on an issue which is of lower-than-first-priority in the eyes of the class being represented.[15]

Ultimately, however, the major drawback would appear to be the lack of organizational support given by these groups to local community organizations. It is often extremely difficult to bring about any observable social change through a given tactic; when that is true, the primary goal must be to foster powerful indigenous organizations through lending the symbolic and real power of a research office to a community group. This cannot be done effectively from several thousand miles distance. This conclusion, however, obviously loses force to the extent that these Washington-based offices represent and support Washington community organizations. It should also be made clear that these groups serve an important function and are valuable tools in fighting injustices facing poor people. However, to the extent that there is presently a shortage of attorneys committed to and working in these areas, it is crucial to use these limited resources in the most effective way.

Existing research offices appear to subsist primarily on foundation grants and private philanthropy, and secondarily on proceeds from publications. Neither of these sources seems particularly reliable (Cahn and Cahn, 1970: 1005, 1007), especially with new IRS regulations penalizing any foundation that supports efforts to achieve legislative reform.[16] Reliance on publishing proceeds for any sort of sustained income, of course, would be very unwise in view of the difficulty of producing or predicting successful sales.

A research office could conceivably be supported, as some law offices are, through membership contributions. Again, it is highly unlikely that such an organization could receive tax-exempt status in the present political climate. On the other hand, a membership organization could ensure that the research office responds to some constituency and in addition might well be a vehicle for community organization, or organization of similar classes from different communities across a state or across the nation. As Marian Wright

Edelman pointed out, action without some sort of a mass base of support is severely handicapped (New Public Interest Lawyers, 1970: 1083). Similarly, a research office could be supported by contributions from local community organizations intended to be served by the office. None of these sources of funding seems particularly stable or generous. All possible sources should be explored. If an office is run on a realistically minimal budget, state and even city research offices should be possible.

The structure of these offices seems to be hierarchical, with a director in charge, often under a board of directors of some variety. The relationships of these offices with constituencies, as has been pointed out, seem diluted, although representation on boards of directors of leaders of community organizations remedies some of the problems. Internally, it is unclear to what extent the traditional dominance of lawyers in decision-making has been broken down. This is a major personal block in most persons with legal training, although it is capable of being overcome. It applies not only to relationships with clients and client organizations but also within the offices themselves, in terms of relationships between lawyers and researchers and legal workers (secretaries). Salary distinctions may be a symptom of the disease.

This problem should be more easily overcome in research offices than in community law offices, because the types of activities undertaken lend themselves far more to including varieties of expertise besides legal training. Consequently, the people manning the office should be drawn broadly from local communities and from other sources of expertise. Finally, twin problems of choice of specialty or agency to deal with, and of community control, present special difficulties. Government of a research office by a board of directors drawn from community groups should go a long way toward curing this problem.

Referral Offices

The third type of organization may be called a referral office, organized to refer callers to people who can give them, at no charge or for a nominal price, whatever social service they happen to need, whether it be legal services, medical care, welfare advice, or whatever.

This type of office needs by no means be as passive as its name implies. In addition to heavy efforts to recruit, say, volunteer lawyers to take occasional cases at no fee and to organize law students to provide time-consuming legal research on such cases, a referral office can broaden its activities to include mastery of all charitable and social services available in a given area inexpensively. Public education on the extent of assistance available to people in trouble can be extended to include information on local activities of community organizations, community projects people might want to join, and so forth. The referral office tackles the critical task of information flow to the community, providing not only general knowledge but also very detailed in-depth knowledge tailored to the caller's particular problem. In addition, affirmative functions can be served as well; for example, a "police complaint center."

Community "switchboards" have been springing up in many communities, offering a wide range of advice to deal with problems from a potential suicide, a drug crisis, a mass arrest situation, to a welfare problem. These switchboards may well offer substantial access to legal services, both through informing the community of free or inexpensive services presently available, and through organizing volunteer lawyer panels to deal with specific kinds of cases. The Selective Service Lawyers' Panel in the San Francisco area has achieved remarkable results in providing highly skilled specialists, often at little or no fee, to persons with draft problems. Similar panels could be established to deal with anything from political defense cases to consumer problems.

Such an office could educate the community regarding its legal rights, practicing what might be called "preventive law." The laws provide an often oppressive framework within which people's lives and property are often jeopardized and sometimes saved. Knowledge of the legal hazards in any area may well be useful in individual cases, even though the legal system as a whole is seriously stacked against urban minorities. Community education, of course, can be closely tied to community organization. And drawing together a concrete knowledge of where to obtain a wide variety of different services can be a source of strength to the community.

Funding such an office should require far less resources than either the community law office or the research office, and may depend far less on the availability of lawyers to serve full time on the staff.

Dedicated volunteer help can provide the major resource, and limited financial help from one or two community organizations should be able to generate enough funds to keep a small office, telephone service, and one or two full-time staff members going on an annual basis. A successful referral office could probably be run for roughly $15,000 per year, and its activities could be as broad as community need and the extent of volunteer time available.

NEW LEGAL INSTITUTIONS

Ombudsman

At the heart of the democratic political theory rests a cardinal premise: the individual rights and dignity of man. In addition, constitutionalism requires that the actions of the government and its officers conform to the rule of law. Such notions are meaningless without a method of redress for the people, even under a constitutional system which is designed to limit and channel the powers of the government. Despite the current availability of judicial checks, legislative review, and executive controls, the Ombudsman plan warrants consideration as an additional institutional device to protect and safeguard the rights of an individual by a less formal process.

If constitutional government is to have any meaning, it is fundamental that an individual who feels wronged by the very agencies set up to help him, or even to regulate his conduct, should be able to appeal quickly, easily, and without cost to a politically independent and impartial office. Thus, conceived, the Ombudsman proposal, in practice, provides a governmental official to whom the average citizen may go to state a grievance and expect results.

The introduction of the Ombudsman as an additional mechanism to handle grievances against governmental agencies, officials, and employees, has been widely suggested.

The Kerner Commission saw the need for the establishment of Ombudsman-like machinery when it recommended "that formal mechanisms for the processing of grievances, many of which will relate to the performance of the city government, be established independent of the local administration" (Report of the National Advisory Commission on Civil Disorders, 1968: 151). The commission found that much of the frustration reflected in the disorders which it was commissioned to investigate resulted "from the lack of accessible and visible means of establishing the merits of grievances against the agencies of local and state government, including but not limited to, the police."

It is doubtful whether there is a governmental institution in the world today which has aroused as much interest as the institution of the Ombudsman. What is the reason for this popularity, and why has interest been focused on it especially since its establishment in Denmark in 1954, even though its Swedish counterpart has been working efficiently since 1809, and in Finland the institution has existed since 1919? It would seem that this popularity derives from a general feeling that it is necessary to devise new ways and means of safeguarding the civil and political rights and freedoms of the individual, in light of the growing interference on the part of the modern state—the welfare state—in the lives of its citizens.

Although the word Ombudsman had its origin in a primitive legal order of decentralized governments of the germanic tribes (Anderson, 1969: 2), the modern embodiment of the Ombudsman is reflected more accurately in the Danish version as provided in the 1953 constitution. The Ombudsman as we now know him is a constitutional officer appointed by parliament to receive, investigate, and report on citizens' complaints of bureaucratic abuse.

Descriptions of the office incorporate the following elements: (1) the Ombudsman is viewed as the people's representative; (2) he is an independent, politically neutral officer; (3) he is appointed by the legislative branch; (4) his main concern is citizen protection against bureaucratic abuse in its various forms; and (5) the position presupposes a law officer of the highest professional standing. Because of my desire to gear this office to the urban crisis, as I have been using that term, I would add that the Ombudsman must have complete acceptance by the minority elements in the community he serves. Without this acceptance, he will have virtually no impact whatsoever upon the real and imagined grievances of the minority community.

The Ombudsman does not act in an adversary fashion as counsel for the complainant, but remains independent of both citizen and government as a mediator or intermediary. He tries to understand both sides of the dispute and bring about a satisfactory resolution of the citizen's complaint. If he finds that the complaint is well founded and the branch of government concerned refuses to remedy the situation, the Ombudsman would be authorized to report the abuse directly and publicly to the legislative body that created his office. While the Ombudsman normally acts in response to a citizen's complaint, he also should have the ability and authority to act on his own motion and should do so, without hesitation.

The Ombudsman is a politically independent, external, critic. He is outside the bureaucratic chain of command, and he does not have the responsibility for formulating, executing, or defending the policies and personnel of any agency. His method of handling grievances against administrative decisions is direct, informal, speedy, and inexpensive. He is independent, impartial, expert, and accessible to the public.

The position of Ombudsman would in no way, of course, solve all of the problems being discussed herein. It is a drop in the bucket, and must be viewed as one of many new devices feasible and necessary to effect a solution. Indeed, William B. Gwyn has described the institution as "an unspectacular and relatively minor governmental reform" (Gwyn, 1968: 67).

To the problems in our society today, however, the role of the Ombudsman can be expanded and adapted to effectively work toward the resolution of problems arising in our urban centers.

While there have been Ombudsman bills presented in the United States Congress, they have been extremely limited and completely inadequate to meet the problems herein discussed. For example, Representative Henry S. Reuss (D.–Wisconsin) included a restriction in his pioneer 1963 proposal, H.R. 7593, that the Administrative Council (Ombudsman) would be authorized to receive complaints only from members of Congress (Congressional Record, 1963a, 1963b).

The name Ombudsman is a catchy one, and, as its familiarity increases, we should be on guard to prevent its improper use. An office of inquiry and information, for example, is not an Ombudsman. While such an office may have expertise and easy accessibility, it lacks independence and impartiality. It would also, I predict, lack acceptance and credibility, since such an office is an integral part of administration, not independent of it.[17]

Some university officials have resorted to creation of Ombudsman-like positions to ease tensions on college campuses. The University of California at Berkeley, viewed by many as having the most active student body in the country, has designated one of its associate deans as an "informal" Ombudsman, hearing student complaints and trying to rectify them (Daily Californian, 1968). Again, however, such roles lack the independence and perhaps the credibility so indispensable to the role of a true Ombudsman.

As mentioned earlier, unrealistic expectations must be avoided. It would be grossly overoptimistic, for example, to expect Ombudsmen to simply cure the urban crisis. Ombudsmen cannot cool the long, hot summers of ghetto violence. They cannot create jobs, provide transportation, or build homes. Ideally, the Ombudsman would promote public confidence in government by providing fast, economical, and impartial redress of grievances in those areas where judicial relief or procedural safeguards are unsuitable or expensive. Especially in the areas of administrative delay, negligence, arrogance, or inefficiency, the Ombudsman may be most valuable. Irritation, frustration, distrust, and fear of coping with an amorphous bureaucracy may well be minimized or eliminated by easy and economical access to the Ombudsman.

Community Law Centers

The operative language and procedures of our legal system are complex and often inscrutable to those unfamiliar with the patterns of legal language or the phenomena of legal procedure. The system, however, depends upon an understanding of its language and procedures for effective operation. The minority communities have not had and do not have sufficient lawyers to advocate positive changes on their behalf or to eliminate instances of abusive exercises of authority. In essence the laws are there to be used, the channels of persuasion are open, but the minority communities do not have sufficient trained representatives who both understand the laws and are knowledgeable about the demands and needs of political and legal persuasion in a highly legalistic system.

Another root of the problem is that the members of minority communities largely labor in ignorance and fear of the law and its processes. Though this condition also exists in the nonminority community under some circumstances, the problem is exacerbated in minority communities which have historically been subjected to discriminatory treatment under a system of laws which unfortunately are too often designed and administered for the advantage of those in power.

One means of combatting these conditions is the development of community law centers to train lawyers who will speak and act for minority communities in their effort to seek orderly change, and to provide those programs in law which will permit the development of a legally enlightened citizenry capable of self-protection against the more obvious abuses of authority.

The need for such centers is obvious. The first Black graduate from an American law school was George Willis Ruffin who finished Harvard in 1869. Today, a century later, there are fewer than 200 Black students currently graduating from American law schools, as contrasted with 10,000 white graduates, according to figures furnished by the National Bar Association. Stanford Law School, where I am employed, graduated its first Black student in 1968, although it has been in existence since 1893. Today, there are approximately 300,000 lawyers in the nation, and less than 3,000 are Black—less than 1%.[18] Statistics gathered by the Equal Employment Opportunity Commission in 1968 reveal that of twenty-six major industry groups in New York City in January 1968, law firms ranked next to last in Black participation; and although 6.7% of the white-collar force in banking firms were Black, Black white-collar workers in law firms constituted just 1.4% of the total employment (Brown, 1970: 385). In 1964-1965 the Black enrollment in American Bar Association-approved law schools was 1.3%. Of the 701 Black students in these law schools during 1964-1965, 267 were enrolled in the 6 predominantly Black law schools (Johnstone and Hopson, 1967).

Numbers alone do not tell the complete story. The bulk of the law schools have curricula which reflect values not entirely consistent with the needs of minority communities. An appropriate response for a law school dedicated to the needs of a minority community would be to emphasize debtors' rights as well as creditors' rights, tenant law as well as landlord law, poverty or urban law as well as a state and gift taxation and traditional property courses. The fields of poverty law and urban law are only now beginning to emerge from the past of contrived inattention. The established law schools largely continue to either ignore these problem areas completely or to use a piecemeal approach. The best answer to these persisting problems is the development of community law colleges, responsive to community needs, which will develop an integrated approach to the education and the law of minority law students and local residents.

The proposal is feasible. The requirements to begin legal studies in California, for example, are surprisingly low. A student can begin the study of law after completing 60 units of college. If he is over 23 years of age and can pass a two-year college equivalency examination administered by the California State Bar, he can enter law school without any formal education.[19] A substantial recruitment program could begin at local colleges and junior colleges, aimed at the many students now capable of completing legal training. The program would inform them of the opportunities and advantages of a career in law, emphasizing the community service aspects of the program. So as not to frustrate expectations, the probability of passing the Bar should be realistically presented.

The formal requirements which must be met to establish a law school are minimal. There are the usual business license forms and permits which must be obtained. The only additional requirement for a corporation is that it have $50,000 in net assets, either real or personal or both, which are devoted to educational purposes. An affidavit certifying the value is filed, and the valuation must be based on an appraisal by a state agency. To meet this need, the school could be established in an existing store or other building estimated to be valued at that sum.

Under state law, the school would operate on a part-time basis—a decided advantage for students working while attending school. All unaccredited law schools have four-year curricula, rather than the normal three-year curricula, to meet California State Bar requirements. Staffing is also easier because of part-time operation. Faculty members can be recruited from members of the local Bar Association. Wages are comparatively low, ranging between $10 and $15 per teaching hour. The administrative personnel would hold the only full-time positions.

What are the prospects for students upon graduation from the proposed community law colleges? The experimental nature of the program compels speculation. It is very likely that the majority of students would be employed in government on the local, state, and national levels. Examples of governmental employment would be the Office of Economic Opportunity legal services, public defender, and similar organizations. Private Bar, business, and teaching, would absorb the balance.

We also have to consider the future of those who complete school but do not pass the Bar examination. It could be expected that the percentage of those who do not pass would be high, since most accredited schools pass about twenty to forty percent of their graduating class through the Bar. Some law schools boast higher percentages passing the Bar, but only after they have had their starting classes reduced sixty to eighty percent prior to graduation.[20] We can reasonably expect that twenty to thirty percent of the students who graduate will pass the Bar. The percentage can be increased by becoming more selective in choosing students who are accepted at the school, but that might defeat the purpose of the schools.

Those who do not pass will occupy a category best described as legal aides capable of doing everything that needs to be done in a law office or agency other than appearing in court or performing those unique services which are presently considered as the practice of law. Because of their legal training, they will hold the LL.B. degree. The career possibilities for a legally trained man who is not a member of the Bar appear to be excellent. Obviously, such a person would be ideally suited for work in a community law office or a research law office.

There is substantial promise that a new community-approved approach to the legal education of members of the minority community for both professional and personal ends will develop a channel for the effective and articulate presentation of legitimate community aspirations. This solution seeks to provide an integrated set of services and to meet the multifarious needs of the community. The purpose of this program is to strive for such a solution, knowing that the social innovations sought can only be brought about by a dramatic improvement over existing approaches.

CONCLUSION

"The Urban Crisis" has become a cliché; the problems themselves have been studied almost to death. We know what the problems are, and the desperate challenge of discovering effective ways to deliver power into the hands of the communities must occupy all our energies. The lawyer's role must be seen and shaped in terms of the environment and actions necessary to achieve substantive changes. These changes cannot be brought about by traditional means. Perhaps the changes we want can only be brought about as Kenneth B. Clark suggests:

> If the ghettos are to be transformed, then forces superior to those which resist change must be mobilized to counteract them. The problem of change in the ghetto is essentially, therefore, a problem of power—a confrontation and conflict between the power required for change and the power resistant to change [Clark, 1967].

To the extent this is true, individual lawyers and the legal profession, public-interest- and social-change-oriented, can use their considerable resources to help effect the necessary redistribution of power that can salvage our cities and our country.

It is impossible to discuss in a paper of this nature all of the many subtle roles the profession and the attorney can play. Certainly, the suggestions contained herein, some of which are already being implemented, are examples of new and imaginative roles for the profession. Each creates its own difficulties and hazards, and none alone can be more than a useful tool for maximizing the professional effort in this area. Many of the problems of our urban areas are not really legal in nature, or are not readily overcome by tinkering with varying forms and structures.

We Americans have taken pride in our government of laws. The framers of the Constitution, for the most part, were lawyers; and lawyers have played an important role in government throughout our history. The relationship between the courts and the other branches of the government is a more significant one in America than anywhere else in the world. When there is severe dissatisfaction and demand for change, it is important that the organized Bar and individual lawyers assume positions of leadership.

In the past, members of the legal profession and too often the organized Bar have not been willing, nor found the time, to provide the leadership of which they are capable. Many members of the legal profession have too often been preoccupied with their particular private interests and have given little attention to interests other than their own and have failed to assume the burden of attempting to accomplish solutions in our society. Assumption of leadership means assumption of a formidable burden. It is far easier to express dissatisfaction than to accomplish solutions in a society encouraging freedom of expression and which allows accommodation of divergent views. Unless lawyers are alert to the needs of the future, change may by-pass them and may produce alterations in the political structure which effect undesirable alterations in the political structure which may, in turn, effect undesirable alterations in the legal profession.

Thus, though our history contains individual good accomplishment by lawyers, in these rapidly changing times the organized Bar is too often involved in "patch-up" thrusts, while not moving firmly ahead with timely solutions and forward-looking remedies. Perhaps this is the nature of the organized Bar, but to the extent it deals with the past, and not the future, it will neither provide nor be looked to for leadership and for solutions.

NOTES

1. The phrase "white power structure" was made popular by the late Dr. Martin Luther King in 1963 during the racial protests in Birmingham, Alabama. Dr. King was referring to those whites who control the political and economic life of a community, and are thus in a position to effect social change.

2. See, for example, *Hamilton v. Alabama,* 376 U. S. 650 (1964) where plaintiff Mary Hamilton successfully challenged a court official who refused to address her by a courtesy title. [For an excellent discussion of the psychological ramifications of this type of treatment upon Blacks, see Poussaint, 1969: 28-29].

3. In a study of the legal profession, it was shown that in a major Eastern city, less than 5% of the Bar reported the median income of its clients to be under $5,000, although more than half the families and single adults reported income of less than $5,000 [see Carlin and Howard, 1965: 381-382; Wald, 1965: 44].

4. Statistics do not completely explain the problem, but they give a reasonable indication of the few resources being put into Legal Aid by the profession. In 1964, 9 cities with a population of 100,000 or more had no Legal Aid offices. That year there were 246 Legal Aid offices and 139 defender offices, a 3% increase since 1949; but only $4 million was being spent in the entire country [see Wald, 1965: 46-48].

5. 42 U.S.C. sections 2701-2981 (1964).

6. The Los Angeles Neighborhood Legal Service Society found that the greatest demand for legal services is in the area of domestic relations (approximately one-third of all cases), followed by "consumer" cases (approximately one-third; see Journal of the State Bar of California 41, 3 [May-June, 1966]: 245). Approximately the same statistics apply to the author's experience with the San Mateo County Legal Aid Society.

7. For example, the LSP office in Watts, California, interviewed 2,395 persons between December 17, 1965, when it opened, and the end of March 1966. Fifty-six percent, or

1,335 of these, were eligible to receive legal services from that office (Journal of the State Bar of California 41, 3 [May-June, 1966]: 244).

8. California Rural Legal Assistance (CRLA), perhaps the most effective LSP office in the country, early decided that it could best serve the poor in the rural areas of California by minimizing the "service function" and concentrating on cases and issues effecting large numbers of its rural clients. With a staff of talented and imaginative attorneys, CRLA has had phenomenal success in changing the laws in such areas as welfare and labor.

9. Of the 408 bankruptcy cases filed by the San Mateo County Legal Aid Society during the fiscal year 1966, 35% of these clients returned to the office within 6 months with additional, or continuing, financial problems. Under the provisions of the Federal Bankruptcy Act, a person can file bankruptcy only once every 6 years (statistics compiled by author).

10. The legal profession can learn a great deal from the medical profession in the use of paraprofessional assistance. Numerous studies by Bar groups have demonstrated that attorneys spend a great deal of their time performing essentially nonlegal tasks, such as interviewing, filling out forms, and the like. They spend a relatively small percentage of their time doing that which they are best equipped to do; i.e., making important legal decisions and implementing them.

11. OEO evaluations of LSP offices indicate that frequent morale problems develop in these offices when the newer attorneys are given the distasteful tasks, such as routine intake, or divorces, while the older attorneys are allowed to handle the more attractive aspects of the office practice (evaluation reports of various Northern California LSP offices prepared by author).

12. This has been a recurring problem for aggressive LSP attorneys, who feel that the local Bar or OEO attempts to control their activities by explicit or implied threats to cut the budget of the office. For example, Terry F. Lenzner, former Director of LSP, stated (November 21, 1970) upon being fired: "My deputy Frank Jones and I were fired yesterday because we sought to run the Legal Services Program in a way that would guarantee the right of the poor to effective representation. . . .The Administration apparently believes in bargain basement justice for the poor, justice tempered by the desire not to step on the toes of anyone who can afford to contribute to a political campaign."

13. Most local Bar Associations have a booklet indicating a "Schedule of Suggested Fees" which their member lawyers are encouraged to follow in setting fees for their clients.

14. According to its application to the Ford Foundation, the Center for Law and Social Policy makes its decisions through a Board of Trustees which establishes "general policies and will determine the substantive legal areas in which the Center's activities will be concentrated" [New Public Interest Lawyers, 1970: 1080]. Decisions to take on particular matters are made by the Center's staff and "all staff attorneys will participate in important decisions such as the decision to take on major litigations or to undertake long-term research projects." (Application of Center for Law and Social Policy to the Ford Foundation, September 20, 1969.)

15. The results of a recently completed survey of Northern California Legal Aid offices indicates that in most instances the attorneys' views of community legal priorities were considerably different from the community views, and in some cases the communities wished the attorneys to establish priorities that were not even legal in nature (unpublished survey of 27 Northern California Legal Services Programs by author, completed November 26, 1970).

16. Internal Revenue Code of 1954, section 4945 (a) (1), and 4945 (e) now provides that a foundation will be forced to pay a 10% tax on any expenditure that supports an attempt to influence legislation by communicating with the general public or with a legislator or his assistants.

17. In East Palo Alto, California, for example, the local sheriff's substation offered free office space to the newly funded LSP office. Though the offer was tempting and would have eased the strain on a tight budget, the Community Action Council (CAC) voted to reject the

offer because, in the words of the CAC chairman, "no black man in his right mind would trust a lawyer who he has to walk through the sheriff's office to see." (Minutes, East Palo Alto Community Action Council, October 23, 1965.)

18. Figures furnished by the National Bar Association.

19. California Business and Professions Code, section 6060.

20. This studied and intentional reduction of class size would run contrary to the purposes of the community law centers, which would have, as an incidental goal, the development of persons with varying degrees of competence and understanding of the legal system.

REFERENCES

ANDERSON, S. V. (1969) Ombudsman Papers: American Experience and Proposals. Institute of Governmental Studies, University of California, Berkeley.

BAMBERGER, C. (1966) J. of the State Bar of California 41, 3 (May-June): 224-231.

BROWN, W. H., III (1970) "Racial discrimination in the legal profession." Judicature 53, 9 (April-May).

CAHN, E. and J. CAHN (1970) "Power to the people or the profession?–the public interest in public interest law." Yale Law J. 79.

– (1964) "The war on poverty: a civilian perspective." Yale Law J. 73.

California, State Bar of (1966) Journal 41, 3 (May-June).

CAPLOVITZ, D. (1967) The Poor Pay More. New York: Free Press.

CARLIN, J. E. and J. HOWARD (1965) "Legal representation and class justice." UCLA Law Rev. 12.

CLARK, K. B. (1967) Dark Ghetto: Dilemmas of Social Power, 200. New York: Harper & Row.

Congressional Record (1963a) July 16: 12069.

--- (1963b) February 11: 2078-2084.

Daily Californian, University of California, Berkeley (1968) July 23: 2.

DE TOCQUEVILLE, A. (1969) Democracy in America. New York: Anchor Doubleday.

DRAKE, J. (1970) "Economic survival for movement lawyers." Southern Legal Action Movement 2, 3 (November).

EDSON, C. (1966) Report to Legal Services National Advisory Committee, Washington, D.C. (September 19).

GWYN, W. D. (1968) "Transferring the Ombudsman," in S. V. Anderson (ed.) Ombudsman for American Government? Englewood Cliffs, N.J.: Prentice-Hall.

JOHNSTONE, Q. and D. HOPSON, Jr. (1967) Lawyers and Their Work. New York: Bobbs-Merrill.

National Advisory Commission on Civil Disorders (1968) Report. New York: Bantam.

New Public Interest Lawyers (1970) Yale Law J. 79 (May).

Newsweek (1967) November 20.

Office of Economic Opportunity (1964) Guidelines for Legal Services Programs 23.

POUSSAINT, A. F. (1969) "The psychology of a minority group with implications for social action," in C. U. Daly (ed.) Urban Violence. University of Chicago Center for Policy Study. Univ. of Chicago Press.

San Francisco Chronicle (1970) December 7: 1.

SILVER, C. R. (1968) "The imminent failure of legal services for the poor: why and how to limit caseload." J. of Urban Law 46.

Stanford Law Review 22 (1970) President's page, preceding p. 433.

WADE, R. C. (1969) "Violence in the cities: a historical view," in C. U. Daly (ed.) Urban Violence. University of Chicago Center for Policy Study. Univ. of Chicago Press.

WALD, P. M. (1965) "Law and poverty." Report to the National Conference on Law and Poverty, Washington, D.C., June 23-25.

Part V

REVIEW, EVALUATION, AND PROGNOSIS

In this section we have attempted to set out two sets of critical evaluations: one of societal trends and the other of research trends in the area of race. In our introductory statement we suggested that a wide variety of strains affected the work of professional and lay students of society. In some measure the authors of the chapters included here both represent and discuss some of these strains. The first two chapters of this section stand in vivid contrast as statements by members of the academic community regarding where we have come from, where we are, and where we may be headed in relation to the crises with which our urban society teems.

The opening chapter by Peter Loewenberg, a historian, represents a distinctive response to academia's confrontation with the new sociopolitical realities of race and change.With Seeman, Bishop, and Grigsby, Loewenberg insists upon the relevance of traditional analytical tools for comprehending the contemporary crises. While the former refer the issues of community and control to the theories of social psychology and methodologies of verification, Loewenberg as a historian knows that "history relativizes social givens," and he invokes and respects historical comparisons for precisely that reason.

As an assessment of current societal trends, Loewenberg's analysis and prescription differ profoundly from those of Harvey Molotch, a sociologist. In an interestingly eclectic piece, Loewenberg hinges his discussion of contemporary problems on ontrapsychic and psychosexual developments peculiar to this age. Paraphrasing Eric Erickson, he points out that "today it is not unusual for a man or woman to undergo the crisis of social and vocational identity formation at an age that was the age of average life expectancy but two centuries ago." Prejudice and racial violence in these circumstances are given a meaning and manifestation unlike those in any previous historical circumstance. Tracing the distinctive character of the contemporary urban scene, its alienation and racial conflict, Loewenberg suggests that, to be effective "the White liberal . . . should not merely engage in moral condemnation . . . he must use the insights of the behavioral and social sciences to understand the needs of the prejudiced person and help him meet these needs without resorting to racial hatred."

Harvey Molotch takes a position opposite that of Loewenberg. Arguing from a social structural position, he contends that managers of social indicators such as public opinion polls have failed to understand and have caused others to misapprehend radical directions developing within the country's various social constituencies. Dismantling the poll data on "backlash" against the activities of students and black people, he indicates that the radical proclivities of black people, working-class and middle-class whites, women, and many others have been strengthened by the increasingly frequent occasions upon which various of these groups confront and fail to affect the acts of large-scale corporate power which negatively determine their lives. (Although Molotch, unlike Loewenberg, does not offer prescriptions for action, the implications are clear.)

The final two chapters, Eugene Grigsby and Councill Taylor frankly and directly take sociology and anthropology to task for failing to give a full, active, and unbiased account of the facts and processes of black urban existence. In a broad-gauged attack on American studies of stratification, poverty, and rebellion, Grigsby finds in them a characterization of black life that supports, in effect, the image of black people liberal whites had sought to offset. Grigsby finally questions whether social science research can serve to aid or impede the aspirations of a people whose nature and sensibilities it has so consistently failed to understand, and he calls for a new agenda for race research.

Councill Taylor, an anthropologist, explores the contribution of American social science to anthropological studies of the black urban experience. Referring in detail to the pioneering work of W. E. B. DuBois on *The Philadelphia Negro,* he finds in it an agenda for contemporary studies. At the same time he acknowledges a deepening hostility in urban black communities to “yet another study” which will result in no perceivable change in their circumstances. Although cautious in his appraisal, Taylor believes that there have been some recent studies which may indeed redeem the social sciences from their past failures.

17

The Radicalization of Everyone?

HARVEY MOLOTCH

□ A CONVENTIONAL WISDOM has emerged in the United States which holds that the American public is going through a "conservative mood"–reflected in a new demand for "law and order" and a willingness to sacrifice previously extant freedoms and social progress in exchange for a restoration of security. This "new majority" has come into being, the argument goes, in response to reckless student and minority group activists who, in their drive for immediate victories, have, in fact, alienated the great bulk of the citizenry from the causes they espouse. Thus, activism has had the consequence of undermining the public constituency which otherwise would support the exercise of civil liberties and would provide the base of public opinion that could generate authentic social reform in matters of race, war, and pollution.

This essay attempts to question this line of argument by offering two major countercontentions: (1) American public opinion has moved in a continuous direction in *support* of, and not in opposition to, the activists' demands; and (2) increasing numbers of persons are themselves adopting the methods and perspectives of the activists.

AUTHOR'S NOTE: *This chapter has been improved by comments from Milton Mankoff and the exchange of views made possible by the Santa Barbara Chapter, New University Conference.*

The argument is thus that the constituency, both passive and active, for social change in America has not declined in recent years, but has, in fact, been continuously expanding.

THE BIG BAD BACKLASH

Any contention that American public opinion has been moving in a leftward direction must deal with the problem of backlash—the layman's social-psychological theory which explains the alleged "hardening" of the public's position on reform and change. Every insurance claim adjuster has learned to be highly suspicious of those people who cry "whiplash" every time their Cadillac is bumped by the Volkswagen behind. We know that people cry whiplash in order to derive maximum advantage from a slight inconvenience imposed by another. The people who cry backlash should be subjected to the same suspicion. It is possible that the backlash cry comes primarily from those who do not wish their lives disturbed by change and who make use of others' demands to gain maximum ideological advantage for their own reactionary position.

The notion of backlash can have two possible meanings, and it is important to distinguish between them. The definition of backlash that emerges from the public talk on the subject, and the one which poses the most serious policy implications for the activists, is as follows: demands on the part of the dissident outgroup for concrete gains are being carried out with such militance that *those who would otherwise be willing to grant those gains become unwilling to do so.* A different interpretation of the meaning of backlash is that those who never had sympathy for the dissident group, and who would always block gains by that group, become increasingly vocal. The evidence would suggest that if there has been backlash in the United States, it has been of the latter sort.

METHODOLOGICAL DIFFICULTIES

Backlash is a theory about public opinion and is thus plagued by the many profound methodological and conceptual issues which the assertion of a "public opinion" contains. What, for example, constitutes the relevant public whose opinion matters? Opinion surveys which eliminate blacks and students from their samples are hardly representative of the public and, thus, methodologically imply that only a certain portion of the public possesses viewpoints to be considered by the backlash argument. That is, those portions of the public which have obviously moved to the left are removed from the reckoning of changes in public attitude.

A more basic difficulty is that the public really holds no consistent political-ideological world view and that public affairs are simply not at all salient ingredients in the daily life routine of the vast majority of persons (compare with Converse, 1964). Thus, the meaning of responses by a national sample to a given survey question can only be explained in terms of the idiosyncratic context of the interview situation (compare with Cicourel, 1964) including the respondent's ability to hastily articulate a bit of the conventional wisdom as communicated through the public media.

THE SURVEY RESULTS

It is therefore on the assumption, made for the sake of the argument, that there is such a thing as a meaningful public opinion which can be measured through survey research, that this analysis proceeds. Misgivings aside, the survey data on backlash simply fail to provide any evidence for the backlash theory—at least in its less trivial interpretation. Instead, there is a more complex situation.

There is indeed a good deal of complaining about the activity of the activists. Certainly, the amount of recorded complaining, by journalists and pollsters, has come to be substantial. People and problems previously unknown to the mass citizenry have been made available to conceptualization of everyday life through activist tactics. Indeed, this is what the activists had in mind; their self-conscious first goal has been to gain attention for their programs and troubles through the staging of public dramas. The affective response to these new people and their new (to the audience) troubles has not been favorable. Thus, as the activists have, seriatim, introduced new confrontation tactics into the public arena, the contemporary affective response to each escalation has been negative.

But in their willingness to grant actual concessions, or at least in their willingness to say that they support the substantive demands of the dissidents, the public has, the evidence would suggest, become increasingly supportive of solutions offered. In other words, a lot of people are willing to talk about their unfavorable attitudes toward the methods utilized by the dissidents and by their activism. But at the same time, greater numbers of persons have come to adopt the world view of the dissidents and to support the goals which they seek.

THE EXAMPLE OF INTEGRATION OPINIONS

Paul Sheatsley (1966), in reviewing findings of Nation Opinion Research Center (NORC) surveys on white attitudes toward integration, can find no evidence for the backlash hypothesis. Over the years when backlash was supposedly in full swing, between 1963 and 1965, the proportion of Americans who found themselves in agreement with the goals of the civil rights movement continuously increased. Sheatsley writes, "With the full expectation of at least some riots in Northern cities, American whites nevertheless (or perhaps for that reason) actually increased their support for integration" (Sheatsley, 1966: 236). Sheatsley further points to Gallup Poll findings which indicate that between the years 1963 and 1965, white Southerners experienced a dramatic positive change in their attitudes toward integration. The Southern case is especially significant given the fact that the South bore the primary brunt of civil rights activism prior to 1965 and was also the site of the most massive white resistance. When Southern white parents were asked in 1963, "Would you yourself have any objection to sending your children to a school where a few of the children are colored?" sixty-percent said yes. Two years later, only thirty-seven percent of Southern parents answered the same question affirmatively.

The Louis Harris Poll documented a similar shift in the attitudes of the American public as a whole on the Civil Rights Act. As Ross et al. (1966) point out, sixty-three percent of the American public favored the Civil Rights Act in 1963, sixty-eight percent in February 1964, and seventy percent by May 1964. These were the years when "black power" emerged and the years in which the first large-scale disorders in the black ghettos began. Similarly, the proportion of Americans who say they would vote for a Negro for President of the

United States (Gallup, April 1969) has increased continuously (with only one wavering point) over the years (1958, thirty-eight percent; 1963, forty-seven percent; 1965, fifty-nine percent; 1967, fifty-four percent; 1969, sixty-seven percent).

When the backlash question is put directly to people, and they are asked whether or not "demonstrations by Negroes have helped more or hurt more the advancement of Negro rights," not only do large numbers of people (usually a majority) think that Negroes are hurting themselves, but most agreement with this proposition comes from those opposed to the substance of Negro demands. The Harris data of 1965, for example, reveal that seventy-four percent of respondents in the South thought the demonstrations were hurting the Negroes' cause, whereas fifty percent of those in the East, and fifty-six percent of respondents in the Midwest thought this to be the case (Erskine, 1968: 659). Similarly, the data of Ross et al. (1966) in Boston revealed that forty-eight percent of their respondents considered that, in the context of the demands for integration of the Boston schools, "The actions Negroes have taken, on the whole, *hurt* their cause." The investigators point out that those who supported the segregationist candidate, Louise Day Hicks, agreed with this kind of statement, "to a considerably greater extent than the anti-Hicks group, with the partial supporters falling between."

On the other hand, the surveys clearly reveal that people continue to express displeasure with certain black leaders, with the "pace of integration" in general, and with the means which the dissidents use to accomplish their goals. But this is a different matter from increasing opposition to those goals; from increasing unwillingness to accept blacks as neighbors, employers, bosses, and brothers-in-law.

This sort of complaining to a survey interviewer does not indicate the degree to which people walk around with their complaint as an important part of their personal lives, the degree to which the trouble is indeed a felt trouble, and the degree to which there are action consequences. One measure, albeit a crude one, of the extent to which the trouble is conceived as an important one, is the extent to which people will put their money where their mouths are. Some evidence on this score was derived by Ross et al. (1966) in their Boston study when they asked respondents to agree or disagree with the following statement: "Although taxes may go up, it is better to lose state aid to Boston public schools than to bring about racial balance in the schools." Although approximately two-thirds of the sample had supported the segregationist, Louise Day Hicks, and were sympathetic to her views, only seventeen percent agreed with the statement. As the researchers conclude, "For the vast majority, lower taxes easily outrank school segregation in importance." Only a minority of even the pro-Hicks group was willing to pay more taxes for segregated schools. Contrast this finding with the results of the recent Gallup survey finding (National Wildlife, 1969) that seventy-three percent of the American people report a willingness to pay higher taxes in order to "fight conservation problems."

THE EXAMPLE OF ANTIWAR PROTEST

The attitude changes on matters of race are paralleled by similar alterations in public attitudes toward the Vietnam War. War protesters, regardless of their methods, have not been popular in the United States. Yet, even as the methods of protest have become increasingly militant, public opinion shifts almost continuously to a stance in support of the war protesters' aims. Between 1965 and 1969, the proportion of Americans who agreed that U.S. involvement in Vietnam was a mistake had increased every year from a low of twenty-four percent in 1965 to almost sixty percent in 1969 (Gallup, May 1968, October 1969). Similarly, the proportion of Americans who favor immediate withdrawal of all troops has increased to an all-time high—thirty-five percent of the population as of February 1970 (Gallup, March 1970). Of course, other events besides protests were occurring during these years: the United States was also reported to be losing the war. But it is clear that militant protest certainly did not prevent a swing toward support of the protesters' position.

It would seem that a similar pattern of survey response is emerging in the case of public attitudes toward the student movement. Perhaps even more than in the case of attitudes toward civil rights, there is wholesale condemnation of the students and their methods. Similarly, significant majorities of the population are willing to endorse severe methods of repressing students, just as large numbers similarly continue to endorse severe punishments for looters in ghetto riots. But the little comparative time data that exist would suggest that, although the generalized harshness of effect remains, there is a softening of attitude and a growing appreciation of the substantive issues which the students are raising.

The California Poll data are particularly relevant to this question, since they reflect attitudes in a state which has experienced the most continuous, dramatic, well-publicized and politicized student disruptions in the United States. Between 1967 and 1970, as campus militance spread from Berkeley to most of the state university and college system, the proportion which thought only a small minority of misfits caused the trouble went down eight points, and the proportion believing that all students should be expelled who challenge authority went down seven points. More significant for the present argument, ten percent more believe students should be given a greater voice in deciding campus rules and regulations, and twelve percent more feel that having police on a college campus makes matters worse instead of better" (Santa Barbara News-Press, 1970b). This evidence does not provide much to go on, but it does suggest that public response to the student movement may be recapitulating the general response to the civil rights movement over the years.

THE OPINION MAKERS

In addition to findings of surveys, there are other sorts of evidence to support the basic hypothesis being advanced. Public opinion may itself be a translation of elite opinion communicated to the masses via the media. In any event, the mass media's opinions are themselves an important factor in any conception of public opinion—whether conceived as a force in shaping the views of large audiences or simply as a reflection of the opinion of an important sector of the elite.

Recently, I spent a good bit of time reviewing the editorial response of the country's major newspapers to the civil rights movement over the past twenty years. The major impression emerging from that inspection is that, whereas the militance of the movement has increased dramatically with time, the response of the media, as expressed through the editorial page, for example, has been consistent. That response has been, "We approve of their ends, but not of their means." Thus, in the immediate postwar years, when the civil rights movement was engaging in court tests of laws and administrative activities perceived as racist, the editorialists were fond of pointing out that "genuine" racial progress could only occur through hard work and "changing the minds and hearts of men." The slogan of the conventional wisdom of the time was that "you can't legislate morality." The same argument was used to deny the efficacy of civil rights legislation designed to effect social change.

This was not the universal response of the media in the postwar era. Rather, it was disproportionately the view of newspapers located in regions where such court tests were being waged and in which such legislation would have its maximum effect—the Southern states. Some newspapers, e.g., the *New York Times,* have tended to be supportive of the civil rights movement consistently when concerned with issues primarily affecting the South. But the more local the issue, the more likely Northern editorialists are to adopt the same critical response that their colleagues in the South adopted.

Nowhere is that response more dramatically shown than in the Northern newspapers' reaction to Martin Luther King. His methods, when applied to the South, were generally accepted in the North, including the tactics of disruption and civil disobedience. The sit-in movement in Southern restaurants and the Selma march received widespread publicity and sympathy from Northern editorialists. When the same tactics were applied in the North, however, the sympathy ended and was replaced by the same stance as had characterized the South at an earlier time. In the years since the death of Martin Luther King, it has become increasingly difficult to realize that this man was not praised by the Northern "liberal" newspapers for his tactics in the North.

Even the *New York Times* shared this general response. The Times was particularly critical of Dr. King's attempt in the last years of his life to broaden his concern to the Vietnam War. King's attempt to relate the cause of war to the causes of racism and to examine the underpinnings of a generally repressive society came under explicit condemnation by the *New York Times.* Thus, in an editorial (April 7, 1967), the Times stated:

> In recent speeches and statements the Rev. Dr. Martin Luther King, Jr. has linked his personal opposition to the war in Viet Nam with the cause of Negro equality in the United States. . . . The political strategy of uniting the peace movement and the civil rights movement could very well be disastrous for both causes. . . . Dr. King can only antagonize opinion in this country instead of winning recruits to the peace movement by recklessly comparing American military methods to those of the Nazis testing "new medicines and new tortures in the concentration camps of Europe". . . . There are no simple or easy answers to the war in Viet Nam or to racial injustice in this country. Linking these hard, complex problems will lead not to solutions but to deeper confusion.

The Times was to alter its opinion both of King and the war. One year later to the day, the Times editorial writers delivered a dramatically laudatory eulogy of Martin Luther King and his life. That editorial, written postmortem, contained the following statements:

> He was an American in the truest sense: for he had a dream. . . . He dreamed for the peace of his countrymen, at home and abroad.
>
> When the United States was honored by his Nobel Peace Prize, he felt obliged to extend his personal philosophy of non-violence from the streets of Selma and Memphis to the rice paddies of the Mecong Delta and the jungles of Viet Nam. He saw the impediments to race and to economic progress at home while a war was waging abroad: "It's inevitable that we've got to bring out the question of the tragic mix-up in priorities. . . . When a nation becomes involved in this kind of war, when the guns of war become a national obsession, social needs inevitably suffer."

The paper further commented on that date, "His militant non-violence accomplished more in his short lifetime than all the violence of the racists, black or white."

The Times was not nearly so enthusiastic about "militant non-violence" when applied to the North prior to King's death. Thus, for example, in a 1963 editorial it remarked:

> The most powerful deterrent to fighting the civil rights battle in the streets is the discovery that direct action tactics rarely accomplish much. . . . Leaders of last summer's unruly equal rights demonstrations in this city—which resulted in more than 700 arrests—have now concluded that mass interference with the rights of others produces little progress toward creating more job opportunities for Negroes or fostering integration.

The editorial then goes on, however, to endorse enthusiastically moves by major corporations to put into effect equal opportunity employment practices.

This inconsistency on the part of the Times is typical of newspaper responses to the civil rights movement and parallels the general public's response to the movement as well. That is, there is continuing denunciation of the tactics which cause intrusions into the normal functioning of the local life-space, but the degree of support for the substantive demands which lie at the base of the activism is certainly not decreased, but instead, would seem to be increased as a result. At the same time, the activists and their leaders, condemned at first, win growing acceptance as their methods pale in comparison to the next cohort of intruders upon the public peace. Thus, Martin Luther King's quite illegal civil disobedience (in the North as well as the South) is conceived as heroism (especially after his death) and Stokely Carmichael's conception of "black power"—a public outrage when first articulated—becomes accepted, or at least relegated to benign neglect. The methods of the left activists are condemned in the locales in which they are practiced and during the time they are practiced. In retrospect, even these condemnations often give way to grudging approval. Simultaneously, the support for the substantive demands contained in the activist programs continues to rise.

THE GROWING RADICAL LEFT

Increasing levels of articulated support for the demands of progressive or "left" causes hardly bespeaks revolutionary fervor in the American heartland. Moreover, such evidence fails to indicate the existence of a growing locus of activist leadership capable of guiding and energizing a successful social movement. Yet there are indeed signs that larger and larger numbers of persons—many from heretofore "establishment" classes and settings—are turning to the radical left for their tactics, politics, and, in still rather rare instances, their self-conception. In the popular term of the day, more Americans are becoming "radicalized."

It would be wise to define the "radical left"—at least for present purposes. The essence of radicalization is that it involves a perception of the connectedness of different aspects of one's social and political worlds; it leads to strategies which are holistic in their attempts to get to the root of a problem. Problems and solutions cease to be viewed as discrete phenomena, but are instead seen as interacting and continuous with the nature of the social order. Seemingly diverse social and personal ills are not viewed serially, to be taken up one after another in bureaucratic problem-solving style. A solution to one problem is a solution to all; a solution to all is necessary for there to be a solution to any. The distinction between members of the American radical left and the American radical right is that whereas the latter locates the root of problems (and thus solutions) in individual or group characteristics (race, attitudes, "evil ways," and the like), the former looks to the social structure and the nature of social institutions as the source of trouble. In other words, the radical-left perspective embodies the sociological imagination (compare with Mills, 1959), and not some variant of a psychological or biological determinist imagination.

A second feature of radicalization is that it is a process—a never-ending process with no real end point. Every so-called radical is capable of further radicalization, just as every nonradical carries a potential for some radicalization. In other words, radicalization involves a change in consciousness which, like any change in consciousness, has boundaries which are infinite. Through changes in one's life conditions and experience, the stage is set for new alterations in what one thinks and does. Radicalism exists only in context of time and place; it exists only in a relative sense. Radical consciousness is held by those on the cutting edge of a given culture; as that culture changes, new opportunities for insight come into being, and as the cutting edge thereby moves, the nature of radicalization must, perforce, move with it. Just as the gaining of the sociological imagination embodies a never-ending search, so the process of radicalization proceeds in the form of a continuously expanding attempt to find connections and to relate features of the social structure to the problems of individuals' lives.

A third characteristic of radicalization is that it involves an activist mentality and life style. That is, people become mobilized to take action on the basis of their beliefs. These actions are public actions and constitute, in the broad sense, political behavior.

Each of these characteristics should be conceived in terms of a continuum, with radicalization constituting the degree to which an individual has moved along the combined continua of the three characteristics. The present concern is with the process by which individuals move along these continua; no assumption is made that the beginnings of the process necessarily result in movement to the end point.

THE SETTING FOR DISILLUSION

The beginnings of the process of radicalization among large numbers of the middle class are rooted in the short-term inability of American society to adapt its institutions to the realities of the postaccumulation era. The wealth and well-being (such as it is) of modern American society comes at the end of a fierce competitive struggle under the aegis of industrial capitalism. The nation continues to experience the existence of institutions which are obsolete given an end to the era which they were designed to serve. All decisions no longer have to be made by the criteria of material growth and profit. Yet policies and priorities continue to function as though this were not true. American foreign policy is based on the premise of dollar diplomacy and continues to be viewed as a means of gaining favorable trade agreements and the ratification of past resource seizures by American corporations (compare with Kolko, 1969; Magdoff, 1969; Horowitz, 1965, 1969). The war in Vietnam, while perhaps not a *direct* result of an attempt to capture resources or to maintain a favorable trade arrangement, springs nonetheless from a frame of reference which is based in a history of imperialist adventures—adventures which were believed to hold important payoffs for the well-being of the American system, at least in terms of corporate profit and general standard of living.

Similarly, domestic government policy continues to foster the disposition and exploitation of internal resources almost purely on the basis of strategic profitability. (Any variation from that system comes only after a heroic battle and constitutes the exception, not the rule.) The result is a physical environment which is aesthetically unpleasurable, an authentic danger to health, and the source of major diseconomies to cities, regions, and the society as a whole.

Finally, the country continues to exist under the rule of persons and institutions which share a social psychology of aggressive Protestant ethicism—an ethos once appropriate to a society struggling for food and shelter for all and material wealth for its elites, but highly inappropriate for a society on the threshold of satisfaction of material want. It is this obsolescence of foreign policy, of methods for resource exploitation, and of the dominant ethos which provides the basis for the radicalization of large numbers of Americans—persons from classes which have heretofore profited, at least to some extent, from the institutions of an industrializing society. As they come into conflict with those institutions, they begin to stand to their fellow citizens as radicals, awaiting the creation of institutions capable of serving their needs and aspirations. To a significant extent, the people of this growing minority emerge from the middle classes of the United States. Should they become dominant, they may cease to hold the position considered (at that later time and place) radical, but they will have moved their society to the substantive left.

THE ENVIRONMENTAL ACTIVISTS

The beginnings of a large activist cadre among middle-class persons organized around the environmental issue may be an important case in point. Some veteran left-radicals view the ecology movements with suspicion and see in them a major cooptation of the left. There are two reasons to suppose that this is not the case. First, the environmental activists seem generally to be persons who were not previously left-radicals—although some of the leadership (probably temporarily) comes from the ranks of left-radicals. Secondly, most serious environmental activists tend to be engaged in actions which, at least to a degree, place them in opposition to dominant institutions. The lessons learned through environmental politics are thus at least as likely to move participants in the direction of the left as to coopt them to the status quo.

Conservation movements have a cyclical history in America, fading and emerging as fad and fancy has moved the upper class from one concern to another (compare with Nash, 1967). But the current group of conservationists are different from their elite forbears. Conservation movements of the past have arisen from two types of motivation: (1) aesthetic concern for the preservation of landscapes and endangered species for a remote posterity; and (2) economic needs of major industrial interests which did not want resources "mismanaged" through profligate small-time profiteering. Such an ardent conservationist as Theodore Roosevelt was a mixture of both: he protected wildlife while simultaneously guarding the long-term interests of cattle corporations, oil prospectors, private utilities, and lumber interests (compare with Ridgeway, 1970).

The concerns of the new conservationists are far more profound. The entire process by which goals are established, resources exploited, and goods distributed has come under attack. Even among those contemporary ecology activists who are interested in maintaining certain species or preserving a given wilderness, the concern is anchored in the ecological notion of interdependence between man and the other species. The concern is not remote but rooted in individual self-interest and an often apocalyptic view of the consequences of environmental disruption. The focus of the current movement is thus by no means upon trivial matters: the activists see themselves as fighting for health, a decent environment, and the continuation of the human species itself. Many have been willing to sacrifice for their cause–in ways which previous environmentalists have not. Some have devised an entire life style around the goals of minimal consumption, minimal intervention in the habitat, and maximum enjoyment of nature. A few (among the young) have come to see themselves as revolutionaries, and are designing a life around this self-depiction.

The new environmentalists emerge at a point in history when two processes are occurring simultaneously. On the one hand, the quality of the environment (at least in certain crucial ways) is, in fact, deteriorating. Air and water pollution, for example, are likely at their highest levels–claims of gasoline companies to the contrary notwithstanding. On the other hand, the sensitivity of middle-class people to their environment and the significance of the environment as a source of personal gratification is increasing.

That the locus of environmental activism should emerge in the more comfortable classes is logical enough. The environment becomes a personal problem only after the more basic problems have been solved. It is a problem which is a luxury, and it comes to large numbers of persons only after large numbers of persons have satisfied their most basic human needs.

That such sorts of discontent should arise out of the success of capitalist industrial development was anticipated in 1874 by Friederich Engels who foresaw the coming contradiction:

> Let us not, however, flatter ourselves overmuch on account of our human victories over nature. For each such victory, nature takes its revenge on us. Each victory, it is true, in the first place brings about the results we expected, but in the second and third places, it has quite different, unforeseen effects which only too often cancel the first. . . . At every step we are reminded that we . . . belong to nature, and exist in its midst, and that our mastery over it consists in the fact that we have the advantage over all other creatures of being able to learn its laws and apply them correctly [Yale, 1970].

Increasing numbers conceive of those laws as *not* being applied correctly; current modes of decision-making, and the premise upon which such decision-making rests, act to preclude such correct application. The goal of increased productivity, blind to all ends except private profit maximization, has led to current environmental difficulties. Echoing Engels, Paul Ray (1968: 18) has pointed out:

> Stocks of energy do not *control,* though we often think of technology as giving us power (and therefore control) over nature. Rather, we are all familiar with the disruptions that ensue, being magnified through an ecosystem. . . . The failure of societies is in the realm of *control* over the environment and over the mutually inconsistent behaviors within society.

In other words, man needs first to define his goals in terms of the pleasures he seeks and the qualities of life he desires and then needs to utilize technology so that there is "*low* utilization to produce effects of major magnitudes with precision" (Ray, 1968: 18).

Technology needs to be the rational harnessing of energy to achieve with maximum efficiency certain life goals *which are determined outside of the context of technology and production.* Thus, for example, if blue skies or sweet air are ultimate goals, there then is no reasonable argument which could dismiss such goals as "impractical." Given the creation of a productive base which is capable of maintaining basic subsistence, the primary historical rationale for determining the use of resources is at its end. Ray (1968: 19) has put the matter succinctly:

> if scarcity no longer is *the* critical problem, then sustenance technology *must* leave the driver's seat, and so must those organizations controlling it. . . . We are not prepared for intellectual paradigms that take affluence for granted and state the *desirable* (rather than necessary) as highest goals.

Capitalism, however imperfectly, has provided the West with a means of attaining the necessary end of sorting out priorities given the overriding need to produce the necessary. It does not, however, contain a means of specifying the desirable or of providing mechanisms to achieve the desired. In fact, capitalism seems to do the opposite—to hinder the vision needed to establish goals and to thwart the achievement of such goals once they are articulated. That is the rub.

There is a great deal of evidence to suggest that the American public is increasingly turning to life visions which challenge the goals of crude energy maximization through the profit system. The consumption of consumer goods as the basis for social organization has been challenged occasionally over the last several decades (for one important version, see Galbraith, 1958), but only recently have there been signs that massive numbers of Americans may be turning to a broader definition of the good life.

Foremost among that evidence has been the sharp rise in public indignation and protest over the pollution of the environment. Pollution has at last arrived on the "most important problem" list of the Gallup poll (Gallup, February 1970). Noteworthy is the fact that, as Gallup has reported, seventy-three percent of the American people report a willingness to pay higher taxes in order to "fight conservation problems" (National Wildlife, 1969). Similarly, a majority feel that "more tax dollars should go for natural resource programs . . . less for space, defense and foreign aid" (National Wildlife, 1970). The environment has become a concern of leading businessmen (Fortune, 1969), as well as a major issue for college students (Gallup, January 1970). A 1956 survey conducted by the California Health Survey reports that sixty-six percent of Los Angeles County residents find that air pollution does "bother" them. In addition, seventeen percent of Los Angeles County residents have considered changing their residence because of air pollution, and nine percent have considered changing their jobs for that reason. The point is that, at least in that metropolitan area, vast numbers of persons are being affected by poor environmental conditions and are aware of and complain about those conditions. Most striking of all is that in California air pollution now ranks about equal (fifty-nine percent versus sixty-three percent) to "crime in the streets" in the proportion of respondents who consider it the "most serious" problem (Gold, 1970). The fact that the crime issue was being actively promoted by California politicians, while the pollution issue was being avoided, adds extra weight to the prominence of pollution as a rather spontaneously generated, directly felt grievance.

These rather flat reports of survey data fail to convey the degree to which at least some members of the higher classes have come to feel environmental degradation as an immediate problem. There is probably no cause which has inspired as much militance among the middle classes as the environmental issue. It began years ago when women in New York suburbs and the South Side of Chicago started climbing trees and chaining themselves to branches in order to provide nature with protection from the axes of highway builders. More recently, there has been the creation of a large number of broad-based voluntary associations which have militant environmental protection as their goal. Such organizations include Ecology Action on the East and West coasts, organizations devised to clean up Lake Erie, Save the Bay in San Francisco, and Get Oil Out (GOO!) in Santa Barbara, California. These groups have created news bulletins and magazines; they have picketed and lobbied; they advertise on a massive basis through the media and through word of mouth; they constitute the beginnings of a national grass roots movement which at base is engaged in a struggle with capitalist institutions over the means of production and the methods of resource exploitation and product distribution. (They usually don't word it that way.) This is no small battle, and the implications could be very significant.

Oil in Santa Barbara. Perhaps the most dramatic head-on clash of an ecological disaster with this emerging life style occurred in Santa Barbara, California, when Union Oil Company's Platform A "blew out"—covering the oceanfront and twenty-five miles of contiguous beach with a coat of crude oil. Almost two years later, oil continues to seep from under the platform—now at a reduced rate of approximately four hundred twenty gallons per day.[1] Numerous bills to end drilling have been introduced into the congressional hopper, but all have become buried or killed in committee. Court cases have been initiated, but without result. Administrative remedies (through hearing and appeal) have been attempted, but have been ignored. It is my argument that the response of Santa Barbara citizens to these events (nonevents) would seem to furnish evidence that the ecological issue serves an as important agency for mobilizing activists and for mass education on the realities of the American political economy. Here and there and in-between, it has even made some radicals. This has happened despite (and, at times, because of) the continuous outpouring of government and oil company rhetoric designed to obfuscate issues and cover up inaction.

The oil spill was remarkable in the degree to which it caused ordinary uninvolved citizens to take some sort of minimal public action. No previous issue has ever brought forth such an outpouring of public sentiment—not even the subsequent 1970 Isla Vista riots and the burning of the Bank of America. The city's newspaper published 372 letters to the editor (almost all condemnatory of the oil companies or the government) during the first February after the spill—and many more hundreds thereafter. A petition campaign to the federal government demanding a halt to all drilling was to gather approximately 200,000 signatures (remarkable in a metropolitan area with a population only slightly above 100,000). The number of persons who attended rallies, participated in anti-oil organizations and wrote their congressmen certainly must have been a significant portion of the adult population of the Santa Barbara area. A public issue, largely without ballyhoo from politicians or any other outside spokesmen, came into existence. This, in itself, is an unusual event in the United States.

Not only was this activated anti-oil constituency remarkable in its size, but it was also noteworthy for its political heterogeneity: it covered most of the spectrum between the area's left and right wings. The most prominent anti-oil organization (Get Oil Out, Inc.–GOO!) had a board of directors which was indeed an unusual coalition. The organization's most important guiding lights were a former state senator who ran one of the nation's blue-blood resort hotels in the Santa Barbara hills and a long-active Santa Barbara civic leader who by birth and demeanor is the daughter of a countess. In addition, the leadership core consisted of a right-wing insurance executive, corporation presidents, traditional conservationist types, a painting contractor, and persons active in the peace movement. At all times, the *Santa Barbara News-Press,* a stodgy old California paper, provided unstinting aid.

This coalition and its surrounding support base was carried through a painful process of failure after failure in its attempt to redress what was conceived of as so obvious and so straightforward a wrong. The initial response of the oil activists had been one of belief in their own efficacy as citizens. Surely after the right individuals had been informed of the facts–the impossibility of drilling safely in the channel's quake-prone and fragile bottom, the nonexistence of any methods of controlling blow-outs or of cleaning the mess they make, the irresponsibility of the so-called regulatory agencies–corrective action would be taken. That corrective action was interpreted everywhere in Santa Barbara and almost without exception as the complete elimination of oil drilling in the ocean.

A recurring theme of industry and government's recent antipollution public relations effort has been to point out that a clean environment, although certainly desirable, is going "to cost you money" (compare with Ridgeway, 1970). This line of argument could conceivably shift attention away from the polluters and pollution control and toward the supposed diseconomies of environmental protection. The industrialist is thereby placed in the role of beneficent protector of the family budget who has heretofore refrained from pollution control only to save the consumer money. The argument, received in the abstract, might indeed have the consequence of depressing a public's appetite for rigorous protection policies, lest production, prices, or taxes rise as a consequence.

But the reality of pollution economics is perhaps very different—and this reality is learned as one is drawn into contact with an actual industry and the economics of an actual case of pollution. Anyone familiar with the political-economic context of off-shore oil pollution could not possibly conclude that pollution contributes in any way to the American people's standard of living. Producing oil in the United States (whether on land or sea) has been a very expensive proposition. The depletion allowance alone is estimated to have amounted to a federal subsidy of over $140 billion (Dugger, 1969: 19). Further, oil company advantages of early tax write-offs and arbitrary import restrictions increase costs to taxpayers still more while at the same time inducing unnecessary exploratory drilling.

Most ironic, however, is the fact that government systems of price supports (primarily through import restriction) artificially inflate the value of domestic oil (approximately one-third above world market prices), thus making off-shore drilling profitable in the first place. That is, drilling in deep water off the Santa Barbara coast is a very expensive proposition compared to the much simpler case of land-based operations. Walter Mead, a prominent petroleum economist, concludes that it actually costs more to get the oil from beneath the oceal floor than the oil is worth (Mead, 1969a; 1969b; Mead with Sorenson, 1970). The only reason the oil companies drill off-shore is that they do, in fact, get more for the oil than it is worth, through inflated domestic prices. The government pays for the oil to be drilled and then cites royalties received by government (a fraction of the subsidy) as a reason why drilling must be continued. Thus, Santa Barbarans (along with other Americans) are paying to degrade the beauty and ecological viability of the off-shore lands.

In other words, drilling for oil in the ocean off the coast of the United States makes no sense whatever. The oil is not needed and the operation constitutes a drag on the national economy. Off-shore drilling is made possible purely through the peculiar machinations of the American political economy. It *is* a planned economy—but planned for strategic profitability rather than for equity or efficiency.

On these matters of the economics of oil, as well as in regard to concerns ecological and aesthetic, the "right people" were easily informed—first through friendly lunches and phone calls (many GOO leaders were on a first-name basis with United States senators and federal officials) and then through official hearings and reports. The fact that this device did not succeed led to a cynicism unusual among persons who usually have both feet firmly planted in the establishment.

There were many interesting manifestations of this rather untypical suspiciousness of the institutions of the establishment. First of all, none of the facts released by the Department of the Interior, any oil company, or any other federal agency came to be taken at face value. Accounts of the volume of oil spillage, damage to wildlife, costs of the clean-up, costs of ending drilling, geological conditions of the channel were almost never accepted. The question was asked—explicitly or implicitly—for whom did the scientist or engineer work, and his expertise came to be evaluated according to the answer given. Thus when oil and interior scientists announced that the only way to stop the continuing seeps was to *accelerate* drilling, the response was bitter skepticism. When scientists, working under grants of the Western Oil and Gas Association, reported minimal effects on the channel's ecology, the conclusions were ignored in favor of contrary evaluations given by scientists not in the employ of oil companies.[2]

At the governmental level, local officials evidenced the same kind of skepticism. Independent scientific environmental boards were created to provide alternative sources of information for local government decision makers. Attempts by the federal government to negotiate compromises were met with a cold shoulder, lest county officials become trapped in lending credibility to new drilling programs which would have no advantage over the old. In one instance, the County Board of Supervisors was invited to participate in a review of what were called new "tough" drilling regulations. The invitation was refused—under the assumption that the "new" rules were simply an excuse for inaction and another public relations stunt. Similarly, federal legislation which would have set aside small portions of the channel in oilfree sanctuaries was either ignored or actively opposed. Again, Santa Barbarans did not want to participate in the nonsolution of their problem.

Finding the more legitimate means blocked—discovering that the legislative process, the courts, and the federal regulatory agencies were directly or indirectly, obviously or subtly, stacked in favor of the oil companies—anti-oil activists were caused to expand their repertory of tactics. With varying degrees of self-consciousness, they came to see themselves in a competitive battle with the federal government and the oil companies. Victory in that battle had little to do with the introduction of bills (which would never pass), entering court suits (which could not win) or mounting evidence (which would never be seriously dealt with). Instead, the real battle was over the manipulation of the historical record—the garnering of the resources needed to create national events which could be "played" by the media and thus be made real to an incipient national constituency. Only by creating public dramas which would embarrass the oil companies and the federal government through public scandal could there be any hope of a policy change. The lesson was learned that "the power to make a reportable event is . . . the power to make experience" (Boorstin, 1962: 10). Making experience, it was hoped, would lead to governmental action.

Thus it was that Santa Barbara's anti-oil movement came to recapitulate over the two years of its existence something approximating the pre-Chicago Convention tactical development of the civil rights and student movements. As with those movements, militance began with actions rather mild. A group began daily pickets of the local wharf serving the oil industries in an attempt to bring in national publicity. A "sail-in" was held one Sunday with a flotilla of local pleasure boats and yachts forming a circle around the offending Union Oil platform, each craft bearing large anti-oil banners. City Hall meetings were packed with citizens reciting demands for immediate and forceful local action.

A dramatic Easter Sunday confrontation (involving approximately five hundred persons) took place between a group of demonstrators and city police. At the conclusion of a rally held at the wharf, an oil service truck unexpectedly began driving up the pier to make a delivery of casing supplies for oil drilling. There was a spontaneous sit-down in front of the truck. Thus a group of Santa Barbarans (some young, some "hippy," but many middle-class adults) was publicly taking the law into its own hands. After much lengthy discussion among police, the truck driver, and the demonstrators, the truck was ordered away and the demonstrators remained to rejoice in their victory. The following day's newspaper editorial, while not supportive of such tactics, found much to excuse—noteworthy, given the paper's long-standing bitter opposition to similar tactics when exercised by dissident Northern blacks or student radicals.

A still more dramatic event occurred when the Sun Oil Company was granted permission to raise still another ugly oil platform adjacent to the leaky lease operated by Union Oil. A group under the leadership of GOO! arranged to have several yachts located around the clock on the very spot where the oil companies were to erect their platform, thus indefinitely delaying its installation. The "fish-in," as it was called, also featured the countess in a helicopter with a line dropped daintily into the ocean below. The legality of the operation was ambiguous. But for at least some participants in the fish-in, legality did not matter, and these participants explicitly said so.

The response of the few students who were involved in the struggle, as well as the students who watched in the wings, seems to have been that of a group who witnessed a confirmation of their worst suspicions. A second sit-in was held at the wharf consisting almost exclusively of college and university students—primarily students from the University of California campus. This sit-in on the wharf successfully blocked all traffic (including oil service vehicles) for a day and a night, but was finally called off when a phalanx of police threatened forcible eviction. It was this group of young people who, after spending the night at the wharf, moved directly to the U.C. Santa Barbara campus where they initiated a new sit-in in front of the campus administration building in protest against the firing of a popular young anthropology professor. Thus began the series of rallies, demonstrations, and confrontations which were to culminate in the weeks and months to come in the burning of a branch of the Bank of America to the ground. I have no information to presume that any of the bank burners were the same individuals who sat at the wharf; but the wharf sit-in was the first act in a politicization sequence which persisted, almost without respite, for the next five months of the school year.

The bank-burning and the oil spill were events widely separated in time and there is no evidence to suppose that their experience with oil and government caused any sympathy for the Isla Vista rioters among Santa Barbara adults. But it does seem that at least in previous moods, many respectable Santa Barbarans held sentiments and gut feelings which, if mixed with a younger body and a younger mind, might have resulted in similarly outrageous behavior.

Some of the city's most eminent citizens discussed, on more than one occasion, blowing up the oil platforms. These remarks were not really serious, one assumes. "I'd like to blow the Goddamned things up" is a sentence which can be easily squeezed between cocktails without implying any serious volition. I also heard a party of elderly prominent citizens discuss, in rather close detail, the actual logistical problems of blowing up an oil platform: how to get out to sea on a calm night; how to destroy the platform without injury to oil workers and without setting off a great conflagration of burning oil. Again, none of these people was serious—but the thought was certainly there.

A widely felt mood was articulated by the President of an international science corporation headquartered in Santa Barbara:

> We are so Goddamned frustrated. The whole democratic process seems to be falling apart. Nobody responds to us, and we end up doing things progressively less reasonable. This town is going to blow up if there isn't some reasonable attitude expressed by the Federal Government—nothing seems to happen except that we lose.

Similarly, a well-to-do widow, during a legal proceeding in a federal district court in which Santa Barbara was once again "losing," whispered in the author's ear:

> Now I understand why those young people at the University go around throwing things. . . . The individual has no rights at all.

Now, almost two years after the initial oil spill, there has been differential adjustment to the continued existence of oil in the Santa Barbara Channel. The national media are gone; those across the country that once opposed oil development in the Santa Barbara Channel are now thinking about other crises. On the local scene, some leaders of GOO! are still weakly hopeful and watch with interest the continuing noises about conservation, pollution, and proposed legislation affecting the Santa Barbara Channel in the United States Congress. Many other local leaders, however, are bitter about their experience but have ceased operating as political activists on the oil issue. Some are involved in the more benign ecological activities—paper recycling, aluminum can collection, and other programs which are less likely to foster a continuous political education. But other GOO! workers have led a bitter fight against developers intent upon subdividing oceanfront ranch land north of the city. Among other results of this campaign, actions by the land developer have led to the arrest of two conservationists (one a 75-year-old Santa Barbara matron) for perjury—a felony charge.

Once again there is a demonstration of the power of vested interests to use government to pursue private profit—this time at the local level.

The tentative lessons which I draw from the Santa Barbara case are as follows. Although environmental activism leads to radical-left insights, adults with comfortable lives in comfortable settings are constrained from carrying out a complete reevaluation of the American system and their place within it. Instead, they tend to seek an avenue in which they can be active without placing themselves in uncompromising opposition to the system. But in terms of the development of their political consciousness, they do get moved. They begin to understand certain things; they see the connectedness of the institutions which surround their lives; they understand what it is like to be frustrated by unresponsive social institutions which will not bend to what is obviously the "right way." In terms of tactics, they become willing to experiment with new methods and to do what is necessary to compete with the politicians and industrial spokesmen in order to gain public attention and to create and manipulate public events. They become especially audacious when they think the chances of success are high. Thus, they show themselves willing to use many of the techniques of the activist left in an attempt to secure their goals. But when the cause is seen as quite futile, they retreat to safe roles in which they can respond to their social conscience, yet at the same time continue their comfortable relationships with each other and with their surrounding society.

At a minimum, however, they have been neutralized as unthinking enemies of the frustrated. They leave the ranks of those who cannot understand the question, "what do those people want?" It may be that as increasing numbers of environmental dissidents become publicly visible, the middle classes and upper-middle classes will conceive their chances of success to be far greater, and thus their willingness to embrace radically different conceptions of their roles in society might increase concomitantly. There is the further possibility that even the "safe" ecology actions will provide a radicalizing experience, simply by generating new forms of consciousness and life style which will free people from old associates and old visions of the way a political economy should be managed. There are also potential conflicts with lumber and mining interests in the recycling programs, and there are inherent future conflicts with land developers, builders, and financiers through wilderness preservation, intelligent zoning, and any action with the potential of inhibiting growth. As these experiences cumulate, a new radical consciousness may well be advanced as a result of environmental activism.

YOUTH

Although young people played only a very minor role in the anti-oil activism of Santa Barbara, the effects of such environmental catastrophes on youth are likely not minimal. Ecology was by no means the first or most important "new left" issue on the American campus, but it is noteworthy that what have been perhaps the three most dramatic campus confrontations in recent years were initiated, at least temporally, by an ecological issue. The events which led to the Isla Vista bank-burning began with the anti-oil sit-in; the Columbia sit-ins were in response to the university's incursion into a community park; and the People's Park rebellion in Berkeley similarly had something of an ecological issue at its core. The presence of such an issue as somewhat central to student activism should heighten our sensitivity to the fact that we are dealing with activist phenomena which are indeed extraordinary to the historical experience. Both in terms of its class origins and the issues raised, the rebellion of American youth is remarkable.

Left activism on American college campuses has by now become so common that it is easy to lose sight of the fact that both by precedent and theory, such a situation is an anomaly. Neither Marxist theory, nor any other, would have predicted it, nor would the preceding youth generation of the 1950s have provided any basis for its anticipation. Middle-class college youth have traditionally provided stability and conservatism, rather than instability to the American system.

The routineness of youth rebellion leads as well to occasional exaggeration of the extent to which the movement has reached all young people. It has not done so—yet. Even among college youth, only a minority consider themselves anything like radical.[3] But it is easily argued that this minority is growing rapidly and that disaffection is spreading from its original elite loci (Ann Arbor, Berkeley, Madison) deep into the academic hinterlands (Kansas, Kent) and beyond to the young working class matriculating in the military services. Should an appreciable proportion of youth become and remain radical, major social change in America will become inevitable. Youth is only one sector of the population (although the most important, and among the largest); college youth is only one sector of the total youth population (although again, the most important and among the largest). As the cohorts of radical youth cumulate into young adulthood, the size of an important radical constituency could become large indeed.

The strains which exist between current institutions and the youth culture are certain to continue building this constituency. The structures which were developed for material accumulation, international domination, and bureaucracy-building are, by their nature, in opposition to the values and modus operandi of the emerging youth culture. Opposition to wars of containment places young people in a continuous life-and-death opposition to the national government and to the local institutions which administer its will; the use of drugs places young people in daily opposition to the police, the courts, and the local elite, as well; the propensity to experiment and to demand humanistic education and administration puts the young in opposition to university administrators, business executives, and those others who administer most of society's institutions.

None of this is news. The news involves the emerging consequences which the development of homogeneous, rather large-scale communities of youthful dissidents creates (compare with Mankoff and Flacks, 1970). Clearly, there are groups arising which are experimenting with a radically new kind of community. These communities (Berkeley, Madison, Isla Vista) are being spoken of as sanctuaries—places where the new society can develop its tactics, its methods for survival, and its bases for interpersonal satisfactions. New institutions are emerging in these communities, such as communes, food co-ops, community gardens, and craft industries of various sorts.

These communities are unique. They bring together similar types of people who gain the potential of joyous communication and culture-building with persons who have similar dispositions: they are real communities. They are ghettos in the sense that they also bring together an alien population, one which is disliked and even despised by those who manage the dominant, surrounding institutions. Generally, they are administered by individuals who do not share the culture which is the ethos of the place; they have colonial status and are ruled from afar. Their people are not taken seriously in governmental circles; they have (as yet) no clout; their residents are seen as less than human beings.

The community of Isla Vista, adjacent to the University of California at Santa Barbara, constitutes one such youth ghetto in the United States. Until recently, it served as the "bedroom community" for what was among the richest and most play-oriented campuses in the West. During the academic year 1969-1970, I observed the development of Isla Vista as a political community, a major rebellion which took place on its streets, and its occupation, first by police, then by National Guard, then by police, and continuously by police. Events in Isla Vista brought out clearly the nature of the relationship which exists between the youth ghetto and the surrounding community. Those relationships are analagous in many ways to the relationships between a black ghetto and its surrounding white "noose"; they are also reminiscent of a village in Vietnam which is controlled by government forces by day, but whose people, food supply, and street safety (especially at night) are in the hands of the Viet Cong.

During the many nights of street fighting in Isla Vista, a clear pattern emerged. The young people are fast on their feet. They know the streets and the buildings intimately. They know which apartment buildings have passageways through courtyards; they know which laundry rooms are open at night and which apartment buildings contain the most sympathetic partisans; they know which back doors are always unlocked.

The police face frustration. They have a sturdy dislike for the long hair, life style, politics, and general reputation of the youth. But they cannot catch anybody. The students in Isla Vista have not been gentle in their dealings with the police. On some nights of street fighting they aggressively began confrontations by stoning police cars; at one point they trapped approximately fifty policemen against the wall of what was then the Bank of America, and continued pelting them with stones and rocks until the police were forced to beat a hasty retreat. Of course, these events isolated from their contexts seem to be without reason or provocation; given the context of police harassment for marijuana smoking, police beatings of students without reason, police harassment of nude bathers on university and surrounding beaches—all of these are the causes of the seemingly senseless aggression of students against police. But for the individual policeman on duty, these antecedent circumstances are not only uninteresting, but may, in fact, be unknown.

Students further police frustration through the development of premeditated community defense tactics. Slingshots and catapults were purchased (or "ripped off") and used to propel rocks at police windshields; nails were driven into roadways, puncturing the tires of police vehicles patrolling after curfew. Street signs were painted over to confuse police—many of whom were assigned from neighboring districts and lacked familiarity with the area.

The police, in their mounting frustration, become willing to expand the repertory ordinarily practiced in a middle-class white community. Groups of policemen went through apartment buildings knocking each door down and beating up the inhabitants. Some students report being lifted off toilets; some were removed from their desks while studying; others found themselves being dragged down flights of stairs by the hair for no reason whatever. The police, almost by the nature of the circumstances, are generally able to catch students who were not involved in any wrongdoing. The fact that all of the students more or less look alike to the police becomes the basis for what the students experience as indiscriminate terror. Increasingly large numbers of students thus come to experience first-hand police brutality, and an important prop of the social order is thereby destroyed.[4]

Students who are illegitimately arrested (and they have numbered in the many hundreds in Isla Vista) then come to experience the nature of the bail system, the cost of an adequate defense, the harassment of adjudication procedures, and the kind of men who are elected or named to judgeships. As one important example, the Santa Barbara County Grand Jury indicted (at the district attorney's request) seventeen Isla Vistans for the bank-burning—which had many hundreds of participants. Of these seventeen, two were in jail during the night in question and were thus eventually dropped from the list. Of the remainder, a majority are well-known student activists who, witnesses report, were nowhere near the bank or (in the case of some) even in Isla Vista during the burning night. One of the indicted had previously served as vice-president of the student body and proved to be the highest vote getter in a recent election for representatives to a newly formed Isla Vista Community Council. Another of the defendants is the past city editor of the student newspaper. Those who know these people (and they are widely known) will simply not believe the prosecution testimony against them. The trial is being run during the summer in an attempt by the court to preclude student attendance, but the students will, upon their return, view the incident as another attempt to destroy their leaders through out-and-out repression.

The consequence of such developments is that, for large numbers of people, the legitimacy of an important institution crumbles. The complaints of others–i.e., the minorities–of the viciousness and illegitimacy of the system become very real. It is much more real to these students than the tracts of socialism from the new left which inspired an earlier generation of student activists. Many of these youth respond not as intellectuals, but on the basis of their own experience. Their ideology, such as it is, is bolstered by a willingness to take risks, to experience the thrill of chase and battle, and a desire for revenge. They are able to use their knowledge of their own community and the relative homogeneity of that community to wage a rather successful battle against the police. They are able, like the Viet Cong, to melt into the populace. With or without intent, they become the first white urban guerrillas.

The emergence of the youth ghetto and the resultant consequences have important implications for the political consciousness of the adult community, as well. Those who previously were not in sympathy with the students, who are prowar and antiblack, may very well be strengthened in their proclivities in that direction. But for other citizens there is an awakening which, while less dramatic than the experience of the students, may be significant nonetheless. First of all, the homogeneity of the Isla Vista youth ghetto, like any youth ghetto, is not quite complete. Adjacent to the student quarter is a family area; sprinkled throughout Isla Vista are older persons from other walks of life who, for one reason or another, like to live near the "swinging" university community. In the course of events in Isla Vista, these people experienced a very dramatic series of events.

Their response to the student activism involved a good bit of vocalization opposing violence and condemning bank-burning. But the burning of the bank, and its attendant activities, took place within a rather narrow piece of geographical turf. The "trashing" of local real estate offices and other targeted "capitalists" occurred on only a few adjacent streets; one could more or less go about one's business anywhere else in Isla Vista during the student disturbances.

Indeed, while the bank was burning, one could purchase an ice cream cone directly across the street and enjoy it in comfort while watching the Bank of America burn to the ground. If one were able to experience routine life across the street from the burning bank, one was certainly safe a block, two blocks, or a half-mile away.

The police response, however, created a very different circumstance. When the augmented police force finally came on the scene, and on the successive nights of police occupation, no one in Isla Vista was safe from excesses of police power. Tear gas canisters were shot off indiscriminately through students' apartment windows and on the lawns of split-level homes in the single family district. A man who turned out to be an assistant district attorney was arrested because he was standing in the driveway of his own home after curfew (7:30 p.m.). An insurance company owner and former member of the Santa Barbara County Grand Jury was arrested for participating in a sit-in protest, and because he lost his balance and touched a policeman's pocket, was charged with assault with attempt to murder a police officer. A father of four children (and a Teamster) was arrested while trying to adjust the television aerial on the top of his home after curfew and was hauled off for arrest in front of his wife and screaming children.

And so it went. Professors, students, businessmen—anyone within the areal unit labeled Isla Vista—was subject to beating and arrest. The number of people radicalized by this experience is beyond my ability to guess. The number is significant enough, however, to have caused some local conservative Isla Vistans to communicate to columnist William Buckley the fact that police actions had caused an entire community to be radicalized. Buckley was convinced of the accuracy of their statements, and, for what I take to be the first time in the history of his syndicated column (or of his writings in general), warned against the dangers of police brutality and indicated that the brutality in Isla Vista was indeed excessive (compare with Santa Barbara News-Press, 1970a). William Buckley's concern is good evidence that his informants, at least, were convinced that a large number of students and adults were radicalized by the continuing occupation of Isla Vista. I would concur.

The consequences upon Santa Barbara citizens were not limited to the students or even the adults who happened to live within Isla Vista's boundaries. One of the important differences between the victims of police terror in a youth ghetto and victims in a black ghetto is the differential social status and credibility which members of the two groups enjoy in the wider society. Thus, when white youths paraded in front of cameras with tales of violence, displaying their wounds, and articulating their grievances, the word was able to get out past the usual boundaries. Perhaps more important were the stories told by the middle-class residents of Isla Vista who happened to get caught up in the various sweeps.

Many of the young people themselves were obviously of the straight variety; some were majors in the physical sciences with no political connections or past history of participation in campus activist events. The young student who was shot to death in front of the Bank of America was himself the prototype of the all-American boy: a "straight" senior honor student who was shot by the police ("accidentally," according to a grand jury report) while trying to defend the temporary structure (which replaced the Bank of America branch building) from its assailants. The handling of the death by police, grand jury, and the media provided another major community irritant.[5] Not only was the news manipulated with insinuations of snipers, and so forth, but the policeman who admitted the killing was not even subjected to a public reprimand or demotion. In response to the brutality allegations, the local district attorney and the sheriff of Santa Barbara County denied the existence of any police wrongdoing and quoted J. Edgar Hoover to the effect that those who meddle in police affairs are the enemies of a stable society.

By this time, a large number of "meddlers" had come to the fore. The local medical society sent doctors as observers; the bar association sent observers; an ad hoc blue ribbon citizens' commission was spontaneously begun and proceeded to take testimony from Isla Vista victims for many days and nights. A group of twenty-four faculty members (along with hundreds of others) surrendered themselves as curfew violators in order to dramatize the brutality of past police activity. First-hand accounts gathered by local reporters, TV cameramen, and various investigators were paraded before the public, providing dramatic visual evidence of wanton police brutality. The communication net about police repression thus came to be expanded as persons who ordinarily know nothing of the machinations of "curbstone justice" and the duplicities of the legal system find themselves in the position of entertaining first-hand accounts from sources which range from the marginally to the completely credible.

It is not being suggested that the majority of Santa Barbara citizens have lost their confidence in the police department, nor that they were undergoing an experience of mass radicalization. I am not particularly impressed by the fact that when it comes to police-student confrontations, the overwhelming public sympathy is with the police. This is to be expected; there is no reason to think bank-burning should be popular with taxpayers. Rather, the extraordinary thing is that any significant minority of middle-class persons should come to lose their belief in the legitimacy of the police and in the American system of justice, and that an even larger group at least becomes neutralized as vocal advocates of police power. That an overwhelming majority of Americans continue to support the police and to reject the idea that the police are routinely brutal (Erskine, 1968: 667) is not as interesting as the fact that some people have come to the belief that the police are, in fact, routinely brutal. The latter fact contains the news.

The legitimacy of a governmental order and its consequent stability is not determined by majority vote. An orderly society requires that virtually everyone have some modicum of belief in the legitimacy of the institutions of social control. A system of order can collapse long before the point is reached where a majority come to disbelieve in the legitimacy of the important institutions. It is thus a very important omen that significant minorities regard the police as pigs—and that the numbers are rising, not declining.

WOMEN

Similar to the student movement in its class origins, women's liberation fosters a similarly critical evaluation of capitalist industrial institutions. The recruiting base of women's liberation seems to be extremely broad, and again, like the ecology and youth movements, appears not to be draining the cadres of the left, but rather to be mobilizing larger numbers of persons against status quo institutions. Again, there may be some sharing of leadership, but the net consequence seems not to be a draining of left resources, but rather to establish still another left constituency, one to be argued with, bargained with, and to be joined in mutually beneficial coalition. As with the ecology movements, a dearth of systematic data precludes anything more than a speculative account of the origins and contemporary forms the women's movements might be expected to take.

But again, there are good reasons to suppose that there is something of a structural contradiction between the aspirations of middle-class women and the nature of contemporary capitalist institutions. Repression of women arises as an issue because the conditions appropriate to the restriction of females to their traditional roles have come to an end. Capitalistic industrial development required vast amounts of human breeding (as did preindustrial agrarian conditions) to constantly replenish a labor supply being killed off by the grueling conditions of early factory life and by the unhealthy conditions of the industrial slum. (The life expectancy in Liverpool and Manchester in 1841 was twenty-six years.) The end of those unhealthy conditions has eliminated the need for such mass breeding; the increasingly mixed benefits of material production further weaken the imperatives of sex-role allocation to promote large labor supplies. The woman's place is no longer (necessarily) in the home raising children.

A second demographic characteristic of capitalism is that whether or not increasing population levels are necessary for economic health (a view once widely held), it does seem clear that an increase in population enhances profits for the capitalist class. That is, although a worker who receives a fixed salary based upon his own productivity is not made more prosperous by the addition of another into the work force, the capitalist stands in a very different position. The surplus labor of each worker accumulates to the advantage of the capitalist class; thus, the larger number of worker-consumers, the greater the size of the surplus labor pie, and the greater the wealth of the capitalist class. As long as the size of the capitalist class remains constant, an orderly increase in population devolves to the advantage of the capitalist who, in effect, takes a constant percentage of the gross—no matter how large. All of this assumes, of course, a complementary relationship between population and productivity growth rates—and it is necessary for such rates to be manipulated (e.g., through birth control) to ensure orderly growth. Where such a relationship is not present, as in certain underdeveloped countries, the situation is not at all optimum (compare with Davis, 1965).

This analysis is quite tentative; regardless of its validity, it is important that a large number of persons believes certain aspects of it. Booming population and booming business are kindred booms in the minds of many. The financial pages of American newspapers are filled with reports of business executives who are publicly licking their chops in anticipation of the new markets which the ripening of bumper baby crops are thought to bring out. Stock market investment letters continue to assess a corporation's potential profitability in terms of the differential effects of various cohort sizes upon different industrial groups. Local and regional elites promote population growth in every way possible to help "build" their community.

In addition to its demographic consequences, the repression of women is functional in that it bolsters the role of exploiter, builder, developer, which capitalist development requires and designates as masculine traits. Thus the services of the male libido are inducted for the purpose of capital accumulation. The passive, trivial, breeding female—invoked as the reason why all of this is done—is the gal behind the man, the prop who "without you, baby, life would have no meaning." She regards her husband with affection and respect in proportion to the number of competitors destroyed, workers exploited, minerals extracted, and acres developed (compare with Schumpeter, 1942). An ideology is constructed which entrenches the female's position by specifying marriage ("sacred bonds") and child-rearing as women's only possible roles. A discriminatory job structure is created which finishes off those women whom the ideology happened to have missed.

A postscarcity society undermines this whole system by questioning the ends for which it was designed and by dangling so many sorts of alternatives to those occupying ascribed roles. The well-ordered, clear-cut division of labor which traditional male-female roles prescribes falls prey to the generalized distaste and growing disutility of such arbitrary categorization systems. The efficiency of sex ascription for increasing productivity is replaced by the higher efficiency of maximizing options and fostering experimentation.

THE NEW WORKING CLASS

Oppenheimer (1970) has argued that there is developing in America a new kind of working class: the intellectuals and quasi-intellectuals. Modern society has created, as never before, large numbers of persons who make their living with their heads and who produce not a material product, but render a service. These individuals include most members of the so-called "helping professions," although those at the top of the most prestigious professions (medicine, law, and the like) are only marginally related to this new phenomenon. The core group consists of the lower-paid "helpers"–teachers, researchers, social workers, and professors–persons who, almost without exception, make their living within large-scale bureaucracies, usually publicly owned (compare with O'Connor, 1970).

The new working class contains individuals who, by virtue of training and personal proclivities, operate with a style and self-concept which is at least somewhat at variance with the goals and methods of the institutions which employ them. The employees celebrate their own "professionalism" with self-policing codes of ethics and a "service norm" (compare with Wilensky, 1964; Parsons, 1949; Greenwood, 1957). Strain in the relationship between the professional and the bureaucracy is brought to the forefront when those who possess control of the organization determine to directly and boldly exercise their formal authority. This is happening more frequently as student and minority activism causes the vociferous right to bring pressure to bear upon public bureaucracies. An attempt is thus made to exercise arbitrary control over individuals whose relatively humanistic and experiential ideologies emphasize creativity and autonomy (especially their own), and who have been rewarded in the past with visions of self-importance in lieu of higher salaries. The beginning of faculty unionization in the largest American colleges and universities (e.g., the California system) is one manifestation of this growing strain.

To be sure, there are large numbers of elite academicians who thrive under these contradictions and who have been only too happy to convert their own work, the work of their students, and, when possible, of their colleagues, to the needs of the corporate state.

They are indistinguishable in terms of the way they carry on their daily affairs from any other members of the managerial elite. But there are others, including some senior scholars, who find themselves at least in sympathy with the complaints raised by the increasing numbers of young academicians who find these circumstances unbearable. These younger academicians, some of whom are alumni of the old SDS, are organizing among themselves in the form of professional radical caucuses and in the national network of academicians which constitutes the New University Conference. Again, the numbers are small, but growing.

THE OTHERS

Of all the obsolescences which developing capitalism has left behind, none is more obvious than the continued existence of an American underclass. The *lumpenproletariat* may be a necessary concomitant of wildly erratic business cycles and a capitalism committed to crude productivity. Those without a place in the industrial machine can be abandoned, perhaps with at least economic justification, in a society perceived to be struggling for basic subsistence capacity. The erosion of such justification under current circumstances lies at the base of the minority groups' movements—movements which were the original stimuli and now the natural allies of middle-class radical movements.

United States domestic policy toward the unproductive continues to be based on a value premise appropriate to a bygone era—the notion that persons not significantly contributing to (or prospering from) material production are, by the very fact, antisocial and thus immoral. It is a system founded upon what Kenneth Boulding terms "malevolence"—the belief that, all other things being equal, the other guy shouldn't have something for nothing. Translated into domestic policy, the assumption results in continuous harassment of those who, for one reason or another, are not making an obvious and direct contribution to the productive apparatus or who lack substantial control of the apparatus. The rage which results on the part of the beneficiaries of this arrangement is a sign of the even greater resentment which is stimulated against the architects and the administrators of this obsolete Henny-Penny morality.

That portion of the working class which consists of minority group members can be easily seen as potential participants in a left-radical coalition. Much more difficult to predict is the political behavior of the white working class—particularly those saved from the lumpenproletariat union contracts. The consensus among sociologists has been that this group is strongly progressive on economic welfare issues, but reactionary on matters dealing with race and nationalism. The recent widespread attention given to hard-hat antistudent sentiment strengthens this characterization in the popular mind. Seymour Lipset (1959) has referred to this general syndrome as "working-class authoritarianism," and his influential writing on the subject continues to dominate relevant scholarly thought.

The evidence is mixed. Voting data cannot be used to support the contention; workers vote overwhelmingly Democrat, and usually (although not always) for the most liberal candidates. Even the survey data, which are the basis for most arguments in support of the Lipset proposition, are not consistent. On the matter of generalized attitudes toward blacks, there is usually found to be a positive relationship between acceptance of black demands and socio-economic status. A summary of poll data on white attitudes toward civil rights demonstrators (Erskine, 1968) reveals five different surveys which provide the same general finding. Paul Sheatsley (1966) reports that measured either by income or education, socioeconomic status is positively related to scores on a "pro-integration scale," constructed as a composite of responses to various integration questions.

There are, however, many factors which make it very unclear what this sort of data might mean. For example, lower-class whites are the ones whose lives are most directly affected by integration; their jobs may be threatened, neighborhood schools racially altered, recreation and leisure locales disrupted (compare with Molotch, 1969). As others have pointed out, it may be quite painless for a securely situated middle-class person to be ideologically magnanimous when dealing in the abstract with such a question raised by a nice-to-impress survey researcher at the door.

At least on the subject of race, there is much evidence, originating with the classic study of LaPiere (1934, 1936), to suggest that expressed ideology and action are two different things. In his study of a changing Detroit community, Mayer (1960) found that more whites who said they would not move in response to change ended up "fleeing" than those who expressed a desire to move—although there still was a positive relationship between moving and stated intention to move. Grier and Grier (1960), in a survey of integrated housing developments across the country, could discern no tendency for integration to be more prevalent among any particular economic, educational, geographical, or ethnic group. Similarly, in my own study of biracial settings in a changing Chicago community (Molotch, 1970b), there was no tendency for successful integration to be centered in any socio-economic group; in fact it was a working-class Fundamentalist church which was among the few integrated institutions, while the well-to-do golf club (like virtually all others) excluded blacks by policy. This demonstrated disparity between articulated attitude and action may be taken either as an index of unreliability of survey research in general, or as a more limited index of the unreliability of surveys as a method of discovering relationships between social class and racial views. In any event, these methodological points can only raise a question; they certainly do not, in themselves, disprove the Lipset authoritarianism hypothesis.

On other issues, such as attitudes toward civil liberties, communism, and so on, there is much survey data to suggest the validity of the Lipset hypothesis (compare with Stouffer, 1955). But there is also important and striking evidence in the opposite direction as well. Richard Hamilton (1968) compared differential class attitudes toward the Korean and Vietnam conflicts. He found that regardless of the war and regardless of the class measure, hard-line solutions are favored by the higher classes. Thus, for example, "bombing China" during the Korean War was favored by thirty-seven percent of the professionals, fifty percent of the self-employed/managers, but only by twenty-nine percent of the unskilled and sixteen percent of the unemployed. Hamilton also reports that it is those who indicate they do the most newspaper and magazine reading who tend to be hard liners. In conclusion, Hamilton dubs the problem to be one of "upper-middle-class authoritarianism," in contrast to the low-status groups which "prove to be strongly favorable to negotiation. It is they who prove to be 'moderate' and conciliatory" (Hamilton, 1968: 445). Some Gallup surveys on attitudes toward the war also support Hamilton's argument.

These very mixed findings are suggestive of the proposition advanced by Michael Mann (1970) that working-class persons respond very differently to questions regarding concrete, everyday experience (where they are consistently "left") than they do to abstract matters dealing with ideology and issues beyond their everyday experience. Thus on such remote matters as constitutional liberties, individuality, the class system, and so forth, their responses tend, according to Mann, to be conservative. Workers support medicare, social security, and almost always side with labor in its disputes with management. Workers tend to be cynical about the legitimacy of the promotion system in their own organization (Blauner, 1964), but agree in the abstract that ability determines who gets ahead (Mizruchi, 1964: 82; Berelson et al., 1954: 58; Lenski, 1963: 165). There may thus be, as Mann argues, "false consciousness" among the workers (or at least pluralistic ignorance), inasmuch as their utterances to survey researchers on abstract questions are so inconsistent with what they think they know from their everyday experience.

The larger implication of these arguments is that once the white workers come to see the relationship between left-radical ideological perspectives and their own practical interests, the relatively fragile and inconsistent attitudes toward nationalism, individuality, and race might crumble. Only recently have there been energetic campaigns by the new left to relate their own ideologies to the practical day-to-day life goals of workers. Recent forays of the student movement into the world of labor politics and strikes (e.g., California farm workers' strike, Follett Publishing Company strike in Chicago, Standard Oil in California) may be groundwork for a creative youth-worker alliance. These are only the barest beginnings, and it would be extremely premature to consider such an alliance as established.

All of this is not to gloss over the fact that there are indeed barriers to effective coalition between the white working class and the various contemporary left movements. White workers conceive of themselves, certainly with encouragement from the politicians, as overburdened by a tax structure which is designed to subsidize the underclass. Students are seen simply as spoiled brats who are inexplicably tearing down the luxurious institutions which the workers' taxes have built. Student actions are kept inexplicable by media which continuously fail to provide a forum for the radical students' perspective and which present student violence as context-free and thus completely mad. To the extent that the working class has access to the media (which is minimal),[6] this is what they get.

Women's movements, to the extent to which they are known, are potential threats to one of the few sources of status gratification which working-class men enjoy. Women's liberation is also, given present arrangements, a threat to working-men's jobs.

The ecology movements are a potential hindrance to the kinds of enjoyments which the more prosperous working-class and lower-middle-class people come to experience when at last they gain some measure of material security. It is the well-to-do working class which buys the motorcycles that tear up the deserts and which make noise in the suburban streets; it is the same social group which disproportionately buys the campers which defile the pristine beauties of the national forests. Again, it is the arriviste workers who have little interest in sailing, but who invest in speedboats and water-skiing equipment that terrorize the quiet gentility of the traditional watersport world. Dune buggies and snowmobiles do similar damage to terrain in other realms.

These somewhat trivial sources of conflict would likely be resolved quite easily in the context of a more general reorientation of American labor away from its present conservative stance. Union membership is becoming available to farm workers, intellectuals, and white-collar professionals. Certain unions are shifting interest to increasingly broad definitions of labor's interests; the United Automobile Workers (granted, not a typical union) is calling attention to ecological problems as posing a danger to the worker—both at the work place and within the larger environment. Demands may shift, as in Europe, away from wage benefits and toward issues involving control of the work place and the quality and disposition of the finished product. The influx of professionals into organized labor and the influence of student participants may yet push labor in such directions. Again, the beginnings are barely visible; the speculative nature of this scenario is well appreciated.

SUMMARY DISCUSSION

There has been no attempt to argue that all of the social groups discussed in this chapter are shortly to become radical revolutionaries. Publics are generally inchoate and, even during periods of profound social change, many lives pass across the earth untouched and ideologically unphased. Instead, the present aim has been to point out that the growing minority of individuals (especially among youth and minorities) who are adopting left-radical perspectives are surrounded by a population—both active and passive—which is increasingly supportive of the content of left-radical programs.

Structural arrangements of the present economy are likely to continue to expand this constituency of the radical left as the ecology movement, the women's movement, and the youth movement bring additional members of the middle classes into that same fold. Meanwhile, those the society has been labeling and punishing as deviant—marijuana smokers, draft evaders, homosexuals, the mentally ill, the poor, and the ex-convicts—can be expected to add their very considerable numbers to the emerging coalition. As a general phenomenon, the "politicization of deviance" reflects the fact that the deviances of an old system—to be destroyed or at least ignored—are emerging as the characteristics of a new majority. Young men who wish to be gentle and humane, women who wish to be efficacious, people who want beauty—members previously of a deviant minority—are the nascent additions to this growing culture.

In the meantime, the emerging left majority are the victims of an increasingly vocal and active right-wing minority—actively attempting to exploit the extant false consciousness of the working class and the actual self-interests of a portion of the national elite. The backlash is an argument and theory which has been used both as a cover for attacks upon dissident activists and a means of undermining the confidence of activists in the efficacy of their tactics. Except as a tactic and expression, it does not exist.

It is the left which has been winning all the arguments in America and which has generated increasing popular support for its programs. Attempts to repress it follow from that success. Whether or not the left majority and its programs for social change emerge as dominant—either through electoral means or some form of revolution—will be determined by factors beyond the scope of the present discussion. But the outcome of this issue will be the definitive test of whether or not public opinion—which has been the subject of this essay—has any real significance for the course of modern American history.

NOTES

1. Approximately half the seepage is said to be captured in tents placed on the ocean floor. For additional details on the Santa Barbara oil spill, see Molotch (1970a) and Ellrodt (1970).

2. The nonneutrality of science was dramatically evidenced when a Berkeley professor of engineering publicly stated that he could not testify on Santa Barbara's behalf in anti-oil court litigation because "my work depends on good relations with the petroleum industry. My interest is serving the petroleum industry. I view my obligation to the community as supplying it with well-trained petroleum engineers. We train the industry's engineers and they help us" (Santa Barbara News-Press, 1969).

3. The Scranton Commission has pointed to a study "indicating that 80 per cent of graduating seniors favored confrontation as a means of social change" (Mankoff and Flacks, 1970).

4. There was a total of 889 arrests during the Isla Vista disturbances and 1,000 pages of testimony on 431 recorded complaints against the police during the final police-student confrontation period (see Potter and Sullivan, 1970; Santa Barbara Citizens Commission, 1970).

5. The Kent State Grand Jury indictments stand as a more dramatic case in point of this dynamic.

6. As suggested to me by Milton Mankoff, it is this minimal connection with the media which may explain Hamilton's data; spared the contrived explanations of war and brutality furnished by the press, workers may have to fall back on their more day-to-day instincts of peaceful coexistence with which they manage the dangers of their own street-life existence (compare with Suttles, 1968).

REFERENCES

BABCOCK, Richard F. (1966) The Zoning Game. Madison: Univ. of Wisconsin Press.

BERELSON, B. R., P. F. LAZARSFELD, and W. N. McPHEE (1954) Voting. Chicago: Univ. of Chicago Press.

BLAUNER, R. (1964) Alienation and Freedom. Chicago: Univ. of Chicago Press.

BOORSTIN, D. J. (1962) The Image. New York: Atheneum Press.

Business Week (1970) "How students see the pollution issue." (February 7): 86-88.

CICOUREL, A. (1964) Method and Measurement in Sociology. New York: Free Press.

CONVERSE, P. E. (1964) "The nature of belief systems in mass publics," pp. 206-261 in D.

CONVERSE, P. E. (1964) "The nature of belief systems in mass publics," pp. 206-261 in D. E. Apter (ed.) Ideology and Discontent. New York: Free Press.
DAVIS, K. (1965) "The urbanization of the human population." Scientific American 213 (September): 41-53.
DUGGER, R. (1969) "Oil and politics." Atlantic (September): 2-26.
ELLRODT, A. G. (1970) "The Santa Barbara oil spill." Bachelor's Paper. Princeton University, Department of Biology. Unpublished.
ERSKINE, H. (1968) "The polls: demonstrations and race riots." Public Opinion Q. 31 (Winter): 655-677.
FLACKS, R. (1967) "The liberated generation." J. of Social Issues (December): 521-543.
Fortune (1969) "The view from the pinnacle." (September): 93-94.
GALBRAITH, J. K. (1958) The Affluent Society. Boston: Houghton-Mifflin.
Gallup Opinion Index: March 1970; February 1970; January 1970; October 1969; April 1969. Princeton: Gallup International, Inc.
GOLD, D. (1970) "Public concern and beliefs about air pollution: a statewide survey." Vol. 3, University of California Research Reports, Project Clean Air.
GREENWOOD, E. (1957) "Attributes of a profession." Social Work 2 (July): 431-455.
GRIER, G. and E. GRIER (1960) Privately Developed Interaction Housing: An Analysis of Experience. Berkeley: Univ. of California Press.
HAMILTON, R. (1968) "A research note on the mass support for 'tough' military initiatives." Amer. Soc. Rev. 33 (June): 439-445.
HOROWTIZ, D. (1969) Empire and Revolution. New York: Random House.
--- (1965) The Free World Colossus. New York: Hill & Wang.
KOLKO, G. (1969) The Roots of American Foreign Policy: An Analysis of Power and Purpose. Boston: Beacon Press.
LaPIERE, R. T. (1936) "Type-rationalizations of group antipathy." Social Forces 15: 232-237.
--- (1934) "Attitudes vs. actions." Social Forces 13: 230-237.
LENSKI, G. (1963) The Religious Factor. New York: Anchor.
LIPSET, S. M. (1959) "Democracy and working-class authoritarianism." Amer. Soc. Rev. 24 (August): 482-502.
MAGDOFF, H. (1969) The Age of Imperialism: The Economics of U.S. Foreign Policy. New York: Monthly Review Press.
MANKOFF, M. and R. FLACKS (1970) "The changing social base of the American student movement: Its meaning and implications," in P. Altbach and R. Laufer (eds.) International Student Protest, The Annals of the American Academy of Political and Social Science (May).
MANN, M. (1970) "The social cohesion of liberal democracy." Amer. Soc. Rev. 35 (June): 423-439.
MAYER, A. (1960) "Russel Woods: change without conflict, a case study of neighborhood racial change in Detroit," in N. Glazer and D. McEntire (eds.) Studies in Housing and Minority Groups. Berkeley: Univ. of California Press.
MEAD, W. J. (1969a) "The economics of depletion allowance." Testimony presented before the Assembly Revenue and Taxation Committee State of California, June 10.
--- (1969b) "The system of government subsidies to the oil industry." Testimony to the U.S. Senate Subcommittee on Anti-trust and Monopoly, March 11.
--- with P. E. SORENSON (1970) "The principal external costs and benefits of marine mineral recovery." Paper prepared for the Offshore Technology Conference, Houston, Texas, April 23. Unpublished.
MILLS, C. W. (1959) The Sociological Imagination. New York: Oxford Univ. Press.

MIZRUCHI, E. H. (1964) Success and Opportunity. New York: Free Press.

MOLOTCH, H. L. (1970a) "Oil in Santa Barbara and power in America." Sociological Inquiry 40 (Winter): 131-144.

--- (1970b) "The politics of expertise." Paper presented at annual meetings of the Pacific Sociological Association, Disneyland.

--- (1969) "Racial integration in a transition community." Amer. Soc. Rev. 34 (December): 878-892.

NASH, R. (1967) Wilderness and the American Mind. New Haven: Yale Univ. Press.

National Wildlife (1970) "New conservation poll." (December-January): 18-19.

--- (1969) "Gallup: 73% of Americans will pay more taxes . . . " (April-May): 18-19.

New York Times. (1967) "Doctor King's Error." (April 7).

O'CONNOR, J. (1970) "The fiscal crisis of the state." Socialist Revolution 1 (January-February, March-April): 34-94.

OPPENHEIMER, M. (1970) "White collar revisited: the making of a new working class." Social Policy 1 (July-August): 27-33.

PARSONS, T. (1949) "The professions and social structure," Essays in Sociological Theory, Pure and Applied. New York: Free Press.

POTTER, R. and J. J. SULLIVAN (1970) The Campus by the Sea Where the Bank Burned Down. Report on the Disturbances at UCSB and Isla Vista, 1968-1970. Submitted to the President's Commission on Campus Unrest. University Bookstore, University of California, Santa Barbara.

RAY, P. H. (1968) "Human ecology, technology and the need for social planning." Amer. Behavioral Scientist (July-August): 16-19.

RIDGEWAY, J. (1970) The Politics of Ecology. New York: Dutton.

ROSS, J. M., T. CRAWFORD, and T. PETTIGREW (1966) "Negro neighbors–banned in Boston." Trans-action (September-October).

Santa Barbara Citizens Commission on Civil Disorders (1970) Report. Santa Barbara: University of California Bookstore.

Santa Barbara News-Press (1970a) July 5: A-11.

--- (1970b) June 16: A-5.

--- (1969) April 12.

SCHUMPETER, J. A. (1942) Capitalism, Socialism and Democracy. New York: Harper & Row.

SHEATSLEY, P. (1966) "White attitudes toward the Negro," pp. 303-324 in T. Parsons and K. Clark (eds.) The Negro American. Boston: Beacon Press.

STOUFFER, S. A. (1955) Communism, Conformity, and Civil Liberties. New York: Doubleday.

SUTTLES, G. (1968) The Social Order of the Slum. Chicago: Univ. of Chicago Press.

WILENSKY, H. (1964) "The professionalization of everyone?" Amer. J. of Sociology 70 (September): 137-159.

YALE, D. R. (1970) Ecology and Revolution. University of Minnesota Chapter, New University Conference. (mimeo)

18

Racism and Tolerance in Historical Perspective

PETER LOEWENBERG

□ URBAN HISTORY is particularly attractive to a generation of scholars impressed with the multiplicity of disciplinary approaches and the explosion of data available to modern research. By concentrating on a single geographically circumscribed unit, such as an urban area, the historian can study the interaction of the manifold aspects, historical, social, and cultural, that condition the institutions and behavior of people in cities (Schorske, 1964).

History relativizes social givens. When put into the perspective of time, institutions and cultural modes are seen as eternally changing and fluctuating in the process of history. Such presumably constant psychological structures as childhood, adolescence, and the family, are not static at all. They have evolved temporally and are still undergoing change. Philippe Ariès (1962) tells us that childhood and the family, as we know them, only began to exist in the sixteenth and seventeenth centuries. Man lived not in the privacy of his home, but in the public places and streets of his town. "The more man lived in the street or in communities dedicated to work, pleasure or prayer, the more these communities monopolized not only his time but his mind" (Ariès, 1962: 375). "This sociability had for a long time hindered the formation of the concept of the family, because of the lack of privacy" (Ariès, 1962: 398). As a consequence, adults played the same games as children, children dressed identically as adults, and both worked at the tasks of life side by side.

The length, intensity, and scope of adolescence as a new stage in growth, with its own subculture, economy, and psychological conflicts, is a development that we are very much in the midst of, and whose implications we experience on campus daily (Loewenberg, 1969). In no other culture has the process of education, vocational apprenticeship, and emotional maturation been so protracted and difficult. Today it is not unusual for a man or woman to undergo the crises of social and vocational identity formation at an age that was the age of average life expectancy but two centuries ago (Erikson, 1963, 1968).

To this prolonged adolescence must be added the biological fact of earlier puberty, which has great psychological implications. Research in Scandinavia, Great Britain, and the United States, indicates that a century ago puberty began at seventeen and that today it averages thirteen and is continuing to decline at a rate of about six months per decade (Comfort, 1970). With puberty beginning earlier and education lasting longer, the structure of adolescent dependency and rebellion is also prolonged and intensified. Today it is often a matter of decades between the time of puberty and social, economic, and emotional autonomy from the parental home for a youth.

The same quality of flux and adaptation exists historically regarding racial and ethnic attitudes. Racial hatred and discrimination toward minority groups has not been universal in history. Medieval Genoa, for example, practiced toleration and met strangers with religious freedom and easy naturalism. One did not have to be a Christian to enjoy full economic and political rights. Jews and Albigensian heretics were not persecuted. "There was no record," writes the Genovese economic historian Robert Lopez (1963: 38), "in the entire history of Genoa, of persecution of the Jews"—a record that few peoples or cities can equal.

The medieval and the early modern European city, just as the polis of the classical world, were cosmopolitan communities. When a man of the Middle Ages said, *"Stadtluft macht Frei"* he was referring to the freedom and peace the city walls offered from the terrors of the feudal countryside. In the free cities of the Middle Ages there was corporate equality, civic autonomy, and a degree of democratic participation in government that exceeded any other area of the world at the time. The king granted to the city of Barcelona, for example, freedom of citizens, officials, messengers, goods, and merchandise, from all toll gatherers or tax gatherers who might wish to impede their movement (Mumford, 1961: 252).

The scale and design of premodern cities was such as to create of necessity the meeting and interaction among inhabitants. There were ample opportunities for people to cross paths, visit, and to mix. The city had squares, watering places, markets, and promenades that provided ample occasion for human contact in the daily round of life. The probabilities were in favor of any resident meeting members of his family, neighbors, and friends as he did his business and ran errands in the streets and market places of the city. This created a sense of community and intimacy that stands in starkest contrast to the impersonality of the city today.

The great German theorist of urban life, Georg Simmel (1903) stressed the impersonal nature and the compulsive quality of modern city life when compared with earlier periods. The large modern city is a place of alienated production for the market place, that is to say for strangers who are unseen and unknown to the producer. Time is the regulator of city life. Without the sharpest punctuality in commitment and performance, the life of the city would become chaos. If all the clocks in Berlin would be off by an hour in varying degrees, says Simmel, then the economy, transportation and private lives of the citizenry would be shattered. The punctuality, calculation, exactitude, required by urban life are directly related to the city's economic system and its intellectual and rational character. They color its life and repress the instinctive, irrational, and impulsive drives which also strive to determine life.

Simmel also emphasized the importance of "the stranger" in the city of Western culture. It is the stranger who provides the necessary quality of criticism and challenge that makes for a creative culture, for "he is not radically committed to the unique ingredients and peculiar tendencies of the group, and therefore approaches them with the specific attitude of 'objectivity.' But objectivity does not simply involve passivity and detachment; it is a particular structure composed of distance and nearness, indifference and involvement" (G. Simmel, 1908: 404). Yet the perspective of the stranger is not one of exclusive advantage. Because of their status as "outsiders" and "others" strangers are depersonalized by the majority. They are not viewed as individuals, but as types or stereotypes, making them subject to "thingification" (G. Simmel, 1908: 407)

With these observations Simmel has fused Karl Marx's alienation of the productive process (Marx, 1961), with Max Weber's concepts of rational economic action and bureaucracy, using the locus of the city as the epitome of modern rational dehumanization and self-alienation. Such dehumanization and objectification of the other, which Simmel saw as the unique quality of city life, was to have a more general historical actualization in the European genocide and racism of the second quarter of the twentieth century.

The cosmopolitan value of the strange was appreciated by the theologian Paul Tillich who also recognized its subversive influence on totalitarianism and racism:

> By its nature, the metropolis provides what otherwise could be given only by traveling; namely, the strange. Since the strange leads to questions and undermines familiar tradition, it serves to elevate reason to ultimate significance. . . . There is no better proof of this fact than the attempts of all totalitarian authorities to keep the strange from their subjects [quoted in Jacobs, 1961: 238].

Contemporary critics of American urban blight concur only in Simmel and Tillich's high valuation of the strange and pluralistic quality of the city. These critics do not share the critique of urban alienation and objectification that is in the tradition of the young Marx (1844) and the post-Freudians such as Wilhelm Reich and Erich Fromm. Today's critics look back with longing to the protean and fecund quality of the nineteenth-century city much as Marx looked back to the nonalienated labor of the medieval craftsman.

Among leading current critics of modern urbanism, Jane Jacobs (1961) romantically praises the spontaneous diversity of the residential, commercial, industrial mix of the American city of the past. Richard Sennett (1969) agrees, suggesting that the "disorder" of a dense uncontrollable city has a positive growth value. In the last decades, he says, the family has appropriated the social functions and contacts that men once sought in the city. The suburbs are a regressive seeking after solidarity and a denial of experiences that foster complexity and disorder. For Sennett it is in the middle-class suburban periphery that the greatest and most self-deceptive alienation of solidarity exists.

THE MODERN CITY

The modern metropolis is an alienating place whose residents lead lives fragmented into vocational, social, family, cultural, and recreational spheres. The city consists of many subcultures, ethnic, social, and geographic. The people who move in one social or regional stratum of the city have little to do with those in any other. The residents of Watts and Bel Air in Los Angeles, the South Side and the Lake Front in Chicago, Harlem and the East Side of Manhattan, have little in common. They appear to share only political boundaries, public utilities, and, in some cases, a transportation net.

Most Americans live in an ethnically circumscribed world of political, social, and vocational relations; many live in residential enclaves as well. There are Polish wards in Milwaukee, Chicago, and Gary (Indiana). Every American city has its Negro district. In the West and Southwest all communities have a Mexican barrio. The larger towns and cities also have Little Tokyos and Chinatowns. The North Shore suburbs beyond Chicago and the "Golden Ghettos" of Long Island contain municipalities inhabited only by middle-class Jews, while neighboring towns, that look alike in every way, are exclusively Gentile. Each has its ethnocentered community agencies, country clubs, and service organizations, in addition to its churches and synagogues. There are also Irish, Italian, and Puerto Rican subcultures in America (Glazer and Moynihan, 1963; Herberg, 1955; Gans, 1962).

Most Americans who live outside the enclaves, barrios, and ghettos, know very little of what is happening inside them. They are stunned and mentally unprepared when racial conflict explodes in the midst of our cities. Each ethnic group is separated from, and suspicious of the others. Each finds comfort and reassurance in being with those who share the same childhood and family patterns (Parsons, 1964), who have similar "signals" of nonverbal communication (Hall, 1959; Lorenz, 1966; Rubenstein, 1970), who know the same intimate worlds of puberty, courtship, and marriage. Karl Deutsch (1952) has found the driving force of modern nationalism in this intimate realm of social communication.

While the modern city is a conglomerate of subcultures and autonomous worlds of social communication, the disparate segments of the city fill vital functions for each other. They are economically and functionally interlocking and interdependent. As currently constituted, the contemporary city needs certain areas for industries and warehouses, where groceries are brought into the city and distributed, where trucks can be parked, and railways can have their marshaling yards. The people who work in the factories that the middle class manages and the upper class owns reside in working-class neighborhoods. There are also districts in every city where the servant and menial classes live, where contractors and employers can go to hire occasional and field labor. As Edward Banfield points out (1968), the central dynamics of urban residential segregation are class and economic, rather than racial.

All cities have an area where the sexual and socially unsanctioned impulses, restricted by middle-class values and banned from the "nicer" residential neighborhoods, find release and come to expression. Here one finds the prostitution, gambling, card games and illegal gaming, pornography, night life, burlesque and stag shows in all their topless and bottomless repetitiveness. In human terms, this means that the upper and middle classes displace their "low" and "dirty" voluptuous feelings from their homes to another part of the city, just as they isolate these emotions from one to another part of their minds.

Man is, as Freud reminded us (1908), born *inter urinas et faeces.* The split between the sensual and the sublime impulses is almost universal among civilized men. The sexual life of modern man is, as a consequence, "severely impaired" (Freud, 1961: 105). "Where they love they do not desire and where they desire they cannot love. They seek objects which they do not need to love, in order to keep their sensuality away from the objects they love" (Freud, 1912: 183). In fact, "the behavior in love of men in the civilized world today bears the stamp altogether of psychical impotence . . . the man almost always feels his respect for the woman acting as a restriction on his sexual activity, and only develops full potency when he is with a debased sexual object; and this in its turn is partly caused by the entrance of perverse components into his sexual aims, which he does not venture to satisfy with a woman he respects" (Freud, 1912: 185). Yet, both in the case of the splitting that constitutes individual man's neurosis and the functional divisions of the modern city, neither the "pure" nor the libidinal can exist without the other. In the case of both man and society, it is healthier to be able to fuse them rather than enforce an unnatural separation and a denial of the instincts and proclivities of man.

The last thirty years of extensive research on the problems of race has made it clear that an understanding of racial hatred demands a comprehension of the irrational and intrapsychic forces in man. It is striking how patterns of racial hatred share certain universal characteristics, be it current Indian attitudes toward the Chinese, Nazi attitudes toward Jews, or white American attitudes toward Negroes, Chicanos, and Orientals.

Racial attitudes are not inborn, they are learned. Hatred and fear of other races is acquired early in life from parents and from the primary figures who socialize a child. These early agents of socialization transmit the cultural norms for intergroup attitudes. A growing child learns social behavior and internalizes social patterns of prejudice by watching and emulating the attitudes and conduct of the significant other people around him. He learns racial prejudice by assimilating the many cues, often nonverbal, and nuances of behavior of the important people in his world—parents, family, nursemaids, and friends. Often something as subtle as a tensing of a handclasp, the rigidity of musculature, or a tremor or inflection of a voice, communicate the message to a child—"this is an object that is different, dangerous, and to be feared." Thus a child learns to respond to people of other racial and ethnic groups by imitating and internalizing the values of the important people in his life.

Racial hatred is often manifested by sexual fears, but it is rooted in fantasies of the forbidden. Prejudice constitutes the link between the child's culturally imposed disgust with his own body and his fears in the social world. Until culture imposes its values, children have no aversion to their bodies. They delight in being naked and in displaying themselves. They take pleasure in the orifices and senses of the body, in its products, odors, and warmth. The child learns a hierarchy of dirtiness, leading from excrement, which is so dirty that it may not be touched, to food which is so hygienic that it may be eaten. What goes into the body is clean, what comes out is filth. The child is taught that his body is a dirt factory which takes in food and expels dangerous contaminants. There are deeply rooted and highly charged contradictory emotions of curiosity, fascination, excitement, and fear released by the teaching of body shame (Kubie, 1965).

The member of the ethnic outgroup is considered to be a dirty foreigner just as sex and pleasure in the body is considered dirty. Thus images of the castrated, mutilated, dirty body, of the evil, inferior, impotent, ugly, and enslaved creature who must be cast off, repressed, and hidden, are subsumed into racial prototypes of great power to mobilize hatred. "Unconscious associations," says Erik Erikson (1959: 30) "of ethnic alternatives with moral and sexual ones are a necessary part of any group formation." Erikson's concepts of positive and negative identity formation clarify this (Erikson, 1959). All of the negative identifications of a lifetime, all those qualities that are most shunned and feared, all those attributes that are most dangerous and tempting, are integrated into a unified identity which is then projected outward onto an ethnic minority.

The process of acquiring the prevailing racial prejudices of our culture begins at an early age and has been measured among children between the ages of three and five-and-a-half. Both Negro and white children had a high or a medium awareness of racial characteristics and of the positive or negative social implications of group membership. Both the Negro and white nursery school children preferred white dolls and fairy-tale characters to dark ones. These preferences applied to actual people as well (Goodman, 1952). Among white children in this age group there was a clear-cut and steady increase in preference for a white playmate (Harding et al., 1969: 20). At age seven, prejudice already interferes with mental functioning. Ethnocentric children are more intolerant of ambiguity, meaning a tendency to impose in a rigid manner certain preconceived sets upon ambiguous perceptual data or upon the solving of reasoning problems. Ethnically prejudiced children have a lower level of abstract reasoning than more tolerant children (Harding, et al., 1969: 37).

Prejudice in children, although it is a reflection of community social norms, is accentuated by a discipline-focused, status-oriented, and emotionally harsh family setting (Adorno, 1950). In one study mothers of highly prejudiced children believe that obedience is the most important thing a child can learn; that a child should never be allowed to set his will against his parents; that a quiet child is better than a noisy one; and that sex play by the child should be punished (Harris et al., 1950).

How can what Myrdal (1944) termed "the American dilemma" of racial prejudice in a democracy dedicated to the equality of men be resolved? For the white liberal the program must remain to fight for full political, social, and economic equality for all Americans regardless of race. However, he should not merely engage in moral condemnation. To be effective he must use the insights of the behavioral and social sciences to understand the needs of the prejudiced person and to help him meet these needs without resorting to racial hatred.

Many Blacks have come to the realization that integration into white culture is bad for their emotional health and self-respect. Insecurity and anxiety are conducive to low self-esteem. Integration as a solution to prejudice fails on both the levels of altering white attitudes and of satisfying Blacks. Whether or not intimate contact with members of another race serves to reduce prejudice is a function of personality structures and social setting. Contact per se does not increase or decrease prejudice. Simply getting to know members of a minority group is not in itself likely to confirm or change racial attitudes. Contact on any other than an absolutely equal dimension will simply reinforce prejudice. A Black menial will be seen as shiftless and stupid by his boss and the Black will see the whites as exploitative and hypocritical.

At this time the policy of integration still carries with it the psychological connotation for both Blacks and whites that it is more desirable to mix with whites and share their social life and values than it is to be barred. Furthermore, while Blacks have taken the initiative by sit-ins and bus boycotts of public services and premises, ultimately it is whites who determine where and for which Black men the cultural, social, and economic institutions of our society shall be opened. Whites must sanction Black admission and participation. Thus psychologically, white superiority over Blacks is still the implication of integration as practiced until now.

BLACK CULTURAL AND POLITICAL IDENTITY

The most significant development in race relations in this decade has been the ascendancy of Blackness as a cultural and political identity. Black identity, or what Frantz Fanon (1963) terms "Negritude," takes the power from white culture to elevate individual Blacks by integration. In rejecting the goal of integration the movement of Negritude has weakened the selective process whereby a few isolated Blacks are invited to participate in America's dominant middle-class social and cultural life. Instead, the Black movement is demanding absorption on a status of full equality and dignity of an entire, largely poor American subculture.

The New Black identity of contemporary militants is still primarily a negative identity. To be wholly Black, a Negro must deny many of the psychological adjustments to a dominant white society that were conditioned in childhood. Thus, Blacks who would define a new consciousness must, as a beginning, define what they are not: They are not white and not middle class and do not wish to emulate either. They are not emasculated Uncle Toms. They are not

Christians and not descended from Europe, ethnically or culturally. These negative identifications are the beginnings of the hard process of hammering out a new authentic Black identity. They clear the Black self-image of past identifications that undermined its holders with the ambivalence of internalized aggression and low self-esteem. Now new inner objects are being shaped out of vigorous encounter with American social reality and internal confrontation with the estranged self.

The movement of Black Power must be placed in the context of the historical development of revolutionary movements. The original minimal demands of Booker T. Washington, and later of the founders of the NAACP, were limited by what seemed possible. But the ground was being prepared. The significant social and economic progress that came with industrial unionism and the war economy in the 1930s and 1940s created new demands leading to legal desegregation and the Civil Rights Movement of the 1950s.

The collapse of the Civil Rights Movement came with the realization that the legal promises of the civil rights legislation would not bring to Blacks either social integration or the middle-class standard of living. This relative deprivation—the proximity or sight and taste of the goal and the realization that one is barred from reaching it—has caused depression and violent militancy. Bitterness and rage have accelerated among Blacks, particularly in the North, as they have made gains that bring them to an expectation of equality, seemingly never to be fulfilled by nonviolent, evolutionary means (Sears and McConahay, 1970).

The violent quality of the Black militant movement arises from the discovery of power among those who have always seen themselves as powerless. The urban Negro has been socialized to new patterns of assertion. What we are seeing is the dramatic consciousness of a reversal of roles: Those who have been the victims of violence enjoy the fantasied prospect of exercising violence on their oppressors. The often bloody language of Black Power advocates is, among other things, a way of testing the surrounding white ambiance: How much can whites be frightened or hurt? Will whites respond (or perhaps overrespond) to threats and intimidation?

White ambivalence vitiates the capacity of the majority to understand Black militancy. They overreact to Black militancy and violence in terms of white fantasies of revenge: that the oppressed will now conquer and rule their oppressors. This, incidentally, is a pure fantasy; no Black group or leader has advocated ruling over whites. Most whites interpret Black threats as final intentions of annihilation rather than tests of reality by those who have until now been powerless. They reason on a primary process level, that an injury can be undone or must be punished by a similar deed inflicted on themselves (Fenichel, 1945: 44). Accordingly, whites preoccupy themselves with protection from Black violence which Blacks in turn misapprehend as a preparation for accelerated white oppression and even genocide, which is not a prospect, at least by conventional historical definitions of the term.

One of the many myths of our culture is that our political "freedom" implies an absence of social controls and that our social and economic choices are made in a vacuum, while other social systems may be "totalitarian" and lack freedom. In fact, we have in our society many forms of social control which have been used to curb discriminatory behavior and diminish prejudice.

LONG-RANGE PROSPECTS

The truly hopeful finding of social science concerning race is precisely that when deliberate efforts are made by social authorities to alter patterns and to build inner personal controls, a significant reduction in prejudice can be achieved. In government-managed interracial housing projects, attitudes of prejudice and stereotyping have been markedly changed (M. Deutsch and Collins, 1951). If psychotherapeutic group work supplements the process of living together and getting to know racial minorities, prejudice can be modified (Bettelheim and Janowitz, 1964: 71-73; Simpson and Yinger, 1958).

When racial interaction is on a level of social equality, personality variables of aggression, authoritarianism, and self-esteem determine the individual's response to experience. People who begin with primitive projective defenses to aggression, and who are socially dissatisfied, will reinforce their existing prejudice after interracial contact. Those who have few aggressive needs and whose attitude toward authority and society are more positive, tend, after a period of interracial contact, to give up or reduce prejudices they hold (Mussen, 1950).

Structural changes involving all the governmental and private levers available to utilize the economic, legal, and educational systems for planned social change toward abolishing prejudiced behavior and attitudes, are the only public policy compatible with the values of a democratic society. The instruments of social control that should be invoked to remove all racial and ethnic barriers include massive intervention by all levels of government, from mayors' commissions to federal contracts, urban renewal, the armed forces, fair and full employment activities, and use of the mass media. The educational system must implement all instrumentalities which exist in its storehouse of available techniques to provide compensatory skills. Curriculum changes toward nonmiddle-class culture, Black, and, in certain areas, other minority cultures, are in order for a truly pluralistic society.

There must be adequate governmental funding and meaningful work for all. Planning on a centralized scale tends to defeat local initiative where it is expressed. On issues of participation, the country town with planning powers vested in local boards was more functional than the metropolis of today. The inhabitants of the modern city must live with decisions made for them by distant centralized planners. A series of new institutions in our cities that permit the exercise of local community control over governmental functions such as police, schools, health, and safety, would go far to give urban populations a sense of power and autonomy over their environment. The actual participation of citizens in the input of governmental decisions concerning their lives would serve to restore two-way contact between the rulers and the ruled, as befits a democracy. It would also raise morale and civic responsibility. It is natural to be cynical and nihilistic when one has no participation in decisions over one's life. But it is much harder to disregard expressed community consensus on what has been defined as the communal interest.

Can equality and radical social change be achieved by Black revolution in this country? We must again look at history. Revolutions have always been made by minorities. But they have only been triumphant when the minority of revolutionaries succeeded in persuading the majority that they too have something to gain from a change. When this conviction was not widespread, the passive majority supported a counterrevolution as in Europe in 1849 and 1933. The hope for America lies not in revolution or its suppression, but in a dynamic social encounter and confrontation with latent forces of racism and destruction.

The uncovering of unconscious ambivalence creating racial tension can come about only from social conflict which forces hidden feelings into consciousness, much on the model of George Sorel's (1961) conceptualization of the revolutionary implications of the bourgeois fear of social violence. What we are witnessing is a kind of social shock therapy which in the process of confrontation brings up all the latent conflicts in both Black and white (Spiegel, 1968: 153-154). In personal psychotherapy we assume that the freeing of bound energies that comes from the resolution and integration of unconscious ambivalences will lead to greater creativity and inner peace. This is often accompanied by intense struggle and hardship. Socially, the resolution of our cultural ambivalences will be difficult and violent. But the exposure and facing of unconscious racial attitudes is no less necessary for our national survival. It will release human energies which are now locked in racial conflict for a new emergence of life and social creativity in our cities.

Perhaps we may find a clue to the resolution of human racial conflict in the studies that have recently been generated in the field of comparative ethology. Students of aggressive animal behavior like Konrad Lorenz have observed that almost all species control intraspecies aggression to avoid killing each other within the species. Even so ferocious an animal as the wolf stops short of slaying its rival in combat. Men are almost the sole exception. Human beings are paid the dubious compliment of being equated with rats as beings who destroy fellow members of their own species (Lorenz, 1966: 132, 237, 240).

The long-range prospect of reducing racial prejudice will only be realized when historical movement takes a postracial dimension as its central axis of hostility. This may come about with biological miscegenation. Of course, this will not curb man's potential for violence and anger. Human hostility and aggression is a concomitant of the more beneficent feelings of love and affection, and will continue to exist as long as there are men. But the conflicts may take different, not necessarily racial, forms. A mere four hundred years ago in Europe, one's very life depended on what liturgy or sacraments one adhered to within Christianity. More recently, we have gone through periods of mutual mass murder in the name of nationalism and race. We appear to be moving into an era where ideology is the paramount rationalization for aggressive destruction. This is hardly more attractive than malice and murder on grounds of race. No one knows what forms intergroup tensions will take next. Although brother has slain brother since the beginning of time and will continue to do so, it may be hoped that in the future he will come closer to knowing why he does so rather than to attribute it to a factor so spurious as race.

REFERENCES

ACKERMAN, N. W. and M. JAHODA (1950) Anti-Semitism and Emotional Disorder: A Psychoanalytic Interpretation. New York: Harper & Bros.

ADORNO, T. W., E. FRENKEL-BRUNSWICK, D. J. LEVINSON, and R. N. SANFORD (1950) The Authoritarian Personality. New York: Harper & Bros.

ALLPORT, G. W. (1954) The Nature of Prejudice. Cambridge, Mass.: Addison-Wesley.

ARIES, P. (1962) Centuries of Childhood: A Social History of Family Life (R. Baldick, trans.). New York: Vintage.

BALDWIN, J. (1964) Notes of a Native Son. New York: Bantam.

--- (1961) Nobody Knows My Name. New York: Dell.

BANFIELD, E. C. (1968) The Unheavenly City: The Nature and Future of our Urban Crisis. Boston: Little, Brown.

BETTELHEIM, B. and M. JANOWITZ (1964) Social Change and Prejudice Including Dynamics of Prejudice. New York: Free Press.

BRODY, E. B. (1966) "Psychiatry and prejudice," in S. Arieti (ed.) American Handbook of Psychiatry 3: 629-642. New York: Basic Books.

BROWN et al. v. Board of Education of Topeka et al. (1954) 347 U.S. 494.

BROWN, C. (1965) Manchild in The Promised Land. New York: Signet.

CASH, W. J. (1941) The Mind of the South. New York: Doubleday.

CHRISTIE, R. and M. JAHODA (1954) Studies in the Scope and Method of "The Authoritarian Personality." Glencoe: Free Press.

CLEAVER, E. (1968) Soul on Ice. New York: McGraw-Hill.

COMFORT, A. (1970) "Early puberty spurs clashes with parents." Los Angeles Times, October 16.

DEUTSCH, K. (1952) Nationalism and Social Communication. Cambridge, Mass.: MIT Press.

DEUTSCH, M. and M. E. COLLINS (1951) Interracial Housing: A Psychological Evaluation of a Social Experiment. Minneapolis: Univ. of Minnesota Press.

ERIKSON, E. H. (1968) "Race and the wider identity," pp. 295-320 in Identity Youth and Crisis. New York: W. W. Norton.

--- (1963) Childhood and Society. New York: W.W. Norton. (second edition)

--- (1959) "Ego Development and Historical Change [1946]" in Identity and the Life Cycle: Selected Papers. New York: International Universities Press.

FANON, F. (1963) The Wretched of the Earth. New York: Grove Press.

FENICHEL, O. (1945) The Psychoanalytic Theory of Neurosis. New York: W. W. Norton.

FREUD, S. (1961) "Civilization and its discontents [1930]," pp. 57-145 in J. Strachey (trans. and ed.) Standard Edition of the Complete Psychological Works 21. London: Hogarth.

——— (1912) "On the universal tendency to debasement in the sphere of love," Standard Edition 11: 177-190.
——— (1908) " 'Civilized' sexual morality and modern nervous illness." Standard Edition 9: 177-204.
GANS, H. J. (1962) The Urban Villagers: Group and Class in the Life of Italian-Americans. New York: Free Press.
GLAZER, N. and D. P. MOYNIHAN (1963) Beyond the Melting Pot: The Negroes, Puerto Ricans, Jews, Italians, and Irish of New York City. Cambridge, Mass.: MIT Press.
GOODMAN, M. E. (1952) Race Awareness in Young Children. Cambridge, Mass.: Addison-Wesley.
GREGORY, D. (1964) Nigger: An Autobiography. New York: E. P. Dutton.
GRIER, W. H. and P. COBBS (1968) Black Rage. New York: Basic Books.
HALL, E. T. (1959) The Silent Language. New York: Doubleday.
HARDING, J., H. PROSHANSKY, B. KUTNER, and I. CHEIN (1969) "Prejudice and ethnic relations." Handbook of Social Psychology 5: 1-76. Reading, Mass.: Addison-Wesley.
HARRIS, D., H. GOUGH, and W. E. MARTIN (1950) "Children's ethnic attitudes: II, relationships to parental beliefs concerning child training. Child Development 21: 169-181.
HERBERG, W. (1955) Protestant, Catholic, Jew: An Essay in American Religious Sociology. New York. Anchor Doubleday.
HERNTON, C. C. (1965) Sex and Racism in America. New York: Doubleday.
HYMAN, H. H. and P. B. SHEATSLEY (1964) "Attitudes toward desegregation." Scientific American 211, 1 (July): 16-23.
JACOBS, J. (1961) The Death and Life of Great American Cities. New York: Vintage.
KARDINER, A. and L. OVESEY (1951) The Mark of Oppression: A Psychosocial Study of the American Negro. New York: W. W. Norton.
KIRSCHT, J. P. and R. C. DILLEHAY (1967) Dimensions of Authoritarianism: A Review of Research and Theory. Lexington: Univ. of Kentucky Press.
KLINEBERG, O. (1935) Race Differences. New York: Harper & Bros.
KOVEL, J. (1970) White Racism: A Psychohistory. New York: Pantheon.
KUBIE, L. S. (1965) "The ontogeny of racial prejudice." J. of Nervous and Mental Disease 141, 4: 265-273.
LEVY, C. J. (1968) Voluntary Servitude: Whites in the Negro Movement. New York: Appleton-Century-Crofts.
LEWIN, K. (1948) Resolving Social Conflicts: Selected Papers on Group Dynamics. New York: Harper & Bros.
LOEWENBERG, P. (1969) "Emotional problems of graduate education." J. of Higher Education XL, 8 (November): 610-623.
LOPEZ, R. S. (1963) The crossroads within the wall," pp. 27-43 in O. Handlin and J. Burchard (eds.) The Historian and the City. Cambridge, Mass.: MIT Press.
LORENZ, K. (1966) On Aggression (M. K. Wilson, trans.). New York: Harcourt, Brace.
McLEAN, H. V. (1946) "Psychodynamic factors in racial relations." Annals of the Amer. Academy of Political and Social Science 244 (March): 159-166.
MALCOLM X (1966) Autobiography. New York: Grove Press.
MARX, K. (1961) Economic and Philosophic Manuscripts of 1844. Moscow: Foreign Languages Publishing House.
MORROW, W. R. (1949) "A psychodynamic analysis of the crimes of prejudiced and unprejudiced male prisoners." Bull. of the Menninger Clinic 13: 204-212.
MUMFORD, L. (1961) The City in History: Its Origins, Its Transformations, and Its Prospects. New York: Harcourt, Brace.

MUSSEN, P. H. (1950) "Some personality and social factors related to changes in children's attitudes toward Negroes." J. of Abnormal and Social Psychology XLV (July): 423-441.
MYRDAL, G. (1944) An American Dilemma. New York: Harper & Bros.
PARSONS, T. (1964) "Social structure and the development of personality: Freud's contribution to the integration of psychology and sociology." Social Structure and Personality. New York: Free Press.
PETTIGREW, T. F. (1964) A Profile of the Negro American. Princeton, N.J.: Van Nostrand.
PINDERHUGHES, C. A. (1968) "The psychodynamics of dissent: a clinical appraisal with emphasis on racial activists," in J. H. Masserman (ed.) The Dynamics of Dissent. New York: Grune & Stratton.
RAAB, E. [ed.] (1962) American Race Relations Today. New York: Anchor Doubleday.
ROSE, A. M. (1956) "Community factors related to desegregation." Amer. J. of Orthopsychiatry 26: 445-449.
RUBENSTEIN, R. L. (1970) "Our tribal society." Morality and Eros. New York: McGraw-Hill. pp. 149-163.
SCHORSKE, C. E. (1964) "The threefold revolution." Colloquium 2 (October).
SEARS, D. O. and J. B. McCONAHAY (1970) "Racial socialization, comparison levels, and the Watts riot." J. of Social Issues 26, 1: 121-140.
SENNETT, R. [ed.] (1969) Classic Essays on the Culture of Cities. New York: Appleton-Century-Crofts.
SHEATSLEY, P. B. (1966) "White attitudes toward the Negro." Daedalus 95, 1 (Winter): 217-238.
SIMMEL, E. [ed.] (1946) Anti-Semitism: A Social Disease. New York: International Universities Press.
SIMMEL, G. (1908) "The Stranger." The Sociology of Georg Simmel (K. H. Wolff, trans.). Glencoe: Free Press. London: Macmillan, 1950. pp. 402-408.
--- (1903) "Die Grossstädte und das Geistesleben," pp. 227-242 in M. Landmann (ed.) Brücke und Tür: Essays des Philosophen zur. Geschichte, Religion, Kunst und Gesellschaft. Stuttgart: K. F. Koehler Verlag.
SIMPSON, G. E. and J. M. YINGER (1958) Racial and Cultural Minorities. New York: Harper & Bros.
SMITH, M. B., J. S. BRUNER, and R. W. WHITE (1956) Opinions and Personality. New York: John Wiley.
SOREL, G. (1961) Reflections on Violence (T. E. Hulme and J. Roth, trans.). New York: Collier.
SPIEGEL, J. P. (1968) "The social and psychological dynamics of militant Negro activism: a preliminary report," in J. Masserman (ed.) The Dynamics of Dissent. New York: Grune & Stratton.
WEST, L. J. (1967) "The psychobiology of racial violence." Archives of General Psychiatry 16 (June): 645-651.

19

An Interpretative Analysis of Recent Work

J. EUGENE GRIGSBY, III

□ THE VALUES PERMEATING American society have been essentially conservative in rationalizing and protecting a privileged class position. These values are hung upon a mythology of America's absolute virtue (civilized ethno-centrism) and deep-seated irrational fears of blacks. In short, American commonsense values about racial violence are the product of centuries of American racism, of white America's concept of itself, and of white America's identification of "The Negro Problem" as such.

Social scientists, among others, have taken the lead in preserving this status quo, veiled in "scientific rigour," by casting minority groups—particularly colored minorities—into second-class status roles. Differences in cultural patterns, linguistic styles, and economic status are labeled pathological, culturally deprived, and deviant. Illuminating "information" presented by social scientists about minority groups has been grossly distorted, factually inaccurate, and seldom reflective of reality—at least the reality as seen from the perspective of the minority groups being described.

AUTHOR'S NOTE: *I want to express my appreciation to Sharon Grigsby and Peter Orleans, who assisted me with the preparation and revision of this article. A special note of thanks is also due to Alicia Lopez, James Pitts, and Frank Odo for helpful comments.*

An example of this naive and inaccurate scientific interpretation can be found in the early writings about the migration of various minority groups to the cities. The literature leads us to believe that as more minority people moved to the cities, they gradually began to sort themselves out into enclaves. It was assumed that such sectioning was done by personal choice for convenience and psychological comfort. Little emphasis was given to the influence of discrimination and blatant racism in this sorting-out process. Indeed, the question of race played a minor role in these early writings about the city. The city was conceptualized as a heterogeneous place which both tolerated and rewarded individual differences. This simplistic description ignored the facts of the city.

Some of these facts are that minority group members, in economic terms, play no significant role other than holders of the most menial and low-paying jobs. Physically, they are required to live in the oldest sections of the city—the zones of transition and areas of physical decay. Psychologically, they are entrapped in the dark ghettos which are presently controlled from without and must often submit to invading outside forces. Politically, they are impotent, although there is some indication that rapid change is being made in this area for the black community. Numerically, it is predicted that blacks will hold a plurality in many major urban centers by 1980. But, in essence, they will be occupiers of empty shells, with no possibility of running city governments and without financial assistance if, as is likely, the current white flight to the suburbs continues. Racial tension and strife recently characteristic of many major cities will continue to be inevitable unless urban analysts and policy makers come to understand the significance of the differences among racial groups.

The purpose of this essay is to review literature which has had a major impact in maintaining the status quo between minority and nonminorities in the United States. Possible alternatives in interpretation will be suggested, and a pluralistic or cultural-difference approach, rather than the traditional dominant culture, subdominant, subculture relationship will be proposed. In reviewing these works, three areas or themes will serve to organize the literature:

(1) American stratification

(2) Definitions or themes of deprivation

(3) Revolt, rebellion; the new direction

AMERICAN STRATIFICATION

Stratification, as seen by social scientists, implies that there exist horizontal relationships among certain social groupings. Sociologists have attempted to demonstrate that all societies are stratified, and in doing so they have identified a hierarchical relationship among strata, with some positions being more desirous than others. The legitimacy of such a hierarchy seems to be solely dependent upon a normative consensus of certain groups of individuals.

Davis and Moore (1945) outline the role of stratification in societies and state that societies must have rewards or inducements which are differentially distributed. These rewards and their distribution become a part of the social order process. As a result of this stratification process, social inequality or the hierarchy becomes a natural device by which societies ensure that the most important positions are conscientiously filled by the most qualified members.

The unqualified use of such terms as "society," "social organization" and "the" permits a writer the hidden assumption in politically crucial context of a homogeneous and harmonious whole (Mills, 1942). It is further assumed that the larger texture of the society will take care of itself, it has a "strain toward consistency" running through it. This, and the cooperation of all, yield equilibrium or social balance. In addition to their descriptive use, such terms become normative. The "social" becomes a good term when it is used in ethical polemics against "individualism," or of an "anti-social" lack of "altruism" or of "anti-social" sentiments (Mills, 1942).

If analyzed in depth, we find that theories of stratification follow very closely the format of a structural whole based on normative consensus (Tumin, 1953; Parsons, 1967). In the case of Davis and Moore, it is never explained who or what determines the importance of a position for the social organization of a society. Implied is the fact that organizing values of a society determine the content, character, and basis of the stratification hierarchy (Bendix and Lipset, 1966).

If this be the basis for stratification, then distinctions among the members of a society must always be made because certain positions will always lie close to the core of that which is socially necessary. The occupants of those positions, therefore, will wear the mantle of power and prestige (Bendix and Lipset, 1966; also see Weber, 1967). Consequently, the needs of the society will be determined by those individuals who hold positions of prestige and importance. If we follow this line of logic, it becomes imperative that social values of all members of the society must coincide with the values of those holding power or prestige within that society, if those who hold power are to remain in powerful and prestigious positions. Thus, the notion of a stratification system in equilibrium is established. Such a system is, by definition, a pattern-maintainence system which makes no allowances for the redefinition of values. In essence we have the establishment of a "legitimate" normative order with a given value system. The implications for this theoretical approach become quite clear when we look at research which has resulted from this normative approach to the study of society.

For example, in the case of Lloyd Warner (1963) who has done several community studies, we find that his conception of stratification results from community judgments made about individuals. These evaluations are the basis for determining an individual's status and were themselves based upon a complex of factors evaluating occupation, social participation, family background, and style of life. Stratification results directly from an estimate of the individual qua individual (Bendix and Lipset, 1966). A major assumption underlying this position is that all individuals or at least a majority of individuals will have equal opportunity to make evaluative judgments of their fellow individuals. If such an assumption were not made, the entire basis for Warner's stratification would come under serious question. Warner himself believed that there existed certain antidemocratic features in this type of stratification, but felt that these antidemocratic tendencies could be counteracted by maintaining an open class system with sufficient opportunities for upward mobility (Warner, 1963). America, in fact, has never been an open class system for colored minorities. The belief in this myth, however, is one of this country's foremost values.

Talcott Parsons also believes in the open class system as the vehicle of upward mobility. When we look at what Parsons (1967) has had to say about stratification, we find that "Stratification is relative to the moral evaluation enjoyed by different social units. Subjective judgements by others are the basis of the formation of strata, of social units or status." Members of society who exemplify the most valued qualities or perform the most valued functions are likely, according to Parsons, to receive the highest ranking. Society's majority determines the value and the nature of rewards. Nowhere in the social science literature is there any description of colored minorities occupying the highest social status or receiving society's highest rewards. It can only be concluded, then, that stratification theory in describing the status quo has the effect of validating it. Even in theory, the minorities are relegated to the bottom rung of the status ladder.

DEFINITIONS OF DEPRIVATION: THE THEORIES OF CULTURAL INFERIORITY

Although stratification theory relegates minorities to the bottom rung of the status ladder, these minorities have been the subject of detailed investigation by members of the social science elite. No minority group in the United States has been more intensively studied than the black American.

Black Families

E. Franklin Frazier's (1948, 1957) works on the black family serve as the underpinnings for almost all works which subsequently discuss black families. The justifications or the rationalizations for the urban ills of contemporary black lower-class Americans can be seen as an elaboration of his ideas (Banfield and Wilson, 1963; Moynihan, 1965; Liebow, 1969; Banfield, 1970). In Frazier's view, the urbanizing of the Negro population since 1900 brought the most momentous change in Negro family life since emancipation. Thousands of migrants, according to Frazier, had been solitary men and women who had led a more or less lawless sex life during their wanderings. But many more illiterate or semiliterate and impoverished Negro families, broken or held together only by a fragile bond of sympathy and habit, sought a dwelling place in the slums of the cities. It was because of the dissolution of the rural folkways and mores of these families, that their children helped swell the ranks of juvenile delinquents (Sherman, 1970).

Daniel Moynihan (1965), in studying the works of Frazier, has done a great deal of writing in an attempt to convince us that Frazier's assessment of the black family was indeed correct. His thesis is that at the heart of the deterioration of the fabric of Negro society is the deterioration of the Negro family. This deterioration, he asserts, is the fundamental source of the weakness of the Negro community at the present time (Moynihan, 1965). In contrast with white families who, according to Moynihan, have achieved a high degree of stability and are maintaining that stability, black families are highly unstable, and in many urban centers these families are approaching complete breakdown.

In Moynihan's view, this condition of the black family has come about as a result of three centuries of injustice. "At this point the present tangle of pathology is capable of perpetuating itself without assistance of the white world." The cycle can be broken only if these distortions are set right. Unless this damage is replaced, all efforts to end discrimination and poverty and injustice will come to little (Moynihan, 1965).

Banfield and Wilson (1963) have obviously read Frazier for they have incorporated into their writing the concept of the weak black family as the root of the black political impotency. The anomaly of the Negro's numerical strength and political weakness, they say, can be explained largely in terms of interrelated factors. These factors are the class structure of Negro society and the character of urban political systems. They emphasize, however, the social disorganization which is characteristic of lower-class Negroes and which is reflected in their high rates of crime, delinquency, desertion, divorce, and illegitimacy. These facts are seen as prime reasons for the Negro's political impotency. These characteristics, they say, are in great part the result of the weakness of the family unit. Indeed, the cultural and economic factors which make lower-class Negro family life so uniquely precarious also make its sense of attachment to the community uniquely weak. Banfield and Wilson point out that the attributes which community life presupposes—education, self-respect, personal skills, a belief in the efficacy of one's own efforts, and a sense of attachment to social entities larger than one's self—are often in short supply (in black communities) because there is no strong family unit to inculcate them (Banfield and Wilson, 1963). Consequently, the social institutions of the city, and especially its government, are often looked upon by Negroes as (at best) remote forces to be ignored or (at worst) hostile forces to be reckoned with (Banfield and Wilson, 1963).

Banfield and Wilson, and Moynihan, like Frazier, have confused statistical patterns or census data for cultural configurations. Aside from the inadequacy of these data, their analyses and interpretations speculate beyond the actual data presented. Such interpretation results in a confusion between demography and social structure. This confusion places in doubt the accuracy of such notions as "family deterioration."

Yet, the theme of black family deterioration becomes a major focal point for much social science research. In the urban setting, black family deterioration continues to be cited as one of the major explanatory variables in accounting for underachievement in education and fragmented community.

Education

Education has been hailed as the prime route by which individuals can achieve both status and prominence in the structure of our society. Since the majority of Americans spend approximately twelve years in formalized educational institutions, it is no wonder, then, that these institutions have a potentially profound influence upon the lives of each participant. Parsons (1959) has stated that the two major problems of schools are: (1) how the school functions to internalize in its pupils both the commitments and capacities for successful performance in their future adult roles, and (2) how the school functions to allocate these human resources within the role structure of the adult society. The school is viewed as a socializing agent which develops in the individual both commitments and capacities which Parsons sees as essential prerequisites for future role performance. It is strongly suggested that the higher the degree of formal education one has the more likely one is to be able to successfully compete within the society. And, indeed, it can be argued that rewards of the society are meted out to those individuals who stick to or internalize the values being fostered by our educational institutions.

Numerous sociological studies (Havighurst, 1958; Baltzel, 1958; Rogoff, 1961; Lundberg, 1968) indicate that high status, prestige, wealth, and success in general, within our society are highly correlated with education. Other studies (K. Clark, 1965; Coleman et al., 1966; U.S. Commission on Civil Rights, 1967) indicate that minority groups within American society have a very low level of academic achievement. The numbers of minority group students who go beyond the high school level throughout the nation is an extremely small percentage (B. Clark, 1960; Coleman et al., 1966). If, in fact, education is the single most important vehicle for achieving status, prestige, and power within our society, and if a majority of minority group peoples do not attain college-degree standings, then it is virtually impossible within the structure of our society for these minority group members to ever achieve status, prestige, and power. And yet basic to the values of the American system is the notion that we must all go as far in the educational system as possible.

Turner (1960) states that an essential underlying assumption is that within the formally open class system which provides for mass education, the organizing folk-norm which defines the accepted mode of upward mobility is a crucial factor in shaping the school system and may be even more curcial than the extent of upward mobility. To Turner, the most conspicuous problem, however, with this form of education as a means of upward mobility, is a conspicuous control problem as well as a problem of ensuring loyalty in the disadvantaged classes toward a system in which the members receive less than a proportional share of society's goods. Turner points out the potential for discontent when the society stresses high educational achievement as a means of success and at the same time systematically excludes members of that society from its educational institutions.

Inequities, however, are not confined to the educational system in this society. One need only look at economic institutions, housing, employment, and in fact every significant aspect of American society. There is a conspicuous absence of minority group members. It does not take much imagination on the part of these members of minority groups to realize that they have been left out of the "mainstream" of the American system. That is, left out of the reward structure of the system. Social scientists have attempted to explain and predict the consequences of this systematic exclusion in several ways.

If we take a look at some of the recent articles attempting to explain the underachievement of black children in education, we find such statements as Miriam Goldberg's (1968):

> Of particular importance from an educational point of view is the unstable family pattern and the strong female dominance in the home, characteristic of a disproportionately large number of Negro families. Beginning with the family the early pre-school years present the child from a disadvantaged home with few of the experiences which produce readiness for academic learning either intellectually or attitudinally. The child's view of society is limited by his immediate family and neighborhood where he sees a struggle for survival which sanctions behavior viewed as immoral in the society at large. He has little preparation either for recognizing the importance of schooling in his own life or of being able to cope with the kinds of verbal and abstract behavior which the school will demand of him.

Frank Riessman (1962), in a similar manner, identifies one of the crucial problems of the urban school today as that of the "culturally deprived child," who, lacking certain attributes which the middle-class child has, is thought, therefore, to be incapable of achievement.

Howard Becker (1952), on the other hand, indicates that perhaps the middle-class teacher is in part responsible for this underachieving. Becker points out that the middle-class teacher, who often neither appreciates nor understands the values of lower-class or culturally deprived children, reacts to values and behavior of these children in a negative manner. Indeed, how a teacher perceives each child is a very crucial factor in terms of that child's behavior.

In a study done by Rosenthal and Jacobson (1968), an experimental situation was set up whereby teachers were told in one group that their pupils were extremely bright and were expected to learn a great deal in the course of the academic year. The teachers in a second group were told that their pupils were dull and would probably just maintain their academic competence during the year and that they should do the best they could with this particular group of kids. At the end of the year the teachers with the bright children did indeed have high achievers; at least there was evidence of some academic advancement. On the other hand, with the teachers who had the "dull" group, the children showed little or no learning achievement. Unknown to the teachers at the time of the experiment was that the children had been randomly selected prior to the beginning of the school year. Thus, the difference in achievement was attributable to teacher expectation and not differences among the children.

Kenneth Clark (1965) indicates that Negro families have been very concerned with the quality of the child's education. He presents some data which show that in Harlem, for example, pupils' IQ tests reflect the quality of teaching and the resulting educational achievement more than the intellectual potential of particular children. In fact it can be demonstrated, according to Clark, that Negro children in ghetto schools tend to lose ground in IQ as they proceed through the schools and to fall farther and farther behind the academic performance standard for their grade level. This same factor was pointed out in the recent Coleman report (1966).

Clark's research indicates that children who are treated as if they are uneducable invariably become uneducable. It is this educational atrophy that is perhaps the single most important variable which prevents lower-class children from achieving in the classroom. Clark's focus is not on the deficiencies of the black family, as was the case with Frazier, Moynihan, and Banfield and Wilson, but rather on the deficiencies of the institutions and the society from within which these families operate. The implications of research which focuses on the deficiencies of the child are logical and clear. It is the child and his family who must be altered. If, on the other hand, it is the institutions surrounding and shaping the culturally deprived which are at fault, then solutions to the problems lie in altering the institutions and not their participants.

For Moynihan (1965), however, the solution for the "pathology" of black families was direct governmental intervention. This is seen as the only means of stemming or turning the tide of the inevitable pathology. For Clark, the solution is just the opposite–intervention into white society. In either case there is very little direct empirical evidence which gives much insight into the phenomenon of the disorganized, deteriorated black family. In fact, much of the literature is based on hearsay and opinion, not hard data.

Cultural Poverty: Sociological Evidence

Elliot Liebow (1969), through participant observation techniques, tries to give us insight into the values, attitudes, and beliefs of street-corner-society black men. Like most of the writers before him, Liebow finds that the lives of these men are fragmented and disorganized; attachments to family and friends are weak. In fact, these men so *desperately* need something to hold onto that they must create, according to Liebow, pseudo-kin relationships—"going for brothers and going for sisters." Personal failures of these men are seen by Liebow as being accounted for by the theory of "manly flaws." Liebow finds few, if any, positive factors or strengths in the lives of these men, except for the fact that they recognize the worth of various types of jobs and are not willing to kill themselves accepting jobs which will barely feed their families and pay off the bills. Liebow, in the tradition of Moynihan, Banfield and Wilson, and Goldberg, describes a way of life which is characterized as a "culture of poverty." The notion of a culture of poverty, originally developed by Oscar Lewis, suggests that poverty has associated with it a very distinctive life style.

Lewis' works (1951, 1959, 1961, 1966a, 1966b) are confined mainly to Spanish-speaking peoples, Puerto Ricans and Mexicans. Even though he has been struck by what he has called "the remarkable similarities in family structure, the nature of kinship ties, the quality of husband-wife and parent-child relations, time orientation, spending patterns, value systems, and the sense of community found in lower class settlements in the slums of London, Puerto Rico, and Mexico City, and in Mexican villages, and among lower class Negroes in the U.S." (Burma, 1970).

In Lewis' writing the characteristics of lower-class Spanish-speaking people can best be summed up as a slow pace in life, not valuing work for work's sake, put off until tomorrow that which you could do today (mañana concept), docility, the father as the strong provider, and extended kinship relationships. Similar concepts are reflected in Heller (1966). Heller describes Mexican-Americans as "regarding the future as both vague and unpredictable, consequently planning for the future, so characteristic of American culture, is therefore not part of their way of life" (Heller, 1966). As a result of this orientation, the Mexican-American culture utilizes the alternative of "being"; that is, the preference for the kind of activity which is a "spontaneous" expression of what is conceived to be "given" in one's personality. This attitude, according to Heller, is antithetical to the "doing" orientation of the American culture and, to a large degree, helps explain the underachievement of this particular group; i.e., the Mexican-Americans.

David Matza (1966) capitalizes on the earlier discussion of Frazier and Moynihan in describing a culture of poor blacks. "The disreputable poor," as he calls them, are "people who remain unemployed, casually or irregularly employed, even during periods approaching full employment and prosperity." For that reason and others he says "they live in disrepute." This classification of poor is different in a variety of ways from those who are deemed deserving. Some of the major differences are "that the disreputable poor cannot easily be reformed or rehabilitated and are resistant and are recalcitrant" (Matza, 1966). In conclusion, Matza describes this group of poor people as being an immobilized segment of society located at the point in the social structure where poverty interacts with illicit pursuits. They are, in the words of Charles Bruce, "the dangerous classes who live in regions of squalid want and wicked woe" (Matza, 1966).

Thus, for Matza, the core of disreputable poverty consists of the dregs—persons spawned in poverty and belonging to families who have been left behind by the otherwise mobile ethnic populations (Matza, 1966). What seemingly emerges as the legitimate basis for such a description lies in the fact that poor people—the objectionable poor, that is—do not share some of the most prized conventional values of the middle class. And for this reason they should be condemned as the dangerous classes (see Banfield, 1970). Matza's solution for getting the dregs out of this intolerable situation is for this group to mobilize for self-help. And yet in his earlier description, this same group was seen as being not easily rehabilitated and reformed. The implications being that the major source of poverty among the poor are the poor themselves, and that somehow the remedy must come from the same source. In short, all they really need to do is to learn the value of work.

Charles Valentine (1968) critically reviews the Frazier-Moynihan thesis and is extremely critical of both the insights and conclusions. In Frazier's case, Valentine points out how the misuse of census data as direct evidence of social disorganization is a glaring violation of the cultural distinction between statistical patterns and cultural configurations. It is Valentine's contention that after reading Frazier's thesis one is left with the feeling that he (Frazier) has not observed the life of slum dwellers at first hand, much less participated in that life.

In reviewing the work of Glazer and Moynihan (1963), Valentine finds that they, like Frazier, have made the same mistake. That is, they have confused demographic data with social structures. The result of such confusion is that nonconformity with middle-class norms is uncritically equated with instability and disorganization. Valentine's harshest criticisms are saved for Moynihan. Moynihan's thesis is seen as an attempt to inject the Frazierian concept of lower-class culture into national policy-making. Such an attempt is viewed as "stratagems to win the attention of those in power by giving them something that can be 'felt as well as understood.' " The result is "pseudo-scholarship used to good advantage in appealing to the ignorance and prejudices of the powerful."

Lee Rainwater (1965) is also critical of the culture-of-poverty thesis. Rainwater argues that a caste system superimposed on blacks has toughened black families, enabling them to function in ghetto life. However, he says this is very different from arguing that "the family is to blame" for the deprived situation ghetto Negroes suffer.

Whites, according to Rainwater, have created a caste system by their great power and with this system forced blacks to bear the burdens of the bottom rung. He argues, contrary to Moynihan, that only palliative results can be expected from attempts to treat directly the disordered family patterns of lower-class blacks. He contends that only a change in the original "inputs" of the caste system, the structural conditions inimical to basic social adaptation, can change family forms. Direct intervention, he says, will serve the unintended function of distracting the country from the pressing needs for socioeconomic reform, and of providing an alibi for the failure to embark on basic institutional changes needed to do something about white and black poverty.

Albert Murray (1970) levels a blistering attack against the Moynihan report. Murray says Moynihan's report is arguing, in effect, that those who have been exploiting Negroes for years should now, upon being shown his statistics, become benevolent enough to set up a nationwide welfare program. Moynihan's report isolates the negative figures on the number of black illegitimate births, broken homes, lack of education, crime, narcotics addiction, and so on. It charts Negro unemployment but not once does it suggest national action to crack down on discrimination against blacks by labor unions and other community institutions. Instead, says Murray, it insists that massive federal action must be initiated to correct the matriarchal structure of the Negro family!

A recent article by TenHouten (1970) challenges the assumptions underlying the Moynihan and Frazier theses that the black family in America is rapidly degenerating and that the "female dominant" family has caused psychological damage to black males. His data suggest that the stereotype of the lower-class black families as matriarchal, pathological, and "approaching a state of complete breakdown," constitutes a sociological mythology.

Implications

By now, some of the ramifications of these approaches should be clear. Social scientists have, perhaps inadvertently, conceived of a stratification system which by definition excludes colored minorities from ever hoping to occupy the upper echelons. In such a schema, upward mobility is only feasible for noncolored immigrants. We may wish to pursue the notion that colored minorities have no desire of partaking in the most beneficial rewards of the society, but there are few within social science circles wishing to pursue such a line of thought.

Instead we have seen research attention focused on these colored minorities precisely because they have not been upwardly mobile. Moynihan (1965) and Banfield (1970) feel that in the case of "disintegrated Black families the only solution is direct governmental intervention." Rainwater and TenHouten, on the other hand, question the appropriateness of intervention and raise a more deep-seated question about the rightful role of social science research as a determinant in policy formulation. TenHouten notes that "research on poor people and black people in American society is itself a weapon and a resource. It is a resource controlled almost entirely by white people. The Moynihan research is a political weapon used to rationalize intervention in black communities, and used for white control of black people." TenHouten feels that this process should be altered: "Research on blacks–on their political ideologies, their community structures, their families, and their personality structures–can serve as either a means of control of blacks by whites, or as a means of resistance to such control." TenHouten hopes that studies such as his own will contribute to a rationale by which black people can resist intervention of the type suggested by Moynihan.

It is perhaps with great justification that these writers express the cautions they have. It is essential to distinguish between social science research and political policy formulation. With policy formulation relying more and more on the social sciences, it becomes extremely important for social scientists to be responsible for the potential implications of their work. Such cautions are well founded in light of the suggested interventionist strategies of Edward Banfield (1970), and Daniel Moynihan (1965). As a result of studying Model City programs around the nation, Banfield makes several recommendations including the following:

(1) Distinguish categorically between those of the poor who are competent to manage their affairs and those who are not, the latter being the insane, the severely retarded, the senile, the lower class (inveterate "problem families") and unprotected children.

(2) Make cash income transfers to the first category (the competent) by means of a negative income tax, the rate structure of which gives the recipient a strong incentive to work. Whenever possible, assist the incompetent poor with goods and services rather than with cash; depending on their incompetence, encourage (or require) them to reside in an institution or semi-institution (for example, a closely supervised public housing project).

(3) Give intensive birth control guidance to the incompetent poor.

(4) Pay "problem families" to send infants and children to day nurseries and pre-schools designed to bring the children into the normal culture.

(5) Regulate insurance and police practices to give potential victims of crime greater incentive to take reasonable precautions to prevent it.

(6) Intensify police patrols in high crime areas; permit the police to stop and frisk and to make misdemeanor arrests on probable causes;

institute a system of "negative bail"—that is, an arrangement whereby a suspect who is held in jail and is found innocent later is payed for each day of confinement.

(7) Abridge to an appropriate degree the freedom of those who, in the opinion of the court, are extremely likely to commit violent crimes. Confine and treat drug addicts.

Some of Banfield's recommendations have already been enacted. President Nixon has recently signed an anticrime law which contains a no-knock clause and a preventative detention clause. Both were recommendations of Banfield. We have already examined some of Moynihan's suggestions and the reactions of other sociologists to them. Such partial or incomplete analyses as Banfield's or Moynihan's become dangerous when the misconceptions and assumptions implicit in them are transmitted directly as weaknesses to action programs and make them self-defeating or impossible of achievement from their inception. In the area of education, such a focus has been reflected in the creation of Headstart programs, Upward Bound programs, High Potential programs, High Horizons programs, English as a Second Language programs, Education of the Culturally Deprived programs, Title I and Title II programs, and so forth. In every case, these programs and other suggestions for change reflect only on the need to change colored minorities who have been characterized as being out of phase with the larger society. It was because of this emphasis on "the problems of black people," the "problems of Mexican-Americans," and "the problems of the poor," that social scientists directed their attention to creating programs for dealing with inappropriately defined problems. Black reactions of revolts and rebellions caught many social scientists off-guard.

RIOTS, REVOLTS, REBELLIONS—THE NEW DIRECTION

Masotti and Bowen (1968) have attempted to create a theoretical schema for the analysis of urban riots. In their introductory statements they try to distinguish between riots and revolutions.

Riots in and of themselves are not seen as attempts to seize power or to overthrow the government. Revolution, on the other hand, is seen as a deliberate attempt to replace one power group with another power group. Regardless of how you define the activities of the sixties, the participants are always seen as being against the functional order of the society—or law breakers. As such it is quite easy to construct a rationalization for the immediate cessation or elimination of this dysfunctional behavior. If, on the other hand, the activities of the sixties were defined as rebellions or revolts and these activities viewed as functional to the participants, one is faced with a totally different frame of reference. Masotti and Bowen touch on this problem but do not fully resolve it. They dismiss the revolution notion and discuss rebellions against civil order. But their discussion centers around the politically value-laden term of riot. To avoid this intrinsically negative context, it is more representative to use the term revolt or rebellion since participants perceived their activities in this way (Hacker and Harmetz, 1966; Jacobs, 1966; Geschmender, 1969; Boskin, 1969).

Masotti and Bowen reject the notion that criminal elements are responsible for rebellious behavior, at least the rebellious behavior described as characteristic of American society in the last ten years. They interpret the revolts of the past decade as attacks by dissatisfied citizens on the civil order of the society. For Masotti and Bowen, dissatisfactions with current states of affairs and a precipitating event are seen as perhaps two important preconditions for rebellious behavior. It is the prediction of these two writers that urban insurrections will not occur in the future because legitimacy of the political system is not being challenged. Indeed, it is their contention that the discontented at the moment are basically satisfied with the political system as it theoretically is supposed to operate but are dissatisfied with the political system as it functionally operates. The Constitution of the United States is seen as a document which has built into it the capacity for the very things the dissenters are asking; equal share, community control, self-protection, self-definition. The problem, however, has always been for the discontented to figure out a way of making the society produce the guarantees contained in this document.

Lipset (1963) states that "polities with a very low degree of wealth are likely to be more politically unstable." And Russett

(1964) states "there exists a positive relationship between unequal land distribution and civil disorder." Lipset's analysis implies that poor people generally are more prone to civil disorder, while Russett suggests that in agricultural societies the presence of large numbers of virtually landless individuals tend toward instability. In both, the implication is that, those who are poor and landless (those who do not occupy a solid position within the class status and power structure), are potentially dangerous to that structure.

It is, perhaps, instructive at this point to note that mob action and violence have marred the urban scene in America since at least the beginning of the eighteenth century. The practical causes of these early rebellions have varied, with labor or religious conflict and opposition to the Civil War draft accounting for some of the worst ones in the nineteenth century. Richard Maxwell Brown (1969) has counted thirty-three major outbreaks of violence from 1900-1949, with twenty-two of them occurring between 1915 and 1919. The disorders of the 1960s have differed from some of the earlier ones in that blacks attacked symbols of "white power" instead of individual whites. And yet, in all documented cases the major victims have been black, and police brutality has been an old and persistent element.

Geschmender (1969) in responding to the U.S. Riot Commission report (1968) reflects on the revolts in terms of their having become, since 1967, a tactic consciously used by some ghetto dwellers to coerce a response from those in positions of power. He points out that the inhabitants of rebellious areas had cause to rebel for a long time, and that rebellious activity invariably was related to long-standing grievances. This suggests that there was a high degree of revolt-potential in the ghetto. Many ghetto residents had serious grievances and were aware that earlier rebellion had brought about a favorable political response. Thus, when a precipitating incident occurred fortuitously, there was a preexisting segment of the population ready to respond with politically motivated riot activity. This pool of persons readily provided the organization and leadership for directed activity (Geschmender, 1969).

The U.S. Riot Commission (1968) identified white racism as being at the root of the revolts this country experienced during the 1960s. Allan Grimshaw (1969) suggests that the Commission, in emphasizing racism and failing to note the explanatory priority of structural variables as contrasted to attitudes, made a tactical error.

The source of the error lies in the fact that by identifying racism as the cause of revolt it is virtually impossible for the nation, or the government in particular, to do anything about alleviating or changing conditions to prevent future riotous activity. Grimshaw states that structural elements of society should have been emphasized as the causes of racism and that racism, in turn, was the parent of the rebellions. Grimshaw cites the works of Marx (see Bottomore, 1963; Weber, 1958; Simmel, 1965; Coser, 1956) as bases for his contention. Each of these individuals, in discussing social conflict, emphasizes the structural features of society, particularly insofar as that structure is related to the distribution and exercise of power.

Using what Grimshaw has called the structural aspects of society, Friedman (1969) gives a definition of structural racism:

> Structural racism is a pattern of action in which one or more of the institutions or organizations of society has the power to throw more burdens on and less benefits to the members of one race than another on an ongoing basis, or to support another institution or organization in so doing, and use their power in this manner. It should be emphasized that in this definition the intentions of the actors, or the formal statements of the relevant norms, laws, values, are irrelevant to the question of whether an institution is acting in a structurally racist manner. What counts is whether its actions *in fact* distribute burdens and rewards in a racially biased fashion.

Grimshaw was perhaps touching the core of the matter in defining structural racism and not individual racism as the root and cause of rebellions.

However, this explanation should not come as a total surprise or shock to any knowledgeable social scientist; one has but to review the works of such writers as Frantz Fanon (1963) in which he describes the role of black people in the United States as that of colonial subjects. He specifically outlines the structural relationship of the dominating or colonial government to the subject people. Unfortunately, Fanon relies heavily upon the Algerian situation as his analogy to contemporary American society, an analogy which does not always apply. However, the emphasis on structural or political subjugation of a group of individuals (i.e., black Americans) perhaps does fit the analogy in terms of French colonial domination of the Algerian people.

Malcolm X (1965: 16) also outlines the relationship of the United States government to colored minorities. Malcolm pointed out that whenever one group or set of institutions controls the economic structure, controls the military structure, controls the educational structure, then that group and that economic institution indeed controls the individual. This is a colonial or master-slave type of relationship, irrespective of all the platitudes about equality of education, equality of opportunity, upward mobility, and whatever other myths this country currently operates under. This analysis also is presented by Stokely Carmichael (Carmichael and Hamilton, (1967) who states "Black people in the United States have a colonial relationship to the larger society, a relationship characterized by institutional racism. That colonial status operates in three areas–political, economic, and social." Martin Luther King (1963) recognized the role of the "white power structure" as the force keeping blacks in their place. Although the very legitimacy of the existing system has been challenged by Carmichael and Hamilton (1967), Masotti and Bowen assert that rioters accept the legitimacy of the existing political order (Masotti and Bowen, 1968).

Black writers have long been aware of what the "American System"–its structural institutions–have been doing to hold colored minorities in subject positions. Why is it then that so few social scientists have yet to fully understand this? Perhaps by looking at specific riot studies, it may become clear why most social scientists still do not fully comprehend what lies behind the slogan "We'd rather do it ourselves."

Watts Rebellion

One particular rebellion, the Los Angeles riot of 1965, is deserving of more specific attention. The Watts riot was one of the most studied and talked about rebellions in the United States (McCone Commission, 1965; Jacobs, 1966; Cohen and Murphy, 1966; Ransford, 1968; N. Cohen, 1970). Although the Watts rebellion was not the most devastating in terms of property damage, it may serve as a prototype for efforts to understand what black dissidents, or blacks in general, want from American society. The result and the incidents surrounding its aftermath are important also as a focal point for assessing the redirection of the planning programs and community organization programs which followed.

Immediately after the revolt, there were many speculations as to who participated, what their characteristics were, what their motivations were, and why it happened in this particular city (see N. Cohen, 1970). Some of the major findings of this report were that the majority of the people who participated in the riots were employed. There apparently was a great deal of support from the educated people within this group. Approximately 71,000 people participated in the rebellion in one way or another, either by actively rebelling or by encouraging and supporting those who did rebel. Most of the participants were long-time residents, contrary to the popular belief that they were recent migrants from the South. It was found that merchants within the revolt area who were doing a heavy credit business were more likely to attribute support for the rebellion to racism, Muslims, and communists. Approximately sixty percent of the people in the revolt area who were surveyed felt that police malpractice was predominant in that area. Cohen suggests that the riot was a revolt against an oppressive system, a reaction against an accumulation of grievances. Even though the revolt was a violent one, says Cohen, "it was not a revolution which challenged the fundamental value system of American society." Cohen overlooked the literature which states that the single most agreed-upon reason for these types of rebellions was to challenge the "fundamental value system of America" (see Jacobs, 1966; U.S. Riot Commission, 1968; Cleaver, 1968; Grier and Cobbs, 1968; Lightfoot, 1968; Lester, 1968; R. M. Brown, 1969).

Paul Jacobs (1966), a journalist writing about the Watts riot, has outlined some conditions which contributed to the explosion in Watts. These conditions included segregated housing, job discrimination, inferior schools, policy harassment and malpractice. He found instances where rebellion and looting in the cities could be as much an act of politics as a desire for material goods; it is a way in which the poor can make a representation to the society, for they have no other kind of representation; it is a way in which the black poor can express their hate of the white world for not giving them their chance to share in the rewards. And, he says, it doesn't matter if the store being looted is owned by a black man or a white man; to them, owning a store is what white people do, not black people.

From the point of view of some of the participants, burning was seen as "selective demolition programs." Jacobs states that the brutal fact about American society today is that a whole constellation of forces, agencies, and circumstances, whose origins are different, whose operations are fragmented, and whose jurisdictions are separate, have converged in the city. The effect of this constellation bearing down upon the ghetto's inhabitants has been disastrous. Unfortunately, Jacobs points out, after all the rioting, little has changed for the people of the nation's ghettos (Jacobs, 1966).

But has so little really changed? Granted physical facts have altered little in most riot areas. Granted that the unemployment rate has not changed significantly in most of these areas. Indeed, in the Watts area, unemployment is significantly higher compared to five years ago (Datson and Profitt, 1970). But have there been some gains, perhaps more subtle, yet perhaps more drastic, gains, that is, for the black community, and for brown communities and the yellow communities? The major gain has been a heightened awareness of self-identification, self-definition, and self-direction.

The New Direction

Today's blacks are in a race to build a sense of black community before a racial Armageddon overtakes them. One example of this process can be found on college campuses throughout the country. Black students, in looking at institutions of higher learning, have identified the academic content of American universities (both black and white) as a body of material which was developed almost entirely from models that simply ignored the experience, culture, and presence of black men here and elsewhere (Harding, 1970).

As we have noted, an overwhelming amount of research regarded as central to the literature of the social sciences deals with issues such as inequality, oppression, and unrest from analytic standpoints completely outside those of the people experiencing them. The recognition of this fact has resulted in the creation of black studies centers which have as their purpose, the study of the history, culture, language, and experience of black people; topics which all too often have not been included in the larger university curriculum. This omission persists in spite of the fact that abrasive criticisms of the American universities were being published many years before the black movement began its offense (Woodson, 1922; Thompson, 1937; Ellison, 1952). The creators of these new centers are not impressed by the black and white (but mostly white) academicians who ask them to respect the great traditions of the university, to do things in its honored style, to submit their black studies programs to the wisdom of their faculties. They are aware of the fact that it is these same faculty members who had not heard of Horace Mann Bond, Charles S. Chestnault, Martin Delang, or Charlie Parker, five years ago, and who thought the black experience was worth one course three years ago (Harding, 1970). These are some of the reasons black studies advocates have fought so arduously to maintain, for the time being at least, autonomous black studies programs. In this fashion, it is likely that, with time, black scholarship will begin more accurately to reflect the actual circumstances, of blacks, both descriptively and analytically.

There can be little doubt that the revolts and rebellions over the last several years have greatly accelerated the desire on the part of blacks to take the lead in controlling their own destinies. This cry for control, however, is not just found on the university campuses. Citizen groups are demanding greater control over neighborhood schools. Shortcomings in urban school services are well known to the clients of these institutions. Growing public conflict over how to upgrade such services has been front-page news across the nation. Conflict marked by the participation of newly aroused but traditionally noninvolved parents, teachers, and students has directed public attention to a search for control of their schools (Gittell, 1969; Rogers, 1969; Lutz, 1970).

Malcolm X once noted "unless you controlled you own land, you were not in control of your own destiny" (Malcolm X, 1965). Except for Black Muslims (Massaquoi, 1970), controlling land has not been a goal of black Americans. Community control, however, may mark the first step in this direction. The question of control over land will undoubtedly be a long-range goal of all ethnic minorities. (Deloria, 1969; Wolfe, 1969; Erickson, 1968). Of more immediate concern will be the desire to control the institutions which daily influence the lives of these minoriteis. Social policy will play a significant role in developing strategies for acquiring this desired control. Social science research has traditionally served as the underpinnings for such policy formulation. We have already seen that much of this research has resulted in misguided policy formulation, the results of which have been disruptive for the nation as a whole.

From the ashes of the sixties must now come a Phoenix of hope for the ethnic minorities to play a greater role in the control of their destinies. One major question must now be answered: "Will future social science research retard or enhance this advancement?"

REFERENCES

ABBOT, D., L. H. GOLD, and E. T. ROGOWSKY (1969) Police, Politics, and Race. Cambridge: Harvard Univ. Press.

BALTZEL, E. D. (1958) Philadelphia Gentlemen. New York: Macmillan.

BANFIELD, E. C. (1970) The Unheavenly City. Boston: Little, Brown.

--- and J. Q. WILSON (1963) City Politics. New York: Vintage Press.

BECKER, H. (1952) "Social class variations in the teacher-pupil relationship." J. of Educational Sociology 25 (April).

BENDIX, R. and S. M. LIPSET [eds.] (1966) Class, Status, and Power. New York: Free Press.

BOSKIN, J. (1969) "The revolt of the urban ghettos, 1964-1967." Annals of the American Academy of Political and Social Science (March).

BOTTOMORE, T. B. [ed.] (1963) Karl Marx. New York: McGraw-Hill.

BROWN, H. R. (1969) Die Nigger Die. New York: Dial Press.

BROWN, R. M. (1969) "Historical pattern of violence in America," in H. D. Graham and T. R. Gurr (eds.) Violence in America: Historical and Comparative Perspectives. A Report submitted to the National Commission on the Causes and Prevention of Violence. New York: Bantam Books.

BURMA, J. H. [ed.] (1970) Mexican-Americans in the United States. New York: Schenkman.

CARMICHAEL, S. and C. HAMILTON (1967) Black Power: The Politics of Liberation in America. New York: Vintage Press.

CLARK, B. R. (1960) "The 'cooling-out' function in higher education." Amer. J. of Sociology 65 (May).

CLARK, K. B. (1965) Dark Ghetto. New York: Harper & Row.

CLEAVER, E. (1968) Soul on Ice. New York: McGraw-Hill.

COHEN, J. and W. S. MURPHY (1966) Burn, Baby, Burn: The Watts Riot. New York: Avon.

COHEN, N. [ed.] (1970) The Los Angeles Riots: A Socio-Psychological Study. New York: Frederick A. Praeger.

COLEMAN, J. S. et al. (1966) Equality of Educational Opportunity. U.S. Department of Health, Education and Welfare. Washington, D.C.: Government Printing Office.

COSER, L. (1956) The Functions of Social Conflict. New York: Free Press.

DATSON, J. and N. PROFITT (1970) "Watts! everything has changed, and nothing." Newsweek 26 (August 24).

DAVIS, K. and W. E. MOORE (1945) "Some principles of stratification," pp. 242-249 in Amer. Soc. Rev. 110, 2.

DELORIA, V., Jr. (1969) Custer Died for Your Sins. New York: Macmillan.

ELLISON, R. (1952) Invisible Man. New York: Signet (eighth printing).

ERICKSON, C. A. (1968) "Uprising in the barrios." Amer. Education (November).

FANON, F. (1963) The Wretched of the Earth. New York: Grove Press.

FRAZIER, E. F. (1957) Black Bourgeoisie. New York: Free Press.

--- (1948) The Negro Family in the United States. New York: Dryden Press.

FRIEDMAN, S. (1969) "How is racism maintained?" Radical Sociology 2, 2 (Fall).

GESCHMENDER, J. A. (1969) "On the political objectives of civil disorders." Social Science Q. 50, 3 (December).

GITTELL, M. (1969) "Community control of education," in R. H. Connery (ed.) Urban Riots, Violence and Social Change. New York: Vintage Press.
GLAZER, N. and D. P. MOYNIHAN (1963) Beyond the Melting Pot. Cambridge: MIT Press.
GRIER, W. H. and P. M. COBBS (1968) Black Rage. New York: Bantam.
GRIMSHAW, A. D. [ed.] (1969) Racial Violence in the United States. Chicago: Aldine.
GOLDBERG, M. (1968) "Factors affecting educational attainment in depressed urban areas," in A. H. Passow (ed.) Education in Depressed Areas. New York: Teachers College Press (seventh printing).
HACKER, F. J. and A. HARMETZ (1966) "What the McCone Commission didn't see." Frontier (March): 10-15.
HARDING, V. (1970) "Black studies and the impossible revolution." J. of Black Studies 1, 1 (September).
HAVIGHURST, R. J. (1958) "Education, social mobility and social change in four societies." International Rev. of Education 4.
HELLER, C. S. (1966) Mexican-American Growth: Forgotten Youth at the Crossroads. New York: Random House.
JACOBS, P. (1966) Prelude to Riot: A View of Urban America from the Bottom. New York: Vintage Press.
KING, M. L. (1963) "Letter from Birmingham jail." Why We Can't Wait. New York: Harper & Row.
LESTER, J. (1968) Look Out, Whitey! Black Power's Gon' Get Your Mama! New York: Dial Press.
LEWIS, O. (1966a) La Vida: A Puerto Rican Family in the Culture of Poverty–San Juan and New York. New York: Random House.
--- (1966b) "The culture of poverty," pp. 19-25 in Scientific American 215, 4.
--- (1964) Pedro Martinez. New York: Random House.
--- (1961) The Children of Sanchez. New York: Random House.
--- (1960) Tepoztlan: Village in Mexico. New York: Holt
--- (1959) Five Families: Mexican Case Studies in the Culture of Poverty. New York: Basic Books.
--- (1955) "Peasant culture in India and Mexico: a comparative analysis," in Marriott Village India Studies in the Little Community. American Anthropological Association, Memoir 83.
--- (1953) Controls and Experiments in Field Work. Berkeley: Kroeber.
--- (1951) Life in a Mexican Village: Tepoztlan Restudied. Urbana: Univ. of Illinois Press.
LIEBOW, E. (1969) Tally's Corner. Boston: Little, Brown.
LIGHTFOOT, C. M. (1968) Ghetto Rebellion to Black Liberation. New York: International.
LIPSET, S. M. (1963) Political Man. New York: Doubleday.
LOVE, J. L. (1970) "La Raza: Mexican Americans in rebellion," in J. H. Burma (ed.) Mexican-Americans in the United States. New York: Schenkman.
LUNDBERG, F. (1968) The Rich and the Super-Rich. New York: Bantam.
LUTZ, F. (1970) Toward Improved Urban Education. Belmont, Calif.: Wadsworth.
MALCOLM X. (1970) "Prospects for freedom in 1965," in R. P. Young (ed.) Roots of Rebellion, the Evolution of Black Politics and Protest Since World War II. New York: Harper & Row.
--- (1965) "Message to the grass roots." Malcolm X Speaks. New York: Ment.
MASOTTI, L. H. and D. R. BOWEN [eds.] (1968) Riots and Rebellions. Beverly Hills, Calif.: Sage.
MASSAQUOI, H. J. (1970) "Elijah Muhammad." Ebony. Special Issue (August)
MATZA, D. (1966) "The disreputable poor," in Smelser and S. M. Lipset (eds.) Social Structure and Mobility in Economic Development. Chicago: Aldine.

McCONE Commission (1965) Violence in the City: An End or a Beginning? A Report by the Governor's Commission on the Los Angeles Riots (December).

MILLS, C. W. (1942) "The professional ideology of social pathologists." Amer. J. of Sociology 49 (September).

MOYNIHAN, D. P. (1965) The Negro Family: The Course for National Action. Washington, D.C.: Government Printing Office.

MURRAY, A. (1970) The Omni-American. New York: Outerbridge & Dienstfreg.

PARSONS, T. (1967) in M. M. Tumin (ed.) Social Stratification. Englewood Cliffs, N.J.: Prentice-Hall.

--- (1959) "The school class or a social system: some of its functions in American society," pp. 297-318 in Harvard Educational Rev. 29, 4 (Fall).

RAINWATER, L. (1965) "Crucible of identity: the Negro lower-class family," in T. Parsons (ed.) The Negro American. Boston: Beacon Press.

--- and W. L. YANCY (1967) The Moynihan Report and the Politics of Controversy. Cambridge: MIT Press.

RANSFORD, H. E. (1968) "Isolation, powerlessness and violence: a study of attitudes and participation in the Watts riot." Amer. J. of Sociology 73 (March).

RIESSMAN, F. (1962) The Culturally Deprived Child. New York: Harper & Row.

ROGERS, D. (1969) 110 Livingston Street. New York: Vintage Press.

ROGOFF, N. (1961) "Local social structure and educational selection," in A. H. Halsey, J. Floud, and C. A. Anderson (eds.) Education, Economy, and Society. New York: Macmillan.

ROSENTHAL, R. and L. JACOBSON (1968) Pygmalion in the Classroom. New York: Holt, Rinehart & Winston.

RUSSETT, B. (1964) "Inequality and instability: the relation of land tenure to politics." World Politics 16 (April).

SCHEER, R. [ed.] (1967) Eldridge Cleaver. New York: Rampart.

SHERMAN, R. B. [ed.] (1970) The Negro and the City. Englewood Cliffs, N.J.: Prentice-Hall.

SIMMEL, G. (1965) in L. Coser (ed.) George Simmel. Englewood Cliffs, N.J.: Prentice-Hall.

STRATTERWHITE, F. J., H. AERMOUREUX, J. JOHNSON, and T. PAGNE (1970) Workshop on the Independent Black Institution. Summary Report. Palo Alto, Calif: Satterwhite.

TenHOUTEN, W. (1970) "The black family: myth and reality." Psychiatry 33, 2 (May).

THOMPSON, C. (1937) "The Harrison-Fletcher bill and Negro separate schools." J. of Negro Education (January).

TUMIN, M. M. (1953) "Some principles of stratification: a critical analysis." Amer. Soc. Rev. 18 (August).

TURNER, R. H. (1960) "Sponsored and contest mobility and the school system," pp. 856-867 in Amer. Soc. Rev. 25 (December).

U.S. Commission on Civil Rights (1967) Racial Isolation in Public Schools. Report. Washington, D.C.: Government Printing Office.

U.S. Riot Commission (1968) Report of the National Advisory Commission on Civil Disorders. New York: Bantam.

VALENTINE, C. A. (1968) Culture and Poverty. Chicago: Univ. of Chicago Press.

WARNER, W. L. (1963) Yankee City. New Haven: Yale Univ. Press (third printing).

WEBER, M (1967) in M. M. Tumin (ed.) Social Stratification. Englewood Cliffs, N.J.: Prentice-Hall.

--- (1958) in H. H. Gerth and C. W. Mills (eds.) From Max Weber. New York: Galaxy.

WOLFE, T. (1969) "The new yellow peril" Esquire (December).

WOODSON, C. G. (1922) The Negro in Our History. New York: Associated.

YOUNG, R. [ed.] (1970) Roots of Rebellion. New York: Harper & Row.

20

Clues for the Future: Black Urban Anthropology Reconsidered

COUNCILL TAYLOR

□ IN THE FOLLOWING PAGES I would like briefly to describe and discuss the progress of ethnographic studies of black people in the urban context.

While contemporary anthropology—especially American anthropology—has not distinguished itself through its conduct of acute and sensitive studies in this area, the mood of the black urban community may now be so hostile as to render such studies all but impossible. These communities are demanding action. It is common knowledge among black people that such studies as have been conducted advance individuals but seldom advance the black community.

The anger which exists is related to at least two central facts: the few urban anthropological studies of black people which have been conducted have seldom if ever been followed by clear policy action reflecting the implicit or explicit recommendations contained in them; black ghetto dwellers have come to distrust the sudden new enthusiasm for social scientific information about their lives, especially since it appears to correspond so neatly with their militant quest for social, racial, and political liberation. They question whether the information will be used for change or more insidious forms of control.

Nonetheless, past research in this area has shown that effective studies can be conducted, and in depth, on the life of the urban black community. Looking at a concrete case from the past provides some clues to the present need. I have divided those past and present studies into reformist, integrationist, and the more recent pluralist types, and have chosen to refer the case of W. E. B. DuBois' neglected *The Philadelphia Negro* (1967) to some of the questions which must be asked of contemporary studies. This study demonstrates keenly analytic, insightful, and humane research. It was by necessity inexpensive, yet breathtakingly exhaustive. Many recent and well-financed studies are pale by comparison. Questions which seem germane today might include:

(a) Who sponsored the research?

(b) What controls were imposed upon the ethnographers?

(c) What were their explicit and implicit objectives?

(d) What were the assumptions of the investigators?

(e) What were the consequences of the research?

(f) Of what utility for social actions were the findings?

While this is not the place to attempt so comprehensive an analysis, projection of some of these questions against representative work may help us to confront more effectively some of the problems of black urban anthropology today.

TYPES OF BLACK URBAN ANTHROPOLOGICAL STUDIES

Existing anthropological research[1] on urban black people can be placed into three categories which correspond roughly to time and implicit or explicit political attitudes. These may be identified as reformist, integrationist, and pluralist studies.

The reformist studies, which appear earliest in the historical record, are addressed to the moral sensibilities of those elements of the white power structure whom the authorities believe may be reached, influenced, and caused to modify their behavior by extending full citizen rights to black people. These studies are also addressed to the black leadership and enjoin them to intensify their efforts among lower-class black people in order to make them more visibly worthy of receiving full citizenship rights. The classic example of this school is *The Philadelphia Negro* (DuBois, 1967), originally published in 1899.

The integrationist studies in a number of ways mirror the strong civil rights movements which dominate much of the middle decades of the twentieth century. Often less explicit in ideological statement than its predecessors, this school of urban anthropologists remains single-mindedly committed to the integrationist goal, even after major voices from the black community have repudiated it. The much-discussed *Black Metropolis* (Cayton and Drake, 1945), represents this viewpoint.

The best of the pluralist studies accept as an existential reality the cultural autonomy of today's ghetto. While rarely, or never, explicitly endorsing the efforts of the black community to expand its autonomy by increasing its control over the political, economic, and educational institutions which shape their lives, these urban anthropologists do seek to document the precise cultural and social patterns and networks which have been created by black people in our cities. These modes of adaptation, these techniques of survival, have been systematically described and analyzed—not in terms of their similarity or dissimilarity to the modes of existence of white middle-class life, but rather in terms of the exigent demands made upon its dwellers by the black ghetto. From one viewpoint, these pluralist studies might be described as representative of the "natural history" approach in anthropology (see Valentine and Valentine, 1970).

REFORMIST BLACK URBAN ANTHROPOLOGY: THE CASE OF DuBOIS

The reformist tendency in black urban anthropology is expressed in its most classic form in the pioneering studies undertaken by W. E. B. DuBois toward the end of the nineteenth century and during the opening years of the twentieth century. Although DuBois was to die in Ghana in 1963, an ardent black nationalist and a committed socialist, during his earliest productive years, despite intimations of the ultimate political and intellectual goals to which he was to dedicate his long life, DuBois at the outset was a militant moral reformer.

G. Peter Murdock (1954) and others have pointed out that American sociology was essentially the intellectual offspring of liberal reformist Protestantism and moral philosophy. Toward the end of the nineteenth century in the United States, the new social sciences commenced to bridge the gap between practical issues of the day and advanced social thought. DuBois, in his autobiography *Dusk of Dawn* (1968), relates his searing encounters with the practical issues of his day in Nashville, Tennessee, in 1885, where "I came in contact with a sort of violence that I had never realized in New England" (DuBois, 1968: 30) and in the "windowless log cabin" of rural east Tennessee, where he taught ex-slaves during his summer vacations from Fisk University, and of which "I saw the hard, ugly drudgery of country life and the writhing of landless, ignorant peasants. I saw the race problem at nearly its lowest terms" (DuBois, 1968: 31). His encounters with advanced social thought took place at Harvard University a few years later where "DuBois received intellectual stimulation from psychologist William James and philosopher George Santayana" and where, under the guidance of other professors, "DuBois' interest in history, economics, and sociology was expanded" (Rudwick, 1969). Perhaps he did not formally embrace anthropology, despite his lifelong work in anthropological fields, because at Harvard "he was instructed that there were obvious differences among the major groups of mankind, and it was self-evident the lowest status belonged to the Negroes. DuBois recalled a visit to a museum, where he viewed a group of skeletons in a progression from a monkey to a white man—with the Negro placed slightly ahead of the chimpanzee" (Rudwick, 1969: 23). "In the fall of 1892, at the University of Berlin, DuBois further expanded his encounters with advanced social thought by studying economics, history and sociology under such figures as Adolf Wagner, Heinrich von Treitschke and Max Weber; from this experience came an assimilation of scientific methods of social research" (Rudwick, 1969: 27). When he completed his studies abroad, he writes that he "dropped suddenly back into 'nigger' hating America" (Rudwick, 1969: 27). His objectives had become clear and he writes in his autobiography:

> I determined to put science into sociology through a study of the condition and problems of my own group. I was going to study the facts, any and all facts, concerning the American Negro and his plight, and by measurement and comparison and research, work up to any valid generalization which I could. I entered this primarily with the utilitarian object of reform and uplift; but nevertheless, I wanted to do the work with scientific accuracy. . . . I easily grasped the idea of a changing developing society rather than a fixed social structure [DuBois, 1968: 51].

In his hopes that through the study of facts with scientific accuracy his objective goals of reform and uplift might be approached, DuBois' aspirations reflected those of other practitioners of the new social sciences of his day. As early as 1877, Lewis Henry Morgan in *Ancient Society* had warned that social and cultural patterns must be adjusted to meet altered conditions. Repudiating the current overemphasis of property values, Morgan predicted that "The time will come, nevertheless, when human intelligence will rise to the mastery over property, and define the relations of the state to the property it protects" (Morgan, 1959).

Nineteenth-century optimism expressed in the publications of the emerging social sciences was derived from the spectacular progress achieved by the biological and natural sciences. Similar progress was expected to result from application of the scientific method to enduring social problems.

In 1868 when DuBois was born, the United States, which had entered the Civil War as a primarily agricultural country, had only commenced its growth toward becoming an industrial giant. Its industrial output did not approach that of the European powers. However, the value of its manufactured products rose from $3.3 billion during DuBois' first year of life to over $13 billion in 1899, the year of the publication of *The Philadelphia Negro.* Thus, by the end of the century, this nation had become foremost among the world's manufacturers, dwarfing the production of such countries as Great Britain and Germany. However, the South was one of the least industrially developed regions in the United States and over ninety percent of black Americans were still tied to the South at the turn of the century. In his autobiography he writes, "The Negro problem was in my mind a matter of systematic investigation and intelligent understanding . . . . The cure for it was knowledge based on scientific investigation" (DuBois, 1968: 58).

DuBois initiated the field of black urban anthropology in 1896, when he undertook a field research project in the slums of Philadelphia's Seventh Ward under the sponsorship of the University of Pennsylvania. DuBois states in his autobiography that the reasons the establishment financed his research were: "Philadelphia wanted to prove . . . by figures" a particular theory:

> that theory was that this great, rich, and famous municipality was going to the dogs because of the crime and venality of its Negro citizens, who lived largely centered in the slum at the lower end of the seventh ward . . . . Of this theory back of the plan, I neither know nor cared. I saw only here a chance to study an historical group of black folk and to show exactly what their place was in the community [DuBois, 1968: 58].

In another work, DuBois expands this explanation:

> It all happened this way: Philadelphia, then and still one of the worst governed of America's badly governed cities [this was published in 1944] was having one of its periodic spasms of reform. A thorough study of causes was called for. Not but what the underlying cause was evident to most white Philadelphians: the corrupt, semi-criminal vote of the Negro Seventh Ward. Everyone agreed that here lay the cancer; but would it not be well to elucidate the known causes by a scientific investigation, with the imprimatur of the University? It certainly would, answered Samuel McCune Lindsay of the Department of Sociology. And he put his finger on me for the task [DuBois, 1944: 44].

In conducting his fieldwork, DuBois utilized the methods which later anthropologists were to systematize over the next several decades: participant-observation, map-making; census-taking; selected informants from various age levels and both sexes, and from each of the social classes which he identified along with a number of voluntary associations. He relied heavily upon sociological methods such as questionnaires and supplemented these techniques with historical documentation and the use of federal, state, and city statistical materials as well as data from libraries and private collections.

As participant-observer, DuBois, with his new bride, settled into a small upstairs flat in the heart of the Seventh Ward; they attended many social functions and church services of the community. For over one year he studied this ward intensively. He personally interviewed several thousand individuals.

For dividing his population into four grades, or social classes, DuBois employed a complex series of criteria incorporating not only economic factors and educational level but also cohesiveness of family structure, visible life style, and moral values. In assessing "functional" character of the community's religious institutions, he drew careful attention to some blatant dysfunctions.

In attempting to assess the achievement of DuBois' work, *The Philadelphia Negro,* it might be instructive to compare some of the problems which concerned DuBois with a recent list of problems proposed by George Spindler (1969) for "the emerging domestic urban anthropology" of "current minorities."

Biculturalism. From the fact that "less than one-third of the Negroes living in this ward were born here, and over one-half were born in the South" (DuBois, 1967: 74), DuBois directs attention to the consequences of the differential enculturation and socialization of large sectors of the population.

Shifts in Leadership Roles. In the course of a functional analysis of the Negro church, DuBois pointed out that

> Negro preachers are often condemned for poor leadership . . . it is said that men with so much power and influence could make striking moral reforms. This is but partially true. The congregation does not follow the moral precepts of the preacher, but rather the preacher follows the standard of his flock [DuBois, 1967: 205].

This passage introduces a lengthy analysis by DuBois of shifts in leadership roles within the religious institutions. In another passage, DuBois describes articulatig mechanisms between the white authorities and the protected black leaders of Philadelphia's organized criminal populations (DuBois, 1967: 312).

Other topics suggested by Spindler and discussed or alluded to by DuBois might include the structural adaptation of communities, the influence of traditional modes of social organization on contemporary social relationships,[2] instrumental values developed in current behavioral choices, the adaptation of ideology to changing external conditions, and the presentation of self and the problem of identity.

In his study, DuBois documented the gravity of unemployment as a problem and the institutionalized racism which restricted most black working men and women to menial jobs; the household conditions which led to problems of health and morality. Intensive critical analysis was made of the Negro church and the corrupt state and local political institutions.

Leaping decades ahead of his period, he recommended the construction of a program which might develop through "the mastery of the art of social organized life" into an increasingly self-sufficient social system, under a progressive and reformed black leadership, within the framework of the wider American society.

Still believing in social reform through social science, DuBois became associated with the Department of Sociology at Atlanta University and there proceeded to launch a program designed to minutely study every phase of the life of the black man in America. During his thirteen-year stay at Atlanta University, he continued this correspondence with prominent overseas scholars such as E. D. Morel, the British Africanist, and Max Weber of Heidelberg. Among the authorities whom he brought to Atlanta were Franz Boas, whom he was later to take as a student, and the black anthropologist-novelist Zorah Neal Hurston. The Atlanta University Publications, produced under DuBois' leadership, continued to express his conviction in the possibility of social reform through sociological research. Separate publications dealt with "Negro morality; the Negro family, urbanization; the effort of Negroes for their own social betterment; Negroes in business; college-bred Negroes; the Negro common school; the Negro artisan; the Negro church; Negro crime."

Each annual report carried this statement of aims:

> This study is a further carrying out of a plan of social study by means of recurring decennial inquiries in to the same general set of human problems. The object of these studies is primarily scientific—a careful search for truth conducted as thoroughly,

> broadly and honestly as the material resources and mental equipment at command will allow; but this is not our sole object; we wish not only to make the Truth clear but to present it in such shape as will encourage and help social reform.

While DuBois was struggling to reduce racism in America by gathering, analyzing, and publishing relevant data, racism continued unabated. For example, while lynching reached its climax in 1892, with the public murder of 235 persons, during the next sixteen years of DuBois' academic career, nearly two thousand persons were publicly killed by mobs, and not one of the murderers was punished.

DuBois, of course, knew that the results of scientific inquiry diffuse very slowly to the general population. In the South, however, he was continually confronted with the raw evidences of racist activities. He describes how he was on his way to the editorial office of the Atlanta *Constitution* where he intended to publish a discussion of a pending case in which a black man was being held on charges of having murdered his landlord's wife. Before he arrived at the newspaper office, not only had the accused slayer been lynched, but his knuckles were on display at a nearby grocery shop. Concluding that social research seemed futile, DuBois shifted his activities from reform efforts to protest activities. Today, the black ghetto community has come to share DuBois' disenchantment with social research.

RECENT STUDIES AND THE CONTEMPORARY BLACK MOOD

Despite his extraordinary intelligence and awareness, DuBois himself could easily fail to conduct adequate research in the present climate of the black community. The mood of today's black ghetto is one of massive hostility toward ethnographic research among its members. Here are a few representative responses of black ghetto residents of an eastern city to a pair of social scientists, both black, who were inquiring into the type of reception which might be anticipated if they were to become the subjects of such studies:

> Perhaps a few years ago I would have said—get your figures and go out there and show the man what's going down—tell him what we need. But now we don't give a shit . . . .

> What can you do for me? I've been studied for over 400 years. I've been analyzed for 100 years. I've been systematized, sterilized, organized, and compromised for the past 50 years. Just what can you do for me?
>
> I'm still economically poor. Socially unequal (in your eyes). Physically abused. And mentally amused. Therefore, what you can do for me is read what you have already written. And Act. Yes, act upon your own recommendations for change.
>
> To continue doing academic gymnastics at the expense of the community is criminal. The social scientist maybe runs down a problem, identifies it—but that isn't enough.

Other ghetto residents, echoing the sentiment that "time had run out" for the outside investigator—that he might have been sought in a more credulous yesterday, but today is too late—admitted to having questions which they would like to pose to social scientists:

> Why does Ford Foundation and others always give money to such studies? I would feel that if money is so readily available to do the studies then this money through proper hustling and know-how could somehow be turned over to the community.
>
> I would ask him how he got through all the bullshit of college and graduate school to become one of those social scientists with a high paying job and then use *his* use of the game to my *own* advantage.
>
> How can the police be used? For us? Or is that what it appears? a contradiction in terms . . . stuff like that for us.
>
> What we wanta know . . . how about—what institutions are strongest within our community at enforcing the status quo? And how they can be altered.
>
> To tell me, what *he* knew, and how *we* could use this knowledge to better the community . . . I wouldn't want large "help" programs to solve the problem. But what could the *community* do, by any means possible, to become independent of the whole power structure.
>
> . . . do some people need conservative and reactionary beliefs for their own security?—and, if they do—what if they're in the way—dig it—do we have to kill them?

Similar views, often expressed more trenchantly, could be multiplied readily by inviting random opinion of the "brothers in the street." The reasons for these views should be very evident. Generation after generation of black refugees, displaced from marginally subsistent regions of the South into the teeming urban ghettos, discover that they are again faced with the stark problems of unemployment and underemployment, wretched housing, chaotic educational systems, inadequate medical facilities, heightened living costs, and personal safety. Had social scientists, their students, or other emissaries from the dominant society streamed into the ghettos and offered their skills to ghetto dwellers engaged in survival struggles, they might have been warmly welcomed. Recent history reveals, however, that social scientists have been conspicuously absent from the ghetto except during those periods when the smoldering explosive conditions nurtured by the circumstances of ghetto life actually erupt and threaten the stability of the wider established order. After each eruption, the floodgates are opened; social scientists, armed with sharper pencils and fatter grants, stream into the ghetto. Compendia of causes and cures were compiled after East St. Louis, 1917; Chicago, 1919; Detroit and Harlem, 1943; and more recently Watts, Newark, and Detroit. But after federal, state, or municipally sponsored hearings, field surveys, investigations, chronologies, interviews, and publications, wanted changes did not come about.

From the vantage point of the ghetto dweller, this is the picture which emerges: long periods of time, relatively unmarked by alien intrusion into the ghetto except by welfare investigators, insurance collectors, law officials, salesmen, landlords, social workers, and preelection office seekers; then, in the aftermath of a major eruption, there suddenly appears an avalanche of social science investigators, including sociologists, criminologists, psychologists, penologists, urbanologists, social psychiatrists, and anthropologists. After a few cursory interviews, some retreat to suburbia or central city; many move from door to door, each with long questionnaire schedules; very few attempt to settle in for longer periods of weeks and months.

In the past, this apparently enormous volume of investigative effort has not yielded a large body of sound published works. Despite a fairly recent upsurge of interest in black communities by social scientists, the number of serious urban anthropologists working in the field is extremely limited. Far fewer anthropologists evidently explored this research area during those periods in the past when their presence might have been more readily tolerated.

The extent of the traditional neglect of urban black America by anthropologists is dramatically indicated by a few figures (see Beckham and Beckham, 1963). Although 1,714 book titles are entered in their master bibliography, fewer than 15, or less than one percent, comprise volumes which focus upon black urban or town experience in the United States.[3]

To be sure, this basic list was published a couple of years before black ghetto revolts and guerrilla warfare transformed American inner cities during successive summers, beginning with Watts in August 1965. Since then, urban anthropologists have been invited by the media and various governmental agencies to offer, for fees, their analyses and prognostications. Numbers have responded. However, this new-found interest is not reflected by a sharply increased number of papers appearing in the *American Anthropologist,* although this is not the case of professional publications in closely related fields. Only a very minute proportion of regularly scheduled papers at the national meetings of this association have concentrated upon problems of urban anthropology in black communities.

This neglect cannot be justified by recourse to basic anthropological theory. In the course of his presidential address on urbanization as a process of cultural and social change delivered to the American Anthropological Association twenty years ago, Ralph Beals drew deserved attention to what he called "the rather blundering way in which some anthropologists have entered upon the study of modern cultures." Beals expresses agreement with a critic who remarked that "if anthropologists keep on as they have begun in the study of modern culture, they will in time reinvent sociology, but unfortunately it will be at least fifty years behind the rest of the field." Noting that much of Robert Redfield's work had been concerned with the problem of urbanism, Beals observed that in Redfield's case, "the primary concern has been with the way in which the influence of urbanism is extended into the urban hinterland and particularly into the folk culture." Hence, although ostensibly interested in a problem of urbanism, both Redfield and his students tended to focus their attention upon what they called folk cultures, and measured in these spatially distant folk communities the relative frequency of urban traits in order to obtain a crude index of what was termed acculturation. Beals points out that these measurements might have been taken within the urban context itself because "in the urban area, immigrants often segregate or are segregated to form new groups. . . . This group formation is enhanced by a degree of enforced residential segregation which is even more highly operative in the case of the rural Negro immigrant" (Beals, 1953: 1-10). However, few anthropologists have followed Beals' suggestion to study the urbanization/acculturation processes as experienced by migrants organized into new segregated groups within the city.

Herskovits (1951) and his students approached the same so-called acculturation problem, but in precisely the opposite direction. Scrutiny was indeed made of black communities in the New World, but as in the case of the Redfield school, primary attention was paid to the folk community. However, instead of examining these communities for frequency of diffused urban traits, these communities were minutely combed for cultural evidences of African heritage. In criticizing this approach, M. G. Smith warned that even

> the study of African heritage in purely cultural terms is not adequately conceived and cannot by itself reveal the processes and conditions of acculturation. Thus, if acculturation, rather than the simple identification of elements as African or other, is the aim of such study, we must study the relevant social and cultural conditions equally and simultaneously [Smith, 1957: 35].

In surveying the meager field of American anthropological studies which concentrates on black urban communities, one is forced to conclude that far too few have been carried out, their research methods have been uneven in quality, their theoretical orientations have been exceedingly weak, and their utility as guides for social action—a major concern of members of the communities studied—almost nonexistent. Anthropologists in Africa, a continent whose population might be grossly comparable to the population of the United States, but whose population is less than ten percent urbanized, have published many times more significant volumes concerning African urbanization than have their American colleagues; and in one-third of the time (e.g., Acquah, 1958; Balandier,

1969a, 1969b; Busia, 1950; Banton, 1957; Epstein, 1958; Kuper et al., 1958; Little, 1965; Miner, 1965; Parrinder, 1969; Porter, 1963; Sofer, 1955).

By contrast with the ten percent of the African population that have been urbanized, today about 15 million black Americans live in metropolitan areas, or sixty-nine percent of the black population of 21.5 million. In 1910, 800,000 blacks–nine percent–lived outside the South. Now almost 10 million, about forty-five percent, live in the North or West (National Advisory Committee on Civil Disorders, 1968: 236-237). However, the largest black urban community in the world, Harlem, has never been the object of a published urban anthropological study.

Those studies which have been conducted have, by and large, not been the works of anthropologists. They have reflected the pluralist, urban-ethnographic bias discussed earlier (see, e.g., Liebow, 1967; McCord et al., 1969; Abrahams, 1964; Slim, 1967; Horton, 1967; Kochman, 1969; Rainwater, 1970; Hannerz, 1969).

These works represent a beginning resurgence of concern and care in investigation, but a great deal more thought must go into such studies if we are to understand and aid black people in urban areas in their quest for liberation.

NOTES

1. By the term urban anthropology I refer more to the method and orientation of investigation than to the disciplinary or professional affiliation of the investigator.

2. DuBois' characterization of the Negro church as a reservoir of African tribal functions as shown in its "religious activity, its social authority and general guiding and co-ordinating work; its family functions . . ." is particularly cogent (DuBois, 1967: 201).

3. Of the fifty-two professional anthropologists whose aid was acknowledged by compilers, none listed is black.

REFERENCES

ABRAHAMS, R. D. (1964) Deep Down in the Jungle; Negro Narrative Folklore from the Streets of Philadelphia. Hatboro, Penn.: Folklore Assoc.

ACQUAH, I. (1958) Accra Survey. London: Univ. of London Press.

BALANDIER, G. (1969a) Ambiguous Africa: Cultures in Collision. New York: World.

--- [H. Weaver, trans.] (1969b) Daily Life in the Kingdom of the Kongo: From the Sixteenth Century to the Eighteenth Century. New York: World.

BANTON, M. (1957) West African City. London: Oxford Univ. Press.

BEALS, R. (1953) "Acculturation" in A. L. Kroeber (ed.) Anthropology Today: An Encyclopedic Inventory. Chicago: Univ. of Chicago.

BECKHAM, R. and M. BECKHAM [compilers] (1963) "A basic list of books and periodicals for college libraries," in D. Mandelbaum, G. Lasker, and E Albert (eds.) Resources for the Teaching of Anthropology. Berkeley: Univ. of California Press.

BUSIA, K. A. (1950) Report on a Social Survey of Sekondi–Takoradi. London: Crown Agents.

CAYTON, H. and S. C. DRAKE (1945) Black Metropolis: A Study of Negro Life in a Northern City. (2 vols.) New York: Harper & Row.

DuBOIS, W. E. B. (1968) Dusk of Dawn: An Essay Toward an Autobiography of a Race Concept [1940]. New York: Schocken.

--- (1967) The Philadelphia Negro [1899]. New York: Schocken.

--- (1944) "My evolving program for Negro freedom," in R. W. Logan (ed.) What The Negro Wants. Chapel Hill: Univ. of North Carolina Press.

EPSTEIN, A. L. (1958) Politics in an Urban African Community. New York: Humanities Press.

HANNERZ, U. (1969) Soulside: Inquiries into Ghetto Culture and Community. New York: Columbia Univ. Press.

HERSKOVITS (1951) Man and His Works: The Science of Cultural Anthropology. New York: Alfred A. Knopf.

HORTON, J. (1967) "Time and cool people." Trans-action 4, 5 (April).

KOCHMAN, T. (1969) "Rapping in the black ghetto." Trans-action 6, 4 (February).

KROEBER, A. L. [ed.] (1953) Anthropology Today: An Encyclopedic Inventory. Chicago: Univ. of Chicago Press.

KUPER, L. ẹt al. (1958) Durban: A Study in Racial Ecology. New York: Columbia Univ. Press.

LIEBOW, E. (1967) Tally's Corner: A Study of Negro Streetcorner Men. Boston: Little, Brown.

LITTLE, K. (1965) West African Urbanization: Voluntary Associations in Social Change. New York: Cambridge Univ. Press.

MANDELBAUM, D., G. LASKER, and E. ALBERT [eds.] (1963) Resources for the Teaching of Anthropology. Berkeley: Univ. of California Press.

McCORD, W. et al. (1969) Life Styles in the Black Ghetto. New York: W. W. Norton.

MINER, H. (1965) Primitive City of Timbuktoo. New York: Anchor-Doubleday.

MORGAN, L. H. (1959) Ancient Society; or Researches in the Lines of Human Progress from Slavery, Through Barbarism to Civilization [1877]. London: Routledge & Kegan Paul.

MURDOCK, G. P. (1954) "Sociology in anthropological perspective" in J. P. Gillin (ed.) For A Science of Social Man: Convergences in Anthropology, Psychology and Sociology. New York: Macmillan.

National Advisory Commission on Civil Disorders, Kerner Report (1968) New York: E. P. Dutton.

PARRINDER, E. G. (1969) Religion in Africa. Baltimore: Penguin.

PORTER, A. T. (1963) Creoledom: A Study of the Development of Freetown Society. New York: Oxford Univ. Press.

RAINWATER, L. (1970) Behind Ghetto Walls. Chicago: Aldine.

REDFIELD, R. (1956) Peasant Society and Culture: An Anthropological Approach to Civilization. Chicago: Univ. of Chicago Press.

RUDWICK, E. (1969) W. E. B. DuBois: Propagandist of the Negro Protest. New York: Atheneum.

SLIM, I. (1967) Pimp, the Story of My Life. Los Angeles: Holloway House.

SMITH, M. G. (1957) "Ethnic and cultural pluralism in the British Carribean," in The Secretary-General, International Institute of Differing Civilizations (INCIDI). Ethnic and Cultural Pluralism in Intertropical Countries. Belgium.

SOFER, C. and R. SOFER (1955) Jinja Transformed: A Social Survey of a Multi-Racial Township. Kampala: East African Institute of Social Research.

SPINDLER, G. (1969) "Review of the American Indian today." American Anthropologist 71, 2: 341-342.

VALENTINE, C. and B. L. VALENTINE (1970) "Making the scene, digging the action, and telling it like it is; anthropologists at work in a dark ghetto," in N. Whitten and J. Szwed (eds.) Afro-American Anthropology: Contemporary Perspectives. New York: Free Press.

BIBLIOGRAPHY

ABBOT, D., L. H. GOLD, and E. T. ROGOWSKY (1969) Police, Politics, and Race. Cambridge: Harvard Univ. Press.

ABERBACH, J. D. and J. L. WALKER (1970) "The attitudes of blacks and whites toward city services: implications for public policy," pp. 519-538 in John P. Crecine (ed.) Financing the Metropolis: Public Policy in Urban Economics. Beverly Hills, California: Sage.

ABRAHAMS, R. D. (1964) Deep Down in the Jungle: Negro Narrative Folklore from the Streets of Philadelphia. Hatboro, Penn.: Folklore Assoc.

ACKERMAN, N. W. and M. JAHODA (1950) Anti-Semitism and Emotional Disorder: A Psychoanalytic Interpretation. New York: Harper & Bros.

ACQUAH, I. (1958) Accra Survey. London: Univ. of London Press.

ADORNO, T. W., E. FRENKEL-BRUNSWICK, D. J. LEVINSON, and R. N. SANFORD (1950) The Authoritarian Personality. New York: Harper & Bros.

Advisory Commission on Intergovernmental Relations (1966) Intergovernmental Relations in the Poverty Program: A Commission Report. Washington, D.C.: Government Printing Office.

ALLPORT, G. (1958) The Nature of Prejudice. New York: Doubleday.

ALTSHULER, A. A. (1970) Community Control: The Black Demand for Participation in Large American Cities. New York: Pegasus.

ANDERSON, S. V. (1969) Ombudsman Papers: American Experience and Proposals. Institute of Governmental Studies, University of California, Berkeley.

ANDIC, F. (1963) "Income of wage earner families and economic development of Puerto Rico, 1941-1953." Caribbean Studies 2, 4: 14-27.

ARON, R. (1969) Progress and Disillusion: The Dialectics of Modern Society. New York: New American Library.

BABCOCK, Richard F. (1966) The Zoning Game. Madison: Univ. of Wisconsin Press.

BACHRACH, P. (1967) The Theory of Democratic Elitism. Boston: Little, Brown.

——— and M. S. BARATZ (1970) Power and Poverty. Fair Lawn, N.J.: Oxford Univ. Press.

BAKER, R. S. (1917) "The Negro goes North." The World's Work (July).

––– (1908) Following the Color Line: American Negro Citizenship in the Progressive Era. New York: Harper & Row.

BALANDIER, G. (1969) Ambiguous Africa: Cultures in Collision. New York: World.

––– [H. Weaver, trans.] (1969) Daily Life in the Kingdom of the Kongo: From the Sixteenth Century to the Eighteenth Century. New York: World.

BALDWIN, J. (1964) Notes of a Native Son. New York: Bantam.

––– (1961) Nobody Knows My Name. New York: Dell.

BALTZELL, E. D. [ed.] (1968) The Search for Community in Modern America. New York: Harper & Row.

––– (1958) Philadelphia Gentlemen. New York: Macmillan.

BANFIELD, E. C. (1970) The Unheavenly City. Boston: Little, Brown.

––– and J. Q. WILSON (1963) City Politics. New York: Vintage Press.

BANTON, M. (1964) The Policeman in the Community. New York: Basic Books.

––– (1957) West African City. London: Oxford Univ. Press.

BARTH, E. A. and B. A. ABU-LABAN (1959) "Power structures and the Negro sub-community." Amer. Soc. Rev. 24 (February): 69-76.

BAYLEY, D. H. and H. MENDELSOHN (1969) Minorities and the Police. New York: Free Press.

BEALS, A. (1962) Gopalpur, a South Indian Village: Case Studies in Cultural Anthropology. New York: Holt, Rinehart.

BEALS, R. (1953) "Acculturation" in A. L. Kroeber (ed.) Anthropology Today: An Encyclopedic Inventory. Chicago: Univ. of Chicago.

BENDIX, R. and S. M. LIPSET [eds.] (1966) Class, Status, and Power. New York: Free Press.

BENEDICT, R. (1934) Patterns of Culture. New York: Houghton-Mifflin.

BERCAL, T. E. (1970) "Calls for police assistance: consumer demands for governmental service." Amer. Behavioral Scientist 13 (May-June-July-August): 681-691.

BERELSON, B. R., P. F. LAZARSFELD, and W. N. McPHEE (1954) Voting. Chicago: Univ. of Chicago Press.

BERREMAN, G. D. (1966) "Concomitance of caste organization," pp. 308-324 in G. DeVos and H. Wagatsuma (eds.) Japan's Invisible Race. Berkeley: Univ. of California Press.

––– (1966) "Structure and function of caste systems," pp. 277-307 in G. DeVos and H. Wagatsuma (eds.) Japan's Invisible Race. Berkeley: Univ. of California Press.

BERUBE, M. R. and M. GITTELL (1969) Confrontation at Ocean Hill-Brownsville: The New York School Strikes of 1968. New York: Praeger.

BERWANGER, E. H. (1967) The Frontier Against Slavery. Urbana: Univ. of Illinois Press.

BETTELHEIM, B. and M. JANOWITZ (1964) Social Change and Prejudice Including Dynamics of Prejudice. New York: Free Press.

BITTLE, W. E. and G. L. GEISS (1957) "Racial self-fulfillment and the rise of an all-Negro community in Oklahoma." Phylon 18: 247-260.

BLACK, D. J. and A. J. REISS, Jr. (1967) "Patterns of behavior in police and citizen transactions," pp. 1-139 in Studies in Crime and Law Enforcement in Major Metropolitan Areas, Vol. 2. Washington, D.C.: Government Printing Office.

BLAUNER, R. (1970) "Black culture: myth or reality?" in N. Whitten and J. Szwed (eds.) Afro-American Anthropology. New York: Free Press.

--- (1964) Alienation and Freedom. Chicago: Univ. of Chicago Press.

BOAS, F. (1911) The Mind of Primitive Man. New York: Macmillan.

BOORSTIN, D. J. (1962) The Image. New York: Atheneum Press.

BORDUA, D. J. and A. J. REISS, Jr. (1966) "Command, control, and charisma: reflections on police bureaucracy." Amer. J. of Sociology 72 (July): 68-76.

BOSKIN, J. (1969) "The revolt of the urban ghettos, 1964-1967." Annals of the American Academy of Political and Social Science (March).

BOWEN, D. and L. MASOTTI (1968) "Spokesmen for the poor: an analysis of Cleveland's poverty board candidates." Urban Affairs Q. 4 (September).

BRIGHAM, C. C. (1923) A Study of American Intelligence. Princeton: Princeton Univ. Press.

BRODY, E. B. (1966) "Psychiatry and prejudice," in S. Arieti (ed.) American Handbook of Psychiatry 3: 629-642. New York: Basic Books.

BROWN, C. (1965) Manchild in the Promised Land. New York: Macmillan.

BROWN, H. R. (1969) Die Nigger Die. New York: Dial Press.

BROWN, R. M. (1969) "Historical pattern of violence in America," in H. D. Graham and T. R. Gurr (eds.) Violence in America: Historical and Comparative Perspectives. A Report submitted to the National Commission on the Causes and Prevention of Violence. New York: Bantam Books.

BROWN, W. H., III (1970) "Racial discrimination in the legal profession." Judicature 53, 9 (April-May).

BRYCE-LaPORTE, R. S. (1971) "Slaves as inmates; slaves as men: a sociological discussion of Elkins' 'Sambo' thesis," in A. Lane (ed.) The Debate on Slavery: Stanley Elkins and His Critics. Urbana: Univ. of Illinois Press.

--- (1969) The American slave plantation and our heritage of communal deprivation." Amer. Behavioral Scientist (March-April): 2-8.

BURMA, J. H. [ed.] (1970) Mexican-Americans in the United States. New York: Schenkman.

BUSIA, K. A. (1950) Report on a Social Survey of Sekondi-Takoradi. London: Crown Agents.

California State Department of Education (1967) "Racial and ethnic survey of California public schools. Part I: Distribution of pupils, Fall 1966." Sacramento.

CAMPBELL, A. and H. SCHUMAN (1968) "Racial attitudes in fifteen American cities," pp. 1-69 in Supplemental Studies for the National Advisory Commission on Civil Disorders. Washington, D.C.: Government Printing Office.

CAPLOVITZ, D. (1967) The Poor Pay More. New York: Free Press.

CAPLOW, T., S. STRYKER, and S. WALLACE (1964) The Urban Ambiance: A Study of San Juan, Puerto Rico. New York: Bedminister Press.

CARLIN, J. E. and J. HOWARD (1965) "Legal representation and class justice." UCLA Law Rev. 12.

CARMICHAEL, S. and C. HAMILTON (1967) Black Power: The Politics of Liberation in America. New York: Vintage Press.

CASH, W. J. (1941) The Mind of the South. New York: Doubleday.

CAYTON, H. and S. C. DRAKE (1945) Black Metropolis: A Study of Negro Life in a Northern City. (2 vols.) New York: Harper & Row.

CLARK, K. B. (1967) Dark Ghetto. New York: Harper & Row.

CLEAVER, E. (1968) Soul on Ice. New York: McGraw-Hill.

COHEN, J. and W. S. MURPHY (1966) Burn, Baby, Burn: The Watts Riot. New York: Avon.

COHEN, N. [ed.] (1970) The Los Angeles Riots: A Socio-Psychological Study. New York: Frederick A. Praeger.

COLEMAN, J. S. (1957) Community Conflict. New York: Free Press.

--- et al. (1966) Equality of Educational Opportunity. U. S. Department of Health, Education and Welfare. Washington, D.C.: Government Printing Office.

COMER, J. (1967) "The social power of the Negro." Scientific American 216 (April): 21-27.

COSER, L. A. (1968) Continuities in the Study of Social Conflict. New York: Free Press.

--- (1956) The Functions of Social Conflict. New York: Free Press.

DAHRENDORF, R. (1968) Essays in the Theory of Society. Stanford: Stanford Univ. Press.

--- (1959) Class and Class Conflict in Industrial Society. Stanford: Stanford Univ. Press.

DELORIA, V., Jr. (1969) Custer Died for Your Sins. New York: Macmillan.

DeREUCK, A. and J. KNIGHT [eds.] (1967) Caste and Race: Comparative Approaches. London: Churchill.

DE TOCQUEVILLE, A. (1969) Democracy in America. New York: Anchor Doubleday.

DEUTSCH, K. (1952) Nationalism and Social Communication. Cambridge, Mass.: MIT Press.

DEUTSCH, M. and M. E. COLLINS (1951) Interracial Housing: A Psychological Evaluation of a Social Experiment. Minneapolis: Univ. of Minnesota Press.

DeVOS, G. and H. WAGATSUMA [eds.] (1966) Japan's Invisible Race. Berkeley: Univ. of California Press.

DISKIN, M. (1969) "Estudio estructural del sistema de plaza en el valle de Oaxaca." America Indigena 29, 4: 1076-1099.

DOLLARD, J. (1937) Caste and Class in a Southern Town. New Haven: Yale Univ. Press.

DONOVAN, J. (1967) The Politics of Poverty. New York: Pegasus.

DREGER, R. M. and K. S. MILLER (1960) "Comparative psychological studies of Negroes and whites in the United States." Psych. Bulletin 57, 5 (September).

DuBOIS, W. E. B. (1968) Dusk of Dawn: An Essay Toward an Autobiography of a Race Concept [1940]. New York: Schocken.

——— (1967) The Philadelphia Negro [1899]. New York: Schocken.

——— (1947) "Three centuries of discrimination." Crisis 54 (December): 362-363.

——— (1944) "The American dilemma." Phylon (second quarter).

——— (1944) "My evolving program for Negro freedom," in R. W. Logan (ed.) What the Negro Wants. Chapel Hill: Univ. of North Carolina Press.

——— (1920) Darkwater: Voices from Within the Veil. New York: Schocken.

ELKINS, S. (1963) Slavery: A Problem in American Institutional Intellectual Life. New York: Grosset & Dunlap.

——— and E. McKITRICK, Jr. (1954) "A meaning for Turner's frontier," Parts I and II. Pol. Sci. Q. 69, 3 & 4.

EPSTEIN, A. L. (1958) Politics in an Urban African Community. New York: Humanities Press.

ERIKSON, E. H. (1968) "Race and the wider identity," pp. 295-320 in Identity Youth and Crisis. New York: W. W. Norton.

——— (1963) Childhood and Society. New York: W. W. Norton.

——— (1959) "Ego Development and Historical Change [1946]" in Identity and the Life Cycle: Selected Papers. New York: International Universities Press.

ERSKINE, H. (1968) "The polls: demonstrations and race riots." Public Opinion Q. 31 (Winter): 655-677.

ETZIONI, A. (1968) The Active Society. New York: Free Press.

——— (1968) "Societal guidance: a key to macro-sociology." Acta Sociologica 4.

——— (1961) A Comparative Analysis of Complex Organizations. New York: Free Press.

EVANS-PRITCHARD, E. E. (1951) Social Anthropology. London: Cohen & West.

FANON, F. (1963) The Wretched of the Earth. New York: Grove Press.

FEAGIN, J. R. (1970) "Home-defense and the police: black and white perspectives." Amer. Behavioral Scientist 13 (May-June-July-August): 797-814.

FRAZIER, E. F. (1967) Negro Youth at the Crossways [1940]. New York: Schocken.

--- (1957) Black Bourgeoisie. New York: Free Press.

--- (1948) The Negro Family in the United States. New York: Dryden Press.

--- (1939) The Negro Family in the United States. Chicago: Univ. of Chicago Press.

--- (1932) The Negro Family in Chicago. Chicago: Univ. of Chicago Press.

--- (1927) "The pathology of race prejudice." Forum 77 (June).

FRIEDMAN, S. (1969) "How is racism maintained?" Radical Sociology 2, 2 (Fall).

FURNIVALL, J. S. (1948) Colonial Policy and Practice. Cambridge: Cambridge Univ. Press.

GALBRAITH, J. K. (1958) The Affluent Society. Boston: Houghton-Mifflin.

GANS, H. J. (1962) The Urban Villagers: Group and Class in the Life of Italian-Americans. New York: Free Press.

GARTH, T. R. (1934) "The problem of race psychology: a general statement." J. of Negro Education 3 (July).

--- (1931) Race Psychology. New York: Whittlesey.

--- (1926) "Race psychology." Scientific Monthly 23 (September).

--- (1923) "A comparison of the intelligence of Mexican and mixed and full blooded Indian children." Psych. Rev. 30: 388-401.

GENOVESE, E. (1966) The Political Economy of Slavery. New York: Random House.

GITTELL, M. (1969) "Community control of education," in R. H. Connery (ed.) Urban Riots, Violence and Social Change. New York: Vintage Press.

GLAZER, N. and D. P. MOYNIHAN (1963) Beyond the Melting Pot. Cambridge: MIT Press.

GREENE, L. J. and C. G. WOODSON (1969) The Negro Wage Earner [1930]. New York: Russell & Russell.

GREER, S. (1969) "Urbanization and social character: notes on the American as citizen," pp. 95-127 in H. J. Schwandt and W. Bloomberg (eds.) The Quality of Urban Life. Beverly Hills, Calif.: Sage.

--- (1966) Urban Renewal and American Cities: The Dilemma of Democratic Intervention. Indianapolis: Bobbs-Merrill.

--- (1962) The Emerging City. New York: Free Press.

GREGORY, D. (1964) Nigger: An Autobiography. New York: E. P. Dutton.

GRIER, W. H. and P. COBBS (1968) Black Rage. New York: Basic Books.

GRIFFIN, J. H. (1960) Black Like Me. New York: Houghton-Mifflin.

GRIMSHAW, A. D. [ed.] (1969) Racial Violence in the United States. Chicago: Aldine.

GROVES, W. E. and P. H. ROSSI (1970) "Police perceptions of a hostile ghetto: realism or projection." Amer. Behavioral Scientist 13 (May-June-July-August): 727-743.

HACKER, F. J. and A. HARMETZ (1966) "What the McCone Commission didn't see." Frontier (March): 10-15.

HAHN, H. (1971) "The public and the police: a theoretical perspective," in H. Hahn (ed.) Police in Urban Society. Beverly Hills, California: Sage.

--- (1970) "Civic responses to riots: a reappraisal of Kerner Commission data." Public Opinion Q. 34 (Spring): 101-107.

--- (1970) "Cops and rioters: ghetto perceptions of social conflict and control." Amer. Behavioral Scientist 13 (May-June-July-August): 761-779.

--- (1969) "Philosophy of law and urban violence." Soundings 52 (Spring): 110-117.

--- (1969) "Ghetto sentiments on violence." Science and Society 33 (Spring): 197-208.

--- and J. R. FEAGIN (1970) "Riot-precipitating police practices: attitudes in urban ghettos." Phylon 31 (Summer): 183-193.

HALL, M. O. (1911) Half a Man: The Status of the Negro in New York. New York: Longmans, Green.

HANNERZ, U. (1969) Soulside: Inquiries into Ghetto Culture and Community. New York: Columbia Univ. Press.

HARDING, J., H. PROSHANSKY, B. KUTNER, and I, CHEIN (1969) "Prejudice and ethnic relations." Handbook of Social Psychology 5: 1-76. Reading, Mass.: Addison-Wesley.

HARDING, V. (1970) "Black studies and the impossible revolution." J. of Black Studies 1, 1 (September).

HARNETT, P. (1969) "Why teachers strike: a lesson for liberals," in M. R. Berube and M. Gittell (eds.) Confrontation at Ocean Hill-Brownsville: The New York School Strikes of 1968. New York: Praeger.

HAYNES, G. E. (1968) The Negro at Work in New York City [1912]. New York: AMS Press.

HELLER, C. S. (1966) Mexican-American Growth: Forgotten Youth at the Crossroads. New York: Random House.

HERNTON, C. C. (1965) Sex and Racism in America. New York: Doubleday.

HERSKOVITS, M. J. (1951) Man and His Works: The Science of Cultural Anthropology. New York: Alfred A. Knopf.

--- (1941) The Myth of the Negro Past. New York: Harper & Brothers.

--- (1926) "On the relation between Negro-white mixture and standing in intelligence tests." Pedagogical Seminary 33: 30-42.

HIEBERT, P. (1969) "Caste and personal rank in an Indian village; an extension in techniques." American Anthropologist 71, 69: 434-453.

HOETINK, H. (1967) Two Variants in Caribbean Race Relations: A Contribution to the Sociology of Segmented Societies. London: Oxford Univ. Press.

HOLDEN, M., Jr. (1969) "The quality of urban order," pp. 431-454 in Henry J. Schmandt and Warner Bloomberg, Jr. (eds.) The Quality of Urban Life. Beverly Hills, California: Sage.

HOROWITZ, D. (1969) Empire and Revolution. New York: Random House.

--- (1965) The Free World Colossus. New York: Hill & Wang.

HYMAN, H. H. and P. B. SHEATSLEY (1964) "Attitudes toward desegregation." Scientific American 211, 1 (July): 16-23.

IRVINE, S. H. (1966) "Toward a rationale for testing attainments and abilities in Africa." British J. of Educational Psychology 36 (February): 24-32.

JACOBS, P. (1966) Prelude to Riot: A View of Urban America from the Bottom. New York: Vintage Press.

JOHNSON, C. S. (1941) Growing Up in the Black Belt. Washington, D.C.: Amer. Council on Education.

--- (1934) Shadow of the Plantation. Chicago: Univ. of Chicago Press.

--- (1930) The Negro in American Civilization. New York: Holt.

KARDINER, A. and L. OVESEY (1951) The Mark of Oppression. New York: W. W. Norton.

KATZNELSON, I. (forthcoming) Race, Politics and Migration. London: Oxford Univ. Press.

--- (1970) "The politics of racial buffering in Nottingham, 1954-1968." Race (April): 431-446.

KELLER, S. (1968) The Urban Neighborhood. New York: Random House.

Kerner Commission (1968) Report of the National Advisory Commission on Civil Disorders. New York: Bantam Books.

KING, M. L. (1963) "Letter from Birmingham jail." Why We Can't Wait. New York: Harper & Row.

KLINEBERG, O. (1963) "Negro-white differences in intelligence test performance: a new look at an old problem." Amer. Psych. 18, 4 (April).

--- (1935) Race Differences. New York: Harper & Bros.

--- (1935) Negro Intelligence and Selective Migration. New York: Columbia Univ. Press.

--- (1928) "An experimental study of speed and other factors in 'racial' differences." Archives of Psychology 15.

KLONOSKI, J. R. and R. I. MENDELSOHN (1966) "The allocation of justice: a political approach." J. of Public Law 14 (Winter): 323-342.

KOCHMAN, T. (1969) "Rapping in the black ghetto." Trans-action 6, 4 (February).

KOVEL, J. (1970) White Racism: A Psychohistory. New York: Pantheon.

KRAFT, J. F. (1966) "The attitudes of Negroes in various cities," pp. 1383-1423 in Vol. 6 of Federal Role in Urban Affairs, Hearings before the Subcommittee on Executive Reorganization, Committee on Government Operations, U.S. Senate, Washington, D.C.: Government Printing Office.

KRAMER, J. (1970) The American Minority Community. New York: Thomas Y. Crowell.

KRAMER, R. (1969) Participation of the Poor: Comparative Community Case Studies in the War on Poverty. Englewood Cliffs, N.J.: Prentice-Hall.

KROSNEY, H. (1966) Beyond Welfare: Poverty in the Supercity. New York: Holt, Rinehart & Winston.

KUBIE, L. S. (1965) "The ontogeny of racial prejudice." J. of Nervous and Mental Disease 141, 4: 265-273.

LANE, A. [ed.] (1971) The Debate on Slavery: Stanley Elkins and His Critics. Urbana: Univ. of Illinois Press.

LESTER, J. (1968) Look Out, Whitey! Black Power's Gon' Get Your Mama! New York: Dial Press.

LEVI-STRAUSS, C. (1967) Structural Anthropology. New York: Anchor Doubleday.

LEVITAN, S. (1967) "Planning the anti-poverty strategy." Poverty and Human Resources Abstracts 2 (January-February).

LEVY, C. J. (1968) Voluntary Servitude: Whites in the Negro Movement. New York: Appleton-Century-Crofts.

LEWIS, O. (1966) La Vida: A Puerto Rican Family in the Culture of Poverty–San Juan and New York. New York: Random House.

——— (1966) "The culture of poverty." Scientific American 215, 4 (October): 19-25.

LIEBOW, E. (1967) Tally's Corner: A Study of Negro Streetcorner Men. Boston: Little, Brown.

LIGHTFOOT, C. M. (1968) Ghetto Rebellion to Black Liberation. New York: International.

LIPSET, S. M. (1963) Political Man. New York: Doubleday.

——— (1959) "Democracy and working-class authoritarianism." Amer. Soc. Rev. 24 (August): 482-502.

LITTLE, K. (1965) West African Urbanization: Voluntary Associations in Social Change. New York: Cambridge Univ. Press.

LITWACK, E. (1961) North of Slavery. Chicago: Univ. of Chicago Press.

LOMAX, L. E. (1962) The Negro Revolt. New York: Signet.

LOVE, J. L. (1970) "La Raza: Mexican Americans in rebellion," in J. H. Burma (ed.) Mexican-Americans in the United States. New York: Schenkman.

LYNCH, O. (1969) The Politics of Untouchability. New York: Columbia Univ. Press.

MALCOLM X (1970) "Prospects for freedom in 1965," in R. P. Young (ed.) Roots of Rebellion, the Evolution of Black Politics and Protest Since World War II. New York: Harper & Row.

——— (1966) Autobiography. New York: Grove Press.

——— (1965) "Message to the grass roots." Malcolm X Speaks. New York: Ment.

MANGIN, W. (1967) "Latin American squatter settlements: a problem and a solution." Latin American Research Rev. 2, 3: 65-98.

MASOTTI, L. H. and D. R. BOWEN [eds.] (1968) Riots and Rebellions. Beverly Hills, Calif.: Sage.

MATTHEWS, D. (1969) "Political science research on race relations," in I. Katz and P. Gurin (eds.) Race and the Social Sciences. New York: Basic Books.

McCONE Commission (1965) Violence in the City: An End or a Beginning? A Report by the Governor's Commission on the Los Angeles Riots (December).

McCORD, W. and J. HOWARD (1968) "Negro opinions in three riot cities." Amer. Behavioral Scientist 11 (March-April): 31-34.

McCORD, W. et al. (1969) Life Styles in the Black Ghetto. New York: W. W. Norton.

McKITRICK, E. [ed.] (1963) Slavery Defended. Englewood Cliffs, N.J.: Prentice-Hall.

McLEAN, H. V. (1946) "Psychodynamic factors in racial relations." Annals of the Amer. Academy of Political and Social Science 244 (March): 159-166.

MEIER, A. (1963) Negro Thought in America, 1880-1915. Ann Arbor: Univ. of Michigan Press.

--- and E. RUDWICK (1969) "A strange chapter in the career of 'Jim Crow.' " The Making of Black America, Vol. II. New York: Atheneum.

MEMMI, A. (1965) The Colonizer and the Colonized. Boston: Beacon Press.

MERTON, R. (1957) Social Theory and Social Structure. Glencoe: Free Press.

MILLER, S. M. and M. REIN (1969) "Participation, poverty and administration." Public Administration Rev. (January-February).

MILLS, C. W. (1942) "The professional ideology of social pathologists." Amer. J. of Sociology 49 (September).

MILLS, C. W. (1959) The Sociological Imagination. New York: Oxford Univ. Press.

MINTZ, S. (1956) "Cañamelar: the subculture of workers on a government owned sugar plantation," pp. 314-417 in J. Steward (ed.) The People of Puerto Rico. Urbana: Univ. of Illinois Press.

MIZRUCHI, E. H. (1964) Success and Opportunity. New York: Free Press.

MOLOTCH, H. L. (1970) "Oil in Santa Barbara and power in America." Sociological Inquiry 40 (Winter): 131-144.

--- (1969) "Racial integration in a transition community." Amer. Soc. Rev. 34 (December): 878-892.

MOORE, W. (1969) The Vertical Ghetto. New York: Random House.

MORGAN, L. H. (1959) Ancient Society; or Researches in the Lines of Human Progress from Slavery, Through Barbarism to Civilization [1877]. London: Routledge & Kegan Paul.

MORRIS, R. T. and V. JEFFRIES (1969) "The white reaction study," pp. 480-601 in N. E. Cohen (ed.) The Los Angeles Riots: A Socio-Psychological Study. New York: Frederick A. Praeger.

MOTON, R. R. (1929) What the Negro Thinks. Garden City, N.Y.: Doubleday, Doran.

MOYNIHAN, D. P. (1969) Maximum Feasible Misunderstanding. New York: Free Press.

--- (1969) "Toward a national urban policy." Public Interest (Fall).

--- (1966) "What is community action?" Public Interest (Fall).

——— (1965) The Negro Family: The Course for National Action. Washington, D.C.: Government Printing Office.

MUSSEN, P. H. (1950) "Some personality and social factors related to changes in children's attitudes toward Negroes." J. of Abnormal and Social Psychology 65 (July): 423-441.

MYRDAL, G. (1957) Rich Lands and Poor Lands. New York: Harper & Brothers.

——— (1944) An American Dilemma. New York: Harper & Bros.

National Advisory Commission on Civil Disorders, Kerner Report (1968) New York: E. P. Dutton.

NELSON, J. M. (1969) Migrants, Urban Poverty and Instability in Developing Nations. Cambridge, Mass.: Center for International Affairs, Harvard University.

NEUWIRTH, G. (1969) "A Weberian outline of a theory of community: its application to the 'dark ghetto'." British J. of Sociology 20 (June): 148-163.

NIE, N. et al. (1969) "Social structure and political participation: developmental relationships." Amer. Pol. Sci. Rev. (September).

NIEDERHOFFER, A. (1967) Behind the Shield. New York: Doubleday.

NIEMEYER, J. (1969) "Final report of the advisory committee on decentralization," in M. R. Berube and M. Gittell (eds.) Confrontation at Ocean Hill-Brownsville: The New York School Strikes of 1968. New York: Praeger.

NISBET, R. A. (1970) Tradition and Revolt. New York: Random House.

OPPENHEIMER, M. (1970) "White collar revisited: the making of a new working class." Social Policy 1 (July-August): 27-33.

ORANS, M. (1968) "Maximizing in Jajmaniland." Amer. Anthropologist 70, 5 (October): 875-897.

——— (1965) The Santal. Detroit: Wayne State Univ. Press.

OTTLEY, R. (1943) New World A-Coming: Inside Black America. New York: Grosset & Dunlap.

PABON, M. (1967) "La integración politica en Puerto Rico," pp. 174-186 in S. Lewis and T. Mathews (eds.) Caribbean Integration. Rio Piedras, Puerto Rico: Institute of Caribbean Studies, University of Puerto Rico.

PARK, R. E. (1950) Race and Culture. Glencoe: Free Press.

PARRINDER, E. G. (1969) Religion in Africa. Baltimore: Penguin.

PARSONS, T. (1967) in M. M. Tumin (ed.) Social Stratification. Englewood Cliffs, N.J.: Prentice-Hall.

——— (1966) "On the concept of political power," in R. Bendix and S. M. Lipset (eds.) Class, Status, and Power. New York: Free Press.

——— (1959) "The school class or a social system: some of its functions in American society," pp. 297-318 in Harvard Educational Rev. 29, 4 (Fall).

PEARL, A. and F. RIESSMAN (1965) New Careers for the Poor: The Non-professional in Human Service. New York: Free Press.

PEASE, W. H. and J. H. PEASE (1963) Black Utopia. Madison: The State Historical Society of Wisconsin.

PETERSON, P. (1970) "Forms of representation: participation of the poor in the community action program." Amer. Pol. Sci. Rev. (June).

PETTIGREW, T. F. (1964) "Negro American intelligence: a new look at an old controversy." J. of Negro Education 33, 1 (Winter).

——— (1964) A Profile of the Negro American. Princeton, N.J.: Van Nostrand.

PFAUTZ, H. W. (1968) "The black community, the community school, and the socialization process: some caveats." Prepared for the Conference on the Community School. Washington, D.C.: The Brookings Institution.

PINDERHUGHES, C. A. (1968) "The psychodynamics of dissent: a clinical appraisal with emphasis on racial activists," in J. H. Masserman (ed.) The Dynamics of Dissent. New York: Grune & Stratton.

PINTNER, R. (1934) "The intelligence of the Negro." J. of Negro Education 3 (July).

——— (1931) Intelligence Testing (second edition). New York: Holt.

PITT, J. P. (1970) "The meaning of black consciousness: community or communion?" Paper given at 55th Annual Meeting of the Assn. for the Study of Negro Life and History. Philadelphia.

PORTER, A. T. (1963) Creoledom: A Study of the Development of Freetown Society. New York: Oxford Univ. Press.

POTTER, D. (1954) People of Plenty. Chicago: Univ. of Chicago Press.

POUSSAINT, A. F. (1969) "The psychology of a minority group with implications for social action," in C. U. Daly (ed.) Urban Violence. University of Chicago Center for Policy Study. Univ. of Chicago Press.

RAAB, E. [ed.] (1962) American Race Relations Today. New York: Anchor Doubleday.

RAINWATER, L. (1970) Behind Ghetto Walls. Chicago: Aldine.

——— (1965) "Crucible of identity: the Negro lower-class family," in T. Parsons (ed.) The Negro American. Boston: Beacon Press.

——— and W. L. YANCY [eds.] (1967) The Moynihan Report and the Politics of Controversy. Cambridge: MIT Press.

RANSFORD, H. E. (1968) "Isolation, powerlessness and violence: a study of attitudes and participation in the Watts riot." Amer. J. of Sociology 73 (March): 581-591.

RAY, P. H. (1968) "Human ecology, technology and the need for social planning." Amer. Behavioral Scientist (July-August): 16-19.

REDFIELD, R. (1956) Peasant Society and Culture: An Anthropological Approach to Civilization. Chicago: Univ. of Chicago Press.

——— and M. SINGER (1954) "The cultural role of cities." Economic Development and Cultural Change 3, 1: 53-73.

REISS, A. J., Jr. (1961) Occupations and Social Status. Glencoe, N.Y.: Free Press.

——— (1961) "Aptitude tests for use in the developing nations." Pittsburgh: American Institute for Research.

RIESSMAN, F. (1962) The Culturally Deprived Child. New York: Harper & Row.

REUTER, E. B. (1917) "The superiority of the mulatto." Amer. J. of Sociology 23, 1 (July).

ROBINSON, A. L., C. C. FOSTER, and D. H. OGILVIE [eds.] (1969) Black Studies in the University. New Haven: Yale Univ. Press.

ROCHE, J. P. (1963) The Quest for the Dream: The Development of Civil Rights and Human Relations in Modern America. New York: Quadrangle.

ROGERS, D. (1970) "The New York City school system: a classic case of bureaucratic pathology," in A. T. Rubinstein (ed.) Schools Against Children: The Case for Community Control. New York: Monthly Rev. Press.

——— (1969) 110 Livingston Street. New York: Vintage Press.

ROSE, A. M. (1956) "Community factors related to desegregation." Amer. J. of Orthopsychiatry 26: 445-449.

ROSENTHAL, R. and L. JACOBSON (1968) Pygmalion in the Classroom. New York: Holt, Rinehart & Winston.

ROSS, J. M., T. CRAWFORD, and T. PETTIGREW (1966) "Negro neighbors–banned in Boston." Trans-action (September-October).

ROSSI, P. H. and R. CRAIN (1968) "The NORC Permanent Community sample." Public Opinion Q. 32 (Summer): 261-272.

RUBENSTEIN, R. L. (1970) "Our tribal society." Morality and Eros. New York: McGraw-Hill. pp. 149-163.

RUBIN, L. (1967) "Maximum feasible participation: the origins, implications, and present status." Poverty and Human Resources Abstract 2 (November-December).

RUDWICK, E. (1969) W. E. B. DuBois: Propagandist of the Negro Protest. New York: Atheneum.

RUNCIMAN, W. G. (1968) "Class, status, and power?" pp. 25-61 in J. A. Jackson (ed.) Social Stratification. Cambridge: Cambridge Univ. Press.

——— (1966) Relative Deprivation and Social Justice: A Study of Attitudes to Social Inequality in Twentieth Century England. Berkeley: Univ. of Calif. Press.

——— (1963) Social Science and Political Theory. Cambridge: Cambridge Univ. Press.

SAFA, H. (1970) "The poor are like everyone else, Oscar." Psychology Today (September).

——— (1967) "The social isolation of the urban poor: life in a Puerto Rican shantytown," pp. 335-352 in I. Deutscher and E. Thompson (eds.) Among the People: Encounters with the Poor. New York: Basic Books.

——— (1965) "The female-based household in public housing: a case study in Puerto Rico." Human Organization 24, 2: 135-139.

SCHATTSCHNEIDER, E. E. (1961) The Semisoverign People. New York: Holt, Rinehart & Winston.

SCHEER, R. [ed.] Eldridge Cleaver. New York: Rampart.

SCHEINER, S. M. (1965) Negro Mecca. New York: New York Univ. Press.

SCHERMERHORN, R. A. (1970) Comparative Ethnic Relations. New York: Random House.

SCHNORE, L. (1967) "Community," pp. 82-150 in N. Smelser (ed.) Sociology. New York: John Wiley.

SCHNORR, A. L. (1966) "How the poor are housed," in W. L. C. Wheaton, G. Milgram, and M. E. Meyerson (eds.) Urban Housing. New York: Free Press.

SCHUMAN, H. and B. GRUENBERG (1970) "City as an explanatory variable in the study of racial attitudes." Amer. J. of Sociology 76.

SCHUMPETER, J. A. (1942) Capitalism, Socialism and Democracy. New York: Harper & Row.

SCHWARZ, P. A. (1963) "Adapting tests to the cultural setting." Educational and Psych. Measurement 23, 4: 673-686.

SEARS, D. O. (1969) "Black attitudes toward the political system in the aftermath of the Watts insurrection." Midwest J. of Pol. Sci. 13 (November): 515-544).

--- and J. B. McCONAHAY (1970) "Racial socialization, comparison levels, and the Watts riot." J. of Social Issues 26, 1: 121-140.

SEEMAN, M. (1967) "On the personal consequences of alienation in work." Amer. Soc. Rev. 32 (April): 273-285.

--- (1959) "On the meaning of alienation." Amer. Soc. Rev. 24 (December): 273-285.

SENNETT, R. [ed.] (1969) Classic Essays on the Culture of Cities. New York: Appleton-Century-Crofts.

SHEATSLEY, P. (1966) "White attitudes toward the Negro," pp. 303-324 in T. Parsons and K. Clark (eds.) The Negro American. Boston: Beacon Press.

SHERMAN, R. B. [ed.] (1970) The Negro and the City. Englewood Cliffs, N.J.: Prentice-Hall.

SHERRARD, T. D. and R. C. MURRAY (1968) "The church and neighborhood community organization," in B. J. Frieden (ed.) Urban Planning and Social Policy. New York: Basic Books.

SIMMEL, E. [ed.] (1946) Anti-Semitism: A Social Disease. New York: International Universities Press.

SIMMEL, G. (1965) in L. Coser (ed.) Georg Simmel. Englewood Cliffs, N.J.: Prentice-Hall.

--- (1908) "The Stranger." The Sociology of Georg Simmel (K. H. Wolff, trans.). Glencoe: Free Press. London: Macmillan, 1950. pp. 402-408.

--- (1903) "Die Grosstädte und das Geistesleben," pp. 227-242 in M. Landmann (ed.) Brücke und Tür: Essays des Philosophen zur. Geschichte, Religion, Kunst und Gesellschaft. Stuttgart: K. F. Koehler Verlag.

SIMPSON, G. E. and J. M YINGER (1958) Racial and Cultural Minorities. New York: Harper & Bros.

SLIM, I. (1967) Pimp, the Story of My Life. Los Angeles: Holloway House.

SMITH, M. G. (1965) The Plural Society in the British West Indies. Berkeley.

——— (1962) "History and social anthropology." J. of the Royal Anthropological Institute 67, 5.

——— (1960) "Social and cultural pluralism." Annals of the New York Academy of Sciences 83: 763-777.

SOREL, G. (1961) Reflections on Violence (T. E. Hulme and J. Roth, trans.) New York: Collier.

SPERO, S. D. and A. L. HARRIS (1931) The Black Worker: The Negro and the Labor Movement, Port Washington, N.Y.: Kennikat.

SPIEGEL, J. P. (1968) "The social and psychological dynamics of militant Negro activism: a preliminary report," in J. Masserman (ed.) The Dynamics of Dissent. New York: Grune & Stratton.

SPINDLER, G. (1969) "Review of the American Indian today." American Anthropologist 71, 2: 341-342.

SPORES, R. (1965) "The Zapotec and the Mixtec at Spanish contact," in G. Willey (vol. ed.) Handbook of Middle American Indians, Vol 3, Part 2. Austin: Univ. of Texas Press.

STOCKING, G. W., Jr. (1968) Race, Culture and Evolution. Glencoe: Free Press.

SUTTLES, G. (1968) The Social Order of the Slum. Chicago: Univ. of Chicago Press.

SWADOS, H. (1962) "The myth of the happy worker," pp. 106-113 in E. Josephson and M. Josephson (eds.) Man Alone: Alienation in Modern Society. New York: Dell.

SZASZ, T. S. (1970) Ideology and Insanity. New York: Doubleday.

TANNENBAUM, F. (1946) Slave and Citizen. New York: Knopf.

TAX, S. (1963) Penny Capitalism. Chicago: Univ. of Chicago Press.

TenHOUTEN, W. (1970) "The black family: myth and reality." Psychiatry 33, 2 (May).

——— (1969) Site Sampling and Snowball Sampling in Biased Nets: Methodology for the Study of Community Organization. Report to U.S. Office of Economic Opportunity. Los Angeles.

——— (1969) Community Leadership Structure in Special Impact Areas: Snowball Sample Analysis. Report to Westinghouse Learning Corporation. Los Angeles.

——— (1969) "Scale gradient analysis: a statistical method for constructing and evaluating Guttman scales." Sociometry (March): 80-99.

——— F. TZUEN-JEN LEI, F. KENDALL, and C. W. GORDON (1970) "Ethnic composition, social contexts, and educational plans of Mexican-American and Anglo high school students." Amer. J. of Sociology.

TERMAN, L. M. (1922) "The great conspiracy or the impulse imperious of intelligence testers, psychoanalyzed and exposed by Mr. Lippmann." New Republic 33 (December 27).

——— (1916) The Measurement of Intelligence. New York: Houghton-Mifflin.

THOMAS, P. (1967) Down These Mean Streets. New York: Knopf.

THOMPSON, C. (1937) "The Harrison-Fletcher bill and Negro separate schools." J. of Negro Education (January).

THOMPSON, E., Jr. (1959) "The plantation as a social system." Plantations of the New World. Pan American Union, Washington, D.C.

THUROW, L. C. (1969) Poverty and Discrimination. Washington, D.C.: The Brookings Institution.

TITTLE, C. and D. TITTLE (1965) "Social organization of prisoners: an empirical test." Social Forces 43, 2.

TOMLINSON, T. M., and D. TenHOUTEN (1970) "Methodology: Negro reaction survey," pp. 127-137 in N. Cohen (ed.) The Los Angeles Riots. New York: Praeger.

TORRANCE, R. P. (1968) "Testing the educational and psychological development of students from other cultures and subcultures." Rev. of Educational Research 38 (February): 71-76.

TUMIN, M. M. (1969) "Some social consequences of research on racial relations," in H. P. Dreitzel (ed.) Recent Sociology No. 1: The Social Basis of Politics. New York: Macmillan.

——— (1953) "Some principles of stratification: a critical analysis." Amer. Soc. Rev. 18 (August).

——— and A. FELDMAN (1961) Social Class and Social Change in Puerto Rico. Princeton: Princeton Univ. Press.

TURNER, J. (1969) "Uncontrolled urban settlement: problems and policies," pp. 507-534 in G. Breese (ed.) The City in Newly Developing Countries. Englewood Cliffs, N.J.: Prentice-Hall.

TURNER, R. H. (1960) "Sponsored and contest mobility and the school system." Amer. Soc, Rev. 25 (December): 856-867.

U.S. Commission on Civil Rights (1967) Racial Isolation in Public Schools. Report. Washington, D.C.: Government Printing Office.

U.S. Department of Commerce (1935) Negroes in the United States, 1920-1932. Bureau of the Census. Washington, D.C.

U.S. Department of Labor (1921) The Negro at Work During the World War and During Reconstruction. Washington, D.C.

U.S. Riot Commission (1968) Report of the National Advisory Commission on Civil Disorders. New York: Bantam.

VALENTINE, C. A. (1968) Culture and Poverty. Chicago: Univ. of Chicago Press.

——— and B. L. VALENTINE (1970) "Making the scene, digging the action, and telling it like it is; anthropologists at work in a dark ghetto," in N. Whitten and J. Szwed (eds.) Afro-American Anthropology: Contemporary Perspectives. New York: Free Press.

van den BERGHE, P. L. (1967) Race and Racism. New York: John Wiley.

VERHAEGEN, P. (1956) "Utilite actuelle des tests pour l'etude psychologique des Autochtones Congolais." Rev. de Psychologie Appliqueem 6: 139-151.

WADE, R. C. (1969) "Violence in the cities: a historical view," in C. U. Daly (ed.) Urban Violence. University of Chicago Center for Policy Study. Univ. of Chicago Press.

--- (1964) Slavery in the Cities–The South, 1820-1860. London: Oxford Univ. Press.

WAGATSUMA, H. (1967) "The pariah caste in Japan: history and present self-image," pp. 118-140 in A. DeReuck and J. Knight (eds.) Caste and Race: Comparative Approaches. London: Churchill.

--- (1966) "Non-political approaches–the influences of religion and education," pp. 88-109 in G. DeVos and H. Wagatsuma (eds.) Japan's Invisible Race. Berkeley: Univ. of California Press.

--- (1966) "Post-war political militants," pp. 68-87 in G. DeVos and H. Wagatsuma (eds.) Japan's Invisible Race. Berkeley: Univ. of California Press.

--- and G. TOTTEN (1966) "Emancipation: growth, and transformation of a political movement," pp. 33-67 in G. DeVos and H. Wagatsuma (eds.) Japan's Invisible Race. Berkeley: Univ. of California Press.

WAGLEY, C. (1960) "Plantation America: a culture sphere," in V. Rubin (ed.) Carribean Studies: A Symposium. Seattle: Univ. of Washington Press.

WALD, P. M. (1965) "Law and poverty." Report to the National Conference on Law and Poverty, Washington, D.C. June 23-25.

WANDERER, J. J. (1969) "An index of riot severity and some correlates." Amer. J. of Sociology 74 (March): 500-505.

WARNER, W. L. (1963) Yankee City. New Haven: Yale Univ. Press.

WARNER, W. L., B. H. JUNKER, and W. A. ADAMS (1941) Color and Human Nature. Washington, D.C.: Amer. Council on Education.

WASHINGTON, B. T. (1902) Up From Slavery. New York: Doubleday, Page.

WASKOW, A. I. (1969) "Community control of the police." Trans-action 7 (December): 4-7.

WAYMAN, J. C. et al. (1967-1969) The War on Poverty in San Diego. La Jolla, Calif.: Western Behavioral Sciences Institute.

WEBBER, M. M. and C. C. WEBBER (1967) "Culture, territoriality and the elastic mile," in H.W. Eldredge (ed.) Taming Megalopolis: What Is and What Could Be. New York: Frederick A. Praeger.

WEBER, M. (1967) in M. M. Tumin (ed.) Social Stratification. Englewood Cliffs, N.J.: Prentice-Hall.

--- (1958) "Science as a vocation," pp. 129-158 in H. H. Gerth and C. W. Mills (eds.) From Max Weber. New York: Grosset & Dunlap.

--- (1958) "Class, status, party," in H. H. Gerth and C. W. Mills (eds.) From Max Weber. New York: Grosset & Dunlap.

WEST, L. J. (1967) "The psychobiology of racial violence." Archives of General Psychiatry 16 (June): 645-651.

WHITTEN, N. E., Jr. and J. F. SZWED (1970) Afro-American Anthropology. New York: Free Press.

WHYTE, W. F. (1955) Street Corner Society. Chicago: Univ. of Chicago Press.

WILENSKY, H. (1964) "The professionalization of everyone?" Amer. J. of Sociology 70 (September): 137-159.

--- (1964) "Varieties of work experience," pp. 125-154 in H. Borow (ed.) Man in a World of Work. Boston: Houghton-Mifflin.

WILLIAMS, O. P. and C. R. ADRIAN (1963) Four Cities. Philadelphia: Univ. of Pennslyvania Press.

WILSON, J. Q. (1968) The Metropolitan Enigma: Inquiries into the Nature and Dimensions of America's Urban Crisis. Cambridge, Mass.: Harvard Univ. Press.

WILSON, J. Q. (1968) Varieties of Police Behavior. Cambridge: Harvard Univ. Press.

--- (1968) "The police and the delinquent in two cities," pp. 173-194 in James Q. Wilson (ed.) City Politics and Public Policy. New York: John Wiley.

--- (1964) "Generational and ethnic differences among police." Amer. J. of Sociology 69 (March): 522-528.

--- (1963) "The police and their problems: a theory." Public Policy 12: 189-216.

WISH, H. (1961) Antebellum. New York: Capricorn Books.

WOOD, R. C. (1959) Suburbia: Its People and Their Politics. New York: Houghton-Mifflin.

WOODWORTH, R. S. (1916) "Comparative psychology of races." Psych. Bull. 13,10 (October 15).

WOLF, E. R. (1966) Peasants. Englewood Cliffs, N.J.: Prentice-Hall.

--- (1957) "Closed corporate peasant communities in Mesoamerica and Central Java." Southwestern J. of Anthropology 13, 1: 1-8.

--- (1956) "Aspects of group relations in a complex society." American Anthropologist 58: 1065-1078.

WOLFE T. (1970) Radical Chic and Mau-Mauing the Flak Catchers. New York: Farrar, Straus & Giroux.

--- (1969) "The new yellow peril" Esquire (December).

WOODSON, (1922) The Negro in Our History. New York: Associated.

--- (1919) The Education of the Negro Prior to 1861. Washington, D.C.: Assn. for the Study of Negro Life and History.

WRIGHT, R. (1940) Native Son. New York: Harper & Brothers.

YOUNG, R. [ed.] (1970) Roots of Rebellion. New York: Harper & Row.

THE AUTHORS

JAMES M. BISHOP is a candidate for the Ph.D. in Sociology at the University of California at Los Angeles. His interests lie in the fields of occupations and professions, organizations, and urban sociology. "Community and Control in a Metropolitan Setting" is his first publication.

ROY SIMON BRYCE-LAPORTE is Associate Professor of Sociology and Director of the Afro-American Studies Program at Yale University. He lived, studied, and taught in the Canal Zone until 1959, and received his Ph.D. from the University of California at Los Angeles in 1968. His interests include comparative and historical studies of race and culture, immigration, adaptation, and qualitative urban sociology. Allyn and Bacon, Inc., will publish his forthcoming book, *A Sociology of Black Captivity–The Slave Plantation.*

MARTIN DISKIN is Assistant Professor of Anthropology in the MIT Department of Humanities. He held an NSF grant in 1964 for a study of the peasant market system in Oaxaca, Mexico, which formed the basis for his article in this volume, "Persistence of Tradition in the Urbanization of the Oaxaca Valley," and also a 1969 article entitled "Estudio estructural del sistema de plaza en el valle de Oaxaca," in *America Indigena.*

WILLIAM RUSSELL ELLIS, JR., is Assistant Professor of Sociology in the School of Environmental Design at the University of California at Berkeley. Prior to joining the faculty at Berkeley, he taught at Pitzer College (1966-1969) and at the State University of New York, College at Old Westbury (1969-1970). He is currently revising his dissertation, "Operation Bootstrap: A Case Study in Ideology and the Institutionalization of Protest," for publication in book form.

GARY GOODPASTER is Associate Professor of Law at the University of Iowa College of Law. During 1969-1970 he was Acting Director of Research and Supervisory Attorney at Colorado Rural Legal Services, Inc. He is author of several articles, of which "The Integration of Equal Protection–Due Process Standards and the Indigent's Right of Access to the Courts," published in the *Iowa Law Review* this year, is the most recent.

JEFFERSON EUGENE GRIGSBY, III, a sociologist, is Acting Assistant Professor in the Urban Planning Program of the School of Architecture and Urban Planning at the University of California at Los Angeles. Prior to his appointment at UCLA he was an instructor in the

Urban Studies Program at Occidental College and a consultant of the Los Angeles City School Office of Urban Affairs, where he was responsible for regional organization and establishment of local high school tutorial programs.

HARLAN HAHN is Associate Professor of Political Science at the University of California at Riverside. An active participant in the American Political Science Association, he is the author of numerous articles dealing with urban violence published in political science journals, and a book, *Urban-Rural Conflict: The Politics of Change.* He will edit Volume VI of Urban Affairs Annual Review to be devoted to Urban Politics and People.

THELTON E. HENDERSON is Assistant Dean at the Stanford School of Law. Prior to taking up his position at Stanford, he was Directing Attorney at the East Bayshore Neighborhood Legal Center in Menlo Park. He received his law degree from Boalt Hall School of Law at the University of California at Berkeley.

IRA KATZNELSON received his Ph.D. from Cambridge University in 1969. He is Assistant Professor of Political Science at Columbia University and Editor of *Politics and Society,* a new quarterly journal in the social sciences. His analysis of comparative aspects of British and American race relations, *Race, Politics and Migration,* will be published by Oxford University Press in 1971.

PETER LOEWENBERG is Assistant Professor of History at the University of California at Los Angeles, and a Research Clinical Associate at the Southern California Psychoanalytic Institute. The author of numerous articles in major historical journals, his most recent publication is "The Psychology of Racism," prepared for inclusion in Gary Nash and Richard Weiss, eds., *The Great Fear: Race in the Mind of America.*

DALE ROGERS MARSHALL is a Lecturer in Political Science at the University of California at Berkeley. She completed her Ph.D. in 1969 at the University of California at Los Angeles. Her dissertation, "Politics of Participation in Poverty," is to be published by the University of California Press. She is the author of "Who Participates in What?", a bibliographic essay on individual participation in urban areas published in *Urban Affairs Quarterly* in 1968.

JANE R. MERCER is Associate Professor of Sociology at the University of California at Riverside and Research Specialist for the State of California Department of Mental Hygiene. Her research dealing with mild mental retardation as well as the evaluation of public school integration has been sponsored and supported by the Public Health Service of the Department of Health, Education and Welfare, and by the Office of Compensatory Education of the California State Department of Education. She is the author of numerous journal articles, and a book, *The Eligibles and the Labeled* (in press).

HARVEY L. MOLOTCH is Assistant Professor of Sociology at the University of California at Santa Barbara, where he teaches courses in urban sociology, the sociology of planning and design, and social change. He received his Ph.D. in 1968 from the University of Chicago. In addition to several published articles, he is the author of *Intervention for Integration* published by the University of California Press. His current research is devoted to the analysis of institutional racism and the comparative study of ecological social movements.

MARTIN ORANS is Associate Professor of Anthropology at the University of California at Riverside. He has published widely in major anthropological journals. His interests lie in the areas of economic development, relations of "material" and "non-material" culture, political organization, kinship and stratification. His book, *The Santal: A Tribe in Search of a Great Tradition,* was published by Wayne University Press in 1965.

PETER ORLEANS is Associate Professor at the University of California at Los Angeles with a joint appointment in the Departments of Sociology and Architecture. He is co-editor of *The New Urbanization* and is presently working on a book, *Urban Morphology: A Sociological Interpretation.* He serves on the editorial board of *Environment and Behavior* and is book review editor for *Urban Affairs Quarterly.*

EDWARD L. PAYNTER is an Acting Instructor in American History at the University of California at Berkeley. During the academic year 1968-1969, he was an Assistant Lecturer in American Studies at the University of East Anglia, Norwich, England. In his recently completed doctoral dissertation he examined the thought of various critics of machine technology, including Walt Whitman, Frank Lloyd Wright, Waldo Frank, and Lewis Mumford. "Value Premises in Race Research: The Evolution of Environmentalism," is his first publication.

LISA REDFIELD PEATTIE is Associate Professor of Urban Anthropology in the M.I.T. Department of City and Regional Planning and Research Associate at the M.I.T.-Harvard Joint Center for Urban Studies. She has authored articles published in the *Journal of the American Institute of Planners,* and *Human Organization.* Her book, *The View from the Barrio,* was published by the University of Michigan Press in 1968.

HELEN ICKEN SAFA is Associate Professor of Anthropology at Livingston College, Rutgers, The State University of New Jersey. Her research has focused on the urban poor, public housing, and social mobility. Listed among her journal articles is "The Case For Negro Separatism: The Crisis of Identity in the Black Community," published in *Urban Affairs Quarterly* in 1968.

MELVIN SEEMAN is Professor of Sociology at the University of California at Los Angeles. He is past president of the Pacific Sociological Association and currently serves on the National Council of the American Sociological Association. A social psychologist, concerned with the study of alienation, he is the author of more than thirty articles appearing in a wide range of social science periodicals.

JOHN STERN received his B.A. from UCLA in 1969, majoring in Philosophy and Mathematics. For the past four years he has done statistical data analysis in Anthropology and Sociology, and has worked as a computer programmer. This is his first publication.

MARVIN I. SURKIN is Assistant Professor of Political Science at Adelphi University. He served on the Executive Committee of the Caucus For A New Political Science from 1967 to 1969. In addition to publishing several journal articles, he is co-editor, with Alan Wolfe, of *An End to Political Science,* published by Basic Books, Inc., in 1970.

COUNCILL TAYLOR is Director of the State University of New York College at Old Westbury, Planting Fields Campus. He is a founding member and member of the Advisory Council of the Institute of the Black World in Atlanta, Georgia. Listed among his numerous publications are two books: *Color and Class: A Comparative Study of Jamaica Status Groups* and *Voluntary Associations in the New Africa.*

DIANA TENHOUTEN received an M.A. in Social Psychology in 1965 at Michigan State University. She has been employed as a Research Sociologist at the U.S. Office of Economic

Opportunity, and is Director of Evaluation, UCLA Hospitals and Clinics. She has published papers on riot ideology and on health care evaluations.

WARREN D. TENHOUTEN is Associate Professor of Sociology at UCLA. Since receiving his Ph.D. from Michigan State University in 1965 he has been actively engaged in research on socialization and racism, as well as in the evaluation of community action programs for the U.S. Office of Economic Opportunity. Among his most recent publications is "The Black Family: Myth and Reality," which appeared in *Psychiatry* in 1970. He is currently working on a book, *The Structure of Social Inquiry,* to be published by Harper and Row, and on a study (with John F. Marsh, Jr.), "Thought, Race, and Opportunity" for the Office of Economic Opportunity.